DIVERSITY AND INTEGRATION IN MYCORRHIZAS

Diversity and Integration in Mycorrhizas

Proceedings of the 3rd International Conference on Mycorrhizas (ICOM3)
Adelaide, Australia, 8–13 July 2001

Edited by

S.E. SMITH

and

F.A. SMITH

Reprinted from *Plant and Soil*, Volume 244, Nos. 1–2 (2002).

KLUWER ACADEMIC PUBLISHERS
DORDRECHT / BOSTON / LONDON

Library of Congress Cataloging-in-Publication Data

A C.I.P. Catalogue record for this book is available from the library of Congress.

ISBN 1-4020-0269-6

Published by Kluwer Academic Publishers,
P.O. Box 17, 3300 AA Dordrecht, The Netherlands

Sold and distributed in North, Central and South America
by Kluwer Academic Publishers,
101 Philip Drive, Norwell, MA 02061, U.S.A.

In all other countries, sold and distributed
by Kluwer Academic Publishers,
P.O. Box 322, 3300 AH Dordrecht, The Netherlands

Printed on acid-free paper

Printed in the Netherlands

Contents

DIVERSITY AND INTEGRATION IN MYCORRHIZAS

Proceedings of the 3rd International Conference on Mycorrhizas (ICOM3)
Adelaide, Australia, 8–13 July 2001

icom 3

Plant and Soil **244:** ix–x, 2002.

Foreword

Proceedings of ICOM3: Diversity and Integration in Mycorrhizas

This book contains chapters that were derived from invited Keynote and Symposium presentations at the 3rd International Conference on Mycorrhizas (ICOM3), held in Adelaide, Australia, from 8–13 July 2001. However, it is more than 'a book of the conference': it also contains chapters by authors who did not speak at the conference, but who were invited in order to provide additional depth and balance. We are grateful to all for the efforts that they have put into their contributions. Not all of those invited to write chapters were able to do so, but we hope that the book provides a flavour both of ICOM3 itself and of current research on mycorrhizal symbioses – a field that is rapidly developing. All chapters were reviewed independently, following the normal procedures of *Plant and Soil*. About 450 delegates from 50 countries attended ICOM3 and Abstracts of all presentations – talks and posters – are available on the ICOM3 Website: *http://www.waite.adelaide.edu.au/Soil_Water/3icom.html*

As organizers of ICOM3, it is not for us to judge its success or impact. We are pleased that informal feedback and published reports have been very positive ('Mycorrhizal Research on the Move', Burleigh and Jakobsen, 2002 and 'ICOM3: A Landmark in Mycorrhiza Research!', Colpaert, 2002). The conference theme was **'Diversity and Integration in Mycorrhizas'** and, to ensure that this theme could be appreciated by all participants, our approach was to bring together major topics in mycorrhizal research that crossed some of the traditional boundaries often reflected in mycorrhizal conferences, such as divisions between different classes of mycorrhizas or divisions based on techniques. We also ensured that a number of up-and-coming younger researchers were invited to present their work, particularly since, in the main, they are the people who are applying novel techniques to unravelling details of symbiotic processes. This approach was evidently well appreciated, as emphasized by Professor David Read in his closing overview of ICOM3. The themes of the conference are reflected in the way in which chapters are grouped in this book. After the scene is set by Bengt Söderström in his thoughtful introductory chapter 'Challenges for mycorrhizal research into the new millennium', the chapters are presented in four groups: 'Molecular diversity and phylogeny', 'Development: from genes to structure', 'Functional relationships' and 'Ecological interactions'. The arrangement of chapters in the book is not exactly the same as that of the presentations at the conference. Some of the chapters lie at the boundaries between themes and others are wide-ranging and cover more than one theme; there is also a connecting loop because of the way molecular tools resulting from studies of 'Molecular diversity and phylogeny' are becoming increasingly important in unravelling 'Ecological interactions'. We hope that in the new arrangement no papers seem entirely out of place. No classification can ever be perfect and – as with mycorrhizal fungal taxonomists – we have done our best.

We are conscious of topics that were covered in concurrent sessions at the meeting and are not included in the book, such as those on 'Problem soils and extreme environments', 'Mycorrhizas in the tropics', the 'Mycorrhizosphere', and 'Mycorrhizas in vegetation restoration', as well as the informal workshops on a wide range of specialist topics. We mention those here to emphasize the wide-ranging influence of mycorrhizas on plant–soil interfaces, and the fact that even an international meeting such as ICOM cannot emphasise diversity without becoming much larger, in which case it would lose the integration that is obviously highly appreciated by participants. The integration is also one of the keys to grasping the complexities of the roles of mycorrhizas in ecosystems.

In conclusion, ICOM3 (and hence this book) would not have been possible without generous financial support from our major sponsors: Kluwer Academic Publishers/*Plant and Soil*, Samuel Roberts Noble Foundation, *New Phytologist*, *Canadian Journal of Botany*, Forest & Wood Products Research and Development Corporation, Grains Research and Development Corporation, the Australian Society of Plant Physiologists (now the Australian Society of Plant Scientists), and the University of Adelaide's Centre for Plant Root Symbioses. We also received financial support from other sources within the University of Adelaide (the Vice-Chancellor's Fund, the Faculty of Science and the Department of Soil and Water) and from AusAID, Biorize, the Cooperative Research Centre for Molecular Plant Breeding, Premiertech, Springer-Verlag, the Australian Academy of Science, the Australian Institute of

Biology and last but not least Mrs Lindsay Harley. We thank the team at Kluwer who have played a major role in bringing the publication to completion and in good time, including Arno Flier for his strong encouragement to take on the project, Pieter Poot as managing editor of the volume, Peter-Paul Delsman as desk editor and particularly Otilia Prins who has been at the interface between authors and editors. We cannot thank individually all those who helped arrange ICOM3: the members of the Organizing Committee and the Australasian Advisory Committee, the organizers of Symposia, Workshops and Concurrent sessions, and many others. Nevertheless we will once again gladly thank the conference organizer Sally Jay, her co-organizer Jenny Blanchard, and Mark Stevens the audiovisual coordinator, for everything that they did to bring ICOM3 into being, and the members and associates of the Mycorrhiza Group in the Department of Soil and Water, who helped us in many ways.

Sally Smith (SE Smith)
Andrew Smith (FA Smith)
Department of Soil and Water and
the Centre for Plant Root Symbioses,
Waite Campus,
The University of Adelaide,
PMB 1, Glen Osmond 5064
South Australia,
AUSTRALIA

References

Burleigh, S and Jakobsen, I 2002. Mycorrhizas on the move. Trends in Plant Science, 7, 6–7.
Colpaert, J 2002. ICOM3: A Landmark in Mycorrhiza Research! Mycorrhiza, 12, 51.

Plant and Soil **244:** 1–7, 2002.

Challenges for mycorrhizal research into the new millennium

Bengt Söderström
Department of Microbial Ecology, Ecology Building, Lund University, SE-223 62 Lund, Sweden

Key words: carbon dioxide, carbon flow, climate change, genomic biology, mycorrhiza

Abstract

Some future directions for mycorrhizal research are discussed. Data from The IPCC Third Assessment Reports 2001 on elevated carbon dioxide the concentration and consequences regarding global climate change, form the basis for arguments that a better understanding of the mycorrhizal activities with regard to controlling carbon flow and plant nutrient flow is essential. Mycorrhiza can be regarded as energy and nutrient switches in the plant roots and it is suggested that more studies on mycorrhizal carbon relations are necessary. The impact of the genomic research era on mycorrhizal research is suggested to be of major importance during the years to come, and it is claimed that stronger interactions between different disciplines will be essential.

Introduction

The importance of mycorrhizal symbioses in nature cannot be overestimated. The significance of these interactions in the nutrition and well-being of the individual plant partner is well established (Smith and Read, 1997) and recent evidence also indicates that they may also have major effects on the structure of the plant community (van der Heijden et al., 1998). Their ubiquitous existence and impact on plant and soil biology make it necessary to consider them in all terrestrial ecological research. Since this form of symbiosis was first described in the mid 19th century, the fundamental questions surrounding these intriguing biological systems have attracted a large number of researchers, as has the question regarding potential applications in agriculture and forestry to enhance plant production and sustainability. Most work has been carried out on the ectomycorrhizal and arbuscular mycorrhizal systems, a fact which is not surprising considering their dominance in nature.

It has been predicted that the 21st century will become the century of biology as the 20th century was the century of physics, the so called 'new biology' being the justification for such a prediction. Bearing in mind the critical importance of the mycorrhizal symbioses in nature and their strong influence on many ecological processes, mycorrhizal researchers are in a very favourable position to develop the field of mycorrhizal research even further. This paper exemplifies some areas of research which have the potential for exciting and important development, and some of the prerequisites for such development are discussed.

The mycorrhizal research community

Mycorrhizal researchers have traditionally been recruited from a number of established disciplines such as botany, microbiology or ecology. This is not surprising when one considers that mycorrhizal symbiosis in itself is a very integral phenomenon. For the advancement of our knowledge and understanding of these complex systems, the international meetings dedicated to the study of mycorrhizas have undoubtedly been of great importance. During the past 20 years, these meetings have grown considerably. Based on the number of abstracts submitted to the international conferences with mycorrhiza as the main theme, it seems that the number of researchers involved in mycorrhizal investigations has increased four times the past 20 years. The North American Conferences on Mycorrhiza (NACOM) had a strong international participation, and in 1979 the 4th NACOM attracted 106 abstracts. Since International Conferences of Mycorrhizas have been organized (ICOM, 1996, 1998, 2001), they have attracted between 400 and 500 contributions and the NACOMs and the

European Symposia on Mycorrhiza (ESM) in the late 1980s and early 1990s each attracted 200–250 abstracts. Compared with other comparable fields of research, this seems to represent normal growth, and mirrors the general increase in societies' scientific effort rather than the specific growth of the mycorrhizal research community. The mycorrhizal research community thus does not seem to have attracted more than the normal number of new researchers in spite of the growing acceptance of the biological importance of such symbioses.

A quick glance at the titles of invited presentations and abstracts submitted to the mycorrhizal conferences (NACOM, ESM and ICOM) indicates that the interest in ectomycorrhiza and arbuscular mycorrhiza over time has varied somewhat and considerably more abstracts were submitted to the last three NACOMs (1987, 1990, 1993) dealing with arbuscular mycorrhizas compared with abstracts on ectomycorrhizal studies. The number of ectomycorrhiza related abstracts has generally been larger at the European meetings (ESM). Fewer AM abstracts were submitted to the first ICOM (1996) while in 1998 the numbers were similar. The number of contributions dealing with other types of mycorrhiza, such as orchid and ericoid mycorrhizas, is consistently very low. Studies describing applications of mycorrhiza in agriculture and forestry tend to account for 20% of all abstracts, while contributions describing the development of new experimental techniques only constitute a few percent. All the symposia on mycorrhiza seem to be very strongly specialised; there have been very few contributions on more general biological subjects related to mycorrhiza research.

Global climate change, carbon and mycorrhiza

During the past 100 years we have seen continuous increase in the carbon dioxide concentration in the atmosphere, as well as an increase in mean temperature. Data on this, as well as predicted future scenarios, are convincingly presented in the IPCC (Intergovernmental Panel on Climate Change) Third Assessment Report — Climate Change 2001 (http://www.ipcc.ch). From the beginning of the 19th century to today, the carbon dioxide concentration has increased from 280 to 360 ppm, a concentration that has not been exceeded during the past 420 000 years and probably not during the past 4 million years. There are today overwhelming arguments that human activity is the main cause for this increase. Seven different but realistic scenarios have been tested by the IPCC working group for the prediction of future carbon dioxide concentrations and these suggest an increase in atmospheric carbon dioxide concentration during the coming 100 years from today's value of 360 ppm to between 550 and 950 ppm. Other greenhouse gases such as methane and nitrous oxide have also been modelled. All readings of the earth's surface temperature show that the average global temperature has increased by about 0.6 °C (± 0.2 °C) over the past 100 years, and it is likely that the 1990s was the warmest decade since recording started. The relationship between the increase in the concentration of greenhouse gases and temperature increase is generally accepted. In the IPCC reports future temperature increases are also modelled according to different scenarios. These models indicate that the expected increase in the Earth's surface temperature will be between 1.4 and 5.8 °C by the year 2100. The predicted climate changes resulting from such a temperature increase are also discussed in the IPCC reports, as well as possible stabilization scenarios.

Scenarios such as those mentioned above were one basis for the Kyoto Protocol to the United Nations Framework Convention on Climate Change in 1997 which, during the summer of 2001 was agreed upon after negotiations in Köln by the majority of the signing members. In this protocol carbon dioxide problems play a very important role and each of the signatories commits itself not only to following agreed CO_2 emission reductions and/or increased sinks, but also to promoting research on carbon cycling, carbon budgeting, etc. Unfortunately, a few major industrial countries have declared that they will not sign the treaty. This, however, does not make it less important.

Of major importance for our understanding of carbon cycling is of course, that reliable carbon budgets can be determined. Much effort has already been devoted to this already during the international IBP and IGBP programmes during the 1970s and 1980s. However, fundamental information is still lacking (Schimel, 1995), including reliable estimates of terrestrial carbon pools and flows to and from these pools. These flows are partly regulated by mycorrhizas, and further research into carbon relations in mycorrhizal symbioses may turn out to be crucial in developing reliable terrestrial carbon budgets and carbon cycling models.

From microcosm studies we know that in ectomycorrhizal symbiosis, 20–30% of the carbon assimilated by the host plant may be consumed by the fungal partner (Bidartondo et al., 2001; Ek, 1997; Rygiewicz and Anderson, 1994; Söderström and Read, 1987), and similar figures have been found in arbuscular mycorrhizal systems (Jakobsen, 1995; Jakobsen and Rosendahl, 1990). These estimates were recently confirmed in an elegant field study, in which large-scale girdling of mature pine trees was carried out and the reduction in soil respiration was used to estimate the contributions from saprophytic processes, root respiration and mycorrhizal respiration (Högberg et al., 2001). It thus seems clear that a substantial proportion of the carbon assimilated by the mycorrhizal host plants goes straight into the soil where part of it is respired to produce energy for biomass build-up, and part of it will be found in fungal biomass in the soil carbon pool.

Mycorrhizal fungi may also more directly control the primary production in that they are the major nutrient scavengers for their host plants (Smith and Read, 1997). However, it does not appear that the size of the plant's energy investment in its fungal partner is related to its nutritional benefits. Smith et al. (2000) showed that a fungus (*Scutellospora calospora*) which formed more biomass and thus consumed more carbon was not as efficient in phosphorus transfer to its host plant as was another fungus (*Glomus caledonium*), which produced considerably less biomass when they were both symbiotic with *Medicago truncatula*. *S. calospora* produced almost twice as much biomass while the phosphorus transfer to the host plant was almost three times greater than with *G. caledonium*. Similarly, Bidartondo et al. (2001) showed that of the four ectomycorrhizal fungi they tested, the isolate that consumed less of the host's assimilated carbon was the one that transferred most ammonium to its host when this nutrient was added to the mycelium. To make the situation even more complex, they also showed in the same experiment that of the four fungi studied, *Paxillus involutus* produced the lowest amount of mycorrhizal connections to its host plant, but was able to extract more carbon from its host per mycorrhizal connection, and also transferred most ammonium per connection. Furthermore, this fungus respired less per unit biomass than did the other fungi. The poor relation between the fungal carbon demand and nutrient supply to the host plant may initiate discussions on possible manipulation of fungi and primary production.

From the examples given above, it is easy to compare mycorrhizal fungi to carbon switches situated in the plant roots with the potential to switch the carbon flow to the soil, where it may accumulate or be oxidized to carbon dioxide, as well as the nutrient flow to the host plant and thereby indirectly controlling the carbon dioxide assimilation. This strongly emphasizes the importance of further studying the carbon relations of mycorrhiza. In a recent review on mycorrhizal fungi and elevated carbon dioxide concentrations, Treseder and Allen (2000) concluded: "Our knowledge indicates that the influence of mycorrhizal fungi on carbon dynamics under global change remains largely unknown, but could be a significant factor in soil carbon sequestration". A challenge for mycorrhizal research is thus to gain a much better understanding of carbon flows in mycorrhizas and ways to control these, and to relate this understanding to research on ecosystem carbon budgets which, in turn, are fundamental for carbon dioxide balance on a global scale. The predicted temperature increase of perhaps as much as 5 °C in the next 100 years is a major threat to our planet.

Mycorrhiza and genomic biology

During the past decade much effort and money have been invested in genome-related research. The launch and successful realization of the HUGO (human genome) project and the subsequent sequencing of the genomes of a range of other organisms have totally changed the foundation on which much biology research is based. Today, the complete genomes of, for example, more than 30 bacterial species, *Caenorhabditis elegans*, *Drosophila*, rat and *Arabidopsis thaliana* have been sequenced, and more organisms are being added to the list. This knowledge will have important, and perhaps even revolutionary, consequences for biological research in the years to come. We also see sequencing facilities becoming common in many research organisations. DNA sequencing is still an expensive task, but the cost has decreased considerably and there is good reason to believe prices will decrease even further. Most of the sequences are, furthermore, freely available on the Internet, an enormous source of information most biologists will have to learn to use. However, the means with which to handle and analyse all this information, bioinformatics, is still under development, but will hopefully soon be easily managed by non-specialists. As an effect of these extensive sequencing programmes we now see the development of functional genomics, transcriptomics, proteomics, metabolomics, etc., sub-disciplines which can all be included in the concept of genomics. What all these new terms illustrate is the extremely rapid development within this field of science, a development which is partly an effect of the many man-hours being invested, but also an effect of novelties being brought out into the open. It is a challenge to every biologist today to learn how to use this new information for his or her particular purposes.

Genome-related research has much potential earlier not evident to biological research. Because of the extensive technological development within this area, there is potential for what may be called 'industrial research', i.e., the potential to run large-scale experiments and analyses since much of the analyses can be automatized. By analysing traditional problems from a genomic perspective, physiological or ecological questions can be elucidated in a new way. It may also be possible to formulate new problems from a genomic analysis, problems that may find physiological or ecological solutions. In fact we may soon also be able to study gene expressions in nature to a much greater extent than has hitherto been possible. The methods being developed today may allow us to perform complex observational and experimental analyses in nature and may also allow new types of experiments to be designed.

One very interesting effect of the growth of genomic biology is that within the broadly defined subject of bioinformatics, new and very promising cooperation between biologists, applied mathematicians and computer scientists has been developed. New courses and educational programmes in computational biology, theoretical biology and bioinformatics are being established at most institutes of higher education all over the world, programmes in which students in biology, mathematics and computer science can meet and work on common projects. Given the complexity of biological systems, this kind of cooperation is necessary if we are to model these systems. This development may well open up new research areas in biology. The possible impact of genomic biology on biology in general has been discussed in a number of reviews, e.g. by Brent (2000).

Genomic biology has already had an important influence on mycorrhizal research, something which is very obvious when studying the programme for ICOM3. Without even attempting a review of genome related mycorrhizal literature, a few examples will be mentioned. The best established methods making use of genome-related molecular biology are those used for identification/classification of mycorrhizal fungi, methods that have almost become routine in many laboratories. The most frequently used technique is PCR-RFLP of the nuclear ribosomal DNA ITS region, and species libraries have been established at many laboratories. It should soon also be possible to identify any fungus to species level by use of only vegetative samples and when PCR-RFLP fails DNA sequencing will be an option. By applying PCR-RFLP it has now been well established, contrary to what was earlier believed, that the formation of fruit bodies is not a good reflection of the species composition and species frequency of the ectomycorrhiza community structure in any ecosystem (Buscot et al., 2000; Erland and Taylor, 1999). Instead it seems that a few species are very common and widely spread, some of which were almost never encountered in fruit body inventories.

Gene expression in mycorrhizal systems has been studied for some 10 years and these studies have recently been reviewed (e.g., Harrison, 1999; Lapopin and Franken, 2001; Martin and Tagu, 1999; Martin et al., 1999; Tagu et al., 2000). However, until now, it has not been possible to study more than one or a few genes simultaneously. Considering the complexity of the symbiosis, with two independent partners having to establish themselves separately, and then, as symbiotic partners, each reacting to environmental stimuli, etc., it is clearly disadvantageous to be restricted to single-gene studies. With the advent of sequencing facilities and the progress made in the large scale genome sequencing programmes, completely new opportunities are now available for global expression analysis using cDNA libraries for microarray analysis. The potential of this method has been widely discussed (e.g., Brown and Botstein, 1999; Harrington et al., 2000; Richmond and Summerville, 2000) and the first results obtained by applying these methods have been published (e.g., yeast: Jia et al., 2000; *Arabidopsis*: Schaffer et al., 2001). An attempt to apply these methods for classification in cancer diagnosis (Kahn et al., 2001) resulted in an editorial comment entitled 'Microarrays — the 21st century divining rod' (He and Friend, 2001). The large-scale gene expression analysis which is today within reach via mRNA profiling and bioinformatics may have an immense impact on research on all complex biological systems, not least the mycorrhizal system.

The first global gene expression data from mycorrhizal ecosystems have recently been published. Voiblet et al. (2001) presented results of an analysis of symbiosis-related genes in *Eucalyptus globus* – *Pisolithus tinctorius* ectomycorrhiza while Johansson et al. (2001) studied *Betula pendula* – *Paxillus involutus* ectomycorrhiza. The first authors performed a more detailed analysis of the differential expression of 486 selected genes in the fungus, host plant and mycorrhiza, respectively, while the latter authors reported a number of differences in the first functional profiling of the 1900 putatively unique transcripts from the fungus, plant and mycorrhiza. Both these papers describe the first steps towards the possible microarray analysis of the systems in order to describe the gene expression during formation of the symbiosis, as indicated by Martin (2001).

At present, progress in genomic research is extraordinarily fast, and we can still only speculate on the consequences of the extensive sequencing programmes in progress. However, putting this progress to good use presents a major challenge for mycorrhizal researches, as discussed in more detail by Martin (2001) and Franken and Requena (2001). As Martin (2001) points out: "While familiar with the practicalities of making a PCR or a biochemical analysis, many scientists lack a 'nuts-and-bolts' appreciation of the pros and cons of functional genomics". This is a major challenge for mycorrhizal research, to get into the fast lane of functional genomics, and not to be left behind in the fast progress of this expanding field.

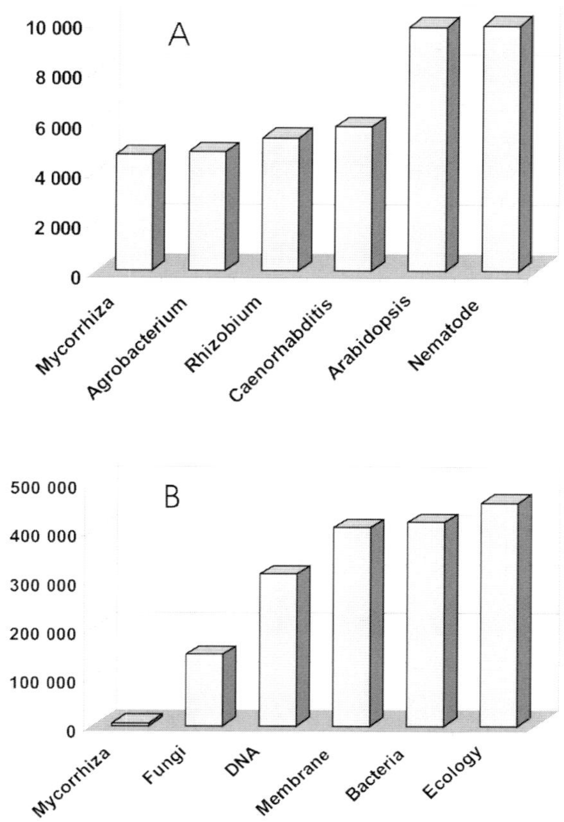

Figure 1. (A,B) Numbers of publications between 1993 and April 2001 in different subject areas, extracted from the BIOSIS database.

The total mycorrhiza research effort

Mycorrhizal symbiotic systems are not only of the utmost ecological importance, for agricultural and forestry production, but also for the sustainability of these and of natural ecosystems. As discussed above, they may also be of primary importance for the global carbon balance which influences climate change. However, they also form superb model systems for studies of interactions between organisms. Awareness of the importance of mycorrhizal symbioses has increased. However, as indicated earlier in this paper it does not seem that the mycorrhizal research community has increased relative to other fields.

Literature databases can be used to obtain an indication of the relative volume of research. A search in the BIOSIS database reveals that the number of articles from 1993 to spring 2001 on 'mycorrhiza' is 4600, and in Figure 1A it can be seen that this is almost the same as the number of papers published on, e.g. *Rhizobium* (incl. *Bradyrhizobium*) or *Agrobacterium*. All publications on nitrogen fixation (incl. *Rhizobium*) amount to 7500. The total volume of research on mycorrhiza thus seems to be equal to that on *Rhizobium*, the most important nitrogen-fixing bacteria (very few papers are published on other nitrogen fixing systems) but is considerably less than the total nitrogen fixation volume. Papers on the use of *Agrobacterium* as a plant transgenic agent (which is probably the major share of the articles with the word '*Agrobacterium*') are equal to mycorrhiza in numbers. The volume of research on *Arabidopsis*, the first plant for which the complete genome was sequenced, and one of the general genetical model organisms, is about twice as large as that on all mycorrhizal research.

A search on more unspecific subjects naturally results in many more publications. Searching on a word like 'fungi' results in 146 000 articles (Figure 1B), indicating that the total volume of research on this group of organisms is about 30 times larger than that on mycorrhiza. Searching for a word like 'membrane' will give 400 000 articles and results from some other wider areas are also shown in Figure 1B.

Considering the breadth of the research in mycorrhiza, as is well illustrated by the abstracts from the conferences discussed above, the volume of mycorrhizal research seems small compared with other more specialised areas of research. However, at the same time it appears that contributions on mycorrhizal research are sparse at other symposia of obvious relevance and it thus seems that reports on the most current advances in mycorrhiza research are not communicated as efficiently as one might expect. In the Proceedings of The 8th International Symposium on Microbial Ecology, Halifax, 2000 (Bell et al., 2000) only 10 pages out of 934 dealt with mycorrhiza. Another example is the XXI International Fungal Genetics Meeting in California in 2001: of the 502 abstracts in the abstract book, only five were mycorrhiza related. Similar figures can be found for botany and plant physiology. A major challenge for mycorrhizal researchers in the years to come will thus be to place mycorrhizal symbiosis more centrally in the expanding research efforts in biology.

Conclusions

Global climate change is of great concern and is of fundamental importance for our planet. Since mycorrhiza may play a critical role in terrestrial carbon exchange processes, a deeper understanding of these processes is urgently needed.

As a consequence of the fast and revolutionary development in genomic and postgenomic research, extraordinary progress in understanding complex biological systems can be expected in the years to come. It is imperative to include this development in all long-term biological research planning. This calls for strong interactions between different disciplines of science.

Acknowledgements

Economical support from the Swedish Research Council and the Royal Physiographical Society in Lund is gratefully acknowledged.

References

Bidartondo M I, Ek H, Wallander H and Söderström B 2001 Carbon sink strength in relation to ammonium and apatite additions of *Suillus pungens, Rhizopogon* section *Amylopogon* spp., and *Paxillus involutus* in ectomycorrhizal association with *Pinus muricata*. New Phytol. 151, 543–550.

Bell C R, Brylinsky M and Johnson-Green P 2000 Microbial biosystems: new frontiers. Proceedings from the 8th International Symposium on Microbial Ecology, Halifax, August 1998. Atlantic Society for Microbial Ecology

Brent R 2000 Genomic biology. Cell 100, 169–183.

Brown P O and Botstein D 1999 Exploring the new world of the genome with DNA microarrays. Nat. Genet. 21, 33–37.

Buscot F, Munch J C, Charcosset J Y, Gardes M, Nehls U and Hampp R 2000 Recent advances in exploring physiology and biodiversity of ectomycorrhizas highlight the functioning of these symbioses in ecosystems. FEMS Microbiol. Rev. 24, 601–614.

Franken P and Requena N 2001 Analysis of gene expression in arbuscular mycorrhizas: new approaches and challenge. New Phytol. 150, 517–524.

Ek H 1997 The influence of nitrogen fertilization on the carbon economy of *Paxillus involutus* in ectomycorrhizal association with *Betula pendula*. New Phytol. 125, 133–142.

Erland S & Taylor A F S 1999. Resupinate ectomycorrhizal fungal genera. *In* Ectomycorrhizal Fungi – Key Genera in Profile. Eds J W G Cairney and S M Chambers. pp 347–363. Springer, Heidelberg, Germany.

Harrington C A, Rosenow C and Retief J 2000 Monitoring gene expression using DNA microarrays. Curr. Opin. Microbiol. 3, 285–291.

Harrison, M J 1999 Molecular and cellular aspects of the arbuscular mycorrhizal symbiosis. Annu. Rev. Plant Physiol. Plant Mol. Biol. 50, 491–503.

He, Y D and Friend S H 2001 Microarrays – the 21st century divining rod? Nat. Med. 7, 658–659.

van der Heijden G A, Klironomos M U, Moutoglis P, Streitwolf-Engel T B, Wiemken A and Sanders I R 1998 Mycorrhizal fungal diversity determines plant biodiversity, ecosystem variability and productivity, Nature 396, 69–72.

Högberg P, Nordgren A, Buchmann N, Taylor A F S, Ekblad A, Högberg M N, Nyberg G, Ottosson-Löfvenius M and Read D J 2001 Large-scale forest girdling shows that current photosynthesis drives soil respiration. Nature 411, 789–792.

IPCC third assessment report — Climate change 2001 Summary for Policymakers — A report from Working Group I of the Intergovernmental Panel on Climate Change. Technical Summary — A report accepted by Working Group I of the IPCC but not approved in detail. http://www.ipcc.ch

Jakobsen I 1995 Transport of phosphorus and carbon in vesicular-arbuscular mycorrhizas. *In* Mycorrhiza. Eds A Varma and B Hock. pp 295–324. Springer, Berlin, Heidelberg.

Jakobsen I and Rosendahl L 1990 Carbon flow into soil and external hyphae from roots of mycorrhizal cucumber plants. New Phytol. 115, 77–84.

Jia M, Larossa R A, Lee J-M, Rafalski A, Derose E, Gonye G and Xue Z 2000 Global expression profiling of yeast treated with an inhibitor of amino acid biosynthesis, sulfometuron methyl. Physiol. Genome. 3, 83–92.

Johansson T, Ahrén D, Le Quéré A, Söderström B, Erlandsson R, Lundeberg J, Uhlén M and Tunlid A 2002 Differentially regulated genes during ectomycorrhiza development: a transcriptome analysis of the ectomycorrhizal root tissue of *Paxillus involutus* and birch (*Betula pendula*). Submitted.

Khan J, Wei J S, Ringnér M, Saal L H, Ladanyi M, Westermann F, Berthold F, Schwab M, Antonescu C R, Petersen C and Meltzer P S 2001 Classification and diagnostic prediction of cancers using gene expression profiling and artificial neural networks. Nat. Med. 7, 673–679.

Lapopin L, and Franken P 2001 Modification of plant gene expression. *In* Arbuscular Mycorrhizas: Molecular Biology and Physiology. Eds Y Kapulnik and D Douds. pp 64–84. Kluwer Academic Publishers, Dordrecht.

Martin F 2001 Frontiers in molecular mycorrhizal research – genes, loci, dots and spins. New Phytol. 150, 499–507.

Martin F and Tagu D 1999 Developmental biology of a plant-fungus symbiosis: the ectomycorrhiza. *In* Mycorrhiza: Structure, Molecular Biology and Function, 2nd ed. Eds. A Varma and B Hock. pp 51–73. Springer, Heidelberg.

Martin F, Laurent P, de Carvalho D, Voiblet C, Balestrini R, Bonfante P and Tagu D 1999 Cell wall proteins of the ectomycorrhizal basidiomycete *Pisolithus tinctorius*: identification, function, and expression in symbiosis. Fungal Gen. Biol. 27, 161–174.

Richmond T and Somerville S 2000 Chasing the dream: plant EST microarrays. Curr. Opin. Plant Biol. 3, 108–116.

Rygiewicz P T and Andersen C P 1994 Mycorrhizae alter quality and quantity of carbon allocated below ground. Nature 369: 58–60.

Schaffer R, Landgraf J, Accerbi M, Simon V, Larson M and Wisman E. 2001. Microarray analysis of diurnal and circadian-regulated genes in *Arabidopsis*. Plant Cell 13, 113–123.

Schimel D S 1995 Terrestrial ecosystems and the carbon cycle. Global Change Biol. 1, 77–91.

Smith F A, Jakobsen, I and Smith S E 2000 Spatial differences in acquisition of soil phosphate between two arbuscular mycorrhizal fungi in symbiosis with *Medicago truncatula*. New Phytol. 147, 357–366.

Smith S E and Read D J 1997 Mycorrhizal Symbiosis. 2nd ed. 605 pp. Academic Press, San Diego, CA.

Söderström B and Read D J 1987 Respiratory activity of intact and excised ectomycorrhizal mycelial systems growing in unsterilized soil. Soil Biol. Biochem. 19, 231–236.

Tagu D, Lapeyrie F, Ditengou F, Lagrange H, Laurent P, Missoum N, Nehls U. and Martin F. 2000 Molecular aspects of ectomycorrhiza development. *In* Current Advances in Mycorrhiza Research. Eds G K Podila and M N Douds. pp 69–89. American Phytopathological Society Press, St. Paul, MN.

Treseder K K and Allen M F 2000 Mycorrhizal fungi have a potential role in soil carbon storage under elevated CO_2 and nitrogen deposition. New Phytol. 147, 189–200.

Voiblet C, Duplessis S, Encelot N and Martin F 2001 Identification of symbiosis-related genes in *Eucalyptus globus — Pisolithus tinctorius* ectomycorrhiza by differential hybridization of arrayed cDNAs. Plant J. 25, 1–12.

Plant and Soil **244:** 9–17, 2002.

Molecular and functional diversity in soil micro-organisms

James I. Prosser
*Department of Molecular and Cell Biology, University of Aberdeen, Institute of Medical Sciences, Foresterhill, Aberdeen, AB25 2ZD, UK**

Received 19 September 2001. Accepted in revised form 15 January 2002

Key words: bacteria, biodiversity, functional diversity, fungi, molecular ecology, ribosomal RNA, rRNA, soil, species diversity

Abstract

Traditional approaches to the study of microbial diversity have relied on laboratory cultivation of isolates from natural environments and identification by classical techniques, including analysis of morphology, physiological characteristics and biochemical properties. These approaches provide information on fine-scale diversity but suffer from bias, resulting from the media and cultivation conditions employed, and from the inability to grow and isolate significant proportions of natural communities in the laboratory. An alternative approach is the amplification of ribosomal RNA and functional genes from nucleic acids extracted directly from environmental samples, with subsequent analysis by 'fingerprinting' methods or by sequencing and phylogenetic analysis. This approach avoids the need for laboratory cultivation and has provided major insights into species and functional diversity of bacterial and archaeal populations. This article reviews molecular approaches to the characterisation of prokaryote diversity in natural environments, their more recent application to fungal diversity and the advantages and limitations of molecular analyses.

Introduction

Micro-organisms are essential for the functioning and sustainability of all natural ecosystems but are frequently ignored due to their small size and consequent methodological difficulties in detecting cells and assessing their activity. Methodology has also severely limited characterisation and analysis of species composition, species diversity and structure of microbial communities. The central role of microorganisms in ecosystem processes, including biogeochemical cycling of nutrients and biodegradation, makes it both unwise and dangerous to ignore any aspect of their ecology. Analysis of diversity is particularly important when considering the ability of ecosystems to respond to changing environmental conditions, the need for conservation of the microbial gene pool and the links between diversity, ecosystem processes, functional and physiological diversity, resilience and sustainability. This article describes the major recent advances in techniques for the characterisation of microbial diversity which provide the potential to address these important issues.

Several approaches are available for assessment of microbial diversity (Atlas, 1984). For those groups with distinctive morphological characteristics, direct observation may be appropriate. This approach is, however, very limited, particularly for analysis of diversity in complex environments, such as soil, and when quantification of measures of diversity is required. Until recently, the only approaches available for assessment of diversity were those based on laboratory cultivation of organisms from environmental samples. Typically, cell suspensions were prepared from environmental samples, dilution series constructed and samples plated onto solid medium. Following incubation, pure cultures obtained from isolated colonies were identified using well-established phenotypic characters. This approach provides information on which species are present, while the relative abund-

* FAX No: +44-1224-273144.
E-mail: j.prosser@abdn.ac.uk

ance of different organisms can be used to determine measures of diversity, e.g. diversity indices, evenness and dominance. The potential limitations of this approach are widely acknowledged and accepted (Torsvik, 1996). Separation of biomass from particulate material will vary between species and with growth form (cells, spores, mycelia), introducing bias. More significantly, it is impossible to design growth media and cultivation conditions that are suitable for all members of the community. Selection of particular groups is inevitable and some organisms may not be able to grow under any of the cultivation conditions utilised. Others will be viable but non-culturable or dormant, requiring lag periods that lead to their being out-competed during incubation or prevent detectable growth within realistic incubation periods. These biases are accentuated when quantification of diversity is required and, for filamentous fungi, are further complicated by difficulties in quantifying and relating biomass derived from hyphae and spores, and consideration of which proportion of mycelial biomass is active. Identification using traditional methods, based on phenotypic characteristics, also raises problems.

Despite these considerable and well-accepted limitations, cultivation-based assessment of microbial diversity has been employed, largely because of the lack of acceptable alternatives, and advances have been made. For example, identification can be accelerated by automated methods, e.g. Biolog, and broad scale measures of diversity, e.g. phospholipid fatty acid (PLFA) profiling (Tunlid et al., 1989), fatty acid methyl ester (FAME) profiling (Cavigelli et al., 1995), DNA hybridisation and re-association (Torsvik et al., 1990), have overcome some of the limitations of cultivation-based techniques. Some of these provide information on functional diversity, e.g. Biolog, which can otherwise be obtained by physiological characterisation of laboratory isolates. They do not, however, enable the fine-scale analysis of diversity that is required to determine the importance of species diversity, and of particular species, on ecosystem function.

Molecular approaches

An alternative approach, which removes many of the above limitations, is the analysis of genes within environmental samples. These genes may be functional genes, i.e. those coding for proteins performing particular metabolic reactions of relevance to ecosystem processes. However, most applications have analysed genes encoding the small subunit (SSU) of ribosomal RNA. Analysis of 16S rRNA genes is now widely used for analysis of bacterial populations, and analysis of 18SS rRNA genes and internal transcribed spacer (ITS) regions is increasingly being used to analyse fungal populations. Ribosomal rRNA genes are ideal for this purpose in that they possess regions with sequences conserved between all bacteria or fungi, facilitating alignment of sequences when making comparisons, while other regions exhibit different degrees of variation, enabling distinction between different groups. These differences provide the basis for a phylogenetic taxonomy and enable quantification of evolutionary differences between different groups. Discrimination of bacteria, using 16S rRNA gene sequences, is greater than that that for fungi, using 18S rRNA sequences, but finer scale information may be obtained by analysis of ITS regions.

The presence of regions of rDNA sequence with different degrees of conservation enables the identification of sequences that are common to all bacteria or fungi, or to specific phylogenetic groups, sometimes to the level of species. These sequences may then be used to design primers for the specific amplification, using the polymerase chain reaction (PCR), of rRNA genes belonging to particular groups or to design specific probes for these groups. These primer sequences provide the basis for analysis of species in natural populations. Two approaches may be adopted, the first based on PCR amplification of rRNA genes and the second involving *in situ* detection of rRNA within cells.

Analysis of amplified genes

The first stages in the analysis of rRNA genes in an environmental sample are cell lysis and extraction of DNA, after which DNA is purified to remove material inhibitory to subsequent enzymatic reactions. PCR amplification is then carried out, using primers specific to the microbial groups of interest. Amplification generates a population of rRNA genes, or gene fragments, of equal size, determined by the particular primers used. This population of gene fragments is considered to be representative of the natural microbial population.

Most information, and fine scale discrimination between groups, is obtained by cloning the amplified rRNA genes and sequencing members of the clone library. Comparison of sequences with those in databases determines which phylogenetic groups are present and, in many cases, enables more detailed identification. This approach is particularly useful for studies of bacteria, as 16S rRNA databases are now extensive and comprehensive. They contain sequences of large numbers of laboratory cultures and also of clones obtained from a range of environments, which are not represented in laboratory cultures. Fungal databases are increasing in size, but are currently less extensive and informative, except for phylogenetic groups that have been the subject of detailed taxonomic study Phylogenetic analysis provides information on the numbers of sequences falling within different groups and may identify novel groups with no sequences of cultured organisms. Sequences obtained in this manner may be used to refine existing primers and probes, increasing their sensitivity and specificity, and to design primers and probes for novel groups of uncultured microorganisms. Finally, if sufficiently large numbers of clones are sequenced, estimates may be obtained of the relative abundance of different groups.

More rapid analysis is achieved using fingerprinting techniques. The most commonly used technique in 16S rRNA studies has been denaturing gradient gel electrophoresis (DGGE, Muyzer et al., 1993), which separates products of the same size, but different sequence, by chemical denaturation. Following staining of gels, banding patterns may be used to compare communities, or to compare the same community following perturbations, and band intensities may be used for semi-quantification of relative abundance (McCaig et al., 1999, 2001). In addition, bands may be excised and genes amplified and sequenced for fine scale analysis. A similar approach is adopted in temperature gradient gel electrophoresis (TGGE), where denaturation results from high temperatures (Felske et al., 1998). A number of fingerprinting techniques involve restriction analysis of PCR products, including terminal restriction length polymorphism (tRFLP, Liu et al., 1997) and amplified ribosomal DNA restriction analysis (ARDRA) (Øverås and Torsvik, 1998). In some cases, database information may be used to predict the banding patterns generated using these techniques by particular rRNA gene sequences, providing some information on the identity of organisms present. PCR-based methods, such as competitive PCR (Jansson and Leser, 1996) and real-time PCR (Heid et al., 1996) are also used to quantify gene copies, and hence cell number or biomass.

Fluorescence in situ *hybridisation (FISH)*

Sequence information can also be used to design and construct fluorescently labelled oligonucleotide probes specific for particular microbial groups. These can then be used to detect cells *in situ* using FISH (Amann et al., 1990). This technique relies on the large number of ribosomes within cells, particularly actively growing cells, which greatly increases sensitivity in comparison to detection of single gene copies. This approach is obviously of great value in determining the location and spatial organisation of different organisms and signal strength may be related to cell activity (but see below). It is also useful in providing supporting evidence for quantitative estimates of relative abundance obtained using PCR-based approaches.

Benefits and limitations

The benefits of molecular techniques have been indicated above, the most important being the ability to characterise diversity without the requirement for laboratory cultivation. In addition, molecular identification, where possible, is more rapid than cultivation-based methods and is becoming the method of choice for commercial identification of many microbial groups.

Molecular techniques avoid problems that necessarily arise from selective media and cultivation conditions and the non-culturability of many cells, but introduce other sources of bias. Lysis efficiency will vary between microbial groups, between spores and mycelia and between regions of mycelia of different ages, influencing DNA extraction yields. Subsequent purification of DNA will reduce yield, potentially also with bias. PCR amplification and use of primers can lead to differences in the proportions of different rRNA genes and sequencing errors, formation of chimeric sequences and introduction of errors during cloning can influence results. *In situ* probing requires semi-permeabilisation of cells, conditions for which will vary within mixed populations. There are also problems associated with molecular-based quantification. Sequence abundance may not necessarily relate directly to cell numbers, for example, through variation in rDNA copy number between different organisms. Considerable care must be exercised when attempting to quantify following PCR amplification and estimation of relative abundance from DGGE gels must be considered semi-quantitative only. As we will see below, phylogenetic information can provide some information on the physiology of organisms, through knowledge of common characteristics of cultivated organisms belonging to particular groups. However, this is not always the case, limiting interpretation of species diversity assessments with regard to ecosystem function.

It is possible to test for many of these potential problems and, in general, cross-referencing and use of several independent techniques indicate that molecular approaches provide true reflection of diversity and community structure within natural assemblages. However, the molecular approach is in its infancy and the importance of these limitations is not yet fully understood, and development of solutions is at an early stage. Molecular techniques have provided enormous advances in our understanding of natural bacterial populations. In the following sections, these will be exemplified by studies of 'total' bacteria and of a more specialised group, ammonia oxidising bacteria, followed by examples of molecular characterisation of 'total' fungal populations. Approaches being developed to assess the mechanisms controlling and driving diversity, and the significance of microbial diversity for ecosystem processes will then be discussed.

Bacterial diversity in soil

Analysis of total bacterial populations can be achieved by PCR amplification of extracted DNA using universal primers, designed to amplify 16S rRNA genes of bacteria and archaea. The following examples typify application of this approach to soil bacterial populations.

In the first, Borneman et al. (1996) analysed 16S rRNA gene clone libraries from East Amazonian soil. Of 124 clones sequenced, 98.4% fell within the *Bacteria* the remainder being mitochondrial DNA. There was considerable diversity, with only 40 of sequences being duplicated, and approximately 25% of sequences fell within five groups that could not be assigned to previously characterised groups within the *Bacteria*. Although they did not obtain isolates, the sequences from extracted DNA differed significantly from those of cultured organisms typically isolated from soil. For example, only 19% and 0.8% of sequences were representative of bacilli and actinomycetes, respectively, and no sequences were found that were representative of pseudomonads, agrobacteria, *Alcaligenes* or *Flavobacterium*. Similar results were found in a study of Brazilian forest and pasture soils (Borneman and Triplett, 1997), with no duplicate sequences in 100 clones analysed, 98 and 2 of which belonged to the *Bacteria* and *Archaea* respectively. No sequence was identical to sequences in databases and 19 were only distantly related to known sequences, suggesting the possibility of new kingdoms. Differences were also found between pasture and forest soils. Kuske et al. (1997), investigating a cinder site (arid, thermal soils with low organic carbon) and a sandy loam soil also found no duplicates in clone libraries and no identity to sequences in databases. Of 56 sequences analysed, 20 were affiliated with known bacterial groups but the remainder fell within five distinct novel groups, different at the division level, which also included novel sequences from other studies. RFLP analysis of 600 cultured isolates from these soils gave no identity with the novel divisions.

While taking into account the potential biases and limitations of 16S rDNA-based techniques, the significance of these findings is remarkable. They suggest that a substantial proportion (sometimes greater than 50%) of soil prokaryotes belong to groups with no representative in laboratory culture. The lack of cultured representatives, and our consequent ignorance of their metabolic and physiological characteristics, prevent meaningful speculation regarding their role in biogeochemical cycling and other environmental processes.

More recent studies have attempted to use molecular techniques, including DGGE analysis of 16S rDNA, to quantify species diversity, evenness and dominance, and to determine the effects of soil management regimes and treatments on prokaryotic diversity (McCaig et al., 1999a, 2001). The high levels of diversity within individual soil samples make it difficult to detect differences between treatments and collector's curves indicate that, even when sequences are grouped at the 95% level of homology, analysis of several hundred clones is required before adequate coverage is achieved.

Diversity of ammonia oxidising bacteria

Autotrophic ammonia oxidising bacteria were traditionally placed in five genera, on the basis of a limited number of phenotypic characters. Analysis of 16S rRNA genes of pure cultures (Head et al., 1993, Purkhold et al., 2000) defined one genus, *Nitrosococcus oceanus*, in the λ-proteobacteria and two genera, *Nitrosomonas* and *Nitrosospira*, in the β-proteobacteria which comprise the majority of strains found in the environment. Ammonia oxidisers grow slowly in the laboratory, with low biomass yields, and isolation and identification are difficult. As a consequence, traditional approaches provided little information on their natural diversity and 16S rDNA-based molecular techniques have proved to be invaluable. Primers specific to β-proteobacterial ammonia oxidisers have been used to amplify and analyse 16S rRNA genes from DNA extracted from a range of environmental samples and the results are typical of those obtained for other microbial groups with specialised function.

Sequence diversity within ammonia oxidisers appears to be high, with very few duplicate 16S rDNA sequences in clone libraries. Sequences frequently differ by only a few bases, possibly through sequencing and cloning errors, but the length of sequence amplified is relatively small (approximately 300 bases) and sequencing of larger stretches or of functional genes (e.g. ammonia monooxygenase) reveals greater diversity. Initial phylogenetic analysis of culture and environmental clone sequences suggested the existence of seven phylogenetic clusters within the *Nitrosomonas* and *Nitrosospira* genera (Stephen et al 1996), but subsequent analysis indicates the possible existence of more (Purkhold et al., 2000). The different clusters are associated with different environments. For example, *Nitrosomonas europaea* cluster 6 and *Nitrosospira* clusters 1 and 5 are obtained from marine environments and *Nitrosospira* clusters 2, 3 and 4 from terrestrial and freshwater environments. Clusters 1 and 5 have no representative in laboratory culture and we cannot therefore predict their distinctive physiological characteristics or ecological role. It is possible, but unlikely, that they are not ammonia oxidisers. Genera previously considered to be restricted to particular environments are now recognised as being more widely distributed. For example, *Nitrosospira* was considered to be absent from marine environments but is now routinely detected.

The ammonia oxidiser type strain *Nitrosomonas europaea* originally isolated from soil, has been the subject of the majority of physiological studies of ammonia oxidation, but sequences related to this strain are rarely encountered and *Nitrosospira* appears to be more important in these environments. This highlights the potential dangers of studies based on laboratory cultivation and the advantages of molecular techniques, which enable us to determine which enrichments and isolates are representative of natural populations. Sequences amplified from directly extracted DNA and from cultures derived from the same environmental samples (Smith et al., 2001) differ significantly in their distribution between ammonia oxidiser clusters and duplicate sequences are found more frequently in enrichment cultures. This indicates laboratory selection of particular groups, but a greater understanding of biases associated with molecular techniques is required before deciding which is the correct picture of natural diversity.

Characterisation of diversity and cluster divisions enabled assessment of the influence of environmental conditions on ammonia oxidiser diversity and community structure. For example, differences have been observed in ammonia oxidiser communities in polluted sediment beneath fish cages (McCaig et al., 1999b), in marine aggregates (Phillips et al., 1999) and in acid soils (Bruns et al., 1999; Stephen et al., 1998). These studies demonstrate the ability of 16S rRNA methods to detect changes in community composition with changes in environmental conditions. They are providing the basis for an understanding of the mechanisms controlling diversity and, importantly, the relationship between species diversity, functional diversity and their impact on process rates.

Fungal studies

Molecular techniques have been used to detect and analyse fungal populations for several years and their specific application to mycorrhizal fungi will be described elsewhere in this volume. In many cases, these techniques have been used to characterise strains isolated on laboratory media or material extracted from hyphae. Studies such as those described above, in which fungal populations are characterised by analysis of DNA extracted from bulk samples, are more rare. This approach is, however, now increasing and several examples will be described. The major advantage is the ability to assess diversity without the requirement for laboratory growth and isolation of fungi. Some studies are based on small subunit rRNA (18S rRNA) but this region possesses less useful information than 16S rRNA genes of prokaryotes. To obtain greater specificity, others have analysed ITS regions. A major limitation, in comparison with bacterial studies, is the relative lack of sequence information in databases but this problem is decreasing and molecular identification, where possible, is more rapid.

Kowalchuk et al., (1997) used nested PCR and DGGE analysis to characterise fungal populations infecting *Ammophila arenaria* which are believed to lead to degeneracy in marram stands. Primers were originally designed to amplify 18S rRNA genes from pure cultures and were not specific, but there was little interference from other soil eukaryotic populations in the relatively specialised environment represented by plant roots. Distinct bands were produced from pure cultures of a range of fungi, but band position in itself does not necessarily provide phylogenetic information, which requires excision and sequencing of bands. Sequences of representative bands from DGGE analysis of healthy and degenerating roots (field samples) were sometimes identical or closely related to those of cultivated fungi, including some isolated from marram roots, but others were only distantly related to known fungal species or showed no clear affiliation with known fungal groups. Also, sequences of several fungi commonly isolated from these systems were not detected and there were indications of differences in populations from roots of healthy and degenerate marram stands.

Primer sets with greater specificity were developed by Smith et al. (1999) for analysis of fungal populations in the wheat rhizosphere. Two of the primer sets were suitable but with one, EF4–EF3, showing bias towards basidiomycetes and zygomycetes and a second, EF4–fung5, with bias towards ascomycetes. The 39 clones obtained using the EF4–fungs primers contained no basidiomycete sequence, while 22 clones derived from EF4–EF3 primers contained only two representatives of the ascomycetes. The 61 clones contained representatives of 24 species but in many cases sequences showed little similarity to those in databases, preventing identification. This reflects the relative paucity of sequence information in the databases at the time, but may also indicate the existence of novel groups that have not yet been cultivated. Banding patterns obtained by TGGE analysis of PCR products indicated bias and differences were found between fungal communities in bulk and rhizosphere soils in microcosms and following incubation of soil for 5 and 10 days. This approach, therefore, provides a means of studying changes in fungal communities in soil with the potential for identification by sequencing bands of interest from TGGE or DGGE gels.

Borneman and Hartin (2000) also developed fungal specific primers for analysis of two avocado grove soils, considered to be suppressive and conducive to infections by *Phytophthora cinnamomi*. Both primer sets were specific and successfully amplified 18S rRNA from pure cultures but differences were found in environmental clones sequences amplified by these primers and those of Smith (1999), emphasising the need for care in interpretation of results and potential primer bias and selectivity. To compare fungal population in disease-suppressive and disease-conducive soils, cloned 18S rRNA genes were sorted into 10 clone types and relative abundances determined for each soil. Significant differences were seen between soils, with four clone types absent from the disease-conducive soil. Comparison of sequences representative of each clone group with those in databases indicated that the most dominant genera differed significantly from those genera isolated using cultivation-based techniques.

Although primer specificity is desirable, in certain situations with low non-fungal diversity, less specific primers may be useful. For example, Vainio and Hantula (2000) analysed fungi colonizing Norway spruce stumps. The lack of non-fungal eukaryotic contaminants enabled comparison of wood colonisers isolated in pure culture and those detected by molecular techniques. In many cases, sequences obtained from extracted DNA and pure cultures were identical, but differences were found in 4 of 17 samples, suggesting selection of non-dominant strains during cultivation. 18S rRNA genes provide good coverage of fungal taxa but do not enable fine scale analysis. This can be obtained by analysis of ITS sequences and primers have been developed for amplification of ITS regions from extracted DNA. For example, Viaud et al. (2000) analysed ITS regions amplified from DNA extracted from soil and from fungi cultivated from the same soil samples. Amplified genes from 67 cultures and 51 clones were grouped on the basis of RFLP patterns and 58 representative ITS sequences were compared with those in databases. At the level of precision used in this study, only one of the cultivated fungi was represented in the clone library. In addition, all cultivated fungi were ascomycetes, except for one basidiomycete, while clone libraries contained a wider range of taxa.

Links between species diversity, physiology and ecosystem processes

Molecular techniques, as illustrated above, are capable of characterising the diversity, species composition and community structure of microorganisms in natural environments. While this information is of value, the most important questions concern the significance of this diversity for ecosystem function. Molecular studies have demonstrated differences in species composition in different environments and differences associated with different treatments. To understand the basis of these differences, we need to determine the characteristics of different components of natural communities and their respective ecosystem functions. Molecular techniques can contribute to such questions and, for unculturable organisms are essential. Examples of a number of approaches are given below.

Analysis of functional gene diversity

Although species composition can be characterised by analysis of 16S rRNA gene sequences, this may not tell anything about the functional diversity of the community, or even which functions can be carried out. Metabolic processes of ecological relevance may be distributed among several phylogenetic groups, while a single group is likely to be capable of performing many functions. This is particularly pertinent when considering, for example, transformations of organic carbon that will be carried out by many heterotrophic microorganisms. Genes for many processes have now been sequenced in a number of organisms. It is, therefore, possible to design primers for the amplification of specific genes, which can then be analysed, as described above for 16S rRNA genes. The diversity of gene sequences within an environment can be characterised and differences in relative abundance of functional genes belonging to different phylogenetic groups can be assessed. For some groups, and some functional genes, groups formed on the basis of functional gene analysis will be related directly to phylogenetic groups. However, this will not always be the case, for example, when functional genes are plasmid-borne or where genes have been transferred horizontally. An important issue in this approach is the choice of functional gene. The success of an organism in a particular environment will depend on a range of properties, rather than the product of a single gene. Functional genes must therefore be chosen with care.

Analysis of RNA

RNA may be used in two ways as an indicator of activity. Detection of mRNA for a particular gene can be used to assess gene expression and determine the response of a particular function to changes in environmental conditions. This may also provide information on which species are responding, if functional and taxonomic genes are related. This approach has been limited by the short half-life of mRNA and difficulties in extracting sufficient quantities for analysis before degradation.

The second approach assumes that more active and faster growing organisms possess more ribosomes and, consequently, higher levels of ribosomal RNA. Studies on the influence of growth rate on ribosomal levels have been restricted to a few organisms and the relationship is not simple, varying between different organisms with, for example, different rates of ribosome degradation following cessation of activity through starvation. Nevertheless, (semi-) quantification of RNA is used to indicate levels of activity of different components of a community and analysis of RNA provides significantly greater sensitivity, as cells typically contain tens of thousands of ribosomes. RNA may be extracted from soil and hybridised with specific, hierarchical probes to determine the relative abundance of different members of the community (Purdy et al., 1997). Alternatively, reverse transcriptase-PCR (RT-PCR) may be used to amplify rRNA genes using RNA, rather than DNA, as a target. Comparison of DGGE profiles following PCR amplification of rRNA genes from both extracted DNA and RNA distinguishes active and inactive populations on the basis of ribosome abundance and can be used to determine which members of the community become active under particularly conditions.

Stable isotope probing

Stable isotope probing (SIP) has enormous potential for determining which components of a microbial community are capable of incorporating a particular substrate, labelled with a heavy isotope. SIP was first used to determine which methanol-utilisers were active in soil. Following incubation of soil with ^{13}C-labelled methanol, DNA was extracted and ^{13}C and ^{12}-DNA separated by density gradient centrifugation. ^{13}C-labelled DNA will be derived only from organisms that have grown and assimilated methanol into DNA and these organisms can be detected by amplification of 16S rRNA genes or functional genes from ^{13}C-DNA, followed by sequence analysis. This approach therefore enables identification of organisms actively metabolising specific compounds *in situ*, with no requirement for laboratory growth. Radajewski et al. (2000) used this technique to distinguish methanol utilisation by two bacterial groups, the α-proteobacterial and *Acidobacterium* lineages, detecting sequences from uncultured organisms. One current limitation of this approach is its sensitivity, requiring levels of growth that enable incorporation of sufficient DNA for separation and detection of ^{13}C-DNA, although sensitivity can be increased by analysis of RNA, rather than DNA. In common with all methods involving substrate amendment, sampling and analysis must be carried out before significant turnover of the substrate occurs.

FISH combined with microautoradiography

The use of FISH, using rRNA probes, has been described above for cell detection in environmental samples. Microautoradiography provides a means of detecting individual active cells and involves incubation of cells in the presence of radiolabelled substrates, which are incorporated only by such cells. Total populations are visualised microscopically and active cells are detected by microautoradiography of material on the same slide. This is similar to combined use microautoradiography and immunofluorescence using specific antibodies, but antibodies are usually prepared from pure cultures of cultivated organisms.

Combining autoradiography with 16S rRNA probing provides greater discriminatory power, more targeted identification and detection of active cells belonging to phylogenetic groups with no known cultivated representative. This approach, termed STAR-FISH, has been used to detect the activity of marine plankton (Ouverney and Fuhrman, 1999) and cells in activated sludge Lee et al. (1999).

Links with traditional approaches

The most straightforward approach to determining the links between species diversity and ecosystem function is to isolate organisms considered to be important in ecosystem function and to assess their potential ecological impact through laboratory-based, physiological studies. Molecular studies can overcome one important limitation in this approach, by identifying which species are important. For example, rRNA-based fingerprinting techniques can determine the environmental factors that lead to changes in relative abundance of particular members of a community. These studies can now be used to inform traditional enrichment/isolation approaches, indicating which cultivated organisms represent those of relevance to ecosystem change. This prevents extensive physiological characterisation of those organisms that are selected by laboratory media and incubation conditions, but are of less importance in the environment. Molecular techniques should also aid in the discovery of conditions suitable for isolation of those organisms that are currently unculturable. Molecular ecological studies will indicate the conditions favouring their growth, informing cultivation conditions to be employed, and molecular analysis of enrichments will determine whether they are present initially and which cultivation conditions lead to their elimination or selection.

Conclusions

The application of molecular techniques has revolutionised our view of the diversity of bacteria in natural environments and is beginning to explain the forces leading to and driving changes in diversity and community structure. Early studies indicate that a similar approach can impact significantly on analysis of fungal diversity in the soil. The molecular approach eliminates a major limitation of traditional studies, the reliance on culturability, but introduces new limitations and biases, the significance of which will become apparent as studies evolve. However, comparative studies employing a range of techniques indicate that they provide a more accurate view of diversity in natural environments. They also allow more rapid identification of individual strains and of relative abundance of different groups. This, in turn, enables greater replication and a degree of quantification that provides the basis for more comprehensive application and testing of theoretical and conceptual models of microbial ecology, including links between diversity and ecosystem function. A combination of techniques is still required and traditional, cultivation-based approaches will retain an important role. Nevertheless, molecular approaches provide the potential for significant advances in our understanding of fungal diversity in natural ecosystems and offer a springboard for the application of genomics to environmental microbiology.

References

Amann R I, Krumholz L and Stahl D A 1990 Fluorescent-oligonucleotide probing of whole cells for determinative, phylogenetic, and environmental studies in microbiology. J. Bacteriol. 172, 762–770.

Atlas R M 1984 Diversity of microbial communities. Adv. Microbial Ecol. 7, 1–47.

Borneman J and Hartin R J 2000 PCR Primers that amplify fungal rRNA genes from environmental samples. Appl. Env. Microbiol. 66, 4356–4360.

Borneman J and Triplett E W 1997 Molecular microbial diversity in soils from Eastern Amazonia: Evidence for unusual microorganisms and microbial population shifts associated with deforestation. Appl. Env. Microbiol. 63, 2647–2653.

Borneman J, Skroch P W, O'Sullivan K M, Palus, J A, Rumjanek N G, Jansen J L, Nienhuis J and Triplett E W 1996 Molecular microbial diversity of an agricultural soil in Wisconsin. Env. Microbiol. 62, 1935–1943.

Bruns M A, Stephen J R, Kowalchuk G A, Prosser J I and Paul E 1999 Comparative diversity of ammonia oxidizer 16S rRNA gene sequences in native, tilled, and successional soils. Appi. Env. Microbiol. 65, 2994–3000.

Cavigelli M A, Robertson G P and Klug M J 1995 Fatty acid methyl ester (FAME) profiles as measures of soil microbial community structure. In Collins HP, Robertson G P and Klug M. J (Eds) The Significance and Regulation of Soil Biodiversity. pp 99–113 Kluwer Academic Publishers, Dordrecht, The Netherlands.

Felske A, Akkermans A D L and De Vos W M 1998 Quantification of 16S rRNAs in complex bacterial communities by multiple competitive reverse transcription PCR in temperature gradient gel electrophoresis fingerprints. Appl. Env. Microbiol. 64, 4581–4587.

Head, I M, Hiorns W D, Embley T M, McCarthy A J and Saunders J R 1993 The phylogeny of autotrophic ammonia-oxidizing bacteria as determined by analysis of 16S ribosomal RNA gene sequences. J. Gen. Microbiol. 139, 1147–1153.

Heid C A, Stevens J, Livak K J and Williams P M 1996 Real time quantitative PCR. Genome Res 6, 986–994

Jansson J K and Leser T D 1996 Quantitative PCR of environmental samples. Molecular Microbial Ecology Manual 2.7.4, 1–19.

Kowalchuk G A, Gerards S and Woldendorp J W 1997 Detection and characterization of fungal infections of *Annophila arenaria* (marram grass) roots by denaturing gradient gel electrophoresis of specifically amplified 18S rDNA. Appl. Env. Microbiol. 63, 3858–3865.

Kuske C R, Barns S M and Busch J D 1997 Diverse uncultivated bacterial groups from soils of the arid Southwestern United States that are present in many geographic regions. Appl. Env. Microbiol. 63, 3614–3621.

Lee N, Nielsen P H, Andreasen K H, Juretschko S, Nielsen J L, Schleifer K-H and Wagner M 1999 Combination of fluorescent *in situ* hybridization and microautoradiography – a new tool for structure-function analyses in microbial ecology. Appl. Env. Microbiol. 65, 1289–1297.

Liu W T, Marsh T L, Cheng H and Forney L J 1997 Characterization of microbial diversity by determining terminal restriction fragment length polymorphisms of genes encoding 16S rRNA. Appl. Env. Microbiol. 63, 4516–4522.

McCaig A E, Glover L A and Prosser J I 1999a Molecular analysis of eubacterial community structure and diversity in unimproved and improved upland grass pastures. Appl. Env. Microbiol. 65, 1721–1730.

McCaig A E, Phillips, C J, Stephen J R, Kowalchuk G A, Harvey S M, Herbert R A, Embley T M and Prosser J I 1999b Nitrogen cycling and community structure of β-subgroup ammonia oxidising bacteria within polluted, marine fish-farm sediments. Appl. Env. Microbiol. 65, 213–220.

McCaig A E, Glover L A and Prosser J I 2001 Numerical analysis of grassland bacterial community structure under different land management regimes using 16S rDNA sequence data and DGGE banding patterns. Appl. Env. Microbiol. 67, 4554–4559.

Muyzer G, de Waal E C and Uitterlinden A G 1993 Profiling of complex microbial populations by denaturing gradient gel electrophoresis analysis of polymerase chain reaction-amplified genes coding for 16S rRNA. Appl. Env. Microbiol. 59, 695–700.

Ovreas L and V Torsvik 1998 Microbial diversity and community structure in two different agricultural soil communities. Microbial. Ecol. 36, 303–315.

Ouverney C C and Fuhrman J A 1999 Combined microautoradiography—16S rRNA probe technique for determination of radioisotope uptake by specific microbial cell types *in situ*. Appl. Env. Microbiol. 65, 1746–1752

Phillips C J, Smith Z, Embley T M and Prosser J 11999 Phylogenetic differences between particle-associated and planktonic β-proteobacteria ammonia oxidising bacteria in the North Western Mediterranean Sea. Appl. Env. Microbiol. 65, 779–786

Purdy K J, Nedwell D B, Embley T M and Takii S 1997 Use of 16S rRNA-targeted oligonucleotide probes to investigate the occurrence and selection of sulphate-reducing bacteria in response to nutrient addition to sediment slurry microcosms from a Japanese estuary. FEMS Microbiol. Ecol. 24, 221–234.

Purkhold U, Pommerening-Roser A, Juretschko S, Schmid M C, Koops H P and Wagner M 2000 Phylogeny of all recognized species of ammonia oxidizers based on comparative 16S rRNA and amoA sequence analysis. Implications for molecular diversity surveys. Appl. Env. Microbiol. 66, 5368–5382.

Radajewski S, Ineson P, Parekh N R and Murrell J C 2000 Stable-isotope probing as a tool in microbial ecology. Nature 403, 646–649.

Smit E, Leeflang P, Glandorf B, Van Elsas J D and Wernars K 1999 Analysis of fungal diversity in the wheat rhizosphere by sequencing of cloned PCR-amplified genes encoding 18S rRNA and temperature gradient gel electrophoresis. Appl. Env. Microbiol. 65, 2614–2621.

Smith Z, McCaig A E, Stephen J R, Embley T M and Prosser J I 2001 Species diversity of uncultured and cultured populations of soil and marine ammonia oxidising bacteria. Microbial. Ecol. 42, 228–237.

Stephen J R, Kowalchuk G A, Bruns M A, McCaig A E, Phillips C J, Embley, TM and Prosser J I 1998 Analysis of β-subgroup ammonia oxidiser populations in soil by DGGE analysis and hierarchical phylogenetic probing. Appl. Env. Microbiol. 64, 2958–2965.

Stephen J R, McCaig A E, Smith Z, Prosser J I and Embley T M 1996 Molecular diversity of soil and marine 16S rDNA sequences related to β-subgroup ammonia oxidising bacteria. Appl. Env. Microbiol. 62, 4147–4154.

Torsvik V, Salte K, Sorheim R and Goksoyr J 1990 Comparison of phenotypic diversity and DNA heterogeneity in a population of soil bacteria. Appl. Env. Microbiol. 56, 776–781

Torsvik V, Sorheim R and Goksoyr J 1996 Total bacterial diversity in soil and sediment communities – a review. J. Ind. Microbiol. 17, 170–178.

Tunlid A, Hoitink H A J, Low C and White D C 1989 Characterization of bacteria that suppress *Rhizoctonia* damping-off in bark compost media by analysis of fatty acid biomarkers. Appl. Env. Microbiol. 55, 1368–1374.

Vainio E J and Hantula J 2000 Direct analysis of wood-inhabiting fungi using denaturing gradient gel electrophoresis of amplified ribosomal DNA. Mycol. Res. 104, 927–936.

Viaud M, Pasquier A and Brygoo Y 2000 Diversity of soil fungi studied by PCR-RFLP of ITS. Mycol. Res. 104, 1027–1032.

Plant and Soil **244:** 19–28, 2002.

Fungal diversity in ectomycorrhizal communities: sampling effort and species detection

Andy F. S. Taylor
*Department of Forest Mycology and Pathology, Swedish University of Agricultural Sciences, PO Box 7026, SE-750 07 Uppsala, Sweden**

Key words: community structure, ectomycorrhizas, fungal distribution, species richness, vertical distribution

Abstract

A number of recent review articles on ectomycorrhizal (ECM) fungal community diversity have highlighted the unprecedented increase in the number of publications on this ecologically important but neglected area. The general features of these species-rich, highly dynamic and complex communities have been comprehensively covered but one aspect crucial to our assessment of diversity, namely the sampling of ECM communities has received less attention. This is a complex issue with two principal components, the physical sampling strategy employed and the life cycle traits of the ECM fungi being examined. Combined, these two components provide the image that we perceive as ECM diversity. This contribution will focus primarily on the former of these components using a recent study from a pine forest in central Sweden to highlight some sampling problems and also to discuss some features common to ECM communities. The two commonly used elements of diversity, species richness and community evenness, present rather different problems in the assessment of ECM diversity. The applicability of using current measures of abundance (number or percentage of root tips colonised) to determine community evenness is discussed in relation to our lack of knowledge on the size of individual genets of ECM fungi. The inherent structure of most ECM communities, with a few common species and a large number of rare species, severely limits our ability to accurately assess species richness. A discussion of theoretical detection limits is included that demonstrates the importance of the sampling effort (no. of samples or tips) involved in assessing species richness. Species area abundance plots are also discussed in this context. It is suggested that sampling strategy (bulk samples *versus* multiple collections of single tips) may have important consequences when sampling from communities where root tip densities differ. Finally, the need for studies of the spatial distribution of ECM on roots in relation to small-scale soil heterogeneity and of the temporal aspects of ECM community dynamics is raised.

Introduction

Recent years have witnessed an unprecedented increase in the number of studies examining the species composition of ectomycorrhizal (ECM) communities below ground. While this can be primarily attributed to rapid progress in the ability to characterise the fungi colonising individual root tips, there is also an increasing awareness that soil organisms, in particular mycorrhizal fungi, can have a significant influence upon aboveground organism dynamics and on ecosystem processes as a whole (Copley, 2000; Read, 1998; van der Heijden et al., 1998).

* FAX No: 18309245. E-mail: Andy.Taylor@mykopat.slu.se

In a recent review on diversity in ECM communities, Horton and Bruns (2001) gave an excellent and comprehensive account of the general characteristics of ECM communities. It is, therefore, not my intention here to give an overview of ECM fungal diversity but rather to focus on a number of interconnected issues relating to the sampling of ECM communities. There has so far been little attention given to the most suitable sampling methods for assessing ECM diversity but it is fundamental to how we perceive community diversity. Given the high species richness and the apparent non-random distribution of species, it is critical that we establish sampling protocols that can accurately determine ECM diversity.

One purpose of the present paper is to examine the relationship between what we know about ECM community structure and what it is hypothetically possible to determine about that structure with different sampling efforts. I will give a brief summary of how we define diversity and then use data from a recent study carried out in Sweden to illustrate a number of points relating to sampling effort and species detection. This data will also be used to discuss some general features that appear to be common to many ECM communities. Although highly relevant for comparing differences between ECM communities, I have not included a discussion on diversity indices in relation to ECM communities. This topic is, by itself, sufficiently large and complex to warrant a separate investigation. The general use of these indices has been covered elsewhere (see Krebs, 1989; Magurran, 1988). The primary aim of the following discussion is to demonstrate that we must exercise caution in our interpretation of diversity data, particularly when comparing the species richness of two communities.

Defining community diversity

Most studies that have examined ECM communities have done so over rather small areas, usually < 1 – 2 ha (see Bruns, 1995; Horton and Bruns, 2001). It therefore seems practical and convenient to use this as the spatial scale at which to discuss sampling of ECM communities. It should be apparent that changes in the scale at which we define communities will have affects on both our ability to accurately determine the species within that area and upon the potential number of species that can occur within the defined area.

Currently, sampling of ECM communities belowground usually involves taking bulk samples of soil with subsequent analysis of the mycorrhizal fungi present on the root tips. Characterisation of the fungi may be done using morphology or molecular techniques or now more commonly a combination of both. A joint approach to characterising the ECM on roots involving a limited amount of morphotyping followed by molecular characterisation of individuals from the morphotypes distinguished has been advocated by Horton and Bruns (2001). Once characterised, community diversity is usually reported as two components, species richness, the number of species in the community and community evenness, a measure of the abundance of each species in the community (Magurran, 1988).

Most studies involve single sampling events, which effectively give a snapshot in time of the community. Under these conditions, it may be important to distinguish two different aspects of species richness; apparent and cryptic species richness. Apparent species richness is that which can be distinguished on the roots at a particular moment in time for a given sampling effort. Cryptic species are species either below the detection limits of the study at that time or those existing only as persistent propagules (spores and sclerotia). These species may only become apparent after temporal changes in species abundance or, in the case of the propagules, after perturbation events e.g. wildfire (Baar et al., 1999; Taylor and Bruns, 1999).

Community evenness is usually reported as the number or percentage of root tips colonised by individual species. A small number of studies have estimated the biomass of species based of estimates of the amount of fungal material associated with the mycorrhizal tips (e.g. Stendell et al., 1999) but our knowledge of the interspecific differences in the extraradical phase of the symbiotic association is such that these estimates may have limited value. However, biomass provides an effective estimate of the mass of roots colonised by individual ECM species and, by derivation, root-associated fungal biomass.

There is a problem inherent within current sampling techniques to using the concept of evenness. For the purpose of defining evenness, we assume that within bulk samples, root tips represent individuals, but there is a high likelihood that mycorrhizal tips within a sample identified as supporting the same fungal species are colonised by the mycelium of one individual. We are therefore effectively sampling parts of the same individual. Horton and Bruns (2001) have discussed some of the problems associated with defining individuals within ECM species and, in addition, give an interesting discussion on the use of frequency *versus* abundance and demonstrate that both can give biased views of a species importance by emphasising different aspect of a species autecology.

The case study

The study (hereafter referred to as the Riddarhyttan study) was carried out in a 50-year-old *Pinus sylvestris* L. stand ca. 100 km SW of Uppsala in central Sweden. It is a typical dry, lichen-heath boreal pine forest on well-drained sand. In 2000, the site was visited on 5 occasions, at two-week intervals, during the main period of sporocarp production. At each visit, the sporocarps from three plots (30 m × 30 m) were collected and identified to species. In addition, 10 soil cores (2.8 cm d.) were taken from each plot during the same period. Each core included the whole of the organic layer. The mycorrhizal tips were extracted from the cores and characterised using a combination of morphotyping (Agerer, 1986–98) and restriction fragment length polymorphism (RFLP) analysis of the ITS region (Kåren et al., 1997). For the purpose of the present discussion, data from the three plots are combined. In total, 1478 sporocarps were collected belonging to 56 species and 5371 mycorrhizal tips were extracted and characterised into 37 morphotypes, of which only 19 could be identified to genus or species level.

It is now well established that the production of sporocarps, the reproductive structures of most ECM fungi, is not a good indicator of the abundance of ECM species belowground (see Dahlberg, 2001). The reasons for this discrepancy are numerous and have been discussed on several occasions (e.g. Erland and Taylor, 2002). At Riddarhyttan, *Cortinarius* was the most abundant genus forming 625 (42.3%) sporocarps representing 25 spp. However, belowground only 1.6% of the mycorrhizal tips examined could be attributed to this genus. Similar patterns have been reported previously (Gardes and Bruns, 1996). It is also common to find greater numbers of species occurring as sporocarps than recorded belowground (e.g. Dahlberg et al., 1997). Horton and Bruns (2001) pointed out this may reflect the common practice, as carried out here, of repeated observations on the sporocarps but only single sampling of the mycorrhizal tips. They suggested that an equally intensive study of the mycorrhizal tips might lead to greater similarity between above and belowground views. It is, however, difficult to see how this effort may be achieved. In the present study, there were almost four times as many root tips sampled as sporocarps. If it were to be done on an area basis, the numbers of root tips would be impossible to handle. There were between 7 and 72 × 10^4 mycorrhizal tips per square metre of forest floor in the present study. Other studies have recorded similar figures (e.g. Dahlberg et al., 1997).

Vertical distribution of species

As mentioned above, there are a number of possible reasons for the disparity between above and belowground views of ECM communities. One factor that has so far received little attention is the vertical distribution of species down the soil profile. A number of studies have demonstrated that ECM species differ in their preference for the organic and mineral soil layers (Danielson and Visser, 1989; Stendell et al., 1999; Taylor and Bruns, 1999). However, the unequal distribution of root tips between the organic and mineral layers in the soil profile must be taken into consideration when making observations on distribution. Recently, Fransson et al. (2000) used a binomial statistical model that incorporated this feature into an analysis of species distribution between organic and mineral soil layers. They showed that *Cenococcum geophilum* mycorrhizas were preferentially found in the organic layers while those of *Tylospora fibrillosa* were associated with the mineral soil. The usefulness of this model was however restricted by the large variation in the data from a single species.

Most studies of ECM communities, including the present one, restrict sampling to the organic horizons. Species that occur further down in the profile are therefore effectively excluded from the analysis. In a recent study carried out in northern Sweden in a mixed coniferous forest growing on a well-developed podsolic soil, 25 ECM taxa were identified on the roots by sequencing of the ITS region. Fifteen of these were found exclusively in the mineral soil (Rosling et al., unpublished). This study illustrates, not just a potential reason for above and below discrepancies, but also highlights a potential problem in determining the species richness of a community. The species restricted to the mineral soil constitute part of the hidden or cryptic community and sampling strategies need to incorporate this phenomenon.

Species rank abundance patterns, theoretical detection limits and sampling effort

In common with most other studies on ECM diversity (Horton and Bruns, 2001), the sporocarp and the mycorrhizal data from the Riddarhyttan pine study comprised a few common species and a long ‘tail’ of rarer species (Figure 1a, b). How does this distribution pattern affect our ability to establish the species richness of a community and to detect differences between communities as a result of perturbations? If we know the relative abundance of a species in a community, it is possible to calculate the probability (p) of finding

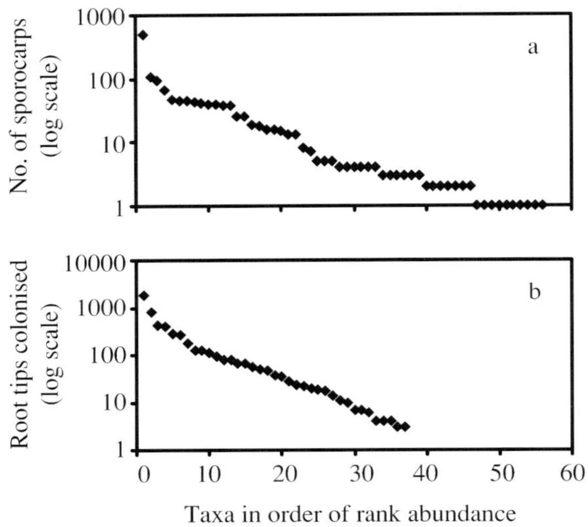

Figure 1. Rank abundance patterns for ectomycorrhizal (ECM) sporocarps (a) and ECM morphotypes (b) using log abundance ($\log_{10}$ no. of sporocarps formed and $\log_{10}$ no. of tips colonised) plotted against ranked abundance. Data from a 50-yr-old pine stand in central Sweden.

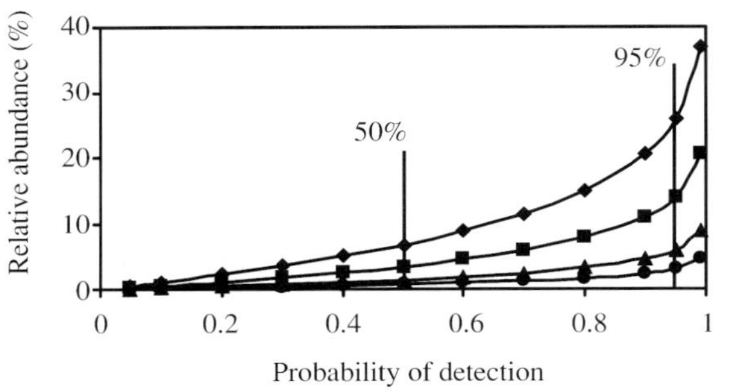

Figure 2. Theoretical detection limits when sampling ectomycorrhizal communties belowground. Relationship between sampling effort (number of tips sampled), probability of detection and the relative abundance of species (percentage of root tips colonized). The vertical lines indicate the relative abundance at each sampling effort (10 tips - ♦, 20 tips - ■, 50 tips - ▲, 100 tips - •) that species must make up in order to stand a 50% and a 95% chance of being detected.

that species in a sample of a given size:

probability of not finding sp. *A* on a tip..........

$$p = 1 - x$$

probability of not finding sp. *A* on *y* root tips...

$$p = (1 - x)^y$$

probability of finding sp. *A* on *y* root tips.......

$$p = 1 - (1 - x)^y, \quad (1)$$

where x is the proportion of species A in the community and y is the number of individuals (in this case root tips) sampled. By changing formula (1) to:

$$x = 1 - \sqrt[y]{1 - p} \quad (2)$$

it is possible to calculate the relative abundance at which species have a given probability of being found in samples. Figure 2 shows the relationships between sampling effort (at 4 different levels; 10, 20, 50 and 100 tips sampled), the probability of detection and the relative abundance of species. The importance of abundance obviously increases with increasing demands on the probability of detection. If we accept an even chance of detecting species in a community, then we have a 50% chance of detecting species that make up 6.69, 3.41, 1.37 and 0.69% of the individual mycorrhizal tips when we sample 10, 20, 50 and 100 tips, respectively. However, with an increase in the level of certainty to 95%, there is a steep increase in the required abundance of species to 25.88, 13.91, 5.82 and 2.95% for the four levels of sampling effort.

How do these levels relate to the relative abundance of ECM species recorded on the root tips at Riddarhyttan? Figure 3 shows a rank abundance plot of the mycorrhizal data using the relative abundance (%) of root tips colonised by each morphotype. The dotted lines represent the lowest percentage of mycorrhizas, at each level of sampling effort, that a species must form in order to have a 95% chance of being recorded in an analysis of the community. For example when 100 tips are sampled, only seven out of the 37 morphotypes occur on or above the dotted line at 2.95% i.e. 30 of the morphotypes have a greater than 5% chance of remaining undetected. The lowest recorded abundance of a morphotype at Riddarhyttan was only 3 tips out of the total of 5371 tips or 0.055%. The calculated number of tips that would need to be sampled in order to stand a 95% chance of detecting this species is 5500, which is remarkably close to the actual number of root tips sampled which was 5371. These theoretical probabilities of detection are based upon the assumption that species are randomly distributed within the community. We still know very little about the spatial distribution of species, but what we do know suggests that species have a clumped distribution rather than being randomly distributed. A clumped distribution would have the effect of decreasing even further our ability to obtain accurate assessments of species richness.

Horton and Bruns (2001) suggested that since most ECM species typically occur in less than 10%

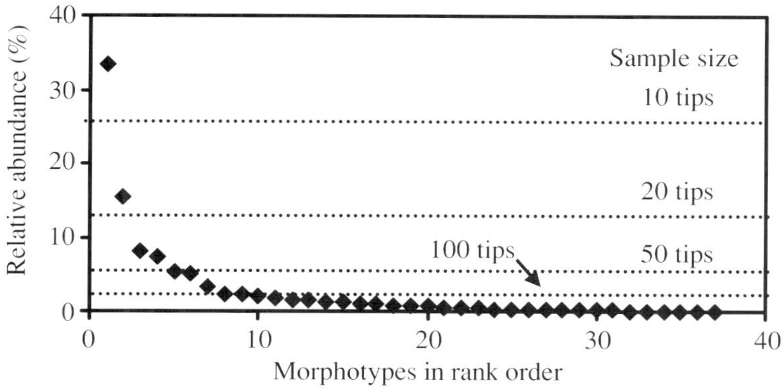

Figure 3. Rank abundance plot, using percentage of root tips colonised, of the ectomycorrhizal morphotypes found on roots of 50-yr-old pine stand in central Sweden. Overlaid on the plot are the 95% theoretical detection probabilities (dotted lines) at four different sample sizes. Morphotypes with a relative abundance that occurs on or above these lines have a 95% or better chance of being detected in four sample sizes.

of soil samples, the distribution of many ECM species is clustered. At Riddarhyttan, almost 84% of the morphotypes occurred in three or fewer samples out of a total of 30. However, the occurrence of species in samples will be greatly influenced by sample size. In addition, low frequency of occurrence in samples, which actually represent a very small proportion of the forest, does not necessarily mean that a species is not randomly dispersed across a site. There is no doubt that where some species occur they form the great majority of the tips in samples, but this is not concomitant with a non-random distribution of individual mycelia of the species across the site. If the phenomenon of clustering within samples is significant then the number of species occurring in individual samples may deviate from a random distribution. This was tested on the Riddarhyttan morphotype data using the Chi-squared test of dispersion (Heath, 1995; Sokal and Rohlf, 1996). This is a simple analysis that examines whether the frequency distribution of organisms in samples is random, clumped or regular. If clustering of tips of a single species influences the numbers of species found in samples it should do so such that the frequency of species in samples should be more regular than expected from a random distribution i.e. the ratio of variance/mean (I – index of dispersal) should be significantly less than 1. The value of I for the Riddarhyttan data was 0.787. This is used to calculate a value of χ^2 using $I*(n\text{-}1)$, giving a value of 22.28, which is well within the expected range of values (16.04 – 45.72) at α_2= 5% for a random distribution. However, the distribution of the number of species in samples will be greatly affected by sample size. Decreasing sample size to the point where it equals that

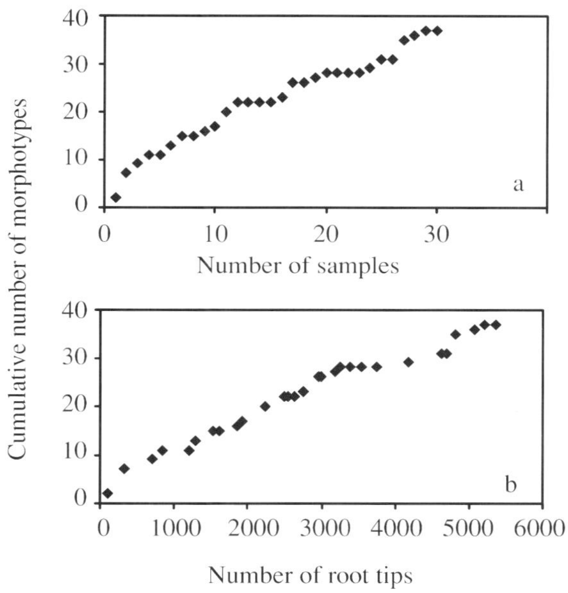

Figure 4. Species area plots for ectomycorrhizal morphotypes found in a study of the mycorrhizal community in a 50-yr-old pine stand in central Sweden. The cumulative number of morphotypes is plotted against number of samples (a) and the cumulative number of root tips (b) examined.

of the mycorrhizal clusters should give a more regular distribution of species within samples. However, this is a very unlikely scenario because in order to achieve this we would have to sample a discrete 3-D block of soil equal in size to the mycorrhizal clusters. Conventional bulk samples invariably constitute a vertical section through the entire organic material and will thus potentially contain many clusters.

There are increasing numbers of published reports on the distribution of genets of individual species (see Horton and Bruns, 2001). Where these include detailed mapping of individuals on a grid system, it is possible to analyse the distribution of genets using the Chi-squared test of dispersion mentioned above. Gherbi et al. (1999) examined the genetic structure of a population of *Laccaria amethystina* (Bolt. ex Hooker) Murr. in a closed 150-year-old beech (*Fagus sylvatica* L.) forest in the Vosges mountains in France. They found a very high density of small genets (>100 per 100 m^2). Analysing the distribution of genets shown in Figure 5 of Gherbi et al. (1999) by assigning each genet to a single 1 m × 1 m square reveals that the distribution of genets is clumped. Increasing the sampling area to 2 × 2 m also gives a very clumped distribution pattern. Gherbi et al. (1999) suggested that *L. amethystina* repeatedly colonises the root system of beech more or less on an annual basis. If this is the case and the main source of inoculum is from spores released from sporocarps, then a clumped distribution pattern of the resultant genets is hardly surprising, since most of the spores will fall in the vicinity of the sporocarps. A similar analysis of the data on *Russula cremoricolor* genets reported by Redecker et al. (2001, Figure 4a) also reveals a clumped distribution. The authors, like Gherbi et al. (1999) for *L. amethystina*, suggested that genets of *R. cremoricolor* were short lived and recolonised from spores. It is, therefore, not surprising that the two species show similar distribution patterns.

There are several reports of much larger genets (> 15–20 m) than the two examples given above (e.g. *Suillus bovinus*, Dahlberg and Stenlid, 1990; *Laccaria bicolor*, Baar et al., 1994; *Pisolithus tinctorius*, Anderson et al., 1998; *Cortinarius rotundisporus*, Sawyer et al., 1999). Sampling in communities primarily composed of large genets should result in a more regular distribution of individual species within samples. However, this would only be the case where the integrity of the mycelium of the genets was more or less intact. Until we are able to examine genet size belowground, we lack knowledge on mycelial integrity and indeed of genet longevity. The apparent existence of large differences in genet size poses a number of problems for sampling ECM communities. If our objective is to assess the abundance of individuals within a species, then the minimum distance between samples must be greater than the size of the individuals being investigated. Many more studies on the genet size and life strategies of ECM fungi, including the same species under different ecological conditions, are required before we can adapt sampling strategies to match individual species.

Species area abundance relationships

It is clear from the above examination of theoretical constraints on detection and the data from Riddarhyttan that the skewed abundance distribution patterns of species within ECM communities must have a considerable influence upon our ability to gain an accurate assessment of species richness within communities. A visual assessment of the accuracy of our observed estimate of species richness may be gained by plotting the cumulative number of species found against the cumulative number of samples. This type of plot may be considered as a species area plot where area is replaced by sample (see Figure 3, Horton and Bruns, 2001). If our estimate of species richness is close to the actual value then the plot of these two parameters should level off indicating that we have sampled sufficient samples to detect most of the species present in the community. There is little tendency for the species area plots for the mycorrhizal data from Riddarhyttan to level off (Figure 4a,b). This is the case when the cumulative number of species is plotted against either the number of samples or the number of tips sampled.

This close relationship between sampling and obtained species richness has some important implication when sampling ECM communities. In particular, when we compare two or more communities with respect to species richness, it is vital that we have some idea of how accurate the assessment of species richness is within each community. For example, if there are two communities, A and B, that are similar in structure but B actually contains fewer species, then it would be easy to assume that the species richness was the same in both communities if we do not consider how sampling effort and species richness is related in these two communities. This is illustrated graphically in Figure 5a. A sampling effort below the line crossing the two community plots would result in the conclusion that the two communities do not differ. Increasing the sampling effort beyond this point would increasingly illustrate that the species richness is actually quite different. Another possible scenario is illustrated in Figure 5b. Here the species in community B are distributed in a more regular pattern than in community A and this results in more species being found in B for a given sampling effort. In this case if we sample below

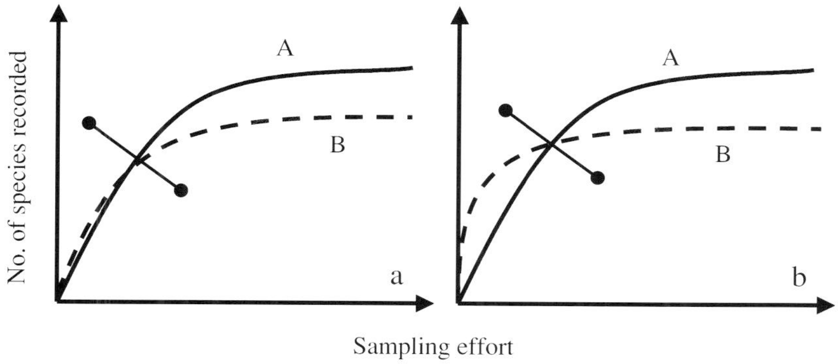

Figure 5. The potential influence of sampling below the asymptote of species area plots on the number of species recorded from ectomycorrhizal communities. (a) Sampling in communties with similar species distributions but which differ in species richness and (b) as (a) but where the species in community B are more regularly distributed than in community A.

the indicated line, then we could reach the conclusion that there were more species in B.

Given the species-rich nature of ECM communities and the high degree of spatial heterogeneity it will, in most cases, be impossible to sample all species within a community. For example, even with intensive sampling over a two-year period, only 22% of the species found fruiting on three Norway spruce sites were found belowground (Peter et al., 2001a). There are a number of statistical techniques that can provide an estimate of the expected number of individuals in communities from which a subsample has been taken (e.g. Jackknife estimate, bootstrap method). These techniques are based on using the community structure (relative abundance of individuals) observed in the subsample to estimate the potential species richness within the community as a whole. However, given the difficulties in defining individuals within ECM communties, the use of these techniques could be questioned.

It is also possible to use the method of rarefaction (Krebs, 1989; Magurran, 1988) to compare the expected number of species in samples of different sizes. This technique was employed by Taylor et al. (2000) to compare ECM diversity in samples of different sizes from *Picea abies* (L.) Karst. and *Fagus sylvatica* L. forests along a N-S transect in Europe. Peter et al. (2001b) also used this method to estimate the number of expected RFLP patterns likely to be found in a standard sample size from a fertilization experiment in a *Picea abies* forest in Switzerland. Although this method is useful, it can only be used to compare expected species richness up to the level of the community (or sample) where the lowest number of individuals were sampled.

There is another aspect to consider in relation to species area abundance relationships and that is the difference between sampling bulk samples containing many mycorrhizal tips and sampling individual mycorrhizal tips at random from the same community. This may be important when comparing two communities that are very similar in all aspects but where the root tip density at one site is significantly lower than the other. Changes in root tip numbers are commonly recorded as a result of perturbations. Fertilization with N often results in a reduction in root tips (Wallenda and Kottke, 1998), while liming, as a counter measure against forest soil acidification, often results in considerable increases in root tip numbers (Kreutzer, 1995). Figure 6 illustrates the potential difference between taking bulk samples and multiple collections of single tips as samples from two communities, A and B. Community B is found on a site where there is a lower density of root tips. Due to the strong relationship between root tips sampled and species found (see Figure 4), taking the same number of bulk samples from each community (Figure 6a) may result in a lower species richness being recorded for B. However, taking individual samples consisting of single root tips from both A and B (Figure 6b), should in theory result in similar species richness being recorded for both communities. This approach was used to good effect recently by Peter et al. (2001b) to investigate the effects of fertilization upon ECM communities. Sampling individual root tips at random to assess ECM diversity may have inherently greater potential for determining both species richness and community evenness than taking bulk samples. It would be very instructive to compare these two sampling strategies.

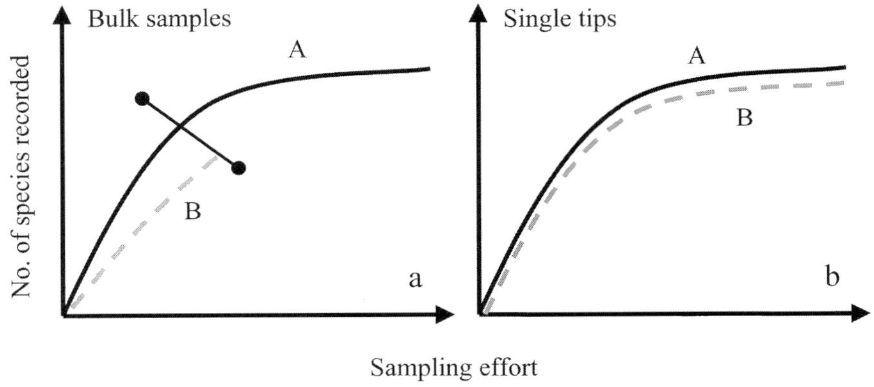

Figure 6. The potential effect of taking bulk samples (a) and individual tips (b) to assess species richness in two similar ectomycorrhizal communities, A and B, where the root tip density in B is significantly less than in A.

Spatial heterogeneity and sampling

Heterogeneity, both spatial and temporal, within soil is high and it has been suggested that this is a significant factor in maintaining the high diversity of ECM communities (Bruns, 1995; Erland and Taylor, 2002) and of other soil organisms (Giller, 1996). With the exception of a small number of studies that have examined ECM distribution at a gross scale in soil (e.g. Harvey et al., 1976, 1978; Kropp, 1982), there are as yet no studies that have specifically examined the microspatial distribution of ECM fungal species belowground in relation to the physicochemical soil environment.

Bebber (1999) recently urged ecologists to get acquainted with geostatistics as a powerful tool in the analysis of spatial and temporal heterogeneity. There have been no studies, as far as I am aware, that have utilised geostatistics to examine soil heterogeneity, both horizontal and vertical, at fine scales (cm – dm) in relation to ECM fungi. Most of what is known about the variance of soil physical and chemical parameters at scales of less than 1–10 m comes from studies of agricultural systems (e.g. White et al., 1987), ecosystems dominated by plants forming arbuscular mycorrhizas (e.g. Dushyantha et al., 1997) or from studies examining the spatial distribution of soil contaminants (e.g. Bringmark and Bringmark, 1998; Olivier and Badr, 1995). A number of studies (e.g. Grundmann and Debouzie, 2000) have examined the distribution of bacteria populations at the millimetre scale and found spatial dependence ranging from 2 to 4 mm for different bacterial groups. Although the mycelial nature of ECM fungi means that they can integrate over a greater volume of soil than bacteria, it is at this level of spatial heterogeneity that studies need to be carried out. Jackson and Caldwell (1993) working in a native sage brush steppe site in Utah demonstrated that soil organic matter, pH, phosphate, potassium, ammonium, and nitrate all showed strong spatial patterns at scales of less than 1 m. At a slightly larger scale (2 m), similar variation was found by Lechowicz and Bell (1991) in a *Quercus / Acer* forest in Quebec. Many of these soil parameters have been linked to the distribution of ECM fungi, at least when assessed by the occurrence of sporocarps (Hansen, 1988; Nantel and Neumann, 1992). It seems likely that the distribution of mycelia and the mycorrhizal tips of ECM species will also be influenced by the same factors. There is a great need for studies that utilise geostatistical techniques for identifying both the structure of variability within individual edaphic dimensions and for quantifying the degree of autocorrelation between samples. Application of these techniques to the analysis of the microdistribution of individual fungal species in relation to soil heterogeneity and to examine the structure of communities sampled in areas of increasing sizes would provide new insights into how individuals are spatially arranged within highly complex and diverse communities.

A fundamental concept within plant and animal ecology is that of the minimum sampling area for recording a certain proportion of individuals within a community. No estimates of minimum sampling areas exist for sampling ECM fungal communities on root tips. Coupled to the concept of minimum sampling area is how samples should be distributed spatially in order to obtain a realistic description of the community occurring on the root tips. Determining the autocorrelation between samples will indicate the necessary distance required between samples when the species abundance within samples is no longer correlated. To obtain a random sample of a community, samples would therefore have to be taken at distances greater than this. Conversely, the same analysis will provide an indication of the predictive power of a sample i.e. how similar will the community be in another sample taken at a set distance from the first sample. Within communities with highly skewed species abundance distributions (i.e. high dominance by one or a few species) the predictive power of a sample of a given size (at least for these species) should be greater than in communties where species are more evenly distributed.

Conclusions

It is clear from the above discussion that sampling effort and sampling strategy can have a major influence upon how we perceive ECM community structure. In addition, our ability to detect the effects of perturbations upon changes in species richness may be limited by the inherent structure of ECM communities. Both these factors, therefore, need to be considered when designing sampling strategies during investigations of ECM diversity. It is also apparent that in order to appreciate the complexity of ECM systems fully, there is a great need for more information on how species are distributed on roots and on the scale and rapidity of temporal shifts in community structure.

Acknowledgements

The author would like to thank the organisers of the 3rd International Conference on Mycorrhizas (ICOMIII) for the invitation to prepare this contribution, the Forestry Research Institute of Sweden for the use of the Riddarhyttan site and for logistical support and the Swedish Forestry companies that financed the project. In addition, thanks are also due to the many people who provided valuable comments during the preparation of this contribution, in particular Björn Lindahl, Roger Finlay, Tom Horton and Martina Peter. Thanks are also due to the anonymous referees for constructive comments.

References

Agerer R 1986–1998 Colour Atlas of Ectomycorrhizae. Schwäbisch-Gmünd, Einhorn-Verlag.

Anderson I C, Chambers S M, Cairney J W G 1998 Use of molecular methods to estimate the size and distribution of mycelial individuals of the ectomycorrhizal basidiomycete *Pisolithus tinctorius*. Mycol. Res. 102, 295–300.

Baar J, Ozinga W A and Kuyper Th W 1994 Spatial distribution of *Laccaria bicolor* genets reflected by sporocarps after removal of litter and humus layers in a *Pinus sylvestris* forest. Mycol. Res. 98(7), 726–728.

Baar J, Horton T R, Kretzer A M and Bruns T D 1999 Mycorrhizal colonization of *Pinus muricata* from resistant propagules after a stand replacing wildfire. New Phytol. 143, 409–418.

Bamforth S S 1995 Interpreting soil ciliate biodiversity. Plant Soil 170, 159–164.

Bebber D, 1999 Spatial autocorrelations. TREE 14(5), 196.

Bringmark E and Bringmark L 1998 Improved soil monitoring by use of spatial patterns. Ambio 27, 45–52.

Bruns T D 1995 Thoughts on the processes that maintain local species diversity of ectomycorrhizal fungi. Plant Soil 170, 63–73.

Copley J 2000 Ecology goes underground. Nature 406, 452–454.

Dahlberg A 2001 Community ecology of ectomycorrhizal fungi: an advancing interdisciplinary field. New Phytol. 150, 555–562.

Dahlberg A and Stenlid J 1990 Size, distribution and biomass of genets in populations of *Suillus bovinus* (L.:Fr.) Ruossel revealed by somatic incompatibility. New Phytol. 115, 487–493.

Dahlberg A, Jonsson L and Nylund J-E 1997 Species diversity and distribution of biomass above- and below-ground among ectomycorrhizal fungi in an old-growth Norway spruce forest in south Sweden. Can. J. Bot. 75, 1323–1335.

Danielson R M and Visser S 1989 Effects of forest soil acidification on ectomycorrhizal and vesicular-arbuscular mycorrhizal development. New Phytol. 112: 41–48

Dushyantha K, Hutchings W and Hutchings M J 1997 The effects of spatial scale of environmental heterogeneity on the growth of a clonal plant: an experimental study with *Glechoma hederacea*. J. Ecol. 85, 17–28.

Erland S and Taylor A F S 2002 Diversity of ectomycorrhizal communities in relation to the abiotic environment. *In* Mycorrhizal Ecology. Eds. Marcel G A van der Heijden & Ian R Sanders. pp 470. Ecological Studies Series. Vol 157. ch 7. Springer-Verlag. Berlin.

Fransson P M A, Taylor A F S, Finlay R D 2000 Effects of optimal fertilization on belowground ectomycorrhizal community structure in a Norway spruce forest. Tree Phys. 20, 599–606

Gardes M and Bruns T D 1996 Community structure of ectomycorrhizal fungi in a *Pinus muricata* forest: above- and below-ground views. Can. J. Bot. 74, 1572–1583

Gherbi H, Delaruelle C, Selosse M –A and Martin F 1999 High genetic diversity in a population of the ectomycorrhizal basidiomycete *Laccaria amethystina* in a 150-year-old beech forest. Mol. Ecol. 8, 2003–2013.

Giller P S 1996 The diversity of soil communities, the 'poor man's tropical rainforest'. Biol. Con. 5, 135–168.

Grundmann G L and Debouzie D 2000 Geostatistical analysis of the distribution of NH_4^+ and NO_2^--oxidizing bacteria and serotypes at the millimetre scale along a soil transect. FEMS Microbiol. Ecol. 34, 57–62

Hansen P A 1988 Prediction of macrofungal occurrence in Swedish beech forest from soil and litter variable models. Vegetatio 78, 31–44.

Harvey A E, Larsen M J and Jurgensen M F 1976 Distribution of ectomycorrhizae in a mature Douglas fir larch forest soil in Western Montana. For. Sci. 22(4), 393–398.

Harvey A E, Larsen M J and Jurgensen M F 1978 Comparative distribution of ectomycorrhizae in soils of three Western Montana forest habitat types. For. Sci. 25(2), 350–358.

Heath D 1995 An Introduction to Experimental Design and Statistics for Biology. UCL Press, University College London, UK. 372.

Horton T R and Bruns T D 2001 The molecular revolution in ectomycorrhizal ecology: peeking into the black box. Mol. Ecol. 10, 1855–1871.

Jackson R B and Caldwell M M 1993 Geostatistical patterns of soil heterogeneity around individual perennial plants. J. Ecol. 81, 683–692.

Krebs C J 1989 Ecological methodology, 2nd ed. Harper & Row, New York. Kreutzer K 1995 Effects of forest liming on soil processes. Plant Soil 168-169, 447–470.

Kropp B R 1982 Formation of mycorrhizae on nonmycorrhizal Western Hemlock outplanted on rotten wood and mineral soil. For. Sci. 28(4), 706–710.

Kårén O, Hogberg N, Dahlberg A, Jonsson L and Nylund J - E 1997 Inter- and intraspecific variation in the ITS region of rDNA of ectomycorrhizal fungi in Fennoscandia as detected by endonuclease analysis. New Phytol. 136, 313–325.

Lechowicz M J and Bell G 1991 The ecology and genetics of fitness in forest plants. II. Microspatial heterogeneity of the edaphic environment. J. Ecol. 79, 687–696.

Magurran A E 1988 Ecological Diversity and its Measurement. Croom Helm, London. Nantel P and Neumann P 1992 Ecology of ectomycorrhizal-basidiomycete communities on a local vegetation gradient. Ecology 73(1), 99–117.

Oliver M A and Badr I 1995 Determining the spatial scale of variation in soil radon concentration. Math. Geol. 27, 893–922.

Peter M, Ayer F, Egli S and Honegger R 2001a Above – and belowground community structure of ectomycorrhizal fungi in three Norway spruce (*Picea abies*) stands in Switzerland. Can. J. Bot. 79, 1134–1151.

Peter M, Ayer F and Egli S 2001b Nitrogen addition in Norway spruce stand altered macromycete sporocarp production and below-ground ectomycorrhizal species composition measured by PCR-RFLP analysis of the ribosomal ITS-region. New Phytol. 149, 311–326.

Read D J 1998 Plants on the web. Nature 396, 69–72.

Redecker D, Szaro T M, Bowman R J and Bruns T D 2001 Small genets of *Lactarius xanthogalactus*, *Russula cremoricolor* and *Amanita francheti* in late stage ectomycorrhizal successions. Mol. Ecol.10, 1025–1034.

Sawyer N A, Chambers S M and Cairney J W G 1999 Molecular investigation of genet distribution and genetic variation of *Cortinarius rotundisporus* in eastern Australian sclerophyll forests. New Phytol. 142, 561–568.

Sokal R R and Rohlf F J 1996 Introduction to Biostatistics. W.H. Freeman and Company, New York. 363 p.

Stendell E R, Horton T R and Bruns T D 1999 Early effects of prescribed fire on the structure of the ectomycorrhizal fungus community in a Sierra Nevada ponderosa pine forest. Mycol. Res. 103, 1353–1359.

Taylor A F S, Martin F and Read D J 2000 Fungal diversity in ectomycorrhizal communities of Norway spruce (*Picea abies* [L.] Karst.) and Beech (*Fagus sylvatica* L.) along north-south transects in Europe. *In* Ed. E-D. Schulze. Ecol. Stud. Vol. 142. pp 343–365. Springer-Verlag, Heidelberg.

Taylor D L and Bruns T D 1999 Community structure of ectomycorrhizal fungi in a *Pinus muricata* forest: minimal overlap between the mature forest and resistant propagule communities. Mol. Ecol. 8,1837–1850

van der Heijden M G A, Klironomos J N, Ursic M, Moutoglis P, Streitwolf-Engel R, Boller T, Wiemken A and Sander I R1998 Mycorrhizal fungal diversity determines plant biodiversity, ecosystem variability and productivity. Nature 396, 69–72.

Wallenda T and Kottke I 1998 Nitrogen deposition and ectomycorrhizas. New Phytol. 139, 169–187.

White R E, Haigh R A and MacDuff J H 1987 Frequency distributions and spatially dependent variability of ammonium and nitrate concentrations in soil under grazed and ungrazed grassland. Fertil. Res. 11, 193–208.

Plant and Soil **244:** 29–39, 2002.

Molecular approaches to ectomycorrhizal diversity studies: variation in ITS at a local scale

Thomas R. Horton

*Department of Forest Science, Oregon State University, Corvallis, OR 97331, USA. Present address: 350 Illick Hall, SUNY - College of Environmental Science and Forestry, Syracuse, NY 13210, USA**

Received 21 August 2001. Accepted in revised form 24 January 2002

Key words: intraspecific variation, ITS, PCR, RFLP, *Tricholoma*

Abstract

In most ectomycorrhizal (EM) community studies involving molecular identification methods there is a poor correspondence between fungi that appear dominant as sporocarps and those that appear dominant on EM roots and the species richness belowground is higher than that above ground. As a consequence, many fungi from root tip samples remain unidentified. In most studies, genetic data from multiple samples of an EM morphotype collected from various sampling locations are compared to genetic data from one to a few sporocarps of each species for identification purposes. The mismatch between above- and belowground species richness may be influenced by these different sampling efforts. To address this, intra-specific variation in the ITS region first investigated in Kårén et al. (1997) is revisited here, but at a spatial scale in which variation is expected to be low. Sporocarps were collected across a 7 km region of the Oregon Dunes National Recreation Area in western North America. ITS–RFLP data are presented for 3-18 sporocarps from each of 44 EM species in 18 genera. A total of 311 sporocarps were analyzed. Fifty-three ITS–RFLP types were observed. Of the 44 species, 38 (86% of total) yielded a single, species specific, RFLP type. No 2 species had the same RFLP type. Polymorphic ITS–RFLP types were observed in six species (14%). The following three species had two ITS-RFLP types with one type dominating: *Inocybe lacera, Laccaria proxima,* and *Rhizopogon subcaerulescens.* The following three species had three RFLP types with no type dominating: *Laccaria laccata, Lactarius deliciosus*, and *Tricholoma flavovirens.* A phylogenetic analysis of ITS sequences in *Tricholoma* revealed that two of the RFLP types in *T. flavovirens* were apparently the result of intra-specific variation, while the third RFLP type was likely a cryptic species. All the other RFLP types observed in *Tricholoma* represented unique phylogenetic species. These results suggest that ITS–RFLP data from single samples (sporocarp or EM) are robust for characterizing most of the species at this scale. However, restriction endonucleases detect a limited amount of existing nucleotide variation and thus have limited value to detect cryptic species. Therefore, additional analyses of sequence data should be added to the RFLP matching technique to identify unknown RFLP types. These data also suggest that polymorphic RFLP types within species do not adequately explain the mismatch between above- and belowground views of EM species richness.

Introduction

A great deal of knowledge about the diversity and ecology of ectomycorrhizal (EM) fungi is available from decades of study (Smith and Read, 1997). Much of the knowledge on fungal diversity resulted from field-based surveys of sporocarps. It can be argued that the presence of a fungus at a site is best assessed by its presence in its vegetative state because many fungi fruit sporadically or cryptically or both. Morphological descriptions of EM roots have provided useful data for identifying the fungi below ground (Agerer, 1987–1996; Goodman et al., 1996–1998; Ingleby et al., 1990), but most species have not been described by this method. Sporocarp surveys and morphological descriptions continue to lay a solid foundation

* FAX No: 1-315-470-6934, E-mail: trhorton@esf.edu

from which EM fungal diversity can be assessed and this foundation increases the utility of molecular techniques in studies of EM fungal diversity and ecology.

The use of the polymerase chain reaction (PCR) has dramatically increased our ability to document the diversity of EM fungi primarily because fungi can be identified directly from vegetative structures, and genetic data is obtained rapidly enough to allow relatively large numbers of samples to be processed. The technique has been applied broadly to identify EM fungi from root tip samples (Agerer et al., 1996; Dahlberg et al., 1997; Eberhardt et al., 2000; Gardes et al., 1991; Gehring et al., 1998; Henrion et al., 1992; Mahmood et al., 1999). In most cases, a combined approach employs morphological sorting of EM root tips with molecular identification using restriction fragment length polymorphism (RFLP) analysis of the internal transcribed spacer (ITS) region of the nuclear rRNA gene repeat.

Horton and Bruns (2001) review the molecular techniques applied to studies of EM communities and what has been revealed from these studies. They suggest that the RFLP typing using ITS sequences is an effective tool for identification purposes, but does have its limitations. For instance, many types remain unknown because no matching RFLP type is observed in available sporocarp databases. This may impact two of the more important findings from these studies: (1) there is a poor correspondence between fungi that appear dominant as sporocarps and those that appear dominant on EM roots; (2) species richness appears higher when analyzing root tip data than when analyzing sporocarp data (Dahlberg et al., 1997; Gardes and Bruns, 1996a; Gehring et al., 1998; Jonsson et al., 1999a,b; Kårén and Nylund, 1997; Mehmann, 1995). A number of issues likely interact to produce these patterns including differential investment in vegetative growth and sexual reproduction, and the production of cryptic sporocarps (Gardes and Bruns, 1996a). However, the pattern may also be influenced by the different sampling efforts applied above and below ground. Many studies analyze ITS–RFLPs from many EM samples of a morphological type (morphotype) and compare these to the ITS–RFLPs from a few sporocarps of each species collected at the research site. If large numbers of sporocarps are sampled for a species, then samples collected at a relatively large spatial scale are included and studies have shown that polymorphic ITS–RFLP types often occur when specimens are collected at such scales (Agerer et al., 1996; Baura et al. 1992; Eberhardt et al., 1999; Farmer and Sylvia, 1998; Gardes et al., 1991; Henrion et al., 1992; Kårén et al., 1997).

The ITS region is expected to show intraspecific variation across large scales because (1) it evolves rapidly, and (2) populations of a species can be reproductively isolated at large scales. Kårén et al. (1997) reported that intra-specific variation was not a factor at a local scale, but only sampled a few sporocarps for each species from herbarium specimens. A broader genetic sampling of sporocarps from a research site will provide a more reliable analysis as to whether ITS polymorphisms are a factor at a local scale. These data will provide additional support for the use of variation in the ITS region for species richness estimates in below ground studies.

The primary objective here is to investigate the use of ITS–RFLP data to rapidly assess the diversity of EM fungi from multiple sporocarps collected at a scale of approximately 10 km^2. Because there is relatively little sequence information in an RFLP pattern, close matches between patterns cannot be satisfactorily analyzed even though they may indicate a high degree of genetic similarity (Bruns et al., 1991; Horton and Bruns, 2001). A secondary objective then is to use a phylogenetic approach using ITS sequence data as a follow up to RFLP analysis in order to clarify species delimitations in *Tricholoma*.

Methods

Site description

The research was conducted at the Oregon Dunes National Recreation Area on the western coast of North America. The coastal sand dune EM plant community includes established forests of *Picea sitchensis, Pseudotsuga menziesii, Tsuga heterophylla*, and on the edges, *Pinus contorta*. Primary succession is continuously occurring in the area as a result of shifting sands covering and killing established forests and exposing uncolonized areas suitable for plant establishment. Primary successional EM plant communities include *P. contorta, Artctostaphylos uva-ursi* and *Salix hookeriana*, with scattered *P. sitchensis* and *P. menziesii* occurring as well. The focus in the current work is on areas dominated by *P. contorta*, but fungi potentially specific to other hosts were also collected. Within this ecosystem, four replicate sites were chosen, each site having similar plant community characteristics. Each replicate site was approximately 500 m × 500 m. The four sites were spread across approximately 7 km of the National Recreation Area.

Sporocarp collections

Sampling of sporocarps occurred weekly in the fall of 1999 and 2000, starting the last week of September and ending the second week of December (the September collection date was omitted in 2000). A time-based sampling approach was employed. Three individuals collected for 45 min at three sites during the peak fruiting periods, with the three sites selected randomly from the four possible sites each week. A total of 21 visits were made over the 2 fall seasons. Epigeous and hypogeous fungi were collected simultaneously during the trips. Hypogeous fungi were located by evidence of fresh mammal digs in the sand.

Data are presented only for species that could be confidently identified based on a morphological concept and for those species that were collected in at least two of the four replicate sites. Sporocarps were studied fresh when possible and were placed on a food dehydrator within 1 week of removal from the field. Macroscopic and microscopic characters were used to identify the fungi and available keys in the OSU mycology collection were consulted (Oregon State University, J. Trappe). Experts in the taxonomy of several difficult genera were consulted (J. Ammirati, M. Seidl, B. Methany, Thom O'Dell for *Cortinarius*; B. Methany for *Inocybe*; K. Shanks for *Tricholoma*; D. Luoma, M. Castellano, E. Cázares, R. Molina for *Rhizopogon*). Many sporocarps could not be identified to the species level because taxonomic treatments are incomplete in general, and because novel species were potentially collected. A large number of *Cortinarius* sporocarps were omitted from this analysis because of uncertain species delimitations. Other species from a variety of genera are not included here because they were collected infrequently.

Molecular approaches

DNA extraction, PCR, and RFLP analysis followed Gardes and Bruns (1996b). The ITS region of the rDNA was amplified using the primer pair ITS–1f and ITS–4b. RFLP patterns were generated with *Hinf*I, *Alu*I, and *Dpn*II. RFLP data were analyzed using Gene Profiler (Scanalytics). These data are not reported here for brevity, but are available from the author. An RFLP type is defined as a composite of data from the three

Table 1. Tricholoma voucher numbers and Genbank accession numbers for ITS sequences

Tricholoma species	Voucher number[a]	Genbank accession number
T. atroviolaceum		AF319432
T. caligatum		AF204813
Tricholoma sp. 2	trh1237	AF458443
T. equestre		AJ236081
T. flavovirens		AB036895
T. flavovirens		AF349689
T. flavovirens6	trh545	AF458449
T. flavovirens6	trh894	AF458450
T. flavovirens6	trh901	AF458451
T. flavovirens7	trh546	AF458452
T. flavovirens7	trh1000	AF458453
T. flavovirens7	trh1023	AF458454
T. flavovirens8	trh670	AF458455
T. flavovirens8	trh652	AF458456
T. focale	trh909	AF462638
T. focale	trh597	AF462639
T. focale		AF319437
T. imbricatum		AF319426
T. imbricatum	trh895	AF462636
T. imbricatum	trh912	AF462637
T. intermedium		AF319434
T. intermedium		AF319434
T. japonicum		AF204810
T. luteomaculosum	trh914	AF458446
T. luteomaculosum	trh1033	AF458447
T. luteomaculosum	trh1187	AF458448
T. magnivelare	trh905	AF458441
T. magnivelare	trh906	AF458442
T. magnivelare		AF309541
T. matsutake		AF309538
T. matsutake		AF204868
T. muricata	trh610	AF458438
T. muricata	trh815	AF458439
T. muricata	trh820	AF458440
T. mutabile	trh916	AF458444
T. mutabile	trh1184	AF458445
T. myomyces		AF319428
T. myomyces		AF319428
T. pardinum		AF319427
T. portentosum		AF241517
T. portentosum		AF241517
T. saponaceum		AF319429
Tricholoma sp. 1	trh567	AF462640
Tricholoma sp. 1	trh883	AF462641
Tricholoma sp. 1	trh899	AF462642
T. ustale		AF204812
T. ustale	trh884	AF458435
T. ustale	trh902	AF458436
T. ustale	trh885	AF458437
T. venenatum		AF319433
T. vernaticum		AF319424

[a] Voucher collection located at SUNY-College of Environmental Science and Forestry Herbarium.

restriction digests and types were considered identical if they matched with all three endonucleases.

Sequence variation in the ITS region was analyzed for 25 putative species in *Tricholoma*, eight of which were collected at the site. The genus *Tricholoma* was chosen because there was a relatively high number of species found at the site and a fairly high number of sequences was available in web databases such as GenBank and EMBL. A phylogenetic analysis was employed to identify unclear species delimitations in the following two cases. RFLP analysis revealed three ITS types in *T. flavovirens*, but it was unclear how much variation in the ITS sequence was represented by these three types. In addition, *Tricholoma* sp. 1 could not be identified with a taxonomic key of western *Tricholoma* species (Shanks, 1997); this may be an unknown species sharing some morphological features with both *T. caligatum* and *T. focale*, but lacking the spicy odor of the former and the orange, viscid cap of the latter (Kris Shanks, pers. comm.). The ITS region of two to three sporocarps of each *Tricholoma* RFLP type from the study area was sequenced. Additional sequences were downloaded from GenBank, selected to increase the diversity of *Tricholoma* spp. covered (Table 1). All available *T. flavovirens* and *T. equestre* sequences are included. While some interesting trends are discussed, the results are not meant to represent a complete phylogenetic analysis of *Tricholoma*. The alignment is available from the author.

PCR products for sequencing reactions were generated with the primers ITS–1f and ITS–4b (Gardes and Bruns, 1996b). The sequencing reactions were conducted on the resultant PCR product with internal primers ITS–2, ITS–4, and ITS–3 (White et al., 1990). Sequences were determined using an ABI Model 373 DNA sequencer (Perkin-Elmer Corporation). DNA sequencing Analysis (version 2.01) and Sequence Navigator software were used to process the raw data. Sequences were aligned by visual estimation using a matrix created in PAUP 4.0b5c (Swofford, 2001). Phylogenetic analyses were conducted with PAUP 4.0b5c using the heuristic search option. Bootstrap values were generated using the fast-bootstrap option and represent 10,000 replicates.

Results

A total of 311 sporocarps in 18 genera and 44 species were analyzed, with a range of 3 – 18 sporocarps per species. Most of the species were collected in both years of the study. Fifty-three ITS–RFLP types were observed (Table 2). Of the 44 species, 38 (86% of total) yielded a single RFLP type for all sporocarps.

While no two species had the same RFLP type, polymorphic ITS–RFLP types were observed in six species (14%). Three species yielded two ITS–RFLP types, with one type dominating: *Inocybe lacera, Laccaria proxima,* and *Rhizopogon subcaerulescens*. Three other species yielded three RFLP types with no type dominating: *Laccaria laccata, Lactarius deliciosus*, and *Tricholoma flavovirens*. In most of these instances, the RFLP variation is the result of variation in only one of the three endonucleases, suggesting limited variation in the ITS sequences. The exceptions were in *Laccaria proxima, L. laccata* and one of the *T. flavovirens* types.

Phylogenetic analysis of *Tricholoma* ITS sequences revealed that all 3 ITS–RFLP types observed in *T. flavovirens* clustered in one clade together with the *T. flavovirens* and *T. equestre* sequences from GenBank. Among these however, three distinct lineages could be distinguished that were separated by 17–20 steps on the tree shown in Figure 1. ITS sequence variation between these lineages was 3.8 to 6.2% in pairwise comparisons (Table 3, Figure 2). Two of the RFLP types from the Oregon dunes, types 6 and 7, show very little sequence variation and are likely to be conspecific. The third *T. flavovirens*, type 8, proved to cluster with the others, but was divergent enough to occur on its own branch supported by a high bootstrap value. The divergence seen between the *T. flavovirens* sequences (and *T. equestre*) is greater than that seen for the *T. caligatum, T. matsutake,* and *T. magnivelare* group.

Although *Tricholoma* sp. 1 has some morphological features similar to both *T. caligatum* and *T. focale,* the phylogenetic analysis clearly does not support a close relationship to *T. caligatum*. The region in the tree with *T. focale* and *Tricholoma* sp. 1 has low overall support, but at present, the unknown species seems most likely to belong to the *T. focale* species group.

Discussion

Unique, species specific ITS–RFLP types were observed in 38 out of 44 species. Six morphologically defined species (14%) yielded multiple ITS–RFLP types. These numbers are very similar to those reported in a larger scale study by Kårén et al. (1997).

Table 2. ITS–RFLP variation in sporocarps of ectomycorrhizal species

	Species	ID[a]	# of plots	# of sporocarps sampled	# of RFLP types
1	*Amanita muscaria*	trh	4	7	1
2	*Bankera fuligneo-alba*	trh	4	10	1
3	*Boletopsis subsquamosus*	trh	3	41	
4	*Boletus edulis*	trh	4	5	1
5	*Boletus piperatus*	trh	2	3	1
6	*Boletus subtomentosus*	trh	4	6	1
7	*Chroogomphus rutilus*	trh	4	6	1
8	*Chroogomphus vinicolor*	trh	3	5	1
9	*Cortinarius aurantiobasis* complex	ja	4	9	1
10	*Cortinarius aureifolius*	pbm	3	7	1
11	*Cortinarius muscigenus*	ms	4	8	1
12	*Cortinarius semisanguineus*	ja	4	6	1
13	*Hydnellum scrobiculatum*	trh	4	6	1
14	*Inocybe lacera*	pbm	4	18	2 (17:1)[b]
15	*Inocybe sambucina*	pbm	2	4	1
16	*Laccaria bicolor*	trh	2	3	1
17	*Laccaria laccata* var. *pallidifolia*	trh	4	11	3 (6:4:1)[b]
18	*Laccaria proxima*	trh	4	9	2 (6:3)[b]
19	*Lactarius deliciosus*	trh	4	10	3 (4:4:2)[b]
20	*Lactarius rufus*	trh	4	7	1
21	*Leccinum manzanitae*	trh	4	6	1
22	*Phellodon melaleucus*	trh	4	9	1
23	*Phellodon niger*	trh	3	6	1
24	*Rhizopogon fuscorubens*	rm	3	3	1
25	*Rhizopogon occidentalis*	dl	4	5	1
26	*Rhizopogon subcaerulescens*	ec	3	4	2 (3:1)[b]
27	*Rhizopogon vulgaris*	mc	4	11	1
28	*Russula cascadensis*	trh	4	5	1
29	*Russula cremoricolor*	trh	3	4	1
30	*Russula pectinoides*	trh	4	5	1
31	*Sarcodon scabrosus*	trh	4	13	1
32	*Suillus albidipes*	trh	2	4	1
33	*Suillus brevipes*	trh	4	7	1
34	*Suillus tomentosus*	trh	4	6	1
35	*Suillus umbonatus*	trh	4	8	1
36	*Thelephora americana*	trh	4	8	1
37	*Tricholoma flavovirens*	trh	4	13	3 (6:4:3)[b]
38	*Tricholoma focale*	trh	3	8	1
39	*Tricholoma imbricatum*	kms	3	4	1
40	*Tricholoma luteomaculosum*	kms	2	2	1
41	*Tricholoma magnivelare*	trh	4	5	1
42	*Tricholoma mutabile*	kms	4	10	1
43	*Tricholoma* sp. 1	kms	3	9	1
44	*Tricholoma ustale*	kms	4	12	1

[a] Initials of individual who identified the species: trh = T.R. Horton, kms = K. M. Shanks, dl = D. Luoma, ja = J. Ammarati, mc = M. Castellano, ms = M. Seidl, pbm = P.B. Matheny, rm = Randy Molina, ec = Efrén Cázares.
[b] Number of collections with each RFLP type.

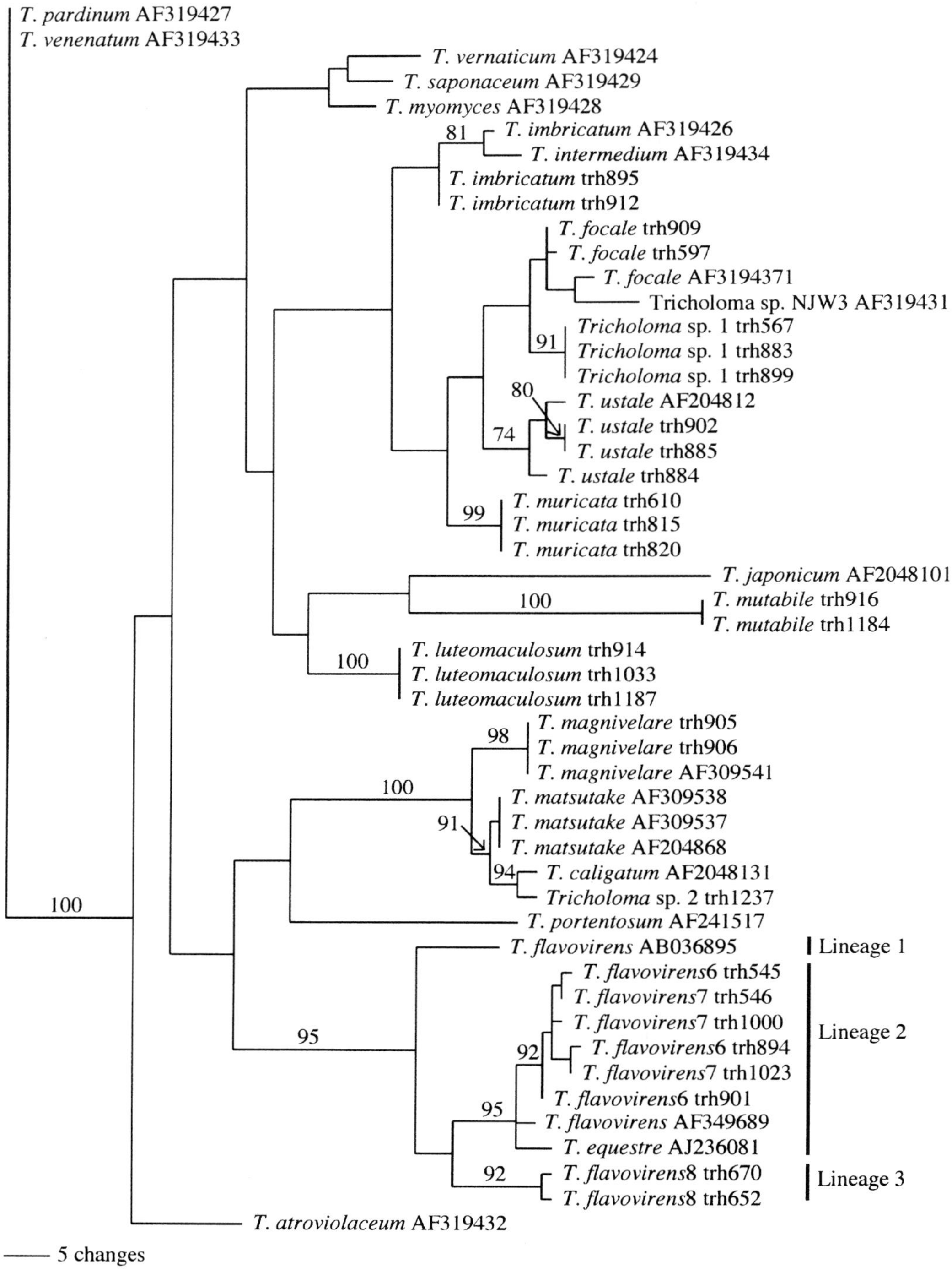

Figure 1. One of 3432 most parsimonious trees for ITS sequence data from *Tricholoma*. Areas that were difficult to align and the entire 5.8s gene were omitted from the analysis. A total of 431 characters were used in the data set, with 241 characters being constant, 50 characters being variable and not parsimony informative, and 140 characters being variable and parsimony informative. Numbers on branches refer to bootstrap values ≥70 from 10,000 replications using the fast bootstrap option.

```
   ITS1 →
1  GGTCATTTAGAGGAAGTAAAAGTCGTAACAAGGTTTCCGTAGGTGAACCTGCGGAAGGATCATTATTGAAATAAGCTTGGTTCGG
2  ..................................................................................G..
3  ..................................................................................G..
4  .....................................................................................
5  ..................................................................................G..
6  ..................................................................................G..
7  --............................................- ..................................G..
8  -----------------.....................................................- ..........G..
9  -.................................................................................G..
10 ...........................C......................................................G..
11 ......................................................................- ..........G..

1  TTGTTGCTGGCTCTGAGGGGCATGTGCACACCTAGCACCAACTTTCTT-ACCACCTGTGCACTTTTTGTAGACTTTGGAAAA----
2  .............................................TC.T.................- ..................
3  ......................................................................................
4  ......................................................................................
5  ......................................................................................
6  .............................................TC.T.................- ..................
7  .............................G....A...................................................
8  ..................................A...................................................
9  ..........T...T...................A...............................................AAAA
10 ..............T...................A..........Y.................YY.................AAAA
11 ..............T....- ........G....A.G....................................G........AACA

1  -TACCT-CTTGGTTT-GAGGACTGTCGTGCGCAA-GCC---GGCTGTCCTTGCACTTTCC-AGTCTATG-TTTTATAATATACACC
2  .....................................................................................G
3  .....................................................................................G
4  ......................................................................................
5  .....................................................................................G
6  .....................................................................................G
7  .........A...............T...................T.......G................................
8  ................................C.A..........T........................................
9  A....CA..C.....T..........A..........TTCT....T............T.C.........................
10 A....CA..C.....T..........A..........TTCT....T............T.C.........................
11 A........C...................................T........T..............T..........C.....

1  TATAGTATGTTTCAGAATGTCATTTAATGGGCTTAACTGCCTTTAATTACCATA--CAACTTTCAACAACGGATCTCTTGGCTCTC
2  ......................................................................................
3  ......................................................................................
4  ......................................................................................
5  .................................C....................................................
6  ......................................................................................
7  ......................................................................................
8  ......................................................................................
9  ..................A...............G.T.......T.A.TT.-..TAT.............................
10 ..................A...............G.T.......T.A.TT.-..TAT........NNNNNNNNNNNNNNNNNNNNN
11 ...........................C........T.......T.A.-..-..TAC.............................
                                                          5.8s gene →
```

Figure 2. ITS (ITS1, 5.8s, ITS2) sequences for 11 sporocarps of *Tricholoma flavovirens* shown in Figure 1 and Table 3. The location of the 5.8s gene was identified following (Baura et al., 1992). Gaps introduced for alignment are indicated (–). Unknown nucleotides are indicated (N).

The species that yielded multiple RFLP types were not clustered in a single genus, but occurred in 5 of the 18 genera: *Laccaria* (2), *Lactarius* (1), *Inocybe* (1), *Rhizopogon* (1), and *Tricholoma* (1). In at least three genera there appears to be a higher probability of observing intraspecific RFLP polymorphisms: *Laccaria* spp. (seen here and in Gardes et al., 1991; Henrion et al., 1992), *Cortinarius* spp. (seen in Kårén et al., 1997), and *Inocybe* spp. (seen here). Note that despite the high probability for intraspecific RFLP polymorphisms in these latter three genera, they are less frequently encountered in EM samples than species of Thelephoraceae, Russulaceae and resupinate fungi not belonging to Thelephoraceae based on sequence analyses of unknown RFLP types (for references see Horton and Bruns, 2001).

RFLP data alone do not adequately quantify the nucleotide variation in the ITS to draw conclusions

```
1  GCATCGATGAAGAACGCAGCGAAATGCGATAAGTAATGTGAATTGCAGAATTCAGTGAATCATCGAATCTTTGAACGCACCTTGCG
2  ......................................................................................
3  ......................................................................................
4  ......................................................................................
5  ......................................................................................
6  ......................................................................................
7  ......................................................................................
8  ......................................................................................
9  ......................................................................................
10 NNNNNNNNNNNNNNNNNNNNNNNNNNNNNNNNNNNNNNNNNNNNNNNNNNNNNNNNNNNNNNNNNNNNNNNNNNNNNNNNNNNNNN
11 ......................................................................................
                                                                                 ITS2→

1  CTCCTTGGTATTCCGAGGAGCATGCCTGTTTGAGTGTCATAAAATTCTCAACCTTTTCAATCTTTATTGGTTGAATTAGGTTTGG
2  .....................................................................................
3  .....................................................................................
4  .....................................................................................
5  .....................................................................................
6  .....................................................................................
7  .....................................................................................
8  .....................................................................................
9  ........................................G...................................G...C....
10 NNNNNNNNNNNNNNNNNNNNNNNNNNNNNNNNNNNNNNNNNNNNNNNNNNNNNNNNNNNNNNNNNNNNNNNNNNNNNNNNNNNNN
11 ........................................G.......................................C....

1  ATGTGGAGTCTTTTGCTGGCTTCTTT----AAAGAGGTCGGCTCTCCTTAAATTTATTAGCAGGA-CCTTTGTGGCCCTCTC-AG
2  .....................................................................................
3  .....................................................................................
4  .........G...........................................................................
5  .........G...........................................................................
6  .........G...........................................................................
7  ..............................-..........................................T...........
8  ..............................-..................CCC.....................T...........
9  ..............................-..............................G..GA.......T......---..
10 NNNNNNNNNNNNNNNNNNNNNNNNNNNNNNNNNNNNNNNNNNNNNNNNNNNNNNNNNNNNNNNNNNNNNNNNNNNNNNNNNNNNN
11 ..A.......................TAGA...AG.A..A.....................G...........T....C.CTC..

1  C-GTTAGTGTGATAATTATCTACGCTATTTGCTGAA-GGCAACTCATAAAATAGGGTTTC-CAGCTTCTAATCGTCTCTTGTAGAC
2  ....................................NNNNNNNNNNNNNNNNNNNNNNNNNNNNNNNNNNNNNNNNNNNNNNNNNN
3  ....................................NNNNNNNNNNNNNNNNNNNNNNNNNNNNNNNNNNNNNNNNNNNNNNNNNN
4  ......................................................................................
5  ........................................................NNNNNNNNNNNNNNNNNNNNNNNNNNNNNN
6  .......................................................NNNNNNNNNNNNNNNNNNNNNNNNNNNNNNN
7  ....................................A-................................................
8  -T..................................A.C...............................................
9  TGC...........G............CC.......AT......G.A....A....G..TA...........T.............
10 NNNNNNNNNNNNNNNNNNNNNNNNNNNNNNNNNNNNNNNNNNNNNNNNNNNNNNNNNNNNNNNNNNNNNNNNNNNNNNNNNNNNNN
11 CT..................................A-.......-C.....G...G..TA................T........

1  AATTTATATGACAATTTGACCTCAAATCAGGTAGGACTACCCGCTGAACTTAAGCATATCAATAAGCGGAGGA
2  NNNNNNNNNNNNNNNNNNNNNNNNNNNNNNNNNNNNNNNNNNNNNNNNNNNNNNNNNNNNNNNNNNNNNNNNN
3  NNNNNNNNNNNNNNNNNNNNNNNNNNNNNNNNNNNNNNNNNNNNNNNNNNNNNNNNNNNNNNNNNNNNNNNNN
4  .........................................................................
5  NNNNNNNNNNNNNNNNNNNNNNNNNNNNNNNNNNNNNNNNNNNNNNNNNNNNNNNNNNNNNNNNNNNNNNNNN
6  NNNNNNNNNNNNNNNNNNNNNNNNNNNNNNNNNNNNNNNNNNNNNNNNNNNNNNNNNNNNNNNNNNNNNNNNN
7  .................................................NNNNNNNNNNNNNNNNNNNNNNNN
8  .........................NNNNNNNNNNNNNNNNNNNNNNNNNNNNNNNNNNNNNNNNNNNNNNNN
9  .........................NNNNNNNNNNNNNNNNNNNNNNNNNNNNNNNNNNNNNNNNNNNNNNNN
10 NNNNNNNNNNNNNNNNNNNNNNNNNNNNNNNNNNNNNNNNNNNNNNNNNNNNNNNNNNNNNNNNNNNNNNNNN
11 .........................................................................
```

Figure 2. (continued.)

about species delimitations. A number of confusing relationships in the genus *Tricholoma* remained unclear after the RFLP analysis but were clarified with the addition of sequence comparisons and phylogenetic ana-

Table 3. Percent pairwise distances between taxa. Data are mean character differences (adjusted for missing data), representing ITS–1, 5.8s, and ITS–2 sequences. Percent differences range from 0.14 to 2.13% within clades and 3.85 – 6.25% across clades (see Figure 1)

	1	2	3	4	5	6	7	8	9	10	11
1. *T. flavovirens*, rflp 6, trh545	–	–	–	–	–	–	–	–	–	–	–
2. *T. flavovirens*, rflp 6, trh894	0.65	–	–	–	–	–	–	–	–	–	–
3. *T. flavovirens*, rflp 6, trh901	0.33	0.33	–	–	–	–	–	–	–	–	–
4. *T. flavovirens*, rflp 7, trh546	0.14	0.82	0.49	–	–	–	–	–	–	–	–
5. *T. flavovirens*, rflp 7, trh1000	0.63	0.65	0.33	0.47	–	–	–	–	–	–	–
6. *T. flavovirens*, rflp 7, trh1023	0.79	0.16	0.49	0.63	0.48	–	–	–	–	–	–
7. *T. flavovirens* AF349689	0.99	1.48	1.15	1.13	1.43	1.60	–	–	–	–	–
8. *T. equestre* AJ236081	1.50	1.85	1.52	1.65	1.96	2.13	1.35	–	–	–	–
9. *T. flavovirens*, rflp 8, trh652	5.07	5.10	4.76	5.21	5.57	5.75	4.69	5.71	–	–	–
10. *T. flavovirens*, rflp 8, trh670	5.90	6.25	5.90	5.90	6.23	6.25	5.63	5.92	0.63	–	–
11. *T. flavovirens* AB036895	4.25	4.60	4.26	4.39	4.78	4.96	3.85	4.97	5.03	5.18	–

lysis. For instance, *Tricholoma* sp. 1 is more closely related to *T. focale* than *T. caligatum,* despite having some morphological features common to both species. In addition, the phylogenetic analysis provided insights into the level of variation represented by the three RFLP types in *T. flavovirens. Tricholoma flavovirens* type 8 is most likely a unique species that at present is morphologically indistinguishable from the other collections under this name. Although *T. flavovirens* is the name used in North America for this fungus with its distinctive lemon-yellow lamellae and stipe (Bessette et al., 1997; Shanks, 1997), there is some confusion regarding whether this name is a synonym of *T. equestre* (Breitenbach and Kränzlin, 1991; Moser, 1983). It is interesting that the one *T. equestre* sequence available in GenBank is within *T. flavovirens* lineage 2 (Figure 1). Whether *T. flavovirens* lineage 2 is synonomous with *T. equestre* cannot be assessed in the present analysis. However, these data strongly suggest that *T. flavovirens* and *T. equestre* group represent a species complex requiring more taxonomic attention. This point is particularly relevant in light of a recent report of poisoning from ingestion of *T. equestre,* an otherwise choice edible fungus (Bedry et al., 2001).

Evidence for multiple biological species within a morphologically defined species has been shown in *Cortinarius rotundisporus* and *Hebeloma* spp. (Aanen et al., 2000; Sawyer et al., 1999). While a phylogenetic analysis was not conducted on any other polymorphic taxa here, direct sequence comparisons suggest multiple species were recovered in *Lactarius deliciosus* and *Inocybe lacera* (data not shown). That multiple species are being discovered in some of these taxa should not come as a surprise. For example, multiple varieties of *Lactarius deliciosus* have been described in western North American (Hesler and Smith, 1979; Methven, 1997), and Bessette et al. (1997) suggest the name is misapplied in North America. If ITS variation observed in some taxa is actually a reflection of cryptic species under one morphological concept, then the ITS–RFLP typing method becomes even more robust. These results highlight that even where polymorphisms are observed, inaccurate biological species concepts may be at least as important as intraspecific variation. This highlights the fact that many EM groups require more taxonomic attention including Cortinariaceae, *Rhizopogon*, Russulaceae, and Thelephorales.

While some of the poor correspondence between above- and belowground data appears to be a function of intraspecific variation even at a local scale, these and other data suggest a greater contribution from other factors including differential investment in vegetative growth and sexual reproduction, and the production of cryptic sporocarps (Gardes and Bruns, 1996a). Sporocarp production for most species is sporadic at best, and it can take years to approach a complete species list at a site (Luoma, 1991; O'Dell et al., 1999), yet many of the species observed fruiting over a 10 year period are likely to occur belowground in a vegetative state during this time (unless the site was recently disturbed or undergoing primary succession). This suggests that even with an increased sampling effort of EM sporocarps, many EM types will remain unidentified with the sporocarp RFLP matching approach alone. ITS–RFLP patterns continue to be a cost effective and highly useful tool for grouping EM fungal species, especially from root tip samples. However, ITS–RFLP data provide limited taxonomic information. Sequence databases, including aligned sequences, are available for a number of DNA regions that are useful at various taxonomic levels (see Horton and Bruns, 2001). As ITS sequences are deposited in GenBank or EMBL, the region becomes increasingly useful for identifying unknown EM types at the species or species-group level. Direct sequencing of the ITS or other region followed by blast searching in GenBank or a phylogenetic analysis is highly recommended to increase the taxonomic information of unknown RFLP types.

Acknowledgements

I thank Dr. Smith and Dr. Carney for the opportunity to present this paper at ICOM3. I thank Dan Segotta for facilitating my work at the Oregon Dunes National Recreation Area, Sara Ashkannejhad and Kevin Hood for many hours of fieldwork, and the following for help identifying fungi: Dan Luoma, Efrén Cázares, Joe Ammirati, Michelle Seidl, P. Brandon Methany, Kris Shanks, Thom O'Dell, Mike Castellano, Randy Molina. I thank two anonymous reviewers and Annette Kretzer for constructive editorial advice. This research was funded by the NRI Competitive Grants Program/USDA award 99-35107-7843 and support from the US Forest Service, PNW Research Station.

References

Aanen D K, Kuyper T W, Boekhout T and Hoekstra R F 2000 Phylogenetic relationships in the genus Hebeloma based on ITS1 and 2 sequences, with special emphasis on the *Hebeloma crustiliniforme* complex. Mycologia 92, 269–281.

Agerer R 1987–1996 Colour Atlas of Ectomycorrhizae. Schwäbisch Gmünd: Einhorn-Verlag Eduard Dietenberger.

Agerer R, Kraigher H and Javornik B 1996 Identification of ectomycorrhizae of *Hydnum rufescens* on Norway spruce and the variability of the ITS region of *H. rufescens* and *H. repandum* (Basidiomycetes). Nova Hedwigia 63, 183–194.

Baura G, Szaro T M and Bruns T D 1992 *Gastrosuillus laricinus* is a recent derivative of *Suillus grevillei*: molecular evidence. Mycologia 84, 592–597.

Bessette A E, Bessette A R and Fischer D W 1997 Mushrooms of Northeastern North America. Syracuse University Press, Syracuse, New York, USA. 582p.

Bedry R, Baudrimont I, Deffieux G, Creppy G, Pomies J P, Dupon M, Gabinski C and Chapalain J C 2001 Wild-mushroom intoxication as a cause of rhabdomyolysis. N. Engl. J. Med. 345, 798–802.

Breitenbach J and Kränzlin F 1991 Fungi of Switzerland: Vol. 3. Mykologia Luzern. Luzern, Switzerland. 361 p.

Bruns T D, White T J and Taylor J W 1991 Fungal molecular systematics. Annu. Rev. Eco. Syst. 22, 525–264.

Dahlberg A, Jonsson L and Nylund J-E 1997 Species diversity and distribution of biomass above and below ground among ectomycorrhizal fungi in an old-growth Norway spruce forest in south Sweden. Can. J. Bot. 75, 1323–1335.

Eberhardt U, Oberwinkler F, Verbeken A, Rinaldi A C, Pacioni G and Comandini O 2000 *Lactarius* ectomycorrhizae on *Abies alba*: morphological description, molecular characterization, and taxonomic remarks. Mycologia 92, 860–873.

Eberhardt U, Walter L and Kottke I 1999 Molecular and morphological discrimination between *Tylospora fibrillosa* and *Tylospora asterophora* Mycorrhizae. Can. J. Bot. 77: 11–21.

Farmer D J and Sylvia D M 1998 Variation in the ribosomal DNA internal transcribed spacer of a diverse collection of ectomycorrhizal fungi. Mycol. Res. 102, 859–865.

Gardes M and Bruns T D 1996a Community structure of ectomycorrhizal fungi in a *Pinus muricata* forest: above- and below-ground views. Can. J. Bot. 74, 1572–1583.

Gardes M and Bruns T D 1996b ITS–RFLP matching for identification of fungi. Meth. Mol. Biol. 50, 177–186.

Gardes M, White T J, Fortin J A, Bruns T D and Taylor J W (1991) Identification of indigenous and introduced symbiotic in ectomycorrhizae by amplification of the nuclear and mitochondrial ribosomal DNA. Can. J. Bot. 69, 180–190.

Gehring C A, Theimer T C, Whitham T G and Keim P 1998 Ectomycorrhizal fungal community structure of pinyon pines growing in two environmental extremes. Ecology 79, 1562–1572.

Goodman D M, Durall D M, Trofymow J A and Berch S M, eds. 1996–1998 A Manual of Concise Descriptions of North American Ectomycorrhizae. Mycologue Publications and Canada-B.C. Forest Resource Development Agreement, Canadian Forest Service, Victoria, B.C.

Henrion B, Le Tacon F and Martin F 1992 Rapid identification of genetic variation of ectomycorrhizal fungi by amplification of ribosomal RNA genes. New Phytol. 122, 289–298.

Hesler L R and Smith A H 1979 North American species of *Lactarius*. University of Michigan Press. Ann Arbor, Michigan, USA. 841 p.

Horton T R and Bruns T D 2001 The molecular revolution in ectomycorrhizal ecology: peeking into the black-box. Mol. Ecol. 10, 1855–1871.

Ingleby K, Mason P A, Last F T and Fleming L V 1990 Identification of ectomycorrhizas. HMSO, London.

Jonsson L, Dahlberg A, Nilsson M-C, Kårén O and Zackrisson O 1999a Continuity of ectomycorrhizal fungi in self-regenerating boreal *Pinus sylvestris* forests studied by comparing mycobiont diversity on seedlings and mature trees. New Phytol. 142, 151–162.

Jonsson L, Dahlberg A, Nilsson M-C, Zackrisson O and Kårén O 1999b Ectomycorrhizal fungal communities in late-successional Swedish boreal forests, and their composition following wildfire. Mol. Ecol. 8, 205–215.

Kårén O, Hogberg N and Dahlberg A 1997 Inter- and intraspecific variation in the ITS region of rDNA of ectomycorrhizal fungi in Fennoscandia as detected by endonuclease analysis. New Phytol. 136, 313–325.

Luoma D L 1991 Annual changes in seasonal production of hypogeous sporocarps in Oregon Douglas-fir forests. *In* Conservation and Management of Native Plants and Fungi. Eds. T N Kaye, A Liston, R M Love, D L Luoma, R J Meinke and M V Wilson. pp 249–253. Native Plant Society of Oregon, Corvallis.

Mahmood S, Finlay R and Erland S 1999 Effects of repeated harvesting of forest residues on the ectomycorrhizal community in a Swedish spruce forest, New Phytol. 142, 557–585.

Mehmann B 1995 Coincidence between molecularly or morphologically classified ectomycorrhizal morphotypes and fruitbodies in a Spruce forest. Biotechnology of Ectomycorrhizae: Molecular Approaches. Eds. V P Stocchi, P Bonfante and M Nuti. pp 41–52. Plenum Press, London, UK.

Methven A S 1997 The Agaricales of California 10: *Lactarius*. Mad River Press, Eureka, California, USA. 78 pp.

Moser M 1983 Keys to Agarics and Boleti. Roger Phillips. London.

O'Dell T E, Ammarati J F and Schreiner E G 1999 Species richness and abundance of ectomycorrhizal basidiomycete sporocarps on a moisture gradient in the *Tsuga heterophylla* zone. Can. J. Bot. 77, 1699–1711.

Sawyer N A, Chambers S M and Cairney J W G 1999 Molecular investigation of genet distribution and genetic variation of *Cortinarius rotundisporus* in eastern Australian sclerophyll forests. New Phytol. 142, 561–568.

Shanks K M 1997 The Agaricales of California 11: Tricholomataceae II – *Tricholoma*. Mad River Press, Eureka California, USA., 22 p.

Smith S E and Read D J 1997 Mycorrhizal Symbioses 2nd ed., pp 605. London, Academic Press.

Swofford D L 2001 PAUP*: Phylogenetic Analysis Using Parsimony, Version 3.1.1. Computer program distributed by the Smithsonian Institution, Washington, DC.

White T J, Bruns T D, Lee S B and Taylor J W 1990 Amplification and direct sequencing of fungal ribosomal RNA genes for phylogenetics. *In* PCR Protocols: A Guide To Methods And Applications. Eds. M A Innis, D H Gelfand, J J Sninsky and T J White TJ. pp 315–322. Academic Press, London.

Plant and Soil **244:** 41–53, 2002.

Ericoid mycorrhizal fungi: some new perspectives on old acquaintances

Silvia Perotto[1], Mariangela Girlanda & Elena Martino
Department of Plant Biology, University of Torino and Centro di Studio sulla Micologia del Terreno, C.N.R., Viale Mattioli 25, 10125 Torino, Italy. [1]*Corresponding author**

Key words: biodiversity, ericoid fungi, group I introns, host range, metal tolerance

Abstract

Many ericaceous species colonize as pioneer plants substrates ranging from arid sandy soils to moist mor humus, in association with their mycorrhizal fungi. Thanks to the symbiosis with ericoid mycorrhizal fungi, ericaceous plants are also able to grow in highly polluted environments, where metal ions can reach toxic levels in the soil substrate. For a long time this mycorrhizal type has been regarded as an example of a highly specific interaction between plants and fungi. More recent studies have been challenging this view because some ericoid mycorrhizal endophytes seem also able to colonise plants from very distant taxa. A molecular approach has allowed the investigation of genetic diversity and molecular ecology of ericoid mycorrhizal fungi, and has revealed that ericaceous plants can be very promiscuous, with multiple occupancy of their thin roots. The molecular analysis of sterile morphotypes involved in this symbiosis has also led to deeper understanding of the species diversity of ericoid fungi. Genetic polymorphism of ericoid fungi is wider than previously thought, and often increased by the presence of Group I introns in the nuclear small subunit rDNA.

Introduction

Most plants belonging to Ericales are able to associate symbiotically with soil fungi to form a distinctive type of mycorrhiza, termed ericoid mycorrhiza. This association was initially investigated in members of the family Ericaceae, more abundant in the northern hemisphere (Bonfante and Gianinazzi-Pearson, 1979; Pearson and Read, 1973), but a morphologically similar mycorrhizal association was described also in the family Epacridaceae, widely distributed in the southern hemisphere (Ashford et al., 1996; Read, 1996; Reed, 1989). This is not surprising since phylogenetic analyses indicate that Ericaceae and Epacridaceae, traditionally considered as two separate families of the order Ericales, are closely related and epacrids may represent a sub-clade of the Ericaceae (Crayn et al., 1998).

The morphology of ericoid mycorrhiza is highly conserved in different plant species (Figure 1). The epidermal cells of the fine ericaceous hair roots harbour more or less dense coils of fungal mycelium (Bonfante and Gianinazzi-Pearson, 1979; Perotto et al., 1995) which appear to remain enclosed within single root cells (see also Figure 3). As in all endosymbioses, the intracellular fungal symbiont is separated from the plant cytoplasm by a plant-derived membrane, which invaginates to follow fungal growth and coil formation. In the infected cells, the plant nucleus moves to a central position.

Although ericoid mycorrhizal plants are commonly found as understorey vegetation in boreal and mediterranean forests, ericaceous shrubs can become dominant in many natural and semi-natural heathland communities. This happens especially in environmental conditions where a slow decomposition of the plant litter occurs, resulting in acidic soils rich in recalcitrant organic matter but low in available mineral nutrients such as N and P (Cairney and Burke, 1998). The survival of ericaceous plants under such nutrient-stressed conditions is thought to depend on the formation of mycorrhizal symbiosis, and the evolution of the association is regarded as having been driven by the selective advantages conferred by fungal infection (Straker, 1996). The saprotrophic potential of ericoid fungi and their ability to degrade complex and recal-

* FAX No: +39-11-6707459. E-mail: silvia.perotto@unito.it

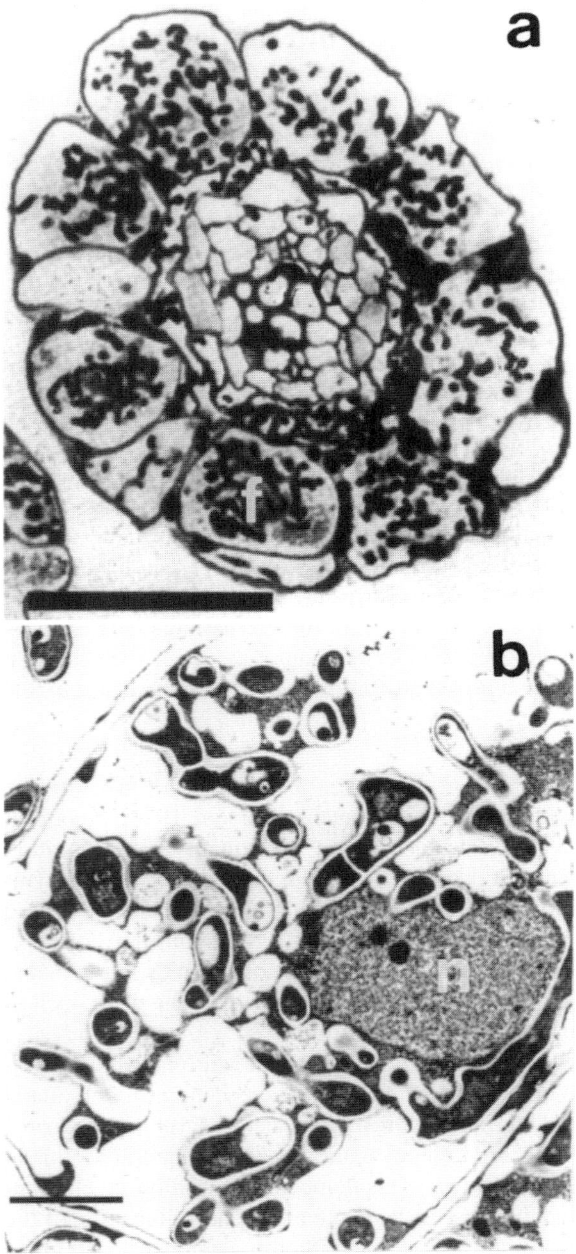

Figure 1. Colonization of *Calluna vulgaris* hair roots by ericoid mycorrhizal fungi. (a.) Transverse section of a hair root to show the general structure of the root. Cells of the outer epidermal layer harbour fungal coils (f) formed by a sterile mycorrhizal morphotype (mycelium G1). The collapsed cortical cells are visible just underneath the epidermis, surrounding the small central cylinder. Bar is 100 μm. (b.) Ultrastructure of an epidermal cell colonised by the sterile mycorrhizal morphotype PSIV. The fungal hyphae form a dense coil that surrounds the plant nucleus (n), which occupies a central position in the cell. Bar is 1 μm.

citrant polymeric substrates have been demonstrated over the years by several authors (e.g., see Bending and Read, 1996, 1997; Cairney and Burke, 1998; Leake and Read, 1991; Perotto et al., 1995; Varma and Bonfante, 1994). Based on these results, it is widely accepted that the major benefit conferred to the ericaceous host plant by mycorrhizal infection is the enzymatic degradation of organic polymers in the soil, and the transfer of some of the resulting products to the root (Smith and Read, 1997). Thanks to their mycorrhizal status, host plants can access otherwise unavailable organic N and P (Nasholm and Persson, 2001).

In addition to their general role in nutrient uptake, it has also been demonstrated that ericoid mycorrhizal fungi confer to their host the ability to compete successfully with other plant species. For example, *Calluna vulgaris* is the most common ericaceous species in the oceanic north west of Europe, where it can form almost pure plant communities (Rodwell, 1991). The superior ability of mycorrhizal *C. vulgaris* to compete over *Nardus stricta* was demonstrated in pot cultures under different nutrient conditions (Genney et al., 2000), and suggested to depend on allelopathy rather than competition for nutrients (see also Pongè, 1998).

Despite the widespread occurrence of ericaceous plants, little progress has been made in the understanding of the molecular bases of plant–fungus interactions during ericoid mycorrhiza formation and functioning. This is in striking contrast with the achievements in the study of arbuscular and ectomycorrhiza, where the use of model systems and the application of functional genomics have led to the discovery of entire sets of genes and proteins regulated during symbiosis (e.g., see Gianinazzi-Pearson, 1996; Van Buuren et al., 1999; Voiblet et al., 2001; and papers in this issue). Delays in this aspect of ericoid mycorrhizal research can be attributed to the difficulty of performing biochemical and molecular studies on this association, due to the minute size of mycorrhizal hair roots, but also to the limited consideration for ericaceous plants in agro-forestry. However, interest in this association has been increasing recently because of the ability of ericoid mycorrhizal plants to grow on polluted sites contaminated by heavy metals, and their potential applications in bioremediation. Some of the tolerance mechanisms, which will be reviewed later, were originally investigated by Bradley et al. (1982), but have been analysed in more detail for both the fungus and the plant (Martino et al., 2000a,b; Sharples et al., 2000a,b).

The development and use of molecular techniques has contributed greatly to gain insights on fungal diversity as well as the molecular ecology of mycorrhizal fungi. These techniques have been applied extensively to the study of ericoid fungi, and the derived knowledge has modified substantially our view of the ericoid mycorrhizal symbiosis, until recently thought to be an endomycorrhizal symbiosis with a high degree of specificity, restricted to few genera of plants and fungi (Straker, 1996). The aim of this contribution is to review some aspects where progress in the study of ericoid mycorrhizal fungi has allowed a better understanding of their physiology (in particular the mechanisms of metal tolerance), or have opened new perspectives on their ecology.

Ericoid fungi: a good model system to study the molecular bases of fungus–heavy metal interactions

The molecular mechanisms that govern interactions between metals and organisms are extremely complex, and depend on the organism involved as well as on environmental factors. It should be remembered, however, that many metals play essential functions in the organisms, and uptake must therefore involve recognition of metal species. Insufficient levels of essential metals can result in stress responses just as severe as those resulting from excess metals, and therefore cells must have developed mechanisms to avoid both extremes (Tomsett, 1993). Some heavy metals have no known functions and are therefore toxic at all concentrations. Molecular recognition allows organisms to differentiate between essential and non-essential ions and, if necessary, to partition them in different ways.

The mechanisms that allow survival of mycorrhizal fungi under heavy metal pollution are relatively poorly understood. Recent reports indicate that ecto- and endomycorrhizal isolates tolerant to high concentrations of heavy metals are selected in contaminated soils (Colpaert et al., 2000; Weissenhorn et al., 1993). Strains of the ericoid mycorrhizal species *Oidiodendron maius* isolated from Cd/Zn/Al-polluted sites also displayed a better ability to grow *in vitro* on media containing these metals when compared with isolates from non polluted sites (Lacourt et al., 2000; Martino et al., 2000a,b). Similarly, populations of arsenate resistant *Hymenoscyphus ericae*, another well known ericoid mycorrhizal partner, have been isolated from As/Cu-contaminated mine soils (Sharples et al., 2000b,c).

There are two major strategies that fungi can adopt to protect themselves against heavy metal toxicity. Avoidance restricts entry of metal ions into the cytoplasm, and relies on decreased uptake or increased efflux of metal ions, or on their immobilization outside the cell. Sequestration occurs to reduce cytoplasmic concentration of free metal ions, either through the synthesis of chelating compounds or by compartmentalisation into the vacuole (Gadd, 1993; Leyval et al., 1997). These general mechanisms of metal tolerance have also been found in mycorrhizal fungi, but it is still unclear if they also regulate transfer to the plant.

In mycorrhizal fungi, as in other fungi, the cell wall is a major binding site for heavy metals. It has been suggested that adsorption to the fungal wall is an important mechanism that reduces metal supply to the host (see Leyval et al., 1997). In ericoid mycorrhiza, the protective effect of the fungus reported by Bradley et al. (1981) has been ascribed to the fact that *Hymenoscyphus ericae* displays strong affinities for metallic cations. In addition, the excretion of a loosely adhering extrahyphal slime has been shown by Denny and Ridge (1995) to correlate in different fungal strains with the tolerance and the amelioration of zinc toxicity. Strains of the ericoid mycorrhizal species *O. maius* derived from polluted and unpolluted soils were investigated for their ability to bind zinc ions. Significant differences were found in the amount of metal adsorbed to the fungal mycelium, probably due to different binding capacity of the cell wall (Figure 2).

Fungi can also interact with metals in the surrounding environment by releasing extracellular metabolites that can modify heavy metals' bioavailability (Gadd, 1993). Mycorrhizal fungi can modify the mobility of metals in their surrounding environment, and solubilization in particular can be very important as it can contribute to bring into soluble forms essential ions that are normally found as insoluble organic and inorganic compounds (Marschner, 1995; Smith and Read, 1997). The efficiency of mycorrhizal fungi in increasing metal availability to plants would appear to be a potential problem in soils that are naturally enriched in toxic metal species or polluted by anthropic activities. By contrast, it is well documented that the ericoid mycorrhizal symbiosis can reduce metal toxicity to the host, allowing plants to survive in soils with potentially toxic amounts of soluble and insoluble metal species (Bradley et al., 1982; Sharples et al., 2000b,c).

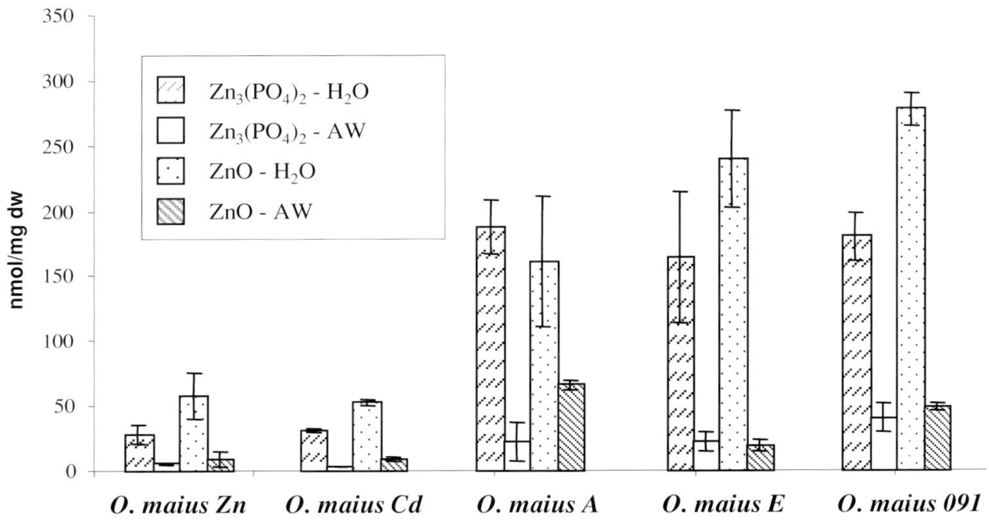

Figure 2. Adsorption of zinc to the mycelium of five different strains of the ericoid mycorrhizal fungus *Oidiodendron maius*. Strains *O. maius* Zn and Cd were isolated from polluted sites near Krakow (Poland), whereas the other strains were derived from unpolluted soils in different geographic locations. Fungi were grown on a solid malt medium added with ZnO or $Zn_3(PO_4)_2$. Significant differences in metal binding properties can be observed. Most of the metal could be released by an acidic wash (AW, HCl 0.5 M). Data are the mean of three independent experiments ± standard deviation.

In these conditions, a possible strategy to reduce metal availability to the plants is suggested by the observation that the ability of ericoid fungi to solubilize insoluble inorganic metal compounds is strongly reduced in fungal strains isolated from polluted soils (Martino et al., unpublished results).

In addition to metabolites that directly interact with metals, fungi can also respond to the presence of metals with the release of specific proteins in the surrounding medium. It was somewhat surprising to observe that the presence of zinc in the culture medium sharply increased the secretion and activity of extracellular enzymes that hydrolyse polymeric substances such as the pectin component of plant cell walls (Martino et al., 2000a). The significance of this increased production is unclear, but oligalacturonans may function as better metal chelators than larger polymers, thus protecting the fungus during saprotrophic growth.

When heavy metals cross the cell wall, they are accumulated in the cell by an active, energy dependent process (Morley et al., 1996). Essential metals are transported across the cell membranes by transport systems specific for these elements (Hughes and Poole, 1989). Non-essential metals can be sometimes transported by the same systems specific for essential metals, or they can use transporters for other nutrients. The mechanism of arsenic tolerance in ericoid mycorrhizal fungi has been investigated by Sharples et al. (2000b,c). This element enters the cell through the phosphate transporter, causing mycorrhizal fungi to enhance both phosphate and arsenate uptake. Sharples et al. (2000c) found that active and specific efflux mechanisms are adopted by ericoid mycorrhizal fungi from polluted sites, so as to decrease cellular concentrations of arsenic while retaining phosphate. These studies are just starting to unravel the cellular and molecular mechanisms of fungal metal tolerance, and they will provide important information to understand and exploit the potential of these fungi in bioremediation.

Biodiversity of ericoid mycorrhizal fungi: what is the true taxonomic range?

Information on the diversity of ericoid mycorrhizal endophytes in the Ericaceae and Epacridaceae has been collected over the years by several authors (e.g., Berch et al., 2001; Bergero et al., 2000; Cairney et al., 2000; Chambers et al., 2000; Hambleton and Currah, 1997; Hutton et al., 1994; Liu et al., 1998; Monreal et al., 1999; Perotto et al., 1996; McLean et al., 1999; Sharples et al., 2000a), and these collective data have contributed to change our view on the specificity of the association with regards to the fungal partner.

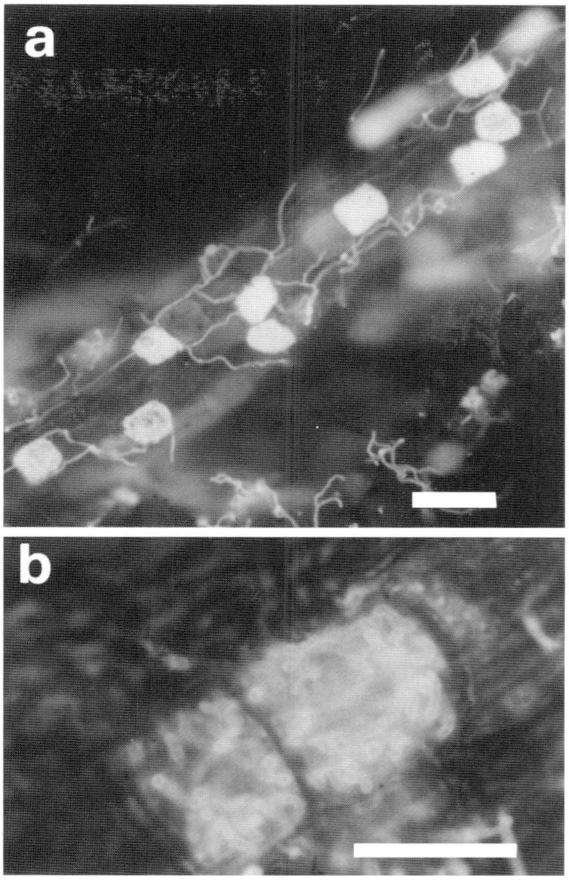

Figure 3. Confocal fluorescence microscopy after staining of a *C. vulgaris* hair root with wheat germ agglutinin-FITC. This lectin binds to the chitin present exclusively in the fungal wall. (a.) Low magnification showing root epidermal cells colonised by fungal coils. The colonised root cells can be well separated from each other and indicate that the fungus does not spread from cell to cell. (b.) The fungal hyphae can be better visualised in two coils shown at higher magnification. Bar is 20 μm.

Fungi recovered from ericoid roots and confirmed to be mycorrhizal in *in vitro* synthesis experiments are mostly sterile when brought into culture. The first two taxa to be recognized as symbionts of ericaceous hosts were both Ascomycetes (see Read, 1996; Smith and Read, 1997). *Hymenoscyphus ericae* (Read) Korf & Kernan is a member of order Helotiales, most frequently found in an anamorphic stage, *Scytalidium vaccinii* Dalpé, Litten & Sigler, proposed by Egger and Sigler (1993). The other known species, *Oidiodendron* spp. (anamorphic fungi), have their teleomorphs mainly in the Onygenales. *H. ericae* and *O. maius* Barron appear to be the dominant fungi in the diverse assemblages of symbionts colonizing the plants investigated to date (Hambleton and Currah, 1997; Johansson, 2001; Monreal et al., 1999; Perotto et al., 1996; Read, 1996; Sharples et al., 2000a; Straker, 1996). It is interesting to note that one or other of the species may be dominant at a particular site. For instance, *C. vulgaris* roots were mostly colonized by *H. ericae* at mine and natural heathland sites in south-west England (Sharples et al., 2000a), whereas *O. maius* was the dominant species in the same plant at a heathland site in northern Italy (Perotto et al., 1996).

Ribosomal DNA sequence comparisons over the past few years are providing new evidence for definition of genetic diversity and taxonomic status of these well-known endophytes. To date, the number of mycorrhizal species within *Oidiodendron* is uncertain, both because of misidentifications of field-collected and historically important strains (Hambleton et al., 1998b) and current re-evaluation of species concepts and boundaries within this genus (Hambleton et al., 1998b; Lacourt et al., 2001). Higher-level taxonomic affiliations and relationships of *Oidiodendron* and related teleomorphs have also been questioned (Hambleton et al., 1998a,b).

Species boundaries are a matter of debate also for *H. ericae*. Vrålstad et al. (2000) found that the ectomycorrhizal symbiont of *Piceirhiza bicolorata* shared approximately 95% ITS1 sequence identity with *H. ericae*, and can be considered part of a single major aggregate with the latter. Several unidentified sterile ericoid symbionts from different ericaceous hosts were also found to segregate with *H. ericae* in sequence based phylogenetic trees (Monreal et al., 1999; Sharples et al., 2000a). In view of the range of genetic diversity observed in such clusters it is presently unclear if a single species is indicated under the name *H. ericae* as currently applied. It is more likely that the *H. ericae* aggregate may correspond to a generic level comprising more than one species showing mycorrhizal behaviour. This generic entity appears to be quite separate from other members of the genus *Hymenoscyphus* (T. Vrålstad, personal communication).

In spite of the likely dominance of a few defined mycobionts, other ascomycetous taxa have been recognized among ericoid symbionts. *Acremonium strictum* W. Gams, an anamorphic fungus with affinities in the Hypocreales (Glenn et al., 1996) formed ericoid mycorrhiza *in vitro* with *Gaultheria shallon* (salal), although association with the host was somewhat atypical (Xiao and Berch, 1996). *Phialocephala fortinii* Wang & Wilcox, another anamorphic fungus with likely affinities in Helotiales based on the small subunit (SSU) rDNA sequences (LoBuglio et al., 1996), was associated with ericaceous roots. However, its status as an ericoid mycorrhizal symbiont is dubious, as it appeared to form intracellular microsclerotia instead of typical hyphal coils in *in vitro* experiments with *Menziesia ferruginea* (Stoyke et al., 1992).

Molecular data are leading to increasing recognition that ericoid fungi encompass a wider spectrum of taxa than once thought, so that the number of mycorrhizal species is currently uncertain. Different authors (e.g., Hambleton and Currah, 1997; Hutton et al., 1994; Liu et al., 1998; McLean et al., 1999; Monreal et al., 1999; Perotto et al., 1996) have hypothesised that sterile mycelia with ericoid mycorrhizal behaviour represent a heterogeneous group of so far mostly unidentified taxa. Recent comparisons of ITS sequences have indeed indicated that several sterile isolates represent taxa having no close relative among species currently included in GenBank or EMBL databases, as indicated by the low percentage similarity of their closest matches in BLAST or FASTA searches. Isolates obtained from *C. vulgaris* in England were found to have highest similarity (approx. 80%) with *Perrotia* spp. (fam. Hyaloscyphaceae, ord. Helotiales) (Sharples et al., 2000a). An isolate from salal was found to be close to *Phialophora finlandia* Wang & Wilcox (Monreal et al., 1999) an ectendo-mycorrhizal species. Interestingly, *in vitro* inoculation of salal seedlings with this species provided some evidence that this species may form typical ericoid mycorrhizae (Monreal et al., 1999).

Relationships of symbionts of ericaceous and epacridaceous hosts have also been the subject of recent research, as endophytes from Epacridaceae may readily form typical ericoid mycorrhiza in Ericaceae and *vice versa*, suggesting close relationships among the fungi (Liu et al., 1998; McLean et al., 1998; Read, 1996; Steinke et al., 1996). This relatedness can also be found by DNA sequence analysis. For example, Bergero et al. (2000) found that one of the dominant sterile mycorrhizal morphotypes obtained from *Erica arborea* plants in a mediterranean forest in Italy had its closest relative (94% ITS sequence similarity) in an epacrid endophyte obtained from *Astroloma pinifolium* (Epacridaceae) in Australia (McLean et al., 1998).

Despite the relatedness among some sterile morphotypes, the occurrence of *H. ericae* and *Oidiodendron* spp. in the epacrids is so far supported mainly by molecular data. Mycorrhizal isolates from *Epacris impressa* formed either distinct clusters or clustered with *Hymenoscyphus ericae* or *Cistella grevillei* (Hyaloscyphaceae, Helotiales) (McLean et al., 1999), although the low degree of sequence identity and the limited range of taxa included in this study may have led to incorrect interpretations, as discussed by Sharples et al. (2000a). Chambers et al. (2000), based on ITS sequences, found that a fungal isolate from *Woollsia pungens* (Epacridaceae) was to be identified as an *Oidiodendron* species.

Ericoid fungi are also likely to encompass members of basidiomycetes. Early circumstantial evidence of basidiomycete fungi forming ericoid mycorrhiza in roots of *Rhododendron*, *Calluna* and *Vaccinium* (Bonfante-Fasolo, 1980; Englander and Hull, 1980; Peterson et al., 1980) were paralleled by recent observation of hyphae with clamp connections and dolipore septa forming typical mycorrhizal coils in roots of *E. arborea* in Italy (Bergero et al., 2000). These associates, however, could not be isolated and their identity remains to be positively ascertained. In addition, direct microscopic observations of the morphological diversity of ericoid fungi in field collected *E. arborea* roots, together with failure to bring many of them into culture, indicate the occurrence of a variety of unculturable mycobionts (Bergero et al., 2000). These results are in contrast with the general opinion that ericoid symbionts have good saprotrophic abilities and are able to grow on common culture media (Leake and Read, 1991). The use of PCR techniques to amplify fungal DNA directly from mycorrhizal roots may help unravelling their taxonomic nature.

The genetic diversity and the phylogeny of ericoid mycorrhizal fungi are discussed in more detail in another paper of this same issue (Berch et al., 2001), and we will focus therefore on some features of ericoid mycorrhiza that may be of ecological and evolutionary relevance.

Ericoid mycorrhizal roots are composite structures with multiple fungal occupants

The morphology of mycorrhizal roots described earlier indicates that epidermal root cells could potentially function as separate units, challenged and colonized by a variety of fungi resident in the rhizosphere (Figure 3). ITS-RFLP analysis of ericoid fungal isolates from different mycorrhizal plant species has demonstrated that multiple occupancy is a common phenomenon in ericaceous roots. We have observed that the mycorrhizal roots of *C. vulgaris* from a heathland site in northern Italy harbour quite a diverse community of mycorrhizal fungi that formed typical mycorrhizal structures after resynthesis trials with axenic plants (Perotto et al., 1996). Similar investigations carried out on ericaceous (Bergero et al., 2000; Hambleton and Currah, 1997; Monreal et al., 1999) and epacridaceous (Chambers et al., 2000) plants also indicate the simultaneous presence of fungi with different ITS-RFLP patterns in the same root system.

Intraspecific genetic polymorphism can be revealed by sensitive techniques such as RAPD. When applied to mycological studies, this technique has been successfully used to identify races or even individual genets of filamentous fungi (Smith et al., 1992). RAPD was used to investigate ericoid fungal isolates derived from a single plant of *C. vulgaris*, either identified as *O. maius* or as sterile morphotypes that were distinguishable for their ITS-RFLP profiles (Perotto et al., 1996). RAPD analyses showed a high intraspecific polymorphism in the *O. maius* isolates, and a lower variability within populations of sterile mycelia. Many isolates of *O. maius* showing polymorphic RAPD bands were derived from the same plant of *C. vulgaris*, thus indicating that the roots of ericaceous plants are a complex mosaic where different populations of mycorrhizal fungi coexist, each represented by a variable number of genets. Multiple occupancy seems therefore to be a common phenomenon in ericoid mycorrhiza, and it has also been reported for example in arbuscular (Clapp et al., 1995) and ectomycorrhizal roots (Brand, 1992). This promiscuous situation may have important functional implications for the plant. The simultaneous association with many and diverse symbiotic fungi may represent an important strategy to broaden the array of functions in the colonization of difficult substrates. The hypothesis is more accreditable if such diversity mirrors a functional diversity, and different species of ericoid mycorrhizal fungi are indeed able to perform different physiological functions. For example, they can hydrolyse complex substrates, or they do so to a different extent (e.g., Leake and Read, 1991; Perotto et al., 1997; Varma and Bonfante, 1994). Even individuals within the same species have been shown to perform differently. For example, Cairney et al. (2000) found that isolates of the ericoid fungus *H. ericae* display different abilities to utilize inorganic and organic nitrogen sources. Different abilities of *O. maius* strains to tolerate high concentrations of heavy metals were demonstrated by Martino et al. (2000a,b), even when they were derived from the same site (Lacourt et al., 2000).

Group I introns in ericoid fungi: a fascinating evolutionary puzzle and a potential problem for molecular studies based on PCR-RFLP

Analysis of the nuclear ribosomal genes has revealed an unusual feature for many ericoid fungal isolates. Amplifications using universal primers designed on the SSU rDNA have yielded fragments which were often much larger in size than expected (Egger et al., 1995; Perotto et al., 2000; Vrålstad et al., 2000) due to the insertion of Group I introns. Group I introns form a structural and functional group characterized by four conserved regions which play a role in the formation of secondary structures. They are widespread but irregularly distributed in lower eukaryotes, especially algae and fungi (Gargas et al., 1995; Johansen et al., 1996). Group I introns occur at several locations along the chloroplast and mitochondrial genome, but seem to be restricted to the rDNA genes in the nuclear genome. They have been shown to splice both *in vitro* and *in vivo* due to the autocatalytic properties of the intron RNA, and are able to insert into intronless copies of the same gene with a process called *homing* (Cech, 1990).

In fungi, nuclear Group I introns have been found both in Ascomycetes and Basidiomycetes. Although it has not yet been clarified if they play a specific biological function in their fungal host, they represent a fascinating evolutionary puzzle and a suitable system for studying lateral movement of genetic elements. Horizontal transmission of Group I introns among fungal taxa has been discussed since their discovery on account of their scattered distribution and presence/absence in related taxa (e.g., Hibbett, 1996; Holst-Jensen et al., 1999; Nishida et al., 1998). Phylogenetic analysis has also indicated that horizontal transmission of Group I introns may have occurred between organelles (Turmel et al., 1995) and between fungi and plants during the intimate contacts established during pathogenic or symbiotic interactions (Nishida and Sugiyama, 1995). In ericoid symbiosis, like in all endosymbioses, the fungal and plant cytoplasm are only separated by a thin interface (Perotto et al., 1995), and so a possibility exists that some introns may be transmitted from the fungus to the host, although this aspect has never been investigated.

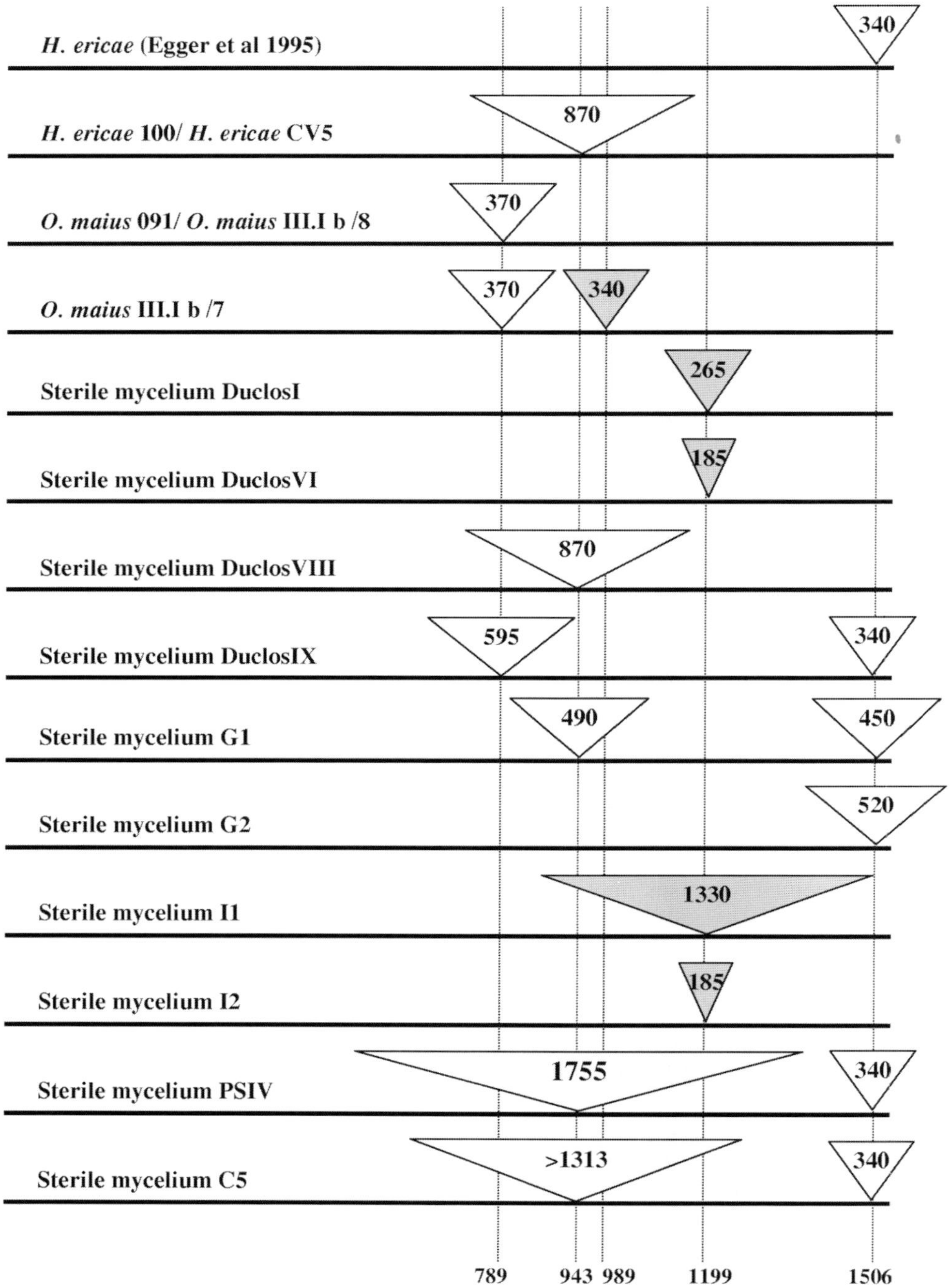

Figure 4. Diagram showing the position of DNA insertions in the SSU rDNA of ericoid isolates. The size of DNA insertions and their exact position (indicated below the diagram and relative to *Escherichia coli* rDNA) were established on the basis of sequence information. White triangles indicate typical Group I introns, whereas dark grey triangles indicate members of a novel class of ribosomal introns. Modified from Perotto et al. (2000).

Following an earlier report of a Group I intron in *H. ericae* (Egger et al., 1995), many additional introns have been found at different locations in the SSU rDNA of a large collection of ericoid fungi (Figure 4) including several isolates of *H. ericae, O. maius* and sterile mycelia (Perotto et al., 2000). The analysis of ericoid mycorrhizal fungi has also allowed the identification of a new class of ribosomal introns that lack the consensus sequences typical of Group I introns. The presence of four distinct and well conserved sequence regions in this new class of introns, together with their occurrence at specific sites different from those colonised by the more typical Group I introns (Figure 4), indicate that they represent a distinct population of insertional elements (Perotto et al., 2000).

Heterogeneity among isolates of the same species, expressed by the presence/absence of specific introns, was found for both *O. maius* and *H. ericae* (Perotto et al., 2000; Vrålstad et al., 2000). The optional presence of Group I introns in the rDNA of ericoid fungi may be a problem in molecular studies that rely on RFLP profiles of either conserved or variable rDNA regions. Since optional introns can be present or absent in the ribosomal genes of strains belonging to the same species, analyses based solely on RFLP profiles may result in misleading information on the diversity of these fungi.

Host range specificity in ericoid mycorrhiza: tight or loose friends?

As already mentioned, ericoid mycorrhiza was until recently considered an example of a highly specific association between few genera of plants and mycorrhizal fungi. The increase in our knowledge on the genetic diversity of ericoid mycorrhizal fungi, described in the previous paragraphs as well as in Berch et al. (2001), has contributed to modify our perception at least with respect to the fungal partner. Experimental evidence is accumulating to suggest that the range of plant species able to associate with ericoid mycorrhizal fungi may be also much wider than previously thought. Duckett and Read have reported in 1995 that *Hymenoscyphus ericae*, the best studied ericoid mycorrhizal symbiont, is also capable of forming *in vitro* intracellular coils in the rhizoids of liverworts. Further support to this observation derives from molecular studies that allowed Chambers et al. (1999) to assign a mycelium isolated in nature from the leafy liverwort *Cephaloziella exiliflora* to the genus *Hymenoscyphus*.

Other recent observations suggest that the host range of ericoid mycorrhizal fungi may be extended also to ectomycorrhizal plants in nature. Ericoid and ectomycorrhizal plants often coexist in natural ecosystems such as boreal or mediterranean forests, where ericaceous plants constitute the understorey vegetation of dominant ectomycorrhizal tree species (Read, 1991). Strong genetic similarities between fungi forming ericoid and ecto-mycorrhiza have been revealed by Vrålstad et al. (2000) in a boreal forest. Following ITS sequence comparison, the ericoid mycorrhizal fungus *H. ericae* was the closest relative of the mycobiont of *Piceirhiza bicolorata*, an ectomycorrhizal morphotype common in post-fire sites. Although the ability of some of these ectomycorrhizal isolates to form ericoid mycorrhiza with an ericaceous host has given so far inconclusive results (Vrålstad, 2001), these data demonstrate that the *H. ericae* aggregate groups together genetically related fungi, able to form distinct mycorrhizal types.

The possible sharing of mycorrhizal fungal isolates among ericoid and ectomycorrhizal plants was also investigated in a mediterranean forest by Bergero et al. (2000). These authors have demonstrated that several fungi (including *Oidiodendron* spp. and sterile morphotypes) were associated with both types of mycorrhizal plants. Fungal associates isolated from ectomycorrhizal *Quercus ilex* root tips formed ericoid mycorrhiza with *Erica arborea in vitro*, and molecular analyses indicate that some of these *Q. ilex* fungal associates were conspecific with mycorrhizal endophytes naturally occurring in *E. arborea* roots (Bergero et al., 2000). The physical and nutritional relationships between these ericoid fungal strains and the ectomycorrhizal plant require further investigation, but it is interesting to note that one ericoid mycorrhizal morphotype could infect *in vitro Q. ilex* seedlings with structures that resembled ectoendomycorrhizal infection (Bergero et al., 2000). The only other fungal species that may be able to form, at least *in vitro*, both ericoid (Monreal et al., 1999) and ecto- or ectoendomycorrhiza (Wilcox and Wang, 1987), depending on the host plant, is *Phialophora finlandia*.

The recent results just described open the exciting possibility that the same ericoid fungus may connect plant species with a different mycorrhizal status. Several studies have shown that the root systems of different plants become interconnected via the hyphae of shared mycorrhizal fungi, by means of which nutrients can be translocated (Arnebrant et al., 1993; Bethlenfalvay et al., 1991; Francis and Read 1984; Graves et al., 1997; Newman and Eason 1993; Simard et al., 1997; Whittingham and Read, 1982). Hyphal links have been reported for ectomycorrhizal fungi (Brownlee et al., 1983) and arbuscular fungi (Francis and Read 1984), but they have also been hypothezised in some more complex situations involving plants with different mycorrhizal status (Smith et al., 1995; Taylor and Bruns 1997). In the case of ericoid mycorrhizal fungi, the possibility of physical links between ericaceous and ectomycorrhizal plants (or liverworts) *via* their fungal associates remains an open question, as well as the nature of functional relationships with these hosts (Read, 2000). It also remains to be established whether or not the same fungus can form different types of mycorrhizal symbioses with distinct hosts in natural conditions. The classical bioassay to establish the mycorrhizal nature of a fungal isolate is the inoculation *in vitro* onto axenic seedlings. Although this assay remains an important test to elucidate the mycorrhizal potential of a fungal isolate, the conditions used are very different from those found in nature and the results must be interpreted with caution (Read, 1996).

Although interesting from a biological point of view, due to the more complex molecular mechanisms involved in the establishment of a tight symbiosis, the type of association of ericoid fungi with different host plants may not necessarily be relevant from a physiological point of view. It has been suggested that root–fungus associations might diverge from classical mycorrhizal types and yet function physiologically as a mycorrhiza under natural conditions (Jumpponen and Trappe 1998; Jumpponen et al., 1998). A study by Abuzinadah and Read (1989) showed that *Oidiodendron* enhances the growth of *Betula pendula* on a medium containing proteins as sole N source without producing a classical ectomycorrhizal infection.

The association of ericoid mycorrhizal fungi with ectomycorrhizal plants may be ecologically relevant in some stressed conditions. For example, genetic relatedness between ericoid and ectomycorrizal fungal associates has been revealed in forest environments subjected to fire (Bergero et al., 2000; Vrålstad et al., 2000). It could be hypothesised that one mycorrhizal host could thus provide an efficient source for biotrophic infection of other plant species. A role as a reservoir of ericoid fungi has also been suggested for liverwort rhizoids by Duckett and Read (1995).

In conclusion, recent biochemical, physiological and molecular studies have allowed us to gain better insights on ericoid mycorrhizal fungi. Identification of some mechanisms involved in metal tolerance provides some clues to understand the success observed for *H. ericae* and *O. maius* in colonizing polluted soils. However, the large number of recent publications on the genetic diversity of ericoid mycorrhizal endophytes indicates that the spectrum of fungal taxa has been largely underestimated, and that other fungi with potentially important functions may still to be discovered.

The molecular ecology of ericoid mycorrhizal fungi, and in particular the range of potential interactions with plants of different mycorrhizal status, is a more delicate issue still open to debate, where contrasting scenarios can be envisaged. Most of the results so far obtained indicate the possibility of a continuum from loose non-mycorrhizal to mycorrhizal associations between ericoid fungi and host plants, possibly related to environmental conditions and to the host plant species. This scenario is mainly based on ITS sequence similarities between isolates colonizing different host plants and on *in vitro* cross-inoculation available only for some fungal isolates. However, other data in the literature indicate that an opposite scenario cannot be discounted, where mycorrhizal specificity may be extremely high and even related to species or subspecies levels. Douglas et al (1989) have observed that an isolate of *Oidiodendron maius* from ectomycorrhizal Sitka spruce could not form ericoid mycorrhiza on an ericaceous host, which could normally harbour other fungal isolates of the same species. Similarly, ericoid mycorrhizal isolates belonging to the *H. ericae* aggregate only formed ericoid mycorrhiza on the ericaceous host when inoculated on ericaceous and tree species, whereas only ectomycorrhizal associations with tree species were observed in a similar cross-inoculation trial when ectomycorrhizal field isolates belonging to the same *H. ericae* aggregate were used (Vrålstad, 2001). These contrasting results indicate that although we have made considerable progress on the molecular analysis of ericoid mycorrhizal fungi, there are still exciting and unanswered questions that need to be addressed. In this process, it will be important to evaluate the results of genetic analyses relative to those from *in vitro* studies to understand the relationships of ericoid fungi with potential host plants in nature.

Acknowledgements

We thank Trude Vrålstad for discussion and for sharing some unpublished information on isolates of the *H. ericae* aggregate. The research described in this paper was funded by CNR 'Target Project on Biotechnology' (Subproject 2), by MIPA Project n.145, and by the Italian MURST.

References

Abuzinadah R A and Read D J 1989 The role of proteins in the nitrogen nutrition of ectomycorrhizal plants. V. Nitrogen transfer in birch (*Betula pendula*) grown in association with mycorrhizal and non-mycorrhizal fungi. New Phytol. 112, 61–68.

Arnebrant K, Ek H, Finlay R D and Soderstrom B 1993 Nitrogen translocation between *Alnus glutinosa* (L.) Gaertn seedlings inoculated with *Frankia* sp. and *Pinus contorta* Doug. ex Loud seedlings connected by a common ectomycorrhizal mycelium. New Phytol. 130, 231–242.

Ashford A E, Allaway W G and Reed M L 1996 A novel role for thick-walled epidermal cells in the mycorrhizal hair roots of *Lysinema ciliatum* R. Br. and other Epacridaceae. Ann. Bot. 77, 375–382.

Bending G D and Read D J 1996 Nitrogen mobilization from protein polyphenol complexes by ericoid and ectomycorrhizal fungi. Soil Biol. Biochem. 28, 1603–1612.

Bending G D and Read D J 1997 Lignin and soluble-phenolic degradation by ectomycorrhizal and ericoid mycorrhizal fungi. Mycol. Res. 101, 1348–1354.

Berch S M 2001 Molecular diversity and phylogeny of ericoid mycorrhizal fungi. Plant Soil, 244, 55–66.

Bergero R, Perotto S, Girlanda M, Vidano G and Luppi A M 2000 Ericoid mycorrhizal fungi are common root associates of a Mediterranean ectomycorrhizal plant (*Quercus ilex*). Mol. Ecol. 9, 1639–1650.

Bethlenfalvay G J, Reyes-Solis M G, Camel S B and Ferrera-Cerrato M 1991 Nutrient transfer between the root zones of soybean and maize plants connected by a common mycorrhizal mycelium. Physiol. Plant 82, 423–432.

Bonfante P 1980 Occurrence of a Basidiomycetes in living cells of mycorrhizal hair roots of *Calluna vulgaris*. Trans. Br. Mycol. Soc. 75, 320–325.

Bonfante P and Gianinazzi-Pearson V 1979 Ultrastructural aspects of endomycorrhiza in the Ericaceae. I. Naturally infected hair roots of *Calluna vulgaris* L. Hull. New Phytol. 83, 739–744.

Bradley R, Burt A J and Read D J 1981 Mycorrhizal infection and resistance to heavy metal toxicity in *Calluna vulgaris*. Nature 292, 335–337.

Bradley R, Burt A J and Read D J 1982 The biology of mycorrhiza in the Ericaceae VIII. The role of the mycorrhizal infection in heavy metal resistance. New Phytol. 91, 197–209.

Brand F 1992 Mixed associations of fungi in ectomycorrhizal roots. *In* Mycorrhizas in Ecosystems. Eds. D J Read, D H Lewis, A H Fitter and I J Alexander. pp 142–147. CAB International, Wallingford, Oxon, UK.

Brownlee C, Duddrige J A, Malibari A and Read D J 1983 The structure and function of mycelial systems of ectomycorrhizal roots with special reference to their role in forming inter-plant connections and providing pathways for assimilate and water transport. Plant Soil 71, 433–443.

Cairney J W G and Burke R M 1998 Extracellular enzyme activities of the ericoid mycorrhizal endophyte *Hymenoscyphus ericae* (Read) Korf & Kernan: their likely roles in decomposition of dead plant tissue in soil. Plant Soil 205, 181–192.

Cairney J W G, Sawyer N A., Sharples J M and Meharg A A 2000 Intraspecific variation in nitrogen source utilisation by isolates of the ericoid mycorrhizal fungus *Hymenoscyphus ericae* (Read) Korf & Kernan. Soil Biol. Biochem. 32, 1319–1322.

Cech T R 1990 Self-splicing of Group I introns. Annu. Rev. Biochem. 59, 543–568.

Chambers S M, Williams P G, Seppelt R D and Cairney J W G 1999 Molecular identification of *Hymenoscyphus* sp. from rhizoids of the leafy liverwort *Cephaloziella exiliflora* (Tayl.) Steph. in Australia and Antarctica. Mycol. Res. 103, 286–288.

Chambers S M, Liu G and Cairney J W G 2000 ITS rDNA sequence comparison of ericoid mycorrhizal endophytes from *Woollsia pungens*. Mycol. Res. 104, 168–174.

Clapp J P, Young J P W, Merryweather J W and Fitter A H 1995 Diversity of fungal symbionts in arbuscular mycorrhizas from a natural community. New Phytol. 130, 259–265.

Colpaert JV, Vandenkoornhuyse R, Adriaensen K and Vangronsveld J 2000 Genetic variation and heavy metal tolerance in the ectomycorrhizal basidiomycete *Suillus luteus*. New Phytol. 147, 367–79.

Crayn D M, Kron K A, Gadek P A and Quinn C J 1998 Phylogenetics and evolution of epacrids: a molecular analysis using the plastid gene *rbcL* with a reappraisal of the position of *Lebetanthus*. Austral. J. Bot. 46, 187–200.

Denny H J and Ridge I 1995 Fungal slime and its role in the mycorrhiza amelioration of zinc toxicity to higher plants. New Phytol. 130, 251–257.

Douglas G C, Heslin M C and Reid C 1989 Isolation of *Oidiodendron maius* from *Rhododendron* and ultrastructural characterization of synthesized mycorrhizas. Can. J. Bot. 67, 2206–2212.

Duckett J G and Read D J 1995 Ericoid mycorrhizas and rhizoid-ascomycete associations in liverworts share the same mycobiont isolation of the partners and resynthesis of the associations *in vitro*. New Phytol. 129, 439–447.

Egger K N and Sigler L 1993 Relatedness of the ericoid endophytes *Scytalidium vaccinii* and *Hymenoscyphus ericae* inferred from analysis of ribosomal DNA. Mycologia 85, 219–230.

Egger K N, Osmond G and Goodier J L 1995 Sequence and putative secondary structure of group I intron in the nuclear encoded ribosomal RNA genes of the fungus *Hymenoscyphus ericae*. Biochim. Biophys. Acta 1261, 275–278.

Englander L, Hull R J 1980 Reciprocal transfer of nutrients between ericaceous plants and *Clavaria* sp. New Phytol. 84, 661–667.

Francis R and Read D J 1984 Direct transfer of carbon between plants connected by vesicular-arbuscular mycorrhizal mycelium. Nature 307, 53–56.

Gadd G M 1993 Interactions of fungi with toxic metals. New Phytol. 124, 25–60.

Gargas A P, DePriest T and J W Taylor 1995 Position of multiple insertions in SSU rDNA of lichen-forming fungi. Mol. Biol. Evol. 12, 208–218

Genney D R, Alexander I J and Hartley S E 2000 Exclusion of grass roots from soil organic layers by *Calluna*: the role of ericoid mycorrhizas. J. Exp. Bot. 51, 1117–1125.

Glenn A E, Bacon C W, Price R, and Hanlin R T 1996 Molecular phylogeny of *Acremonium* and its taxonomic implications. Mycologia 88, 369–383.

Gianinazzi-Pearson V 1996 Plant cell responses to arbuscular mycorrhizal fungi: getting to the roots of the symbiosis. Plant Cell 8, 1871–1883.

Graves J D, Watkins N K, Fitter A H, Robinson D and Scrimgeour C 1997 Intraspecific transfer of carbon between plants linked by a common mycorrhizal network. Plant Soil 192, 153–159.

Hambleton S and Currah R S 1997 Fungal endophytes from the roots of alpine and boreal Ericaceae. Can. J. Bot. 75, 1570–1581.

Hambleton, S, Currah R and Egger K 1998a Phylogenetic relationships of ascomycetous root endophytes of the Ericaceae inferred from 18S rDNA sequence analysis. *In* Abstracts of the Second International Conference on Mycorrhiza. Eds. U Ahonen-Jonnarth, E Danell, P Fransson, O Kårén, B Lindahl, I Rangel and R Finlay. pp. 78. Swedish University of Agricultural Sciences, Uppsala.

Hambleton S, Egger K N and Currah R S 1998b The genus *Oidiodendron*: species delimitation and phylogenetic relationships based on nuclear ribosomal DNA analysis. Mycologia 90, 854–869.

Hibbett D S 1996 Phylogenetic evidence for horizontal transmission of Group I introns in the nuclear ribosomal DNA of mushroom-forming fungi. Mol. Biol. Evol. 13, 903–917.

Holst-Jensen A, Vaage M, Schumacher T and Johansen S 1999 Structural characteristics and possible horizontal transfer of Group I introns betwen closely related plant pathogenic fungi. Mol. Biol. Evol. 16, 114–126.

Hughes M N and Poole R K 1989 Metals and Microorganisms. Chapman and Hall, London.

Hutton B J, Dixon K W and Sivasithamparam K 1994 Ericoid endophytes of Western Australian heaths (Epacridaceae). New Phytol. 127, 557–566.

Johansen S D, Muscarella E and Vogt V M 1996 Insertion elements in ribosomal DNA. *In* Ribosomal RNA: Structure, Evolution, Processing and Function in Protein Biosynthesis. Eds. R A Zimmermann and A E Dahlberg. pp. 89-110. CRC Press, Boca Raton, FL.

Johansson M 2001 Fungal associations of Danish *Calluna vulgaris* roots with special reference to ericoid mycorrhiza. Plant Soil 231, 225–232.

Jumpponen A and Trappe J M 1998 Dark septate endophytes: a review of facultative biotrophic root-colonizing fungi. New Phytol. 140, 295–310.

Jumpponen A, Mattson K G and Trappe J M 1998 Mycorrhizal functioning of *Phialocephala fortinii* with *Pinus contorta* on glacier forefront soil: interactions with soil nitrogen and organic matter. Mycorrhiza 7, 261–265.

Lacourt I, D'Angelo S, Girlanda M, Turnau K, Bonfante P and Perotto S 2000 Genetic polymorphism and metal sensitivity of *Oidiodendron maius* strains isolated from polluted soils. Ann. Microbiol. 50, 157–66.

Lacourt I, Girlanda M, Perotto S, Del Pero M, Zuccon D and Luppi A M 2001 Nuclear ribosomal sequence analysis of *Oidiodendron*: towards a redefinition of ecologically relevant species. New Phytol. 149, 565–576.

Leake J R and Read D J 1991 Experiments with ericoid mycorrhiza. *In* Methods in Microbiology, vol. 23. Eds. J R Norris, D J Read and A K Varma. pp. 435–459. Academic Press, London.

Leyval C, Turnau K and Haselwandter K 1997 Effect of heavy metal pollution on mycorrhizal colonization and function: physiological, ecological and applied aspects. Mycorrhiza 7, 139–153.

Liu G, Chambers S M and Cairney J W G 1998 Molecular diversity of ericoid mycorrhizal endophytes isolated from *Woollsia pungens* (Cav.) F. Muell. (Epacridaceae). New Phytol. 140, 145–153

LoBuglio K F, Berbee M L and Taylor J W 1996 Phylogenetic origins of the asexual mycorrhizal symbiont *Cenococcum geophilum* Fr. and other mycorrhizal fungi among the ascomycetes. Mol. Phylogen. Evol. 6, 287–294.

Marschner H 1995 Mineral Nutrition of Higher Plants. Academic Press, London.

Martino E, Coisson J D, Lacourt I, Favaron F, Bonfante P and Perotto S 2000a Influence of heavy metals on production and activity of pectinolytic enzymes in ericoid mycorrhizal fungi. Mycol Res. 104, 825–833.

Martino E, Turnau K, Girlanda M, Bonfante P. and Perotto S. 2000b Ericoid mycorrhizal fungi from heavy metal polluted soils: their identification and growth in the presence of heavy metals. Mycol. Res. 104, 338–344.

McLean C B, Anthony J, Collins R A, Steinke E and Lawrie A C 1998 First synthesis of ericoid mycorrhizae in the Epacridaceae under axenic conditions. New Phytol. 139, 589–593.

McLean C B, Cunnington J H and Lawrie A C 1999 Molecular diversity within and between ericoid endophytes from the Ericaceae and Epacridaceae. New Phytol. 144, 351–358.

Monreal M, Berch SM and Berbee M 1999 Molecular diversity of ericoid mycorrhizal fungi. Can. J. Bot. 77, 1580–1594.

Morley G F, Sayer J A, Wilkinson S C, Gharieb M M and Gadd G M 1996 Sequestration, mobilization and transformation of metals and metalloids. *In* Fungi and Environmental Change. Eds. J C Frankland, N Magan and G M Gadd. pp 235–256. University Press, Cambridge.

Nasholm T and Persson J 2001 Plant acquisition of organic nitrogen in boreal forests. Physiol Plant 111, 419–426.

Newman E I and Eason W R 1993 Rates of phosphorus transfer within and between ryegrass (*Lolium perenne*) plants. Funct. Ecol. 7, 242–248.

Nishida H and Sugiyama J 1995 A common Group I intron between a plant parasitic fungus and its host. Mol. Biol. Evol. 12, 883–886.

Nishida H, Tajiri Y and Sugiyama J 1998 Multiple origins of fungal Group I introns located in the same position of nuclear SSU rRNA gene. J. Molec. Evol. 46, 442–448.

Pearson V and Read D J. 1973 The biology of mycorrhiza in the *Ericaceae*. I. The isolation of the endophyte and synthesis of mycorrhizas in aseptic cultures. New Phytol. 72, 371–379.

Perotto S, Peretto R, Faccio A, Schubert A, Varma A, and Bonfante P 1995 Ericoid mycorrhizal fungi: cellular and molecular bases of their interactions with the host plant. Can. J. Bot. 73 sup. 1, S557–S568.

Perotto S, Actis-Perino E, Perugini J and Bonfante P 1996. Molecular diversity of fungi from ericoid mycorrhizal roots. Mol. Ecol. 5, 123–131.

Perotto S, Coisson J D, Perugini J, Cometti V and Bonfante P 1997 Production of pectin-degrading enzyme by ericoid mycorrhizal fungi. New Phytol. 135, 151–162.

Perotto S, Nepote-Fus P, Saletta L, Bandi C and Young J P W 2000 A diverse population of introns in the nuclear ribosomal genes of ericoid mycorrhizal fungi includes elements with sequence similarity to endonuclease-coding genes. Mol. Biol. Evol. 17, 44–59.

Peterson T A W, Mueller C and Englander L 1980 Anatomy and ultrastructure of a *Rhododendron* root-fungus association. Can. J. Bot. 58, 2421–2433.

Ponge J F, Andre J, Zackrisson O, Bernier N, Nilsson M C and Gallet C 1998 The forest regeneration puzzle. BioScience 48, 523–530.

Read D J 1991 Mycorrhizas in ecosystems. Experientia 47, 376–390.

Read, D J 1996 The structure and function of the ericoid mycorrhizal root. Ann. Bot. 77, 365–374.

Read D J 2000 Links between genetic and functional diversity — a bridge too far? New Phytol. 145, 363–365.
Reed M L 1989 Ericoid mycorrhizas of Styphelieae: intensity of infection and nutrition of the symbionts. Aust. J Plant Physiol. 16, 156–160.
Rodwell J S 1991 Mires and Heaths. British Plant Communities, II. Cambridge University Press, Cambridge.
Sharples J M, Chambers S M, Meharg A A and Cairney J W G 2000a Genetic diversity of root-associated fungal endophytes from *Calluna vulgaris* at contrasting field sites. New Phytol. 148, 153–162.
Sharples J M, Meharg A A, Chambers S M and Cairney J W G 2000b Symbiotic solution to arsenic contamination. Nature 404, 951–952.
Sharples J M, Meharg A A, Chambers S M and Cairney J W G 2000c Mechanism of arsenate resistance in the ericoid mycorrhizal fungus *Hymenoscyphus ericae*. Plant Physiol. 124, 1327–1334.
Simard S W, Perry D A and Jones M D 1997 Net transfer of carbon between ectomycorrhizal tree species in the field. Nature 388, 579–582.
Smith M L, Bruhn J N and Anderson J B 1992 The fungus *Armillariella bulbosa* is among the largest and oldest living organisms. Nature 356, 428–431.
Smith J E, Molina R and Perry D A 1995 Occurrence of ectomycorrhizas on ericaceous and coniferous seedlings grown in soils from the Oregon Coast Range. New Phytol. 129, 73–81.
Smith S E and Read D J 1997 Mycorrhizal Symbiosis. 2nd edn. Academic Press, London.
Steinke E, Williams P G and Ashford A E 1996 The structure and fungal associated of mycorrhizas in *Leucopogon parviflorus* (Andr.) Lindl. Ann. Bot. 77, 413–419.
Stoyke G K, Egger N and Currah R S 1992 Characterisation of sterile endophytic fungi from the mycorrhizae of subalpine plants. Can. J. Bot. 70, 2009–2016.
Straker C J 1996 Ericoid mycorrhiza: ecological and host specificity. Mycorrhiza 6, 215–225.
Taylor D L, Bruns T D 1997 Independent specialized invasions of ectomycorrhizal mutualism by two nonphotosynthetic orchids. Proc. Natl. Acad. Sci. USA 94, 4510–4515.
Tomsett BA 1993 Genetic and molecular biology of metal tolerance in fungi. *In* Stress tolerance of fungi. Ed. D H Jennings. pp. 69–95. Marcel Dekker, New York.
Turmel M, Cote V, Otis C, Mercier J P, Gray M W, Lonergan K M and Lemieux C 1995 Evolutionary transfer of ORF-containing Group I introns between different subcellular compartments (chloroplast and mitochondrion). Mol. Biol. Evol. 12, 533–545.
Van Buuren M, Maldonado-Mendora I E, Trien A T, Blaylock L A and Harrison M J 1999 Novel genes induced during an arbuscular mycorrhizal (AM) symbiosis between *M. truncatula* and *G. versiforme*. Mol. Plant Microbe Interact. 12, 171–181.
Varma A and Bonfante P 1994 Utilization of cell wall related carbohydrates by ericoid mycorrhizal endophytes. Symbiosis 16, 301–313.
Voiblet C, Duplessis S, Encelot N and Martin F 2001 Identification of symbiosis-regulated genes in *Eucalyptus globulus–Pisolithus tinctorius* ectomycorrhiza by differential hybridization of arrayed cDNAs. Plant J. 25, 181–191.
Vrålstad T 2001 Molecular ecology of root-associated mycorrhizal and non-mycorrhizal ascomycetes. Dissertation submitted to the University of Oslo for the degree Doctor scientiarum N. 146, Unipub forlag, Oslo, Norway.
Vrålstad T, Fossheim T and Schumacher T 2000 *Piceirhiza bicolorata* — the ectomycorrhizal expression of the *Hymenoshypus ericae* aggregate? New Phytol. 145, 549–563.
Weissenhorn I, Leyval C and Berthelin J 1993 Cd-tolerant arbuscular mycorrhizal (AM) fungi from heavy metal polluted soils. Plant Soil 157, 247–256.
Whittingham J and Read D J 1982 Vesicular arbuscular mycorrhiza in natural vegetation systems III. Nutrient transfer between plants with mycorrhizal connections. New Phytol. 90, 277–284.
Wilcox H E, Wang C J K 1987 Ectomycorrhizal and ectoendomycorrhizal associations of *Phialophora finlandia* with *Pinus resinosa*, *Picea rubens* and *Betula alleghaniensis*. Can. J. For. Res. 17, 976–990.
Xiao G and Berch S M 1996 Diversity and abundance of ericoid mycorrhizal fungi of *Gaultheria shallon* on forest clearcuts. Can. J. Bot. 74, 337–346.

Plant and Soil **244:** 55–66, 2002.

Molecular detection, community structure and phylogeny of ericoid mycorrhizal fungi

S.M. Berch[1,3], T.R. Allen[2] & M.L. Berbee[2]
[1]*British Columbia Ministry of Forests Research Branch, Research Branch Laboratory, PO Box 9536, Victoria BC V8W 9C4, Canada,* [2]*Botany Department, University of British Columbia, 6270 University Blvd., Vancouver BC V6T1Z4, Canada.* [3]*Corresponding author**

Received 21 August 2001. Accepted in revised form 20 March 2002

Key words: Ericaceae, ericoid mycorrhizal fungi, Epacridaceae, *Gaultheria shallon*, phylogeny, salal

Abstract

Through traditional culturing and molecular characterization, we have determined that five putative species and 2 polyphyletic assemblages of fungi produce ericoid mycorrhizae in *Gaultheria shallon*, other Ericaceae and Epacridaceae. Using phylogenetic analysis of ITS2 sequences in GenBank, we have confirmed that most of these fungi occur in North America, Europe, and Australia. The low recovery rate of culturable ericoid mycorrhizal fungi from *Gaultheria shallon* may partly be explained by the fact that most mycorrhizal root segments contain an unculturable basidiomycete, revealed by direct amplification, cloning, and sequencing of LSU fungal DNA from root. Molecular characterization and phylogenetic analysis are powerful tools in revealing the geographic distribution and identity of ericoid mycorrhizal fungi.

Introduction

Almost two decades ago, David Read had proposed that ericoid mycorrhizal fungi allow ericaceous plants to access nutrients locked in the organic matrix that would otherwise be unavailable (Read, 1983). So, when growth check was observed on conifer seedlings growing in dense stands of the Ericaceous shrub, salal (*Gaultheria shallon* Pursh), on northern Vancouver Island, Weetman et al. (1989) suggested that ericoid mycorrhizal fungi might be involved. The Salal Cedar Hemlock Integrated Research Project (SCHIRP) was established over a decade ago to understand and solve this problem (Prescott, 1996; Prescott and Weetman, 1994). A component of SCHIRP is the investigation of the role of ericoid mycorrhizal fungi of salal in conifer growth check.

Using traditional fungal culturing, Xiao and Berch (1996) showed that at least four different fungal groups formed ericoid mycorrhizae at the SCHIRP site at Pt. McNeil, Vancouver Island. Two groups sporulated in culture, *Acremonium strictum* W. Gams and *Oidiodendron maius* Barron, and two non-sporulating groups were described as Unknown 1 (exemplified by UBC S9) and Unknown 2 (exemplified by isolate UBC S246). Xiao (1994) described Unknown 2 as resembling Read's isolates 100 and 101 of *Hymenoscyphus ericae* (Read) Korf & Kernan. *Acremonium strictum*, a fungus of wide distribution, was found to form mycorrhizae by Xiao (1994) but has not been isolated or detected in salal roots by subsequent workers. Its abundance and role in the roots of salal remains to be clarified. For all ericoid mycorrhizal fungi from salal, the percent recovery of mycorrhizal fungi from roots was low, around 3 – 5% per fungal taxon (Table 1).

To identify sterile isolates from the SCHIRP site, Monreal et al. (1999) took advantage of high variability of the internal transcribed spacer regions ITS1 and ITS2 of ribosomal RNA repeats. The ITS regions were analyzed with restriction enzymes that digested the amplified DNA into discrete fragments. The resulting restriction fragment length polymorphisms (RFLPs) were unique banding patterns that allowed for cat-

* FAX No: 250-952-4119.
E-mail: shannon.berch@gems7.gov.bc.ca

Table 1. Frequency of confirmed ericoid mycorrhizal fungi isolated from 1120 fine root segments of salal during first study[a]

Fungal taxon	Representative isolates[b]	Number of isolates	Percentage of root segments producing this fungus
Acremonium strictum	S220	41	4
Oidiodendron maius	S4, S18, S45, S80	59	5
Unknown 1	S9	36	3
Unknown 2	S246	41	4

[a]Xiao (1994).
[b]Deposited at UAMH.

Table 2. Frequency of confirmed ericoid mycorrhizal fungi isolated from 800 fine root segments of salal during this study

Fungal taxon	Representative isolates	Number of isolates	Percentage of root segments producing this fungus
Capronia sp.	UBCtra1046	10	1
Hymenoscyphus ericae[a]	M8, UBCtra 241	80	10
Hymenoscyphus ericae[a]	M5, UBCtra175	9	1
Oidiodendron maius	UBCtra133	49	6
Phialophora finlandia	UBCtra323	27	3
Rhytismatales	UBCtra300, UBCtra1018C	2	<1
Unknown 1	S9, UBCtra1128, UBCtra180	25	3
Unknown 2	S246, UBCtra270	23	3
	UBCtra290	19	2
	UBCtra466	9	1
	UBCtra179	6	1

[a]Taxa differentiated due to differences in RFLP patterns (Monreal et al., 1999).

egorization of sterile isolates into four groups capable of forming ericoid mycorrhizae (*Oidiodendron*, Unknown 2, and two new groups, Unknown 3 and 4). Then, using sequencing of the ITS2 region, Monreal et al. (1999) determined that Unknown 3 (isolate M8) and Unknown 4 (isolate M20) were similar to *Hymenoscyphus ericae*.

We returned to the same SCHIRP site, isolated more fungi from salal roots and confirmed their ability to form ericoid mycorrhizae. We found some of the fungi isolated by Xiao (*Oidiodendron maius*, Unknown 2) and by Monreal (Unknown 3). In addition, we found a number of isolates not identifiable as any of these fungi that were nonetheless able to form mycorrhizae on back-inoculation (Table 2). Again percent recovery of ericoid mycorrhizal fungi from mycorrhizal root segments was low.

Hymenoscyphus ericae has been consistently reported to be the major ericoid mycorrhizal fungus of Ericaceae such as *Calluna* and *Erica* in Europe since 1973 (Pearson and Read, 1973; Read, 1983, 1986; Strandberg and Johansson, 1999). *Scytalidium vaccinii* Dalpé, Sigler & Litten, which was found through molecular characterization to be the anamorph of *H. ericae* (Egger and Sigler, 1993), has also been isolated from the roots of *Vaccinium vitis-idaea* L. in Finland (Currah et al., 1999) and from *Vaccinium angustifolium* in Quebec, Canada (Dalpé, 1989). Fungal isolates from ericoid mycorrhizae, including those of *H. ericae*, are usually sterile in culture, or form arthroconidia as in *Scytalidium vaccinii* that might not be easily recognizable to the non-taxonomist, thus making their identification difficult using classical taxonomic approaches. Although *H. ericae*-like sterile or arthroconidial cultures have been isolated from Ericaceae in North America for awhile, only recently have Hambleton et al. (1999) induced apothecium production in a culture isolated from *Ledum groenlandicum* Oeder in Alberta, thus confirming the occurrence of *H. ericae* in Canada. Given this recalcitrance of *H. ericae* to sporulate in culture (Straker, 1996), molecular characterization is clearly the best way to identify this fungus. Molecular characterization can also be done directly on roots from the field, which allows for efficient and complete surveys of fungal communities.

Around the globe, research groups have begun to reveal a diversity of ericoid mycorrhizal fungi associated with ericaceous and epacridaceous plants (Chambers et al., 1999; Monreal et al., 1999; Perotto et al., 1996; Xiao and Berch, 1996), but the number of taxa of ericoid mycorrhizal fungi is still uncertain (Straker, 1996). The objective of this research was to use molecular characterization to determine the diversity and distribution of ericoid mycorrhizal fungi at two scales: in the fine roots of salal at the SCHIRP site on northern Vancouver Island, and of ericoid mycorrhizal fungi globally.

Methods

Ericoid mycorrhizal fungus isolation

Roots of salal (*G. shallon*) were collected over a number of years (Monreal, 1997; Xiao, 1994; this study) from the SCHIRP installation on northern Vancouver Island, British Columbia (50° 60′ N, 127° 35′ W). Roots were kept cool, transported to the University of British Columbia, and held at 4 °C until cleaned. Fine roots were then removed from the rhizomes, soaked in cool tap water, washed gently to remove soil and plant debris, surface sterilized in 30% hydrogen peroxide for 30 – 60 s, and rinsed 3 times in sterile distilled water. The fine roots were cut into 0.5 cm segments and plated on potato dextrose agar (PDA).

Mycorrhiza synthesis

Each fungal isolate was tested for mycorrhizal synthesis on *G. shallon* in Petri dish growth chambers as described by Xiao and Berch (1995). Seeds of *G. shallon* were sterilized in 30% hydrogen peroxide for 1 min. and placed on water agar for germination. Seedlings were transplanted to a low nutrient medium, modified Melin-Norkrans agar (MMN), in plastic Petri dishes, and inoculated with a fungal isolate. Mycorrhizal colonization was confirmed with light microscopy. Fungal isolates that formed mycorrhizae were re-tested by culturing the fungus from sterilized roots of colonized plants and inoculating new plants. Isolates that did not immediately form mycorrhizae were maintained on salal in Petri dish growth chambers and checked for 1 year.

Molecular characterization

Genomic DNA from each fungal isolate was extracted from young mycelium (Lee and Taylor, 1990). Following the PCR amplification procedures of Monreal et al. (1999), the ribosomal DNA regions of the ITS1 and ITS2 were amplified with the fungal specific primer combination ITS1-F and ITS4 (White et al., 1990). The genomic DNA of each isolate was characterized by RFLP with three enzymes: *Msp*1, *Rsa*1, and *Cfo*1 following the manufacturer's instructions (New England Biolabs). The isolates were grouped by RFLP banding patterns. The ITS2 regions of selected isolates from each group were sequenced with primer ITS3. To find matching sequences in the GenBank database, ITS2 sequences were subjected to BLAST searches. Matching sequences obtained during the BLAST searches were combined with previously collected sequences of fungal isolates from the same field site to reconstruct phylogenetic relationships.

Fungi detected by direct amplification and cloning of fungal DNA from root

Total genomic DNA was extracted from 1 cm segments of sterilized field collected roots. Genomic DNA from sterile leaves of *G. shallon* was used as negative controls for DNA amplification and cloning. Total fungal DNA was amplified with primers ITS1-F and ITS4, or ITS1-F (Gardes and Bruns, 1993) and TW13 (5′-GGTCCGTGTTTCAAGACG-3′) (Tom White, personal communication), to target a 1200 bp region of the ribosomal RNA repeat including the ITS region and about 500 bp of the LSU. PCR products were cloned using Invitrogen TOPOTMTA Cloning kit following the manufacturers' instructions. Cloned material was characterized with three restriction enzymes *Msp*1, *Cfo*1, and *Rsa*1. For phylogenetic analysis, the ITS2 and 500 bp of the LSU region were sequenced from selected clones for each RFLP group.

Sequence analysis

Sequences from clones and cultures were manually aligned in SeqPup with sequences retrieved from GenBank and analyzed with PAUP 4.0b5. GenBank accession numbers for all isolates used in our analyses are listed in Appendices 1 (ITS2) and 2 (LSU).

Results

Phylogenetic tree of ericoid mycorrhizal fungi of salal

To reconstruct phylogenetic relationships for salal's ericoid mycorrhizal fungi, we combined isolates and clones from all our studies (Monreal, 1997; Xiao, 1994; this study). The highly diverse ITS2 region of the rDNA was used to determine relationships of fungal groups at species level. Our sequences of ericoid mycorrhizal fungi and mycorrhiza-inhabiting clones from salal comprise a number of groups (Figure 1) which are discussed in the following section. All of the groups found in salal roots have now been reported from elsewhere in the world and all groups of ericoid mycorrhizal fungus reported from elsewhere in the world have now also been found in the roots of salal on a single site on northern Vancouver Island (Figure 2).

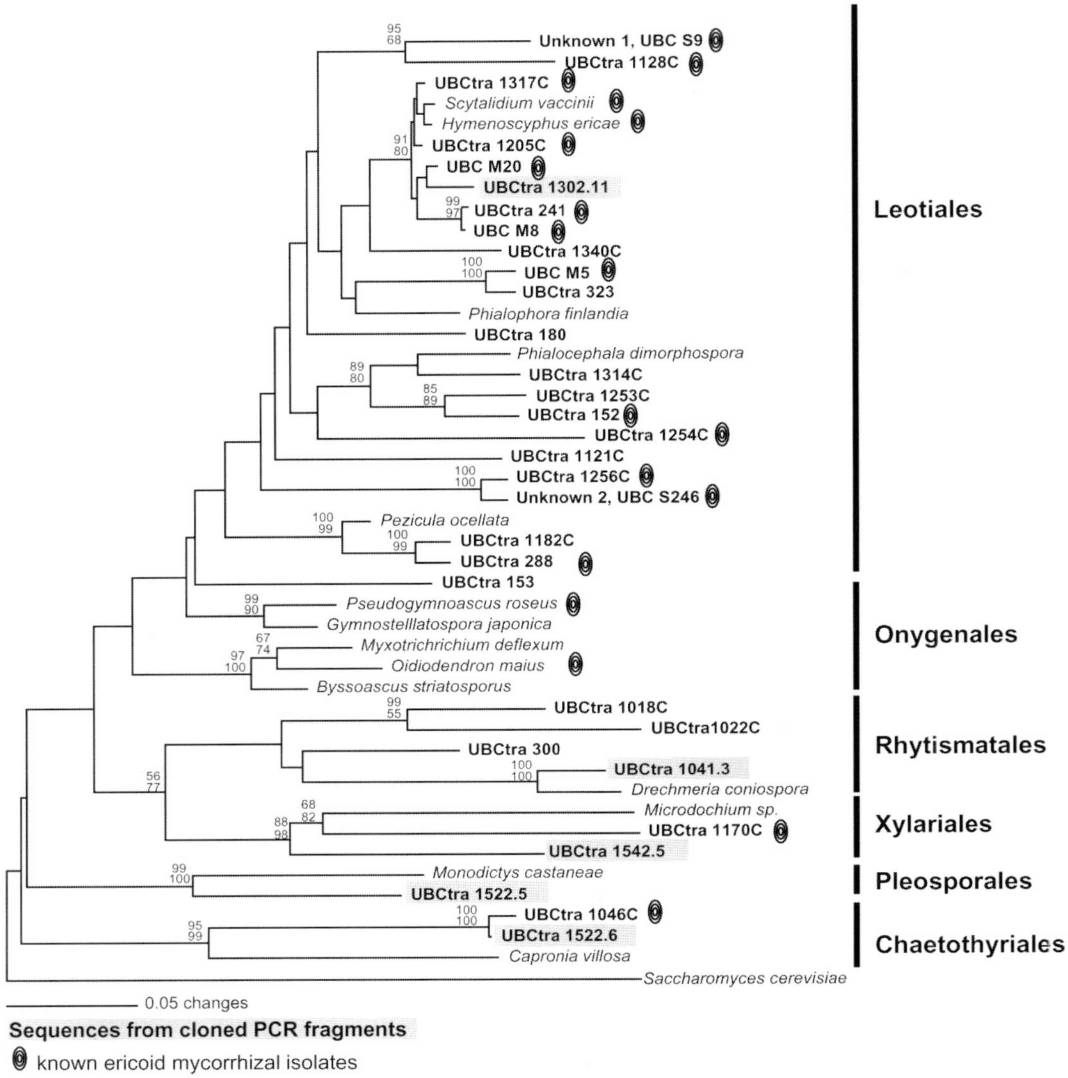

Figure 1. Relationships of cultured fungi and fungi detected by PCR amplification and cloning directly from ericoid mycorrhizal roots of *Gaultheria shallon*. Distances for this neighbour-joining tree were calculated using a Jukes-Cantor correction and were based on an alignment of about 350 nucleotides of the 5.8S and ITS2 rDNA regions. Branch lengths are proportional to base pair differences. Percentages from 500 bootstrap replicates are shown with neighbour-joining on top and fast parsimony without branch swapping on the bottom (values <50% not shown).

Phylogenetic tree of ericoid mycorrhizal fungi available in GenBank

Included in the phylogenetic analysis (Appendix 1) were ITS2 sequences from GenBank from June, 2001, selected to represent the diversity of fungal root endophytes of Ericaceae (Hambleton et al., 1998; Monreal et al., 1999; Sharples et al. 2000; this study) and Epacridaceae (Chambers et al., 2000; McLean et al., 1999). Also included were sequences of fungi from leafy liverworts in the Antarctic (Chambers et al., 1999) and from oak ectomycorrhizae in Italy (Bergero et al., 1999) that clustered with known ericoid mycorrhizal fungi.

The majority of ITS2 sequences sampled from GenBank belong to the Leotiales and many cluster in a species complex around *Hymenoscyphus ericae* (Figure 2). Sequences of this complex originate from Ericaceae in western Canada and eastern USA (Monreal et al., 1999; this study) and Europe (Chambers et al., 2000; Sharples, 2000), from Epacridaceae in Australia (McLean et al., 1999), and from the leafy

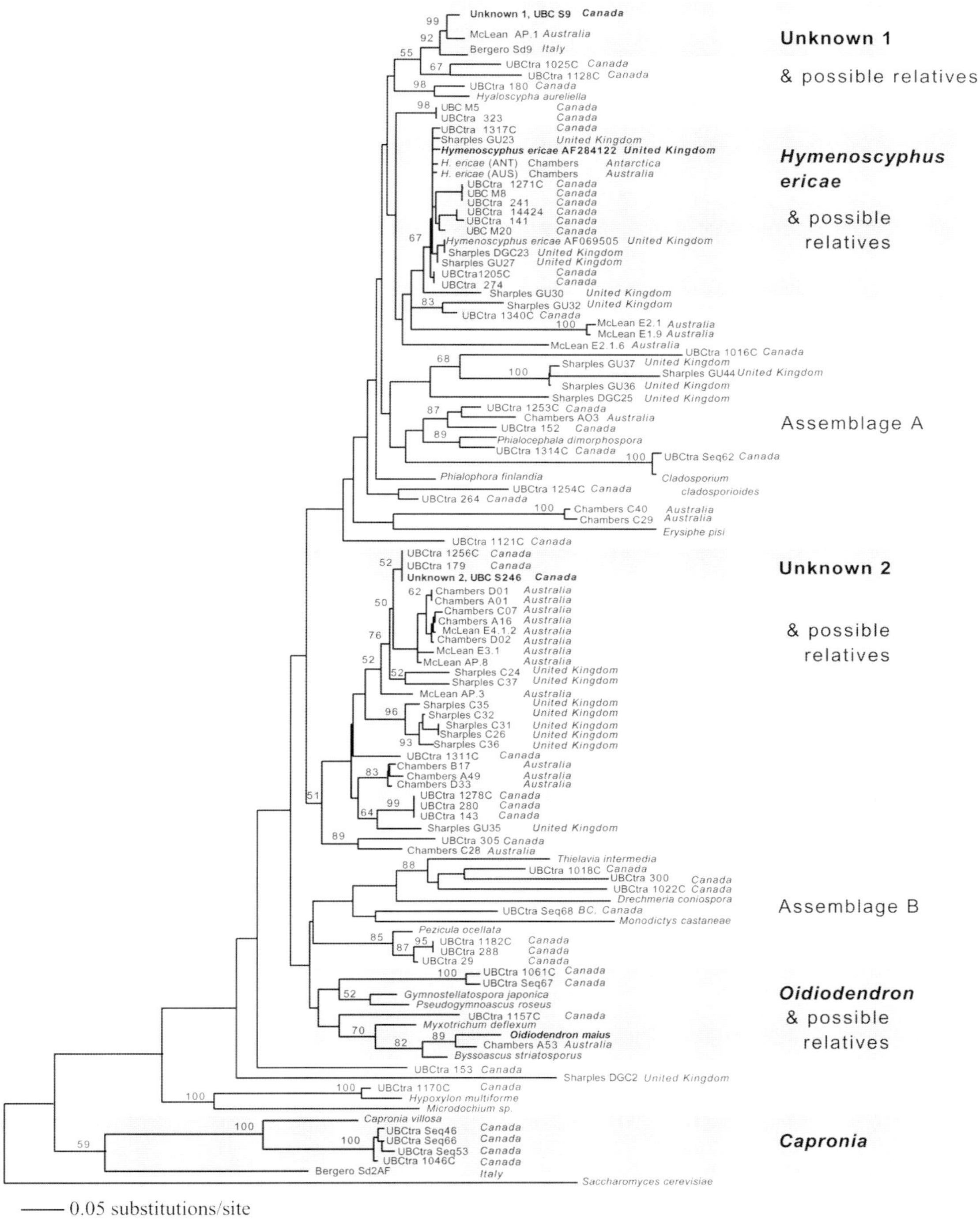

Figure 2. Tree comparing the relationships among ITS2 rDNA sequences of fungi in GenBank either from roots of Ericaceae and Epacridaceae or from other substrates but related to the ericoid mycorrhizal fungi. Tree illustrates main taxa and geographic origin of the sequences analysed. All isolates designated as originating from Canada are from a single research site on Vancouver Island, British Columbia. Distances for this neighbour-joining tree were calculated using a Jukes-Cantor correction and were based on an alignment of about 350 nucleotides of the 5.8S and ITS2 rDNA regions. Branch lengths are proportional to base pair differences. Percentages from 500 bootstrap replicates are reported (values <50% not shown).

liverwort, *Cephaloziella exiliflora,* in Australia and Antarctica (Chambers et al., 2000) (Figure 2).

Close relatives of the taxon originally recognized by Xiao (1994) as Unknown 1 from *Gaultheria shallon* in coastal British Columbia, Canada, have now also been reported from Epacridaceae in Australia (Chambers et al., 2000; McLean et al., 1999) and from ectomycorrhizae of *Quercus ilex* L. in Italy (Bergero et al., 2000) (Figure 2). These fungal isolates cluster in the Leotiales but their identity is still unknown and will remain so until isolates sporulate or more taxa are sequenced. Chambers et al. (2000) recognized that their isolate A03 grouped with *H. ericae* but sequence identity of A03 to *H. ericae* was low (<81%). Our analysis confirms this isolate, A03, clusters in Leotiales (as part of Assemblage A), but is not closely related to *H. ericae* (Figure 2).

Assemblage A consists of isolates known to form ericoid mycorrhizae with salal (e.g. UBCtra152) and sequences of DNA cloned directly from salal roots (e.g. UBCtra Seq62), a sequence from the roots of *Woollsia pungens* (Cav.) F. Muell. in Australia (Chamber et al., 2000), and sequences from *Calluna vulgaris* in Devon, UK (Sharples et al., 2000) (Figure 2). Assemblage A may be a paraphyletic or polyphyletic grouping of fungi, many of them root-inhabiting, that includes *Phialocephala dimorphospora* Kendrick, a potential root pathogen of black spruce (Yamasaki et al., 1998), and *Phialophora finlandia* which forms ectomycorrhizae on *Pinus strobus* L. (Ursic and Peterson, 1997).

Except for Canadian sequences from salal ericoid mycorrhizae, there are no ITS2 sequences identical to Unknown 2 in GenBank. However related fungal isolates have often been found in epacrid roots in Australia (Chambers et al., 2000, McLean et al., 1999), salal roots in BC, and *Calluna vulgaris* in the UK (Sharples et al., 2000) (Figure 2).

Outside of the Leotiales, three main groups have members known to form ericoid mycorrhizae. *Oidiodendron maius* and related taxa have now been reported to form mycorrhizae with *Gaultheria shallon* in British Columbia, Canada (Monreal, 1997; Xiao, 1994; this study), with *Vaccinium corymbosum* in Quebec, Canada (Dalpé, 1989; Hambleton et al., 1998), and with *Calluna vulgaris* in Devon, UK (Sharples et al., 2000). One isolate in the *Oidiodendron* group, A53, from Woollsia *pungens* in Australia did not form mycorrhizae after back inoculation on *Vaccinium macrocarpon* Ait. (Chambers et al., 2000).

A group of fungi related to *Capronia villosa*, Chaetothyriales, is now known to form ericoid mycorrhizae with *Gaultheria shallon* in British Columbia (this study) and with *Erica arborea* L. in Italy (Bergero et al., 2000). Fungi with very similar sequences have also been isolated from ectomycorrhizae of *Quercus ilex* L. (Figure 2) (Bergero et al., 2000).

Finally, Assemblage B may be a chance assemblage of unrelated taxa. It contains sequences from isolates shown to form ericoid mycorrhizae with *Gaultheria shallon* in British Columbia and sequences cloned directly from *G. shallon* roots, some of which may not come from mycorrhizal fungi. GenBank contains no other ericoid mycorrhizal fungus sequences belonging to this cluster of fungi. Another cluster in this assemblage groups with *Thielavia intermedia* Stchigel & Guarro in the Sordariales and *Drechmeria coniospora* (Drechsler) Gams & Jansson in the Hypocreales. A second cluster groups weakly with *Monodictys castaneae* (Wallr.) S. Hughes in the Pleosporales. The third groups with *Pezicula ocellata* (Pers.:Fr.) Seaver in the Leotiales.

Fungi detected by direct amplification and cloning of fungal DNA from root

Comparison of LSU sequences of representatives from each RFLP group found one group of basidiomycetes and several groups of ascomycetes (Figure 3; Appendix 2). In total, of 281 screened clones from 16 root segments, 193 had the DNA insert representing the basidiomycete and 88 had inserts from ascomycetes. The ascomycetes found in direct PCR amplification were also cultured from adjacent root segments, but the basidiomycetes were never obtained in cultures from the roots.

Discussion

Isolation of fungi from mycorrhizae of Ericaceae and Epacridaceae frequently produces sterile cultures that can be identified by ITS2 sequencing to belong to a relatively small number of groups. Of the 5 putative species and 2 assemblages with members forming ericoid mycorrhizae thus far reported in the literature and in GenBank, four are in the Leotiales (*Hymenoscyphus ericae*, Unknown 1, 2, and Assemblage A), one is in the Onygenales (*Oidiodendron*) and one is in the Chaetothyriales (*Capronia*). The seventh (Assemblage B) is polyphyletic with representatives from

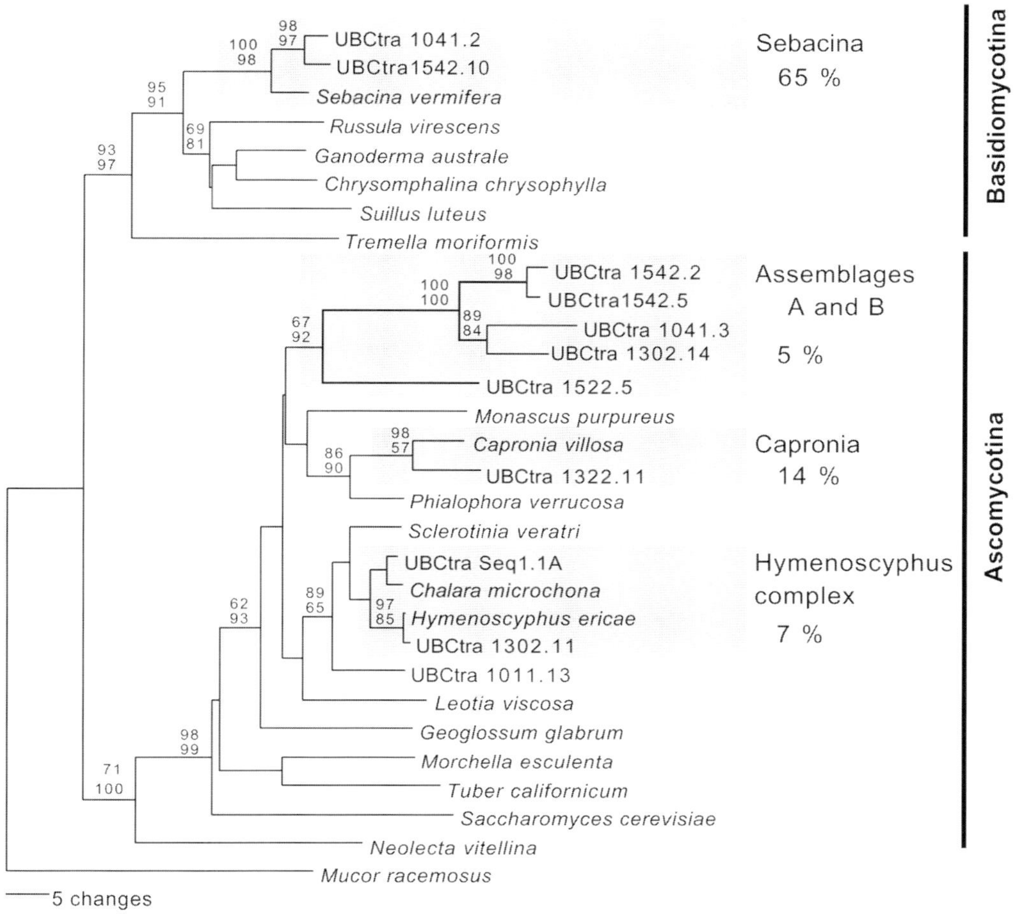

Figure 3. Tree comparing relationships among LSU rDNA sequences of fungi from salal roots at the SCHIRP installation on northern Vancouver Island illustrating the relative proportion of clones from each of the four main fungi detected from among a population of PCR amplified DNA fragments. Distances for this neighbour-joining tree were calculated using a Jukes-Cantor correction and were based on an alignment of LSU rDNA sequences that were about 500 nucleotides in length. Branch lengths are proportional to base pair differences. Percentages from 500 bootstrap replicates are shown with neighbour joining on top and fast parsimony without branch swapping on the bottom (values <50% not shown).

different orders. All but Assemblage B are widespread taxa, found now in three or four continents and forming mycorrhizae with various Ericaceae and Epacridaceae. Some of these fungi have also been reported from atypical habitats such as rhizoids of liverworts and ectomycorrhizae of oak.

However, we have determined that the success rate in isolating ericoid mycorrhizal fungi from *Gaultheria shallon* is quite low, in the range of 1 – 10%. We found this puzzling until we sequenced the LSU region and discovered that 65% of the mycorrhizal root segments of salal that we examined contained a basidiomycete that we could not culture. If this pattern is similar for other ericoid mycorrhizal plants, then it is possible that we are ignoring an important component of these fungus–plant systems when we concentrate on fungi that can be isolated and grown in culture. The strength of direct amplification and cloning of fungal DNA from root is that non-culturable root inhabitants can be detected; the weakness of course is that we cannot currently determine whether these fungi are really mycorrhiza formers.

Using phylogenetic reconstruction, we have determined that the basidiomycete found repeatedly in the fine roots of salal is related to *Sebacina vermifera* (Auriculariales), which has been reported to be an orchid and ectomycorrhizal fungus (Warcup, 1988; Weiss and Oberwinkler, 2001). Jelly fungi were previously detected in ericoid mycorrhizae of *Calluna vulgaris* based on ultrastructure of the hyphal septal pore (Bonfante-Fasolo, 1980). The exact identity of our fungus and its role in *Gaultheria shallon* roots remain to be determined.

Acknowledgements

Guoping Xiao, Marcia Monreal, Tony Millar, Katherine Greene, and Firoozeh Nafar all made significant contributions to this research. Forest Renewal BC and the National Sciences and Engineering Research Council of Canada (NSERC) provided funding for this research. Funding assistance by Forest Renewal BC does not imply endorsement of any statements or information contained herein.

References

Bergero R, Perotto S, Girlanda M, Vidano G and Luppi A M 2000 Ericoid mycorrhizal fungi are common root associates of Mediterranean ectomycorrhizal plants (*Quercus ilex*). Molecular Ecol. 9, 1639–1649.

Bonfante-Fasolo P 1980 Occurrence of a basidiomycete in living cells of mycorrhizal hair roots of *Calluna vulgaris*. Trans. British Mycol. Soc. 75, 320–325.

Chambers S M, Liu G and Cairney J W G 2000 ITS rDNA sequence comparison of ericoid mycorrhizal endophytes from *Woollsia pungens*. Mycol. Res. 104, 168–174.

Chambers S M, Williams P G, Seppelt R D and Cairney J W G 1999 Molecular identification of *Hymenoscyphus* sp. from rhizoids of the leafy liverwort *Cephaloziella exiliflora* in Australia and Antarctica. Mycol. Res. 103, 286–288.

Currah R, Niemi M and Huhtinen S 1999 *Oidiodendron maius* and *Scytalidium vaccinii* from the mycorrhizas of Ericaceae in northern Finland. Karstenia 39, 65–68.

Daple Y 1989 Ericoid mycorrhizal fungi in the Myxotrichaceae and Gymnoascaceae. New Phytol. 113, 523–527.

Egger K N and Sigler L 1993 Relatedness of the ericoid endophytes *Syctalidium vaccinii* and *Hymenoscyphus ericae* inferred from analysis of ribosomal DNA. Mycologia 85, 219–230.

Gernandt D S and Stone J K 1999 Phylogenetic analysis of nuclear ribosomal DNA places the nematode parasite, *Drechmeria coniospora*, in Clavicipitaceae. Mycologia 91, 993–1000.

Hambleton S, Egger K N and Currah R S 1998 The genus *Oidiodendron*: Species delimitation and phylogenetic relationships based on nuclear ribosomal DNA analysis. Mycologia 90, 855–870.

Hambleton S, Huhtinen S and Currah R S *Hymenoscyphus ericae*: A new record from western Canada. Mycol. Res. 103, 1391–1397.

Lee S B and Taylor J W 1990 Isolation of DNA from fungal mycelia and simple spores. *In* PCR Protocols: A Guide to Methods and Applications. Eds. M A Innis, D H Gelfand, J J Sninsky and T J White. pp 282–287. Academic Press, San Diego.

McLean C B, Cunnington J H and Lawrie A C 1999 Molecular diversity within and between ericoid endophytes from the Ericaceae and Epacridaceae. New Phytol. 144, 351–358.

Monreal M, Berch S M and Berbee M 1999 Molecular diversity of ericoid mycorrhizal fungi. Can. J. Bot. 77, 1580–1594.

Monreal M A 1997 Molecular identification of ericoid mycorrhizal fungi. Ph.D. Thesis. University of British Columbia, Department of Soil Science,Vancouver, BC.

Pearson V and Read D J 1973 The biology of mycorrhiza in the Ericaceae. I. The isolation of the endophyte and synthesis of mycorrhizas in aseptic culture. New Phytol. 72, 371–379.

Perotto S, Actis-Perino E, Perugini J and Bonfante P 1996 Molecular diversity of fungi from ericoid mycorrhizal roots. Molecular Ecol. 5, 123–131.

Prescott C E and Weetman G F 1994 Salal Cedar Hemlock Integrated Research Program: A Synthesis. Faculty of Forestry, University of British Columbia,Vancouver. 85 pp.

Prescott C E 1996 Salal Cedar Hemlock Integrated Research Program. Research Update #1: December 1996. Faculty of Forestry, University of British Columbia, Vancouver, B.C.

Read D J 1983 The biology of mycorrhiza in the Ericales. Can. J. Bot. 61, 985–1004.

Read D J 1996 The structure and function of the ericoid mycorrhizal root. Ann. Bot. 77, 365–374.

Sharples J M, Chambers S M, Meharg A A and Cairney J W G 2000 Genetic diversity of root-associated fungal endophytes from *Calluna vulgaris* at contrasting field sites. New Phytol. 148, 153–162.

Spatafora J W and Blackwell M 1993 Molecular systematics of unitunicate perithecial ascomycetes: The Clavicipitales-Hypocreales connection. Mycologia 85, 912–922.

Straker C J 1996 Ericoid mycorrhiza: Ecological and host specificity. Mycorrhiza 6, 215–225.

Strandberg M and Johansson M 1999 Uptake of nutrients in *Calluna vulgaris* seed plants grown with and without mycorrhiza. For. Ecol. Manage. 114, 129–135.

Ursic M and Peterson R L 1997 Morphological and anatomical characterization of ectomycorrhizas and ectendomycorrhizas on *Pinus strobus* seedlings in a southern Ontario nursery. Can. J. Bot. 75, 2057–2072.

Warcup J H 1988 Mycorrhizal associations of isolates of *Sebacina vermifera*. New Phytol. 110, 227–232.

Weetman G F, Fournier R, Barker J, Schnorbus-Panozzo E and Germain A 1989 Foliar analysis and response of fertilized chlorotic Sitka spruce plantations on salal-dominated cedar-hemlock cutovers on Vancouver Island. Can. J. For. Res. 19, 1501–1511.

Weiss M and Oberwinkler F 2001 Phylogenetic relationships in *Auriculariales* and related groups – hypotheses derived from nuclear ribosomal DNA sequences. Mycol. Res. 105, 403–415.

White T J, Bruns T, Lee S and Taylor J 1990 Amplification and direct sequencing of fungal ribosomal RNA genes for phylogenetics. *In* PCR Protocols: A Guide to Methods and Applications. Eds. M A Innis, D HGelfand, J J Sninsky and T J White. pp 315–322 Academic Press, San Diego.

Xiao G 1994 The role of root associated fungi in the dominance of salal. Ph.D. Thesis. University of British Columbia, Department of Botany, Vancouver BC.

Xiao G and Berch S M 1995 The ability of known ericoid mycorrhizal fungi to form mycorrhizae with *Gaultheria shallon*. Mycologia 87, 467–470.

Xiao G and Berch S M 1996 Diversity and abundance of ericoid mycorrhizal fungi of *Gaultheria shallon* on forest clearcuts. Can. J. Bot. 74, 337–346.

Yamasaki S H, Fyles J W, Egger K N and Titus B D1998 The effect of *Kalmia angustifolia* on the growth, nutrition, and ectomycorrhizal symbiont community of black spruce. For. Ecol. Manage. 105, 197–207.

Appendix 1. GenBank accession numbers for sequences of isolates included in ITS2 analysis

Accession number	Identity	Plant Source	Location	Mycorrhizal status	Source
	Unknown 1 and possible relatives				
AF081442	UBC S9, (Unknown 1)	*Gaultheria shallon* Pursh	Canada	+	GenBank
AF099089	AP-1	*Astroloma pinifolium* (R.Br.) Benth.	Australia	+	GenBank
AF269067	Sd9	*Quercus ilex* L.	Italy	+	GenBank
AF300741	UBCtra 1025C	*Gaultheria shallon*	Canada	+	Culture from root
AF300746	UBCtra 1128C	*Gaultheria shallon*	Canada	+	Culture from root
AF149071	UBCtra 180	*Gaultheria shallon*	Canada	Ambiguous	Culture from root
U57495	*Hyaloscypha aureliella* (Nyl.) Huhtinen			N/A	GenBank
	Hymenoscyphus ericae and possible relatives				
AF081440	UBC M5	*Gaultheria shallon*	Canada	+	GenBank
AF149083	UBCtra 323	*Gaultheria shallon*	Canada	Ambiguous	Culture from root
AF300748	UBCtra 1317C	*Gaultheria shallon*	Canada	+	Culture from root
AF252834	GU23	*Calluna vulgaris* (L.) Hull	UK	–	GenBank
AF284122	*H. ericae* (Read 100 5183)	*Calluna vulgaris*	UK	+	GenBank
AF069440	*H. ericae* (AUS)	*Cephaloziella exiliflora* (Tayl.) Steph.	Australia	Not tested	GenBank
AF069439	*H. ericae* (ANT)	*Cephaloziella exiliflora*	Antarctica	Not tested	GenBank
AF300750	UBCtra 1271C	*Gaultheria shallon*	Canada	+	Culture from root
AF081435	UBC M8	*Gaultheria shallon*	Canada	+	GenBank
AF149068	UBCtra 241	*Gaultheria shallon*	Canada	+	Culture from root
AF300751	UBCtra 14424	*Gaultheria shallon*	Canada	N/A	PCR product from root
AF149067	UBCtra 141	*Gaultheria shallon*	Canada	+	Culture from root
AF081438	UBC M20	*Gaultheria shallon*	Canada	+	Culture from root
AF069505	*H .ericae* (Read 101 5183)	*Calluna vulgaris*	UK	+	GenBank
AF252833	DGC23	*Calluna vulgaris*	UK	+	GenBank
AF252835	GU27	*Calluna vulgaris*	UK	+	GenBank
AF300749	UBCtra 1205C	*Gaultheria shallon*	Canada	+	GenBank
AF149069	UBCtra 274	*Gaultheria shallon*	Canada	+	Culture from root
AF252836	GU30	*Calluna vulgaris*	UK	+	GenBank
AF252837	GU32	*Calluna vulgaris*	UK	+	GenBank
AF300752	UBCtra 1340C	*Gaultheria shallon*	Canada	–	Culture from root
AF148951	E2-1	*Epacris impressa* Labill.	Australia	–	GenBank
AF099978	E1-9	*Epacris impressa*	Australia	+	GenBank
AF098291	E2-1-6	*Epacris impressa*	Australia	+	GenBank
AF081439	**Scytalidium vaccinii* Dalpé, Sigler & Litten, UAMH5828	*Vaccinium angustifolium* Ait.	US	+	GenBank
AF081437	**H. ericae* Read 101	*Calluna vulgaris*	UK	+	GenBank
AF081436	**H. ericae* UAMH 6735	*Calluna vulgaris*	UK	+	GenBank
AF151089	**H. ericae* Read 100 5183	*Calluna vulgaris*	UK	+	GenBank
AF284124	*UBCtra 1302.11	*Gaultheria shallon*	Canada	N/A	PCR product from root
AF300726	*UBCtra 1011.13	*Gaultheria shallon*	Canada	N/A	PCR product from root
AF284123	*UBCtra 1302.5	*Gaultheria shallon*	Canada	N/A	PCR product from root
AF284127	*UBCtra 1.01	*Gaultheria shallon*	Canada	N/A	PCR product from root
AF300724	*UBCtra Seq1.1A	*Gaultheria shallon*	Canada	N/A	PCR product from root
AF300718	*UBCtra 1086.12	*Gaultheria shallon*	Canada	N/A	PCR product from root
AF300755	*UBCtra Seq68	*Gaultheria shallon*	Canada	N/A	PCR product from root
	Assemblage A				
AF300740	UBCtra 1016C	*Gaultheria shallon*	Canada	–	Culture from root
AF252843	GU37	*Calluna vulgaris*	UK	–	GenBank
AF252842	GU44	*Calluna vulgaris*	UK	+	GenBank
AF252840	GU36	*Calluna vulgaris*	UK	–	GenBank
AF252841	DGC25	*Calluna vulgaris*	UK	–	GenBank
AF300753	UBCtra 1253C	*Gaultheria shallon*	Canada	–	Culture from root
AF072291	A03	*Woollsia pungens* (Cav.) F.Muell.	Australia	+	GenBank
AF149072	UBCtra 152	*Gaultheria shallon*	Canada	+	Culture from root
AF081434	*Phialocephala dimorphospora* Kendrick			N/A	GenBank
AF300754	UBCtra 1314C	*Gaultheria shallon*	Canada	Ambiguous	Culture from root

Appendix 1. Continued

Accession number	Identity	Plant Source	Location	Mycorrhizal status	Source
AJ300335	*Cladosporium cladosporioides* (Fres.) de Vries			N/A	GenBank
AF300730	UBCtra Seq62	*Gaultheria shallon*	Canada	N/A	PCR product from root
AF011327	*Phialophora finlandia* Wang & Wilcox			N/A	GenBank
AF300745	UBCtra 1254C	*Gaultheria shallon*	Canada	+	Culture from root
AF149070	UBCtra 264	*Gaultheria shallon*	Canada	Ambiguous	Culture from root
AF072300	C40	*Woollsia pungens*	Australia	+	GenBank
AF072299	C29	*Woollsia pungens*	Australia	+	GenBank
AF073348	*Erysiphe pisi* DC			N/A	GenBank
AF300747	UBCtra 1121C	*Gaultheria shallon*	Canada	Ambiguous	Culture from root
	Unknown 2 and possible relatives				
AF300737	UBCtra 1256C	*Gaultheria shallon*	Canada	+	Culture from root
AF149077	UBCtra 179	*Gaultheria shallon*	Canada	+	Culture from root
AF081443	UBC S246 (Unknown 2)	*Gaultheria shallon*	Canada	+	GenBank
AF072301	D01/C01	*Woollsia pungens*	Australia	+	GenBank
AF072296	*C01 /D01	*Woollsia pungens*	Australia	+	GenBank
AF072290	A01	*Woollsia pungens*	Australia	+	GenBank
AF072297	C07	*Woollsia pungens*	Australia	+	GenBank
AF072292	A16	*Woollsia pungens*	Australia	+	GenBank
AF097312	E4-1-2	*Epacris impressa*	Australia	+	GenBank
AF072302	D02	*Woollsia pungens*	Australia	+	GenBank
AF099979	E3-1	*Epacris impressa*	Australia	+	GenBank
AF099091	AP-8	*Astroloma pinifolium*	Australia	+	GenBank
AF252838	AC24	*Calluna vulgaris*	UK	+	GenBank
AF252850	AC37	*Calluna vulgaris*	UK	–	GenBank
AF099090	AP-3	*Astroloma pinifolium*	Australia	+	GenBank
AF252847	AC35	*Calluna vulgaris*	UK	–	GenBank
AF252846	AC32	*Calluna vulgaris*	UK	+	GenBank
AF252845	AC31	*Calluna vulgaris*	UK	+	GenBank
AF252844	AC26	*Calluna vulgaris*	Devon, UK	+	GenBank
AF252848	AC36	*Calluna vulgaris*	UK	–	GenBank
AF300739	UBCtra 1311C	*Gaultheria shallon*	Canada	Ambiguous	Culture from root
AF072295	B17	*Woollsia pungens*	Australia	+	GenBank
AF072293	A49	*Woollsia pungens*	Australia	+	GenBank
AF072303	D33	*Woollsia pungens*	Australia	+	GenBank
AF300738	UBCtra 1278C	*Gaultheria shallon*	Canada	Ambiguous	Culture from root
AF149073	UBCtra 280	*Gaultheria shallon*	Canada	Ambiguous	Culture from root
AF149075	UBCtra 143	*Gaultheria shallon*	Canada	Ambiguous	Culture from root
AF252839	GU35	*Calluna vulgaris*	UK	+	GenBank
AF149076	UBCtra 305	*Gaultheria shallon*	Canada	Ambiguous	Culture from root
AF072298	C28	*Woollsia pungens*	Australia	+	
	Assemblage B				
AJ271588	*Thielavia intermedia* Stchigel & Guarro			N/A	GenBank
AF300727	UBCtra 1018C	*Gaultheria shallon*	Canada	+	Culture from root
AF149079	UBCtra 300	*Gaultheria shallon*	Canada	–	Culture from root
AF300728	UBCtra 1022C	*Gaultheria shallon*	Canada	Ambiguous	Culture from root
AF106018	*Drechmeria coniospora* (Drechsler) Gams & Jansson			N/A	GenBank
AJ238678	*Monodictys castaneae* (Wallr.) S.J. Hughes			N/A	GenBank
AF141199	*Pezicula ocellata* (Pers.:Fr.) Seaver			N/A	GenBank
AF300744	UBCtra 1182C	*Gaultheria shallon*	Canada	–	Culture from root
AF149074	UBCtra 288	*Gaultheria shallon*	Canada	–	Culture from root
AF149087	UBCtra 29	*Gaultheria shallon*	Canada	–	Culture from root
	Oidiodendron and possible relatives				
AF300742	UBCtra 1061C	*Gaultheria shallon*	Canada	+	Culture from root
AF300743	UBCtra Seq67	*Gaultheria shallon*	Canada	N/A	PCR product from root

Appendix 1. Continued

Accession number	Identity	Plant Source	Location	Mycorrhizal status	Source
AF062818	*Gymnostellatospora japonica* Udagawa, Uchiyama & Kamiya			N/A	GenBank
AF062819	*Pseudogymnoascus roseus* Raillo			N/A	GenBank
AF300736	UBCtra 1157C	*Gaultheria shallon*	Canada	Ambiguous	Culture from root
AF062814	*Myxotrichum deflexum* Berk.			N/A	GenBank
AF081424	*Oidiodendron maius* Barron, S4	*Gaultheria shallon*	Canada	+	Culture from root
AF062799	**Oidiodendron maius*	*Vaccinium corymbosum* L.	Canada	+	Culture from root
AF072294	A53	*Woollsia pungens*	Australia	–	GenBank
AF062817	*Byssoascus striatosporus* (Barron & Booth) von Arx			N/A	GenBank
	Other taxa				
AF149078	UBCtra 153	*Gaultheria shallon*	Canada	Ambiguous	Culture from root
AF252849	DGC26	*Calluna vulgaris*	Devon, UK	Not tested	GenBank
AF300729	UBCtra 1170C	*Gaultheria shallon*	Canada	+	Culture from root
AJ246219	*Hypoxylon multiforme* (Fr.:Fr.) Fr.			Not tested	GenBank
AJ279481	*Microdochium* sp.			Not tested	GenBank
	Capronia				
AF050261	*Capronia villosa* G.F. Samuels			N/A	GenBank
AF300731	UBCtra Seq46	*Gaultheria shallon*	Canada	N/A	PCR product from root
AF300732	UBCtra Seq66	*Gaultheria shallon*	Canada	N/A	PCR product from root
AF300733	UBCtra Seq53	*Gaultheria shallon*	Canada	N/A	PCR product from root
AF300734	UBCtra 1046C	*Gaultheria shallon*	Canada	+	Culture from root
AF284129	*UBCtra 1041.4	*Gaultheria shallon*	Canada	N/A	PCR product from root
AF300725	*UBCtra 1086.11	*Gaultheria shallon*	Canada	N/A	PCR product from root
AF284128	*UBCtra 1322.11	*Gaultheria shallon*	Canada	N/A	PCR product from root
AF269068	Sd2	*Quercus ilex*, *Erica arborea* L.	Italy	+	Culture from root
	Outgroup				
Z73326	*Saccharomyces cerevisiae* Meyen ex Hansen			N/A	GenBank

*Sequences have at least a 95% sequence similarity to previous sequence and were not included in phylogenetic trees.
N/A = not applicable, meaning that cultures were not available.

Appendix 2. GenBank accession numbers for sequences of isolates included in LSU analysis

Accession number	Identity	Source
	Basidiomycetes	
AF284135	UBCtra 1041.2	PCR product from root
AF284136	UBCtra 1542.10	PCR product from root
AF202728	*Sebacina vermifera* Oberw.	GenBank
AF041548	*Russula virescens* (Schff.) Fr.	GenBank
X78780	*Ganoderma australe* (Fr.:Fr.) Pat.	GenBank
U66430	*Chrysomphalina chrysophylla* (Fr.: Fr.) Clé.	GenBank
AF042622	*Suillus luteus* (Fr.) S. F. Gray	GenBank
AF075493	*Tremella moriformis* Smith: Berk.	GenBank
	Ascomycetes	
AF284131	UBCtra 1542.2	PCR product from root
AF284130	UBCtra 1542.5	PCR product from root
AF284132	UBCtra 1041.3	PCR product from root
AF284125	UBCtra 1302.14	PCR product from root
AF284133	UBCtra 1522.5	PCR product from root
AF033394	*Monascus purpureus* Went	GenBank
AF050261	*Capronia villosa* Samuels	GenBank
AF284128	UBCtra1322.11	PCR product from root
AF300719	*UBCtra 1322.2	PCR product from root
AF050283	*Phialophora verrucosa* Medlar	GenBank
AF113739	*Sclerotinia veratri* Cash & R.W. Davidson	GenBank
AF300724	UBCtra Seq1.1A	PCR product from root
AF222467	*Chalara microchona* W. Gams	GenBank
AF284122	*Hymenoscyphus ericae* (Read) Korf & Kernan	Culture UAMH
AF284124	UBCtra 1302.11	PCR product from root
AF300726	UBCtra 1011.13	PCR product from root
AF113737	*Leotia viscosa* Fr.	GenBank
AF113738	*Geoglossum glabrum* Pers.	GenBank
U42669	*Morchella esculenta* (L.) Pers.	GenBank
AF127120	*Tuber californicum* Harkn.	GenBank
Z73326	*Saccharomyces cerevisiae* Meyen ex Hansen	GenBank
U42695	*Neolecta vitellina* (Bres.) Korf & J.K. Rogers	GenBank
AJ271061	*Mucor racemosus* Fresen.	GenBank

*Sequences with at least a 95% sequence similarity to previous sequence were not included in phylogenetic trees.

Plant and Soil **244:** 67–73, 2002.

Molecular identification and phylogeny of arbuscular mycorrhizal fungi

Dirk Redecker
*Institute of Botany, University of Basel, Hebelstrasse 1, CH-4056 Basel, Switzerland**

Received 21 August 2001. Accepted in revised form 12 December 2001

Key words: arbuscular mycorrhiza, evolution, Glomales, molecular identification, phylogeny

Abstract

The fossil record and molecular data show that the evolutionary history of arbuscular mycorrhizal fungi (Glomales) goes back at least to the Ordovician (460 million years ago), coinciding with the colonization of the terrestrial environment by the first land plants. At that time, the land flora only consisted of plants on the bryophytic level. Ribosomal DNA sequences indicate that the diversity within the Glomales on the family and genus level is much higher than previously expected from morphology-based taxonomy. Two deeply divergent lineages were found and described in two new genera, *Archaeospora* and *Paraglomus*, each in its own family. Based on a fast-growing number of available DNA sequences, several systems for molecular identification of the Glomales within roots have been designed and tested in the past few years. These detection methods have opened up entirely new perspectives for studying the ecology of arbuscular mycorrhiza.

A problem for morphology-based taxonomy: One fungus in two different families?

The taxonomy of arbuscular mycorrhizal (AM) fungi has traditionally been based on the morphology of the spores. Families and genera were mainly distinguished by the hyphal attachment and mode of formation of the spore, whereas the substructure of the spore walls played an important role in species identification (Gerdemann and Trappe, 1974; Morton, 1988; Walker, 1992). The resulting three-family structure Glomaceae, Acaulosporaceae, Gigasporaceae (Morton and Benny, 1990) was initially supported by rDNA data (Simon et al., 1993a). However, these data also indicated a large genetic diversity within the genus *Glomus* that was not reflected by the few morphological characters of the spores.

The closest non-mycorrhizal relative to arbuscular mycorrhizal fungi was found to be *Geosiphon pyriforme* (Gehrig et al., 1996). This fungus forms a peculiar symbiosis with cyanobacteria, intracellularly harboring *Nostoc* symbionts. It also produces spores that resemble those of *Glomus* species very closely, even in spore wall ultrastructure (Schüßler et al., 1994). The *Geosiphon* symbiosis, therefore, appeared to mirror a possible ancestor to arbuscular mycorrhiza that could have occurred in terrestrial ecosystems before the origin of land plants.

The discovery of an arbuscular mycorrhizal fungus forming spores that formally belong to two different genera and families was obviously not easily compatible with morphology-based taxonomy. Spores of *Acaulospora gerdemannii* and *Glomus leptotichum* were observed to be formed on the same hypha (Morton et al., 1997). As no evidence could be found that one of these spore types may represent a sexual stage, the two spore types were termed 'synanamorphs'. Molecular biological analysis showed no evidence that they originate from two different organisms (Redecker et al., 2000b). However, their 18S rDNA sequences proved that this dimorphic organism neither belongs to *Glomus*, nor to *Acaulospora*, nor to any of the established families (Sawaki et al., 1998). Instead it belongs to one of several ancestral, deeply divergent lineages within the Glomales (Redecker et al., 2000b).

* FAX No: +41-61-267-2330. E-mail: dirk.redecker@unibas.ch

Deeply divergent lineages of the Glomales

Ac. gerdemannii/G. leptotichum was not the only AM fungal species found to fall outside the previously known phylogenetic groupings. To date, four other species were reported in two major lineages. Some of these species are dimorphic, one forms only spores of the acaulosporoid type, and others produce only *Glomus*-type spores (Redecker et al., 2000b).

The acaulosporoid spore type of a second dimorphic fungus had been previously described as *Glomus gerdemannii* (Rose et al., 1979). Its actual glomoid morph remained undetected until recently. Spores of both its two synanamorphs are very similar in structure to *Acaulospora gerdemannii/Glomus leptotichum*. rDNA sequences confirmed that they are closely related species. Because of their large phylogenetic distance to all previously known glomalean genera, a new genus was established and the two dimorphic fungi were renamed *Archaeospora leptoticha* and *Ar. gerdemannii*, respectively (Morton and Redecker, 2001).

A third species in this lineage forms rather inconspicuous, small, hyaline spores of the acaulosporoid but not the glomoid type. It was previously known as *Acaulospora trappei*. Ribosomal sequence data showed that it is related to *Ar. leptoticha* and *Ar. gerdemannii* (Redecker et al., 2000b). It was established as the type species of the new genus *Archaeospora* as *Ar. trappei* (Morton and Redecker, 2001).

The molecular data allowed to re-evaluate morphological characters of the acaulosporoid spores of *Archaeospora* that had appeared as mere anomalies before, and had not been assigned great significance. Spores of *Ar. leptoticha* and *Ar. gerdemannii* have inner flexible walls, but these are quite different in their fine structure from the germinal walls of *Acaulospora* species. They do not appear to form a 'germination orb' during germination that allows the germination hypha to penetrate the spore wall. Instead, whenever it was observed, germination appears to occur through the hyphal attachment or its remnants, as in some *Glomus* species. The acaulosporoid spores of *Archaeospora* are separated from the subtending hypha by a short 'pedicel' not found in *Acaulospora* (Morton and Redecker, 2001).

The glomoid spores of *Archaeospora* do not have morphological characters that distinguish them from their distant relatives in *Glomus* sensu stricto (e.g. *G. intraradices*). The same is true for *Glomus occultum* and *Glomus brasilianum*, members of another, deeply divergent lineage which has no close affiliation to *Archaeospora* or *Geosiphon*. *G. occultum* and *G. brasilianum* were placed in the new genus *Paraglomus* in the family Paraglomaceae. However, like *Archaeospora*, *Paraglomus* is distinguished from *Glomus* by rDNA sequences, unique fatty acids and the mycorrhizal morphology. The mycorrhizae of these two genera do not stain at all or only weakly with the standard dyes (Trypan Blue, Acid Fuchsin, etc.) that are routinely used to detect and quantify AM colonization. This is one reason why these fungi have not been extensively studied in their symbiotic interactions. In fact, they may be interesting objects for study, because their large phylogenetic distance to other AM fungi might be correlated with physiological differences. Based on 18S rDNA sequences, *Paraglomus* seems to be the most ancestral lineage within the Glomales (Figure 1; Schwarzott et al., 2001), although there is only low bootstrap support for this.

The relationship of *Geosiphon pyriforme* to the basal mycorrhizal lineages and their branching order has been unclear. The exact branching order of *Geosiphon, Paraglomus*, *Archaeospora* and the clade with the previously known three families was not resolved well (Redecker et al., 2000b). An updated sequence of *Geosiphon* in the databases places this fungus closer to *Archaeospora leptoticha/gerdemanni* (Figure 1; Schwarzott et al., 2001). This finding is quite surprising because currently there are no other data suggesting this relationship except for the small subunit sequence. The sequence signature used to define *Archaeospora* (Morton and Redecker, 2001) is rather similar to the updated *Geosiphon* sequence at the respective site.

Based on the available rDNA data, it appears that *Geosiphon* is derived from mycorrhizal lineages and that its symbiosis with cyanobacteria is not an ancestral precursor to AM. In the long term, the definition of the Glomales will have to be modified to accomodate non-mycorrhizal fungi, because it is likely that the transition from mycorrhizal to other nutrition modes occurred more than once, as has already been shown for ectomycorrhizal fungi (Hibbett et al., 2000).

Other new results of molecular phylogenetics

The genus *Sclerocystis* had been the subject of some taxonomic wrangling. It was unclear whether the formation of complex sporocarps merited a separate genus (Almeida and Schenck, 1990; Wu, 1993). Most

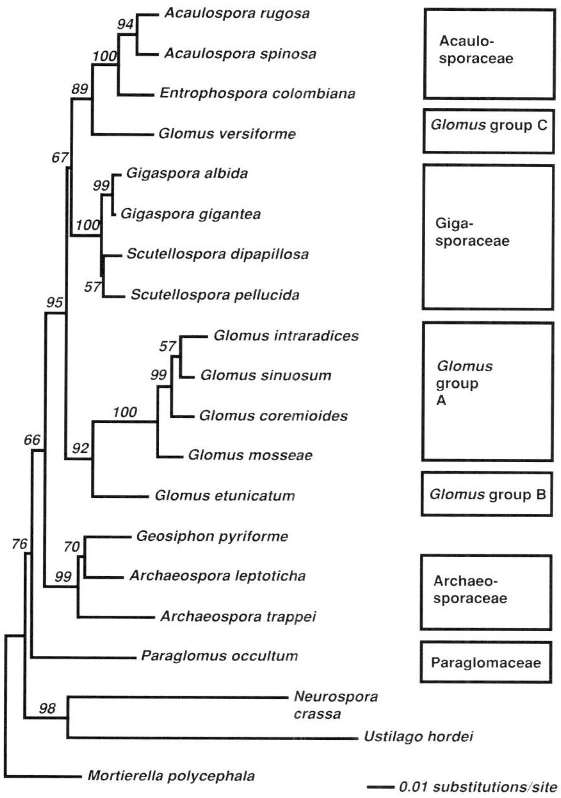

Figure 1. Phylogenetic tree of the Glomales (neighbor-joining analysis of 18S rDNA sequences). Numbers on the nodes denote bootstrap values from 10 000 replicates. Boxes on the right indicate families and the *Glomus* subgroups proposed by Schüßler et al. (2001).

authors agreed that the component spores of the sporocarps showed all characteristics of *Glomus* spores, but as we now know the spore type is formed by several unrelated lineages. There were also several intermediate sporocarp morphologies between *Glomus* and *Sclerocystis*. Using 18S rDNA sequences, Redecker et al. (2000c) resolved the issue and found that the complex sporocarps were formed by members of *Glomus* group A (see Figure 1). From the phylogenetic standpoint, there was no reason anymore to keep *Sclerocystis*. Therefore, the authors placed the last remaining species *S. coremioides* (sensu Almeida & Schenck) into *Glomus*.

Within the Gigasporaceae, *Gigaspora* seems to be the advanced and not the primitive genus. Its few species form a very narrow clade compared to the large variation within *Scutellospora* (Schwarzott et al., 2001). There is even the possibility that *Scutellospora* may be paraphyletic. If this is true, the complex system of germinal walls and the germination shield of *Scutellospora* were lost during evolution of *Gigaspora*.

In previous analyses, *Glomus versiforme* grouped within a weakly supported monophyletic group of *Glomus* species, comprising *G. intraradices, G. mosseae* and *G. etunicatum* (Gehrig et al., 1996; Redecker et al., 2000b; Schüßler, 1999). The recently updated sequence in the databases places it in its own clade with some affinity to the Acaulosporaceae (Figure 1; Schwarzott et al., 2001). Several other *Glomus* species were reported to be in this lineage (Schwarzott et al., 2001).

The fossil history of the Glomales

It has been hypothesized previously that AM fungi played a substantial role in the colonization of terrestrial habitats by early land plants (Pirozynski and Malloch, 1975). The earliest evidence for the AM symbiosis were fossil arbuscules found in specimens of the early land plant *Aglaophyton* from the Devonian Rhynie Chert (400 Myr ago, Remy et al., 1994). Fossil spores and hyphae resembling those of glomalean fungi had been detected in plant material from this site by the 1920s (Kidston and Lang, 1921).

Aglaophyton is thought to be in an intermediate position between bryophytes and vascular plants and was, therefore, already higher evolved than the putative first terrestrial plants (Taylor and Taylor, 1993). Most likely the earliest land plants colonized terrestrial ecosystems in the Ordovician and were similar to today's hornworts and liverworts. Only microfossil evidence has been found for these plants: spores and cellular scraps ('cuticle fragments', Graham, 1993; Gray et al., 1982). Similar microfossils were actually found in connection with hepatic thalli from later epochs and, therefore, confirmed this notion (Edwards et al., 1995). In agreement with fossil findings, molecular and morphological cladistics both support liverworts and hornworts as the oldest branches of the land plant lineage that evolved from green algae (Kenrick and Crane, 1997).

Spores and hyphae strongly resembling today's AM fungi were recently found in 460-million-years-old dolomite rocks from the Ordovician of Wisconsin (Redecker et al., 2000a). This is the earliest evidence for the existence of fungi. The oldest generally accepted fungal fossils found so far had been Ascomycete–like hyphae in arthropod frass from the

Silurian (Sherwood–Pike and Gray, 1985) and the numerous groups of fungi found in the Rhynie Chert (400 Myr ago, Kidston and Lang, 1921; Remy et al., 1994; Taylor et al., 1994, 1999). The Ordovician spores are remarkably similar to today's *Glomus* type spores and have a size of 40–95 μm. Their hyphae are non-septate and 3–5 μm wide.

These findings prove that glomalean fungi were present at a time when the land flora only consisted of plants of the bryophyte grade. Together with the majority of land plants, today's hornworts and liverworts do form associations with AM fungi, although they do not meet all the criteria of AM, because they do not have true roots (Read et al., 2000; Schüßler, 2000). Interestingly, extant mosses do not form AM (Read et al., 2000).

Although a direct connection between those early glomalean fungi and the land plants on the bryophyte level was not demonstrated, their co-occurrence and the proven ability of extant hornworts and liverworts to form an AM symbiosis suggest that they may have been associated. These findings give strong support to the notion that AM fungi were instrumental in the success of early land plants.

New perspectives for ecology: Molecular identification from roots

It has long been recognized as a major problem of ecological studies on AM that the fungi cannot be easily identified within roots. Traditionally, spores were used to determine which species was present and active. However, it is well known that spore formation is highly dependent on physiological parameters and often not correlated with root colonization. Species determination from spores also has many pitfalls for non-experts, when spores are immature, parasitized or lack distinctive morphological characters. Moreover, spore morphology does not reflect mycorrhizal diversity when we consider that various unrelated lineages within the Glomales form very similar spores of the *Glomus* type, and that acaulosporoid spores occur in both Acaulosporaceae and Archaeosporaceae

Therefore, several approaches were taken to develop molecular tools that would allow species identification of AM fungi independent from spore formation. A few authors designed PCR primers from random sequences, most used a systematic approach, comparing sequences of known function. A randomly amplified fragment from *Gl. mosseae* was used to generate specific primers for this species (Lanfranco et al., 1995) and to establish a quantitative PCR system to measure root colonization (Edwards et al., 1997). Most authors, however, used the systematic approach with ribosomal sequences. Provided a sufficient database of sequences and knowledge of phylogeny, this approach allows to tailor the specificity of probes or PCR primers to species, genera or any other level of taxonomy.

Simon et al. (1992) conducted the first pioneering studies of molecular phylogeny and detection of AM fungi. The authors designed the primer VANS1, based on only three 18S subunit sequences from the Glomales, and a set of other primers specific for subgroups of the Glomales. They also proposed an identification system using Single Strand Conformation Polymorphism (SSCP) (Simon et al., 1993b). As more and more sequences became available from a broader taxon sample from the Glomales, it became clear that the VANS1 primer did not amplify from all glomalean fungi (Clapp et al., 1999; Schüßler et al., 2001). Nevertheless, it was used in some studies (Chelius and Triplett, 1999; Clapp et al., 1995; Di Bonito et al., 1995)

Using the group-specific PCR primers for a portion of the 18S rDNA designed by Simon et al. (1993b) and a subtraction hybridization technique to enrich fungal sequences, Clapp et al. (1995) performed the first molecular study of a field population of AM fungi. Their PCR results for *Acaulospora* and *Scutellospora* species were mainly in agreement with spore counts but interestingly there was a strong discrepancy between strong root colonization by *Glomus* and absence of sporulation. An influence of canopy type on this cryptic colonization was also detected. Moreover, concurrent colonization of the same 5 cm root length by all three genera was demonstrated. These results underlined the need to apply molecular methods in order to obtain more reliable data of AM fungal populations in roots.

Other authors targeted different parts of the ribosomal genes, e.g. the large subunit (Kjøller and Rosendahl, 2000; van Tuinen et al., 1998) or the Internal Transcribed Spacers (Redecker, 2000). van Tuinen et al. (1998) employed a nested PCR procedure to circumvent the problem of PCR inhibitors present in many root samples. These authors designed primers specific for four inoculant species to study AM fungal populations in a greenhouse experiment. Root fragments of 1 cm length were analyzed in this study, and again, the pieces were usually colonized by more than one fungus. In this study, a complex pattern of fungal species interactions was detected. The colonization by the two species from the Gigasporacae was significantly enhanced in combination with certain others, suggesting synergistic interactions among species. A similar approach was used to study AM fungal populations in heavy-metal-polluted soils (Turnau et al., 2001).

A set of five primers targeted at major phylogenetic groups within the Glomales was designed to amplify parts of the 18S ribosomal subunit, the ITS and the 5.8S subunit (Redecker, 2000). These groups also include *Archaeospora* and *Paraglomus*. In addition to the difficulties to detect them by staining, these AM fungi had not been detectable with previously designed molecular tools.

Kjøller and Rosendahl (2000) developed a specific primer for a part of the ribosomal large subunit from members of *Glomus* group A (see Figure 1). This primer was used in a nested PCR reaction. Single Strand Conformation Polymorphism (SSCP) was used by these authors to screen PCR products for sequence differences, which is less labor-intensive than sequencing. Isolates of *G. mosseae* and *G. caledonium* could be differentiated from each other and *G. geosporum* by this approach, and detected side by side in colonized roots. In samples collected from a *Pisum sativum* field study, this method again showed the dicrepancy between spore and root populations of AM fungi. Some sequence types from roots were not found in trap cultures from the respective field site (Kjøller and Rosendahl, 2001). On the other hand, *G. mosseae* was not detected within roots although it was established in trap cultures. One possible explanation for the latter may be a strong seasonal influence on AM colonization of this species, that may remain undetected by spore counting.

Interestingly, Helgason et al. (1998) showed that a putative *G. mosseae* was the most abundant type of AM fungus found in roots in some arable fields in the UK. In fact, these authors reported that the AM species diversity in cultivated soils was strikingly low when compared to a seminatural woodland. On the other hand, *G. mosseae* was not found in the woodland, suggesting that it is better adapted to the conditions in arable land. The authors attributed the differences in AM species composition to recurring soil disturbance by ploughing in the cultivated field sites. The AM population in roots of bluebell (*Hyacynthoides non-scripta*) in the woodland site was characterized by complex seasonal patterns and a significant influence of the canopy type (Helgason et al., 1999). Helgason et al. (1998, 1999) used a variable portion of the 18S small subunit that was amplified in a single step from roots with the PCR primer AM1. This primer discriminates most AM fungal from plant DNAs but also amplifies a wide range of other fungi. The initial screening of the PCR products was performed by RFLPs, before they were sequenced. The proportion of the clones sequenced from the PCR products was assumed to be proportional to root colonization and used to quantify the respective species.

All these studies demonstrate the power of molecular identification methods to elucidate the species composition of active AM fungal communities. So far it has been difficult to obtain alternative sequence markers to the ribosomal subunits. Due to the numerous micro-organisms living on AM fungal spores, the glomalean identity of novel markers is difficult to establish. Even for ribosomal sequences, the steadily growing number of mislabeled sequences in the databases can cause confusion.

Another problem that has been reported with all sequence markers from the rDNA of the Glomales so far, is the variation all these sequences show within spores and morphospecies (Sanders et al., 1995). There is no easy correlation between sequence identity and species identity, as each spore harbours different sequence types. There is no phylogenetic species concept for the Glomales and several aspects of their genetics continue to be a mystery. Until we have obtained more knowledge in these fields, taxa have to be bundled in groups as shown in the studies above, under the assumption that all taxa in the respective group share at least some ecological or physiological characteristics.

Acknowledgments

I would like to thank Andres Wiemken and Thomas Boller for support and Arthur Schüßler and Rasmus Kjøller for access to unpublished data and interesting discussions.

References

Almeida R T and Schenck N C 1990 A revision of the genus *Sclerocystis* (Glomaceae, Glomales). Mycologia 82, 703–714.

Chelius M K and Triplett E W 1999 Rapid detection of arbuscular mycorrhizae in roots and soil of an intensively managed turfgrass system by PCR amplification of small subunit rDNA. Mycorrhiza 9, 61–64.

Clapp J P, Fitter A H and Young J P W 1999 Ribosomal small subunit sequence variation within spores of an arbuscular mycorrhizal fungus, *Scutellospora* sp. Mol. Ecol. 8, 915–921.

Clapp J P, Young J P W, Merryweather J W and Fitter A H 1995 Diversity of fungal symbionts in arbuscular mycorrhizas from a natural community. New Phytol. 130, 259–265.

Di Bonito R, Elliott M L and Des Jardin E A 1995 Detection of an arbuscular mycorrhizal fungus in roots of different plant species with the PCR. Appl. Environ. Microbiol. 61, 2809–2810.

Edwards D, Duckett J G and Richardson J B 1995 Hepatic characters in the earliest land plants. Nature 374, 635–636.

Edwards S G, Fitter A H and Young J P W 1997 Quantification of an arbuscular mycorrhizal fungus, *Glomus mosseae*, within plant roots by competitive polymerase chain reaction. Mycol. Res. 101, 1440–1444.

Gehrig H, Schüßler A and Kluge M 1996 *Geosiphon pyriforme*, a fungus forming endocytobiosis with *Nostoc* (Cyanobacteria) is an ancestral member of the Glomales: Evidence by SSU rRNA analysis. J. Mol. Evol. 43, 71–81.

Gerdemann J W and Trappe J M 1974 Endogonaceae in the Pacific Northwest. Mycologia Memoirs 5, 1–76.

Graham L E 1993 Origin of land plants. John Wiley & Sons, New York.

Gray J, Massa D and Boucot A J 1982 Caradocian land plant microfossils from Libya. Geology 10, 197–201.

Helgason T, Daniell T J, Husband R, Fitter A H and Young J P W 1998 Ploughing up the wood-wide web? Nature 394, 431.

Helgason T, Fitter A H and Young J P W 1999 Molecular diversity of arbuscular mycorrhizal fungi colonising *Hyacinthoides non-scripta* (bluebell) in a seminatural woodland. Mol. Ecol. 8, 659–666.

Hibbett D S, Gilbert L B and Donoghue M J 2000 Evolutionary instability of ectomycorrhizal symbioses in basidiomycetes. Nature 407, 506–508.

Kenrick P and Crane P R 1997 The origin and early evolution of plants on land. Nature 389, 33–39.

Kidston R and Lang W H 1921 On Old Red Sandstone plants showing structure, from the Rhynie Chert Bed, Aberdeenshire. Part V. The thallophyta occurring in the peat bed; the succession of the plants throughout a vertical section of the bed, and the conditions of accumulation and preservation of the deposit. Trans. Roy. Soc. Edinburgh 51, 855–902.

Kjøller R and Rosendahl S 2000 Detection of arbuscular mycorrhizal fungi (Glomales) in roots by nested PCR and SSCP (single stranded conformation polymorphism). Plant Soil 226, 189–196.

Kjøller R and Rosendahl S 2001 Molecular diversity of Glomalean (arbuscular mycorrhizal) fungi determined as distinct *Glomus* specific DNA sequences from roots of field grown peas. Mycol. Res. 105, 1027–1032.

Lanfranco L, Wyss P, Marzachi C and Bonfante P 1995 Generation of RAPD-PCR primers for the identification of isolates of *Glomus mosseae*, an arbuscular mycorrhizal fungus. Mol. Ecol. 4, 61–68.

Morton J B 1988 Taxonomy of VA mycorrhizal fungi: Classification, nomenclature, and identification. Mycotaxon 32, 267–324.

Morton J B and Benny G L 1990 Revised classification of arbuscular mycorrhizal fungi (Zygomycetes): A new order Glomales and Gigasporineae and two new families Acaulosporaceae and Gigasporaceae with an emendation of Glomaceae. Mycotaxon 37, 471–491.

Morton J B, Bever J D and Pfleger F L 1997 Taxonomy of *Acaulospora gerdemannii* and *Glomus leptotichum*, synanamorphs of an arbuscular mycorrhizal fungus in Glomales. Mycol. Res. 101, 625–631.

Morton J B and Redecker D 2001 Two new families of Glomales, Archaeosporaceae and Paraglomaceae, with two new genera *Archaeospora* and *Paraglomus*, based on concordant molecular and morphological characters. Mycologia 93, 181–195.

Pirozynski K A and Malloch D W 1975 The origin of land plants: a matter of mycotropism. BioSystems 6, 153–164.

Read D J, Duckett J G, Francis R, Ligrone R and Russell A 2000 Symbiotic fungal associations in 'lower' land plants. Phil. Trans. Roy. Soc. London B Biol. Sci. 355, 815–830.

Redecker D 2000 Specific PCR primers to identify arbuscular mycorrhizal fungi within colonized roots. Mycorrhiza 10, 73–80.

Redecker D, Kodner R and Graham L E 2000a Glomalean fungi from the Ordovician. Science 289, 1920–1921.

Redecker D, Morton J B and Bruns T D 2000b Ancestral lineages of arbuscular mycorrhizal fungi (Glomales). Mol. Phylogen. Evol. 14, 276–284.

Redecker D, Morton J B and Bruns T D 2000c Molecular phylogeny of the arbuscular mycorrhizal fungi *Glomus sinuosum* and *Sclerocystis coremioides*. Mycologia 92, 282–285.

Remy W, Taylor T N, Hass H and Kerp H 1994 Four hundred-million-year-old vesicular arbuscular mycorrhizae. Proc. Nat. Acad. Sci. USA 91, 11841–11843.

Rose S, Daniels B A and Trappe J M 1979 *Glomus gerdemannii* sp. nov. Mycotaxon 8, 297–301.

Sanders I R, Alt M, Groppe K, Boller T and Wiemken A 1995 Identification of ribosomal DNA polymorphisms among and within spores of the Glomales: Application to studies on the genetic diversity of arbuscular mycorrhizal fungal communities. New Phytol. 130, 419–427.

Sawaki H, Sugawara K and Saito M 1998 Phylogenetic position of an arbuscular mycorrhizal fungus, *Acaulospora gerdemannii*, and its synanamorph, *Glomus leptotichum*, based upon 18s rRNA gene sequence. Mycoscience 39, 477–480.

Schüßler A 1999 Glomales SSU rRNA gene diversity. New Phytol. 144, 205–207.

Schüßler A 2000 *Glomus claroideum* forms an arbuscular mycorrhiza-like symbiosis with the hornwort *Anthoceros punctatus*. Mycorrhiza 10, 15–21.

Schüßler A, Gehrig H, Schwarzott D and Walker C 2001 Analysis of partial Glomales SSU rRNA gene sequences: implications for primer design and phylogeny. Mycol. Res. 105, 5–15.

Schüßler A, Mollenhauer D, Schnepf E and Kluge M 1994 *Geosiphon pyriforme*, an endosymbiotic association of fungus and cyanobacteria – The spore structure resembles that of arbuscular mycorrhizal (AM) fungi. Bot. Acta 107. 36–45.

Schwarzott D, Walker C and Schüßler A 2001 *Glomus*, the largest genus of the arbuscular mycorrhizal fungi (Glomales), is nonmonophyletic. Mol. Phylogen. Evol. 21, 190–197.

Sherwood–Pike M A and Gray J 1985 Silurian fungal remains: probable records of the class Ascomycetes. Lethaia 18, 1–20.

Simon L, Bousquet J, Levesque R C and Lalonde M 1993a Origin and diversification of endomycorrhizal fungi and coincidence with vascular land plants. Nature 363, 67–69.

Simon L, Lalonde M and Bruns T D 1992 Specific amplification of 18S fungal ribosomal genes from vesicular–arbuscular endomycorrhizal fungi colonizing roots. Appl. Environ. Microbiol. 58, 291–295.

Simon L, Levesque R C and Lalonde M 1993b Identification of endomycorrhizal fungi colonizing roots by fluorescent Single-Strand Conformation Polymorphism Polymerase Chain Reaction. Appl. Environ. Microbiol. 59, 4211–4215.

Taylor T N, Hass H and Kerp H 1999 The oldest fossil ascomycetes. Nature 399, 648.

Taylor T N, Remy W and Hass H 1994 Allomyces from the Devonian. Nature 367, 601.

Taylor T N and Taylor E L 1993 The Biology and Evolution of Fossil Plants. Prentice Hall, Englewood Cliffs, USA.

Turnau K, Ryszka P, Gianinazzi PV and van Tuinen D 2001 Identification of arbuscular mycorrhizal fungi in soils and roots of plants colonizing zinc wastes in southern Poland. Mycorrhiza 10, 169–174.

van Tuinen D, Jacquot E, Zhao B, Gollotte A and Gianinazzi–Pearson V 1998 Characterization of root colonization profiles by a microcosm community of arbuscular mycorrhizal fungi using 25 rDNA-targeted nested PCR. Mol. Ecol. 7, 879–887.

Walker C 1992 Systematics and taxonomy of the arbuscular mycorrhizal fungi (Glomales) – a possible way forward. Agronomie 12, 887–897.

Wu C-G 1993 Glomales of Taiwan: III. A comparative study of spore ontogeny in *Sclerocystis* (Glomaceae, Glomales). Mycotaxon 47, 25–39.

Plant and Soil **244:** 75–83, 2002.

Molecular phylogeny, taxonomy, and evolution of *Geosiphon pyriformis* and arbuscular mycorrhizal fungi

Arthur Schüßler
*Technische Universität Darmstadt, Institut für Botanik, FB10, Schnittspahnstrasse 10, D-64287 Darmstadt, Germany**

Key words: arbuscular mycorrhiza, evolution, *Geosiphon pyriformis*, *Glomeromycota*, molecular phylogeny, taxonomy

Abstract

Geosiphon pyriformis (Kütz.) v. Wettstein is the only known example of a fungus living in endocytobiotic association with a cyanobacterium. The close phylogenetic relationship of *Geosiphon* with some arbuscular mycorrhizal fungi (AMF) and the phylogenetic position of *Geosiphon* are shown in detail. Comprehensive small subunit (SSU) rRNA sequence analyses allow the erection of a new, molecular phylogeny-based taxonomic system for the AMF, including *Geosiphon* (Geosiphonaceae). Within the recently described phylum Glomeromycota (with one class, Glomeromycetes), a system including four orders was proposed. The erection of several new families will also be necessary. Evolutionary implications are discussed, referring to different possibilities of the influence of AMF on the colonization of the terrestrial habitat by plants.

The *Geosiphon pyriformis* endosymbiosis

Geosiphon pyriformis (Kütz.) v. Wettstein is presently the only known example of a fungus living in endocytobiotic association with a cyanobacterium. Originally, interest in *Geosiphon pyriformis* was based mainly on the fact that it can be considered as a 'primitive' photoautotrophic endosymbiosis. The photobiont (*Nostoc punctiforme*) can be separated experimentally and cultured without the fungal partner, which, as far as is presently known, is obligately symbiotic. The study of *Geosiphon* led to fundamental insights into evolutionary steps leading to more derived systems, finally giving rise to the organelles within the eukaryotic cells (Schnepf, 1964). In spite of its potential importance for research on endocytobioses and the biology of the eukaryotic cell, few studies have been done on *Geosiphon*, mainly because the organism is rare in nature and its cultivation is difficult. In the past only five habitats of *Geosiphon* have been documented from Germany and Austria, and presently only one is known in the Spessart mountains in Germany.

Because of the close relationship of the obligate symbiotic fungal partner to the arbuscular mycorrhizal fungi (AMF), it has been suggested recently (Schüßler and Kluge, 2001) that the *Geosiphon* endocyanosis could provide a model system for the study of several fundamental mechanisms and evolutionary questions concerning the ecologically and also economically important arbuscular mycorrhiza (AM) (Smith and Read, 1997).

The *Geosiphon* fungus lives together with the cyanobacterium, *Nostoc punctiforme*, on the surface and in the upper layer of humid, nutrient poor soils. When a fungal hypha comes into contact with *Nostoc* cells, the latter can be incorporated by the fungus at the hyphal tip. The irregular structure formed thereafter swells and forms a unicellular 'bladder', appearing on the soil surface. When mature, these bladders (Figure 1) are about 1–2 mm long and can live for more then 6 months. Inside a bladder the cyanobacteria are physiologically active and multiply. These processes, as well as ecological aspects and the structure of the symbiosis, are described in detail in Schüßler and Kluge (2001) and the references cited therein.

The present paper focuses on the taxonomy and molecular phylogeny of *Geosiphon* and the AMF,

* FAX No: +49-6151-166049.
E-mail: schuessler@bio.tu-darmstadt.de

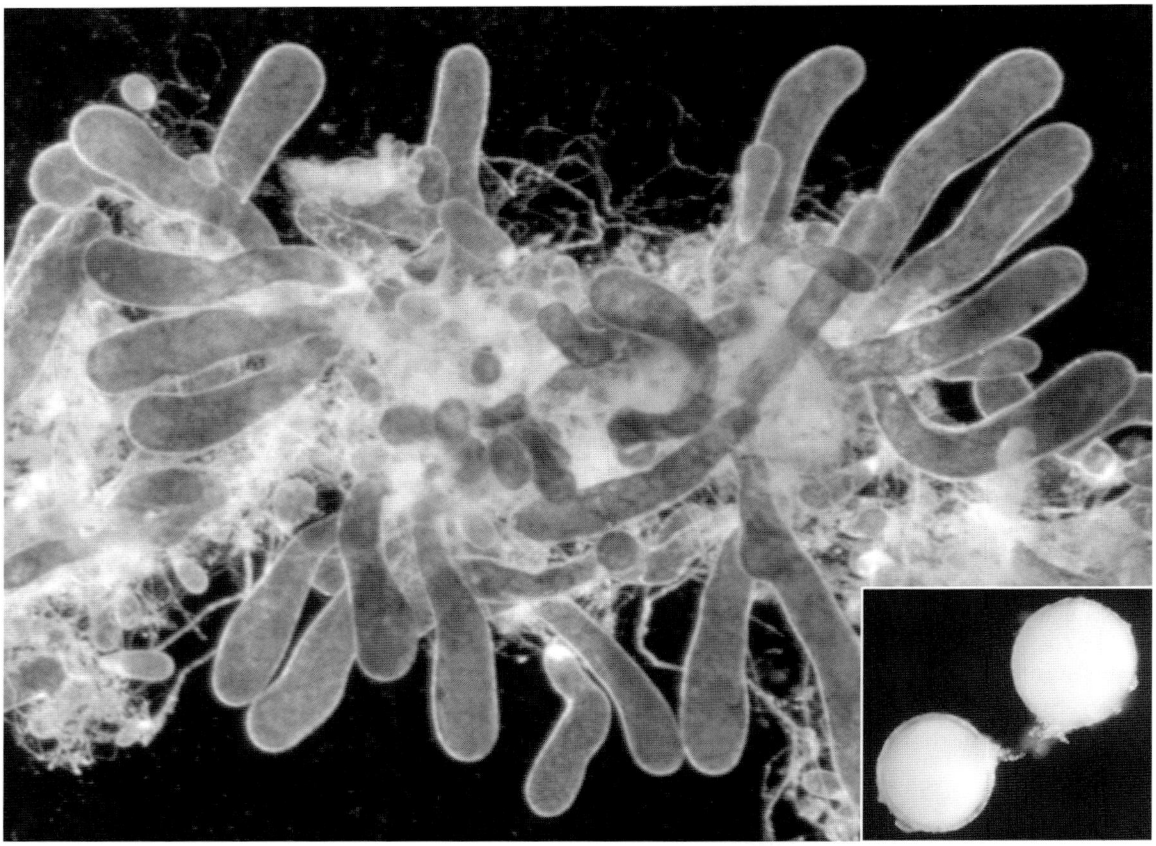

Figure 1. *Geosiphon* bladders and spores, harvested from a laboratory culture and incubated in nutrient solution. Bladders of different size and age are shown. The largest bladders are about 2 mm in length. The insert shows *Geosiphon* spores, which are 250 μm in diameter.

starting with some classical aspects and further discussing new results regarding phylogeny and evolution of *Geosiphon* and the AMF. The new taxonomic system for the AMF and related fungi, based on the proposed new phylum Glomeromycota (Schüßler et al., 2001b), is discussed.

Taxonomy of *Geosiphon pyriformis*

Until the last decade, little was known about the taxonomic position of the fungal partner of *Geosiphon pyriformis*. Nowadays, the application of extensive SSU rRNA sequence analyses reveals the true affiliation of the *Geosiphon* fungus within the group formerly containing only AMF. The status of the AMF within the kingdom Fungi is also clarified by the same molecular methods (see below). However, since some misleading data exist in the literature, published taxonomic studies on *Geosiphon* will be summarised and interpreted here first, before the new data are discussed.

Originally, *Geosiphon* was described as a siphonal alga, *Botrydium pyriforme* (Kützing, 1849). The symbiotic nature of the organism was first recognised by Fritz v. Wettstein (1915), who described it in German as *Geosiphon pyriforme*, a symbiosis between a heterotrophic, siphonal alga and *Nostoc* (Cyanobacteria). This was a valid description, since the description in Latin for the genus is required only from 1 January 1935 (Article 36.1 of the ICBN). However, the orthographically correct name is *Geosiphon pyriformis*, since '*sipho*, *siphonis*' (greek) is male and *pyriformis* adjectival (Articles 23.5 & 32.5 1 of the ICBN). *Geosiphon* thereafter was assumed to be the first known case of a symbiosis between two algae (Drude and Schorler, 1916). Later, the organism was described as typifying its own family, Geosiphonaceae, characterised by chitinous cell walls, within the siphonalean algae (Engler and Gilg, 1924). This description also was introduced before 1935 and therefore was valid.

The fungal ('phycomycete') nature of the macrosymbiont was first recognised by Knapp (1933), who had already documented photographically the incorporation of cyanobacteria by hyphal tips. He described *Geosiphon pyriformis* as a new type of lichens, an 'intracellular phycomycetous lichen'. Therefore the species name referred to the symbiotic consortium. A cultured fungal mycelium was stated to represent the *Geosiphon* fungus. However, this fungus was very probably a contaminant, since the photographs show a hyphal mycelium atypical of *Geosiphon* (and AMF) and it was not possible to re-establish the *Geosiphon* symbiosis by addition of cyanobacteria (Knapp, 1933). This is noted here since in some publications it is wrongly stated that the *Geosiphon* fungus can be cultured without symbiosis partner. There is presently no evidence that this is possible. Due to the description of the symbiosis as a 'phycomycetous lichen', Ciferri and Tomaselli (1957) introduced the generic name '*Geosiphonomyces*' (Geosiphonomycaceae, Geosiphonomycetales, Phycolichenomycetes) in a proposed new microlichen taxonomy as a name specifically for the fungal element. It was subsequently ruled that such changes are illegitimate since the name of a lichen applies to the fungal element (see Article 13.1d of the ICBN) and the name *Geosiphon* was already validly published. Regarding higher taxonomic rankings, Mattick (1954) first proposed the ordinal name 'Geosiphonales', but no Latin was provided and this name was invalid.

The genus name *Geosiphon* is now properly used for the symbiotic consortium as well as for the fungal partner alone. However, since *Geosiphon* does show several unique features not typical for lichens (Hawksworth and Honegger, 1994), the emendation of the genus (see below) refers to the fungal partner only.

Mollenhauer (1992) first speculated that *Geosiphon* might be related to the genus *Glomus* and therefore the AMF. At that times the taxonomy of the AMF was based mainly on the structural characteristics of the spores. A subsequent study comparing morphological and ultrastructural criteria of *Geosiphon* spores with those of AMF (Schüßler et al., 1994), revealed striking similarities between *Geosiphon* and some *Glomus* species. Gehrig et al. (1996) showed by SSU rRNA sequence analyses, that *Geosiphon* is related to the AMF, forming a monophyletic clade with them. Ignoring this knowledge, Cavalier-Smith (1998) described with a brief Latin diagnosis a new class, Geomycetes, containing one new order, Geosiphonales, with the sole family Geosiphonaceae and *Geosiphon* as the sole monospecific genus. Moreover, this 'class' was incorrectly placed within the Ascomycotina. Although this diagnosis is taxonomically completely misleading, it fulfils the minimum requirements for valid publication of a name (ICBN Article 32.2), and therefore should be regarded as valid.

The order Geosiphonales was recently used as a taxon (Benny et al., 2001), but without mention of the class Geomycetes and the Glomomycetes, the latter also very briefly described by Cavalier-Smith (1998) to contain the Glomales (AMF) and Endogonales. Unfortunately many of the taxa within the AMF are not described orthographically correct. Here the correct names, Glomeromycetes, Glomerales, Glomeraceae, etc., are used (Schüßler et al., 2001b). To resolve some of the taxonomic problems an emended diagnosis for *Geosiphon* and the Geosiphonaceae is given below. The class Geomycetes and the order Geosiphonales (Cavalier-Smith, 1998), typified by *Geosiphon*, appear to be valid names, but will not withstand future taxonomic concepts and therefore are not emended here, since the rule of priority is not mandatory for ranks above family (ICBN Article 16, note 2).

Emendation of Geosiphonaceae and *Geosiphon pyriformis*

Geosiphonaceae Engler and Gilg, 1924 (Syllabus der Pflanzenfamilien, Verlag von Gebrüder Borntraeger, Berlin, page 24), emend. Schüßler

Fungi with coenocytic mycelium, when senescent or dead sparsely septate hyphae occur; living mostly hypogeously. Forming asexual spores with a rigid, chitinous structural wall by blastic development of hyphal tip or intercalarly, followed by thickening of structural wall components. Spores globose to subglobose, produced singly or in loose clusters, occurring in soil. Fungi, as far known, biotrophic, forming close symbiotic relationship with photoautotrophic organisms. Presently one species known which produces endocytosymbioses with cyanobacteria. Phylogenetically belonging to an order (Archaeosporales) of the arbuscular mycorrhizal and related fungi (phylum Glomeromycota Walker and Schüßler; class Glomeromycetes Cavalier-Smith). Phylogenetic sequence analyses were described in detail in Schüßler et al. (2001a; Mycol. Res. 105, 5–15), Schüßler et al. (2001b; Mycol. Res. 105, 1413–1421), and Schwarzott et al. (2001; Mol. Phylogenet. Evol. 21, 190–197). Presently containing one species, *Geosiphon pyriformis*.

Genus typicum:
Geosiphon pyriformis F. von Wettstein, 1915 (Österreichische Bot. Z. 65, 145-156) = *Botrydium pyriforme* Kützing 1849 (Species algarum. Lipsiae, Leipzig, p. 486)

Geosiphon pyriformis (Kützing) F. von Wettstein, 1915 (Österreichische Botanische Zeitschrift 65: 145–156), emend. Schüßler

Fungus mostly hypogeous; symbiotic saccules ('bladders') epigaeous. Mycelium coenocytic, or sparsely septate when senescent. As far as known, biotrophic, forming close symbiotic relationship with photoautotrophic organisms. Known to produce endocytosymbioses with cyanobacteria (*Nostoc* sp.). Further bacterial endosymbionts living in the fungal cytoplasm, about 0.5 μm in diameter and ovoid, frequently with a median constriction. Fungus phylogenetically belonging to a taxon at the level of order (Archaeosporales) of the arbuscular mycorrhizal and related fungi (phylum Glomeromycota Walker and Schüßler; class Glomeromycetes Cavalier-Smith). Forming white to slightly brownish chlamydospores, produced singly or in loose clusters, occurring in soil.

Spore wall structure of a rigid, chitinous structural wall. Spores formed by blastic development of hyphal tip or, less frequently, intercalarly, followed by thickening of structural wall components. Wall of the spores consists of a thin, flexible innermost layer, a laminated layer with a thickness of 10–13 μm, and an irregularly shaped evanescent outer layer, extending for some micrometers along the subtending hypha, sometimes lacking on mature spores. Electron microscopic observations show that the laminated layer comprises helicoidally arranged microfibrils and is separated from the evanescent outer layer by a thin (about 0.2 μm) sublayer. Innermost layer about 0.4 μm thick, delimited against the inner wall layer by a thin sublayer. Occlusion of subtending hypha by formation of an amorphous plug comprising thin, septum-like structures in the lumen of the subtending (sporogenous) hypha. Plug draws through the laminated wall of the spore, within the hypha often ends at a thick septum.

Mature spores formed in laboratory cultures on fine sand are globose (diameter 160–320 μm; mean 253 μm, SD = 24 μm, n = 166 spores; >93% of the spore diameters between 220–300 μm). Spores in coarse substrate tend to be more ovoid to irregular in shape. Spores contain many lipid droplets of different size (up to 20 μm in diameter) and 'structured globules' (SGs, diameter 3–6 μm). The latter consist of a highly light-refracting, homogenous outer and a granular interior part. They swell within a few minutes after they have been released in water by squeezing the spore. Electron microscopy shows the SGs to contain slightly bent rods in paracrystalline arrays. The rods have a diameter of about 6 nm, are packed hexagonally with a center to center spacing of about 8 nm.

Germination occurs through the subtending hypha or directly through the spore wall, by one or several germination hyphae. A 'cytoplasmic pole', free of SGs and big lipid droplets, often is visible at the location of hyphal outgrowth.

For further details of spore characters, see Schüßler et al. (1994; Bot. Acta 107, 36–45). Characters of symbiotic interface and compartmentation were described in detail in Schüßler et al. (1996; Protoplasma 190, 53–67). Phylogenetic sequence analyses were described in detail in Schüßler et al. (2001a; Mycol. Res. 105, 5–15), Schüßler et al. (2001b; Mycol. Res. 105, 1413–1421), and Schwarzott et al. (2001; Mol. Phylogenet. Evol. 21, 190–197).

Etymology

Genus: Greek: *ge*, soil; *sipho*, *siphonis*, tube; SPECIES: Latin: *pírus*, *pýrus*, pear; *fórmis*, shaped – referring to the syncytial mycelium and the tube-like, pear shaped appearance of the unicellular bladders formed by this fungus, containing the endosymbiotic cyanobacteria.

Molecular phylogeny of *Geosiphon* and the AMF

After the morphological and ultrastructural investigations, showing that *Geosiphon* forms '*Glomus*-like' spores (Schüßler et al., 1994), further evidence that *Geosiphon* is closely related to the AMF was provided by Gehrig et al. (1996), analysing the small subunit ribosomal RNA (SSU rRNA) genes of *Geosiphon* and *Glomus versiforme*. The phylogenetic trees showed that the AMF include *Geosiphon* as a basal branch, and furthermore that the AMF form a distinct clade not clustering with other zygomycetes. Moreover, it had already been indicated in that study that the AMF share a common ancestor with the Ascomycota and Basidiomycota. However, the number of sequences of AMF and zygomycotan fungi was still too limited to draw final conclusions of the positioning of the AMF

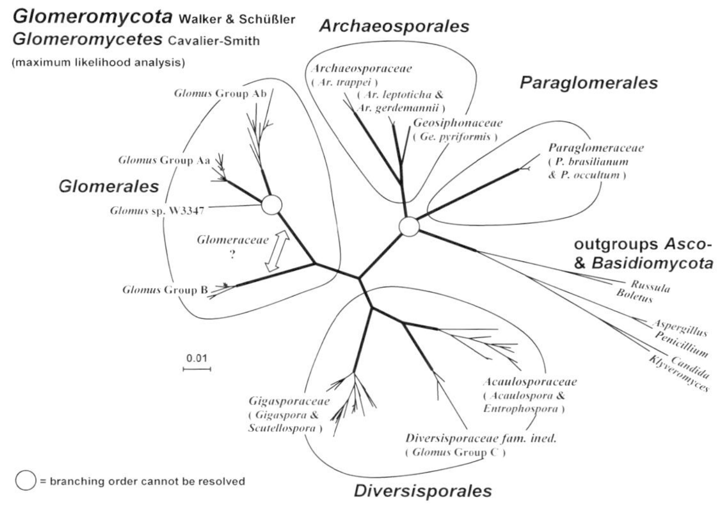

Figure 2. A maximum likelihood tree, which reflects the topology of different neighbour joining and parsimony bootstrapp analyses, showing the phylogeny of the AM fungi (Glomeromycota). A taxonomic system is proposed with four orders within one class.

within the kingdom Fungi and recently the monophyly of the AMF itself was doubted (Morton, 2000).

Based on the increasing amount of molecular data, the natural relationships within the AMF are now becoming clearer. Recently two new families, the Archaeosporaceae and Paraglomeraceae (orthographically corrected here), were described (Morton and Redecker, 2001), representing ancestral lineages within the AMF (Redecker et al., 2000b). Both families include species forming spores with *Glomus*-like morphology. Nevertheless, even after the separation of these two families, the AMF taxonomy failed to reflect a natural phylogeny. The largest genus within the AMF, '*Glomus*', had been indicated to be inadequately defined (Gehrig et al., 1996; Walker, 1992) and represents a conglomerate of fungi that are morphologically hard to separate and which have genetic distances as large as that between the AMF families Gigasporaceae and Acaulosporaceae (Schüßler et al., 2001a). Furthermore it could be shown by analyses of a large data set of near full length SSU rRNA gene sequences that *Glomus* is non-monophyletic and a new family (Diversisporaceae fam. ined.) was proposed (Schwarzott et al., 2001), together with a new family ranking system based on molecular phylogeny. Parts of this system – mainly because of the present lack of convincing morphological characters for a diagnosis – are not yet validly described or named (see Figure 2), but it is intended to form the basis of a new taxonomy for this important fungal group.

With respect to the molecular phylogeny of *Geosiphon*, a monophyletic origin of the AMF and *Geosiphon* was recently confirmed, but it was stated that *Geosiphon* is an ancestral member of the Glomerales, and that the order 'Geosiphonales' represents a sister group to the Glomerales (Benny et al., 2001). However, earlier publications already showed *Geosiphon* to be closely related to a certain AMF species (Kramadibrata et al., 2000; Schüßler, 1999) and recently it was shown that the new monogeneric family Archaeosporaceae (Morton and Redecker, 2001) is paraphyletic (Schüßler et al., 2001b; Schwarzott et al., 2001), since one of its subclades comprises *Geosiphon pyriformis* (Geosiphonaceae).

It is, therefore, evident that the *Geosiphon* fungus represents a member of an ancestral lineage within the AMF, as already hypothesised by Gehrig et al. (1996). From the known sequence data, it can be unequivocally stated that *Geosiphon* is – with respect to the species analysed up to now – most closely related to the dimorphic AMF *Archaeospora leptoticha* (synonyms *Acaulospora gerdemannii*; *Glomus leptotichum*). The SSU rRNA gene of this species is well characterised since it has been sequenced independently in the USA and Japan (Redecker et al., 2000b; Sawaki et al., 1998) and one of the isolates was re-investigated in Germany (Schwarzott et al., 2001). It was shown recently by PCR approaches that *Ar. leptoticha* is commonly found in wild grasses in Japan (Sawaki et al., 2001). The other newly described ancestral AMF family, Paraglomeraceae, is indicated as the most ancestral lineage within the AMF in phylogenetic analyses, but with only low bootstrap support. The species presently known to belong to this group have also been re-investigated to obtain full length sequences (Schwarzott et al., 2001). Comparing the presently known AMF sequences to that of *Geosiphon*, it turned out that *Geosiphon*-specific PCR primers can be designed in several regions of the SSU rRNA gene. Several such primers are developed now to evaluate the occurrence of *Geosiphon* in natural habitats and to answer the yet unsolved question if *Geosiphon* also forms an AM with higher plants or bryophytes (Schüßler et al., 2001a).

A new phylum for the arbuscular mycorrhizal fungi (AMF)

Regarding higher taxonomic levels, recent molecular works indicate that the phylum Zygomycota is polyphyletic and cannot yet be sustained in a reliable phylogenetic classification (Benny et al., 2001; O'Donnell et al., 2001). From a morphological point of view the classification of this phylum also is of doubtful validity and perhaps the majority of the organisms assigned to it are not known to have a sexual stage (Benjamin, 1979). Therefore, the inclusion of the AMF in the Zygomycetes was an artificial grouping. It was recently shown, based on SSU rRNA sequence analyses, that the AMF (together with *Geosiphon*) can clearly be separated in a well-supported monophyletic clade (Schüßler et al., 2001b), which probably diverged from the same common ancestor as the Ascomycota and Basidiomycota. Consequently the AMF (and 'related' fungi, meaning fungi like *Geosiphon*) were removed from the Zygomycota and placed into a new monophyletic phylum, the Glomeromycota Walker and Schüßler (Schüßler et al., 2001b), containing a single class, Glomeromycetes Cavalier Smith.

In addition, a new family- and order-ranking was proposed (Schüßler et al., 2001b; Schwarzott et al., 2001). Several of the new groups have not yet been formally described since no clear morphological characters have been identified up to the present – if existing at all. The results of the studies quoted above are summarised in Figure 2. Within this taxonomic ranking the order Glomerales still includes many of the 'classical' taxa as described by Morton and Benny (1990). The order Diversisporales is proposed to contain the Gigasporaceae, Acaulosporaceae and Diversisporaceae *fam. ined.* The Archaeosporaceae, together with the Geosiphonaceae and the Paraglomeraceae represent families of two further separate orders, Archaeosporales and Paraglomerales. As defined, the suborders 'Glomineae' and 'Gigasporineae' are clearly misleading and should not be used as taxa.

Some earlier descriptions should be emended. The system shown in Figure 2 is based on molecular phylogeny and therefore open for further changes (e.g., new orders or classes) without serious conflict with the proposed taxa. However, the main open questions is how to define the genera and species within the AMF. For some groups, this seems to be possible by specific sequence patterns and specific PCR primers, as proposed for the families Archaeosporaceae and Paraglomeraceae (Morton and Redecker, 2001) and the new taxa shown in Figure 2 (Schüßler et al., 2001b). On the other hand, based on the phylogenetic relationships shown, the first preliminary indications for the identification of morphological features characterising some groups, or genera, appear. This search for morphological characters, in concordance with molecular phylogeny, might solve the problem in future, but needs detailed investigations by AMF taxonomists and molecular biologists working in close co-operation rather than competition. Some recent molecular studies have been partially ignored and on the INVAM taxonomy web pages (http://invam.caf.wvu.edu/), serving as valuable and important sources for taxonomic information about AMF, unfortunately contradicting phylogenetic concepts are shown. In the 'classification tree' shown, the Archaeosporaceae and Paraglomeraceae are placed into the suborder 'Glomineae' (Morton and Benny, 1990), which itself forms a sister group to the 'Gigasporineae'. However, on the same pages this scheme is contradicted by a molecular phylogenetic tree showing that Gigasporaceae and Acaulosporaceae share an immediate common ancestor. This contradicting schemes demonstrate the problems with the suborder concept and the need for a change in the AMF taxonomy, as recently proposed (Schüßler et al., 2001b). As a reference for molecular data of AMF phylogeny, SSU rRNA gene sequence alignments and phylogenetic trees are deposited at http://www.amf-phylogeny.com and will be updated when new data are available.

The taxonomic concept proposed for the AMF, based on SSU rRNA sequence analyses, and the placement of *Geosiphon* within the AMF, are shown in Figure 2 (for details, terminology, and species composition of the *Glomus* groups see Schüßler et al., 2001a; Schwarzott et al., 2001). The Glomeraceae requires emendation and should represent *Glomus* Group A (containing, e.g., *G. mosseae* and *G. intraradices*) or B (containing, e.g., *G. claroideum* and *G. lamellosum*), dependent on the placement of the type species, *Gl. microcarpum*. The remaining *Glomus* Group represents a new family. The description of the Archaeosporaceae also should be emended since the family, as currently described, is paraphyletic. *Ar. trappei* is the type species and therefore *Ar. leptoticha* and *Ar. gerdemannii* probably should be excluded from the Archaeosporaceae, since they form a monophyletic branch together with *Geosiphon*. The Diversisporaceae *fam. ined.* (*Glomus* Group C, containing, e.g., *G. versiforme* and *G. spurcum*) will be described with a formal diagnosis soon. Note that the branches shown in Figure 2 as thick lines are statistically well supported (for details see Schwarzott et al., 2001). At least nine families are proposed in this system, whereas the Diversisporaceae and two additional families are new, and a further family is already indicated by preliminary analyses (unpublished).

Implications for the interpretation of AM evolution

In 1993 the first SSU rRNA sequence based phylogenetic tree of the Glomerales was published (Simon et al., 1993), indicating the AMF as an 'old' fungal group. Since that time it has become increasingly obvious that the earliest land plants, which had no true roots, were already colonised by hyphal fungi. These formed arbuscules strikingly similar to that of modern AM (Remy et al., 1994) and the occurrence of AMF has recently been dated back to at least 450 Myr (Redecker et al., 2000a). The sequence analyses of Simon et al. (1993) indicated *Glomus* as an ancestral genus within the AMF and mycobionts in early endomycorrhizal associations, developing in the Paleozoic in parallel to the first land plants, were said to be formed by '*Glomus*-like fungi'. This was in agreement with '*Glomus*-like' fossil records (Stubblefield et al., 1987) and the discovery of arbuscules in 400-Myr-old *Aglaophyton* fossils (Remy et al., 1994). However, today it is clear that the genus concept of the AMF is misleading (see above). Several families within the AMF form '*Glomus*-like' spores and are not closely related to *Glomus*, as, e.g., the ancestral lineages Archaeosporaceae and Paraglomeraceae (Redecker et al., 2000b; Morton and Redecker, 2001) as well as the Diversisporaceae *fam. ined.* (Schwarzott et al., 2001). If the term '*Glomus*-like' is used as a reference for the morphology of the fossil fungi it should be kept in mind that this is a morphological character referring to a highly divergent conglomerate of AMF genera, and that the 'classical' *Glomus* species are not ancestral within the AMF.

Absorption of water and nutrients (especially the poorly labile element, P) from primeval 'soil' was one of the major barriers to the evolution of land plants, and it is nowadays frequently discussed that AMF may have played a crucial role in exploring these resources in symbiosis with plants. 'Primitive' AM symbioses may still be reflected by AM-like symbioses between bryophyte thalli and AM fungi, and at least some of these fungi also form AM with higher plants (Schüßler, 2000). Furthermore, Pirozynski and Malloch (1975) postulated a partnership between two basically aquatic organisms, an alga and a 'phycomycetous' fungus, as the initial step of land plant evolution. It was argued that AM-like associations with symbiotic fungi were crucial elements in the advance of primitive plants to the land (Malloch et al., 1980). Regarding this hypothesis and the evolution of AM – with respect to *Geosiphon* – it was speculated by Gehrig et al. (1996) that 'a hypothetical association of a *Glomus*-like fungus with a green alga as a step during the evolution of the land plants appears probable. .... *Geosiphon pyriformis* now indirectly supports this hypothesis by the formation of a symbiosis between a fungus belonging to the Glomerales and a photoautotrophic prokaryote. Thus all other possible earlier associations between fungi and photoautotrophic organisms are less speculative but appear probable'. With the present background one can hypothesize that the *Geosiphon* – *Nostoc* endosymbiosis represents an organisation level of 'AMF' symbioses with cyanobacteria (Schüßler and Kluge, 2001). It shows an extant symbiotic organisation level even with bacterial photoautotrophic organisms, perhaps reflecting an early evolutionary stage of an AM-like symbiosis, when 'lower' plants had not yet colonised land and cyanobacteria would have been prominent under the prevailing environmental conditions.

This hypothetical stage of AM evolution is, however, only one of several possibilities (Schüßler et al., 2001b): (i) symbioses similar to *Geosiphon* – *Nostoc* evolved before plants colonized land and the 'AMF' later developed the relationships with the plants; (ii) 'AMF' and plants (algae) already formed symbioses in the aquatic environment, and colonized land together from the beginning; (iii) precursor 'AMF' emerged as saprobes or parasites, later becoming symbiotic with plants when appearing on land; (iv) primitive land-plants first evolved independently from the fungi for some time and later evolved the symbiosis with the 'AMF'.

In this context it is very interesting that recent molecular analyses (Heckman et al., 2001 and literature cited therein) date back the origin of AMF to 1200–1400 Myr, the Basidio- and Ascomycota to about 1200 Myr, green algae to at least 1000 Myr and the first land plants (bryophytes) to at least 700 Myr. Even with more recent calibration points for the molecular clock analyses (rRNA as well as nuclear proteins data sets), a colonization of land by fungi is dated back to at least 900 Myr. This is the time the 'earliest possible mycorrhizal association' took place – probably before land plants evolved. Therefore, terrestrial AMF probably evolved before land plants, and it is well possible that they were already symbiotically living. Altogether, this data give favour to possibility (i) and (ii) noted above. Another interesting aspect regarding fungal symbioses is that most ascomycotan lineages stem from lichen forming ancestors (Lutzoni et al., 2001). Nowadays cyanobacteria symbioses are widespread in very diverse organism groups. The hypothesis that AMF formed symbioses with cyanobacteria (*Geosiphon*-like?) or green algae earlier then AM with land plants may sound very speculative on the first view, but fits well into the data noted above. However, reconciling these different possibilities will require further fossils and more molecular data from extant organisms. A further step would be the discovery of a fossil or recent green algal 'AM' symbiosis. In any case, present knowledge indicates clearly that the formation of the symbiosis with AMF was an early and fundamental step in land colonization by plants, maybe driven by fungi which were already capable of symbiosis formation.

Conclusions

Geosiphon pyriformis is the only known fungus known to form endocyanosis, and recent studies showed that the fungus clearly belongs to the Glomeromycota. The investigations on *Geosiphon* led to a comprehensive molecular phylogeny of the AMF and allowed the proposal of a well founded natural taxonomic concept for this important group of fungi. This is not only of taxonomic interest, but also important for the understanding, planning and interpretation of (eco)physiological, genetic, biodiversity, and other studies. As an example, one may note the relatively normal colonization of the 'mycorrhiza-defective' *rmc* tomato mutant by *Glomus versiforme* (Gao et al., 2001). On the contrary to the other investigated AMF this fungus did not induce the expression of several 'defence' genes of the plant. *G. versiforme* belongs to the Diversisporaceae *fam. ined.* (*Glomus* Group C) – do members of this group behave different compared to the other AMF investigated? Such aspects now can be worked out.

Another example for the importance of understanding the phylogeny could be *Geosiphon pyriformis* itself. Because of the size of the cells and the exposition of the fungal plasmamembrane to the surrounding medium in the symbiotic stage, it has considerable advantages for the study of particular physiological features of AMF, as, e.g., nutrient transport across the fungal plasmamembrane (Schüßler and Kluge, 2001). A further advantage of the *Geosiphon – Nostoc* symbiosis is that only one eukaryote is present in the symbiosis. In contrast to the AM symbiosis, where in the symbiotic stage the plant is always present and the fungal gene expression is difficult to investigate, in *Geosiphon* the fungal mRNA can be isolated specifically because the prokaryotic mRNA lacks a poly(A) tail. Without understanding the molecular phylogeny it would not be uncovered that *Geosiphon* appears to be a very promising model system in AM research.

Acknowledgements

Thanks to the Deutsche Forschungsgemeinschaft for supporting the studies on *Geosiphon* and AMF phylogeny (SCHU1203) and to John David (CABI) for very helpful information.

References

Benjamin R K 1979 Zygomycetes and their spores. *In* The Whole Fungus Vol 2. Ed. B Kendrick. pp 573–622. National Museum of Natural Sciences and National Museum of Canada, Ottawa.

Benny G L, Humber R A and Morton J B 2001 Zygomycota: Zygomycetes. *In* The Mycota VII part A. Eds. McLaughlin, Mc Laughlin, Lemke. pp 113–146. Springer, Berlin, Heidelberg.

Cavalier-Smith T 1998 A revised six-kingdom system of life. Biol. Rev. 73, 203–266.
Ciferri R and Tomaselli R (1957) Prospetto di una sistematica microlichenologica. ATTI Series 5, Vol. XIV (1–3), pp 247–262. Istituto Botanico della Università Laboratorio Crittogamico Pavia, Pavia.
Drude O and Schorler B 1916 Beiträge zur Flora Saxonica – III. Über eine merkwürdige Alge Sachsens [*Geosiphon pyriformis* (Ktz.) F. v. Wettst.]. Sitzungsberichte und Abhandlungen der Naturwissenschaftlichen Gesellschaft ISIS e.V., Dresden, pp 58–61.
Engler A and Gilg E 1924 Geosiphonaceae (Siphonales). Syllabus der Pflanzenfamilien, Verlag von Gebrüder Borntraeger, Berlin, p. 24.
Gao L L, Delp G and Smith S E 2001 Colonization patterns in a mycorrhiza-defective mutant tomato vary with different arbuscular-mycorrhizal fungi. New Phytol. 151, 477–491.
Gehrig H, Schüßler A and Kluge M, 1996 *Geosiphon pyriformis*, a fungus forming endocytobiosis with *Nostoc* (Cyanobacteria), is an ancestral member of the Glomerales: evidence by SSU rRNA analysis. J. Mol. Evol. 43, 71–81.
Hawksworth D and Honegger R 1994 The lichen thallus: a symbiotic phenotype of nutritionally specialized fungi and its response to gall producers. *In* Plant Galls – Organisms, Interactions, Populations. Ed. M A J Williams. pp 77–98. The Systematics Association Special Volume No. 49, Clarendon Press, Oxford.
Heckman D S, Geiser D M, Eidell B R, Stauffer R L, Kardos N L and Hedges S B 2001 Molecular evidence for the early colonization of land by fungi and plants. Science 293, 1129–1133.
Knapp E 1933 Über *Geosiphon pyriformis* Fr. v. Wettst., eine intrazelluläre Pilz-Algen-Symbiose. Ber. Dtsch. Bot. Ges. 51, 210–217.
Kützing F T 1849 *Botrydium pyriformis*. *In* Species algarum. Ed. F A Brockhaus. p 486. Lipsiae, Leipzig.
Kramadibrata K, Walker C, Schwarzott D and Schüßler A 2000 A new species of *Scutellospora* with a coiled germination shield. Ann. Bot. 86, 21–27.
Lutzoni F, Pagel M and Reeb V 2001 Major fungal lineages are derived from lichen symbiotic ancestors. Nature 411, 937–940.
Malloch D W, Pirozynski K A and Raven P H 1980 Ecological and evolutionary significance of mycorrhizal symbiosis in vascular plants (a review). Proc. Natl. Acad. Sci. USA 77, 2113–2118.
Mattick F (1954) XIII. Abteilung: Lichenes. Flechten. *In* A. Engler's Syllabus der Pflanzenfamilien – I. Allgemeiner Teil Bakterien bis Gymnospermen. Eds. H Melchior and E Werdermann. pp 204–218. Gebrüder Borntraeger, Berlin.
Mollenhauer D 1992 *Geosiphon pyriformis*. *In* Algae and Symbiosis: Plants, Animals, Fungi, Viruses, Interactions Explored. Ed. W Reisser. pp 339–351 Biopress, Bristol. Morton J B 2000 Evolution of endophytism in arbuscular mycorrhizal fungi of Glomerales. *In* Microbial Endophytes. Eds. C W Bacon and J H White. pp 121–140. Marcel Dekker, New York.
Morton J B and Benny G L 1990 Revised classification of arbuscular mycorrhizal fungi (Zygomycetes): a new order, Glomerales, two new suborders, Glomineae and Gigasporineae, and two new families, Acaulosporaceae and Gigasporaceae, with an emendation of *Glomeraceae*. Mycotaxon 37, 471–491.
Morton J B and Redecker D 2001 Two new families of Glomerales, Archaeosporaceae and ParaGlomeraceae, with two new genera *Archaeospora* and *Paraglomus*, based on concordant molecular and morphological characters. Mycologia 93, 181–195.
O'Donnell K, Lutzoni F M, Ward T J and Benny G L 2001 Evolutionary relationships among mucoralean fungi (Zygomycota): Evidence for family polyphyly on a large scale. Mycologia 93, 286–296.
Pirozynski K A and Malloch D W 1975 The origin of land plants: a matter of mycotrophism. BioSystems 6, 153–164.
Redecker D, Kodner R and Graham L E 2000a Glomalean fungi from the Ordovician. Science 289, 1920–1921.
Redecker D, Morton J B and Bruns T D 2000b Ancestral lineages of arbuscular mycorrhizal fungi. Mol. Phylogenet. Evol. 14, 276–284.
Remy W, Taylor T N, Hass H and Kerp H 1994 Four hundred-million-year-old vesicular arbuscular mycorrhizae. Proc. Natl. Acad. Sci. USA 91, 11841–11843.
Sawaki H, Sugawara K and Saito M 1998 Phylogenetic position of an arbuscular mycorrhizal fungus, *Acaulospora gerdemannii*, and its synanamorph *Glomus leptotichum*, based upon 18S rRNA gene sequence. Mycoscience 39, 477–480.
Sawaki H, Kojima T, Shinozaki N and Saito M 2001 Diversity of deeply branched arbuscular mycorrhizal fungi, *Archaeospora* spp., and their detection with PCR. Abstract of the Third International Conference on Mycorrhizas 8–13 July 2001, Adelaide, Australia.
Schnepf E 1964 Zur Feinstruktur von *Geosiphon pyriformis*. Arch. Mikrobiol. 49, 112–131.
Schüßler A 1999 Glomerales SSU rRNA gene diversity. New Phytol. 144, 205–207.
Schüßler A 2000 *Glomus claroideum* forms an arbuscular mycorrhiza-like symbiosis with the hornwort *Anthoceros punctatus*. Mycorrhiza 10, 15–21.
Schüßler A and Kluge M 2001 *Geosiphon pyriformis*, an endocytosymbiosis between fungus and Cyanobacteria, and its meaning as a model system for arbuscular mycorrhizal research. *In* The Mycota IX. Ed. B Hock. pp 151–161. Springer, Berlin, Heidelberg, New York.
Schüßler A, Mollenhauer D, Schnepf E and Kluge M 1994 *Geosiphon pyriformis*, an endosymbiotic association of fungus and Cyanobacteria: the spore structure resembles that of arbuscular mycorrhizal (AM) fungi. Bot. Acta 107, 36–45.
Schüßler A, Bonfante P, Schnepf E, Mollenhauer D and Kluge M 1996 Characterization of the *Geosiphon pyriformis* symbiosome by affinity techniques: confocal laser scanning microscopy (CLSM) and electron microscopy. Protoplasma 190, 53–67.
Schüßler A, Gehrig H, Schwarzott D and Walker C 2001a Analysis of partial Glomerales SSU rRNA genes: implications for primer design and phylogeny. Mycol. Res. 105, 5–15.
Schüßler A, Schwarzott D and Walker C 2001b A new fungal phylum, the *Glomeromycota*: phylogeny and evolution. Mycol. Res., in press.
Schwarzott D, Walker C and Schüßler A 2001 *Glomus*, the largest genus of the arbuscular mycorrhizal fungi (Glomerales), is nonmonophyletic. Mol. Phylogenet. Evol., in press
Simon L, Bousquet J, Lévesque R C and Lalonde M 1993 Origin and diversification of endomycorrhizal fungi and coincidence with vascular land plants. Nature 363, 67–69.
Smith S E and Read D J 1997 Mycorrhizal symbiosis, 2nd edn. Academic Press, London, UK.
Stubblefield S P, Taylor T N and Trappe J M 1987 Fossil mycorrhizae: a case for symbiosis. Science 237, 59–60.
von Wettstein F 1915 *Geosiphon* Fr. v. Wettst., eine neue, interessante Siphonee. Österr. Bot. Z. 65, 145–156.
Walker C 1992 Systematics and taxonomy of the arbuscular endomycorrhizal fungi (Glomerales) – a possible way forward. Agronomie 12, 887–897.

Plant and Soil **244:** 85–95, 2002.

The application of genetic approaches for investigations of mycorrhizal symbioses

Susan J. Barker[1,3], Sébastien Duplessis[2] & Denis Tagu[2]
[1]*Faculty of Natural and Agricultural Sciences, The University of Western Australia, Crawley, WA 6009 Australia.* [2]*INRA-Nancy, UMR 1136 Interactions Arbres/Micro-organismes, 54280 Champenoux, France.* [3]*Corresponding author**

Received 28 September 2001. Accepted in revised form 30 April 2002

Key words: arbuscular mycorrhiza, ectomycorrhiza, genetic mapping, genetic variation, mycorrhizal mutants, quantitative trait loci

Abstract

Genetic analyses of mycorrhizal symbioses have been far less common to date than molecular biological investigations. This review aims to address the problem that genetic research approaches are some of the least familiar to non specialists by providing some detailed explanations of the requirements and processes involved, including concepts of genetic variation and genetic mapping. Each section includes examples of research progress which is restricted to studies of arbuscular mycorrhizal (AM) and ectomycorrhizal (EcM) symbioses. Most such research has focussed on AM hosts or EcM fungi. For AM hosts, some early work on natural genetic variation has not been exploited yet, but new research with barley and clover will enable genetic mapping of mycorrhizal associated QTLs for the first time. EcM fungal studies have shown a genetic basis for mycorrhizal capacity and quantitative genetic differences in mycorrhizal capacity. Some recent work with EcM hosts has begun genetic mapping of QTLs associated with mycorrhizal status. Most AM genetic research has focussed on analysis of nodulation-defective mutants for their AM host status. Map-based cloning and characterisation of the first genes shown by these analyses to be essential for establishment of both nodulation and mycorrhizal symbioses are anticipated shortly. Comparisons with molecular and genetic research on plant disease resistance genes and signalling pathways may prove useful as those studies are more advanced and underlying biochemical and evolutionary relationships are likely to exist.

Abbreviations: AM – arbuscular mycorrhiza; EcM – ectomycorrhiza; NILs – near isogenic lines; QTL – quantitative trait loci

Introduction

Attention is being increasingly focussed on evidence that mycorrhizal symbiosis is an underlying component of land plant evolution. As also discussed elsewhere (Barker and Larkan, this volume; Selosse and Le Tacon, 1998), one consequence of this fact, along with the continued presence of mycorrhizas during subsequent plant evolution until the present day, is that the genetic pathways involved in mycorrhizal symbiosis may be fundamental to root biology. Consequently, understanding these pathways may provide insights into more recently evolved interactions between plant roots and other organisms, as the genetic components of the interaction might have been adapted for subsequently evolved processes. Also, the continued selection pressure for the capacity to form a mycorrhiza, which is apparent from the current symbiotic status of most land plant families, may have constrained the capacity of plants to evolve protective responses in roots to non mu-

* FAX No: 61-8-9380-1108. E-mail: sbarker@agric.uwa.edu.au

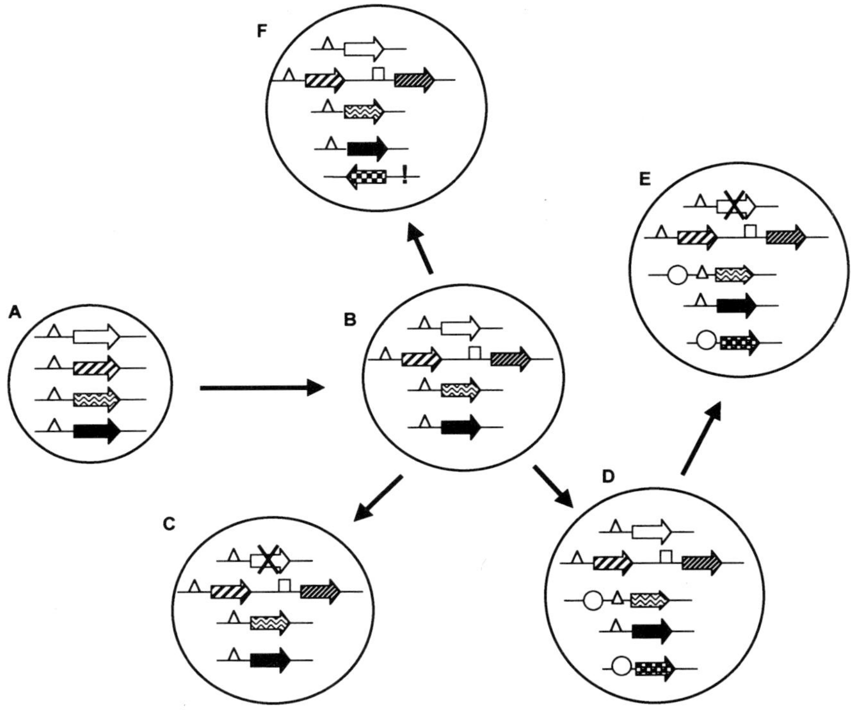

Figure 1. Evolution of gene sets from the vancestral AM host. Genes are shown as block arrows and adjacent DNA sequences as lines. Shading variations indicate different gene functions. A triangle on the line to the left of the gene represents that the gene is regulated by and/or essential for AM symbiosis. Alternative symbols indicate a different role for the adjacent gene. (A) The ancestral gene set required for successful AM symbiosis. (B) Duplication and divergence of role and possibly function of a subset of the AM symbiosis genes. Examples include the phosphate transporter gene family members of potato (Rausch et al., 2001); chitinase genes (Salzer et al., 2000) and possibly, kinase receptor genes and associated signal transduction gene family members (see text). Square indicates divergence of expression of the duplicated gene, similar shading indicates the duplicated gene retains a similar function to the ancestral gene from which it arose. This genome is representative of modern hosts that are solely AM symbionts. (C) Black X indicates loss of expression of an essential AM symbiosis gene such as in mycorrhizal defective mutations (e.g. *rmc* tomato; Barker et al., 1998a). This may also be the genetic basis for 'passive' non AM hosts such as members of the Chenopodiaceae (Koide and Schreiner, 1992; Schwartz-David et al., 2001). (D) Evolution of novel genetic potential, such as the ability to form a nodule primordium in legumes when induced by colonising bacteria, or the ability to form short roots when colonised by ectomycorrhizal fungi, along with new gene functions as required by the new symbiosis, and recruitment of some members of the AM gene set. Circles indicate genes that are responsive to and/or required for the new symbiosis. (E) Black X indicates loss of expression of an essential AM symbiosis gene. This genome is representative of plants that have evolved to be non AM symbionts but that do exhibit another more recently evolved root symbiosis e.g. Orchidaceae. (F) Additional checked gene with ! regulatory element indicates evolution of an inhibitory pathway that actively blocks AM symbiosis. An example may include members of the Brassicaceae, where production of isothiocyanates in the roots may inhibit colonisation by AM fungi (reviewed by Koide and Schreiner, 1992). Note that in C, E and F the model predicts that (many of) the genes necessary for AM symbiosis are still present in the genomes of these species. In some cases e.g. *Lupinus*, anti-AM symbiosis genetic elements of both E and F may be operational (Koide and Schreiner, 1992).

tualistic interactions. However, although most land plants have some kind of mutualistic root symbiosis, recent research on AM symbiosis has shown that it is relatively easy to identify mutant host plants that have normal growth and fecundity under nutrient-rich growth conditions but have lost the capacity to form a normal symbiosis. Therefore, at least some of the genes involved in this most ancient symbiosis are not compensated by other genes and also are not essential for plant viability under optimal nutritional conditions. The fact that some plants are dual hosts for AMs and more recently evolved symbioses, such as EcM or nodules, means that these ideas of common genetic pathways can and are being directly tested (see below). Figure 1 illustrates conceptually, the evolution, gene recruitment and subsequent divergence of plant genomes, from an ancestral simple AM symbiotic host, to the current day possibilities.

The AM fungi are obligate symbionts with biological characteristics that make genetic analysis difficult; they are not or are rarely sexually reproducing (Tommerup and Sivasithamparam, 1990), and they are multinucleate with as yet unknown nuclear ploidy. In contrast, the capacity exists to examine the genetic contributions of EcM fungi to the functioning of the EcM symbiosis because classical genetic manipulations (sexual crosses in controlled conditions, haploid strains, genetic transformation) are available for some model species (Marmeisse et al., 1995). EcM symbiosis has probably evolved, not from pathogenic interactions as usually thought, but from saprophytic fungi. Some fungal EcM species have reverted to a saprophytic life, indicating that probably few genes make the difference (Hibbett et al., 2000). In summary, genetic analysis of AM is focused towards the host, particularly in species that are model annual plants for which collections of data and genotypes are available (*Lycopersicon esculentum* (tomato), *Medicago truncatula* and so on). In contrast, genetic analysis of EcM is more directed to the fungal partner because trees – the main hosts for EcM – are difficult to manipulate genetically (issues include long generation times, outcrossing reproductive requirements, diploid genotypes).

Over the past two decades, there has been increasing attention to the use of molecular biology to dissect mycorrhizal symbioses. However, genetic analysis has been slower to become established and, as discussed by Smith and Read (1997), natural genetic variation within species of the ability to form AM symbioses has been virtually ignored. Only recently has work on AM host or EcM fungal mutations become popular by the use of mutagenesis in laboratory conditions. One of the strengths of taking a genetic approach to analysis of any particular process is that it complements a molecular biology approach without requiring the same type of sophisticated laboratory equipment and expensive consumables. In genetic research, observation of functional changes in a particular organism, or differences between two genetically distinct but related organisms grown under defined conditions, are deductively linked to involvement of one or more specified regions of that organism's genome. Molecular biology usually moves in the other direction, namely from observation of expression of a gene or genes during a response to determination of whether that gene expression is causal or responsive to the function being examined. The two research approaches meet in the emerging field of molecular genetics, where molecular research tools are used to identify the specific genetic regions (loci) involved in the process, and complementation or generation of a mutant phenotype are used to show the role of specific genes in the process being examined. In summary, the genetic approach compares a biological process in different genotypes cultivated under equal conditions, whereas molecular biology compares the behaviour of a given genotype in different biotic or abiotic conditions.

This paper describes the extent of genetic analyses (using both natural variants and induced mutants) that have taken place on AM and EcM symbioses. Some details on the experimental approaches used to validate genetic contributions to the symbiosis are also provided because this aspect of genetic research is probably least familiar to researchers outside the field. This paper is aimed to complement, but not to repeat, information in the solid reviews that have recently been written detailing the use of host mutants in the study of AM (Marsh and Schultze, 2001; Peterson and Guinel, 2000; Stougaard, 2001) or EcM (Marmeisse et al., 1995) symbioses.

Preliminaries – identification of 'useful' genetic variation

Genetic variation is any heritable difference in a particular process or trait of an organism. Some genetic variation, such as for example the Arum or Paris type intra-radicle growth pattern of one AM fungus in plants from different families (reviewed Smith and Smith, 1996), whilst interesting, is not easily investigated in a comparative fashion. This is because there are too many other genetic differences between the plants to enable a precise dissection of the specific host genes responsible for the differences in fungal growth habit. A Mendelian geneticist looks for variation in the trait or process of interest that can be examined by controlled interbreeding between the variant organisms. The use of biometrical trial designs and statistical tests on progeny from these controlled matings (crosses) then allows prediction of the number of different genomic regions (loci) that contribute to the variation in that trait or process. Organisms that are asexual, such as described for the majority of AM fungi examined, cannot be subjected to Mendelian genetic analysis. This is because the experimental approach relies on meiotic cell divisions – thus random assortment of chromosomes and DNA recombination (crossing over) between loci on the same chromosomes – to provide progeny that are the result of a combination between genomes of the two parents.

Genetic variation can be natural or induced by human manipulation such as controlled breeding (sexual crosses) or generation of mutants in the laboratory (genetic transformation, chemical mutagenesis). Apart from being observed in interbreeding organisms, useful genetic variation should provide some insight into the process of interest. So, for example, a mutation that renders a mycorrhizal host incapable of forming the symbiosis, but that also causes multiple additional changes in the host such as reduced root growth, sterile pollen, stubby dwarfed shoots, is not a useful experimental tool. In practice, the geneticist will either work with experimentally generated mutations or distinct breeds or accessions of material. These are tested for variation in the trait of interest and if any variation is found, a cross between variants is followed by segregation analysis of offspring to determine the extent to which genetic characters can be ascribed to be responsible for the variation observed.

Genetic analyses of different plant – microbe interactions showed that some host responses are qualitative (resistance *versus* susceptibility to the infection), whereas others are quantitative (or metrical; a gradient of tolerance to the infection). A qualitative response usually results from the effect of a major (vertical) gene, whereas a quantitative response is due to the combination of the action of several genes which each contribute to the response in an additive way. Mendelian genetic analysis of a quantitative trait is still possible if not too many genes contribute to the phenotype, but interpretation of data requires a strong statistical support.

Genetic variation in AM symbiosis

Interspecific genetic variation in AM symbiosis of a host has been identified, for example: in split root experiments on mechanisms of competition between different AMF species (Pearson et al., 1993); in experiments with the tomato mutant *rmc* that is colonised by some AMF species but not others (Gao et al., 2001); in wild type tomato, where different AMF form either the Arum or Paris colonisation morphology (Cavagnaro et al., 2001). However, intra-species genetic variation in the AM fungi has not yet been shown to result in functional differences between isolates (e.g. Sanders et al., 1996).

In contrast, there have been some attempts to examine variation between AM host cultivars in their capacity for, or responses to, mycorrhization. Until the research with legume nodulation mutants (see below), the majority of research in this area was with the cereals *Zea mays* (maize), *Hordeum vulgare* (barley) and *Triticum aestivum* (bread wheat) (summarised by Gianinazzi-Pearson, 1984; Smith and Read, 1997). It was established that in these species host genotype variation in colonisation occurred. Formal genetic studies with *Triticum aestivum* indicate that mycorrhizal-responsiveness genes may exist on different chromosomes of the 'Cheyenne' cultivar (Hetrick et al., 1995). Also, a recent investigation by Zhu et al. (Y Zhu and S E Smith, pers. comm.) on different *Hordeum vulgare* cultivars has shown significant differences in P acquisition between cultivars for which doubled haploid mapping populations exist (Langridge et al., 1995). Therefore, the way has been paved in that research for a genetic dissection of at least one nutritional component of the functional AM symbiosis.

The importance of AM symbiosis in low input agricultural systems should not be denied. However, plant breeders in general have ignored the symbiosis and beneficial traits like enhanced response to mycorrhization, likely to have been present in older accessions and cultivars that pre-date the heavy phosphate fertilisation era, may well have been lost from modern cultivars (Hetrick et al., 1993; Smith and Read, 1997). In that light, another recent research report demonstrating effects of host genotype on AM colonisation and host growth response, using 43 NILs of *Trifolium repens* (white clover) and with research links into a plant breeding program (Eason et al., 2001) is encouraging. If the world's food supplies can continue to be produced at current or increased abundance, but with lower overall inputs and environmental degradation, an increased awareness and assessment of the role of the AM symbiosis in low input agriculture is urgently required (e.g. Douds and Millner, 1999).

Another approach to genetic analysis of AM symbiosis, namely analysis of mutants, has been more extensively followed. This approach dates from an inspired experiment where neighbouring labs pooled resources and expertise to test the idea whether legume nodulation mutants might also be non mycorrhizal, and indeed, up to 50% of them were (Duc et al., 1989; Gianinazzi-Pearson, 1996). Since that time, a large number of nodulation-defective mutants in a range of legumes have been tested (Peterson and Guinel, 2000). More than 20 loci have been reported amongst the different legume species tested (Marsh and Schultze, 2001) and of these, 17 at least control both nodulation and mycorrhization amongst three different model species; *Pisum sativum* (garden pea) (more than eight loci known), *Medicago truncatula* (three loci) and *Lotus japonicus* (six loci) (Stougaard, 2001). Mapping and possible functions of some mycorrhizal-specifying genes are discussed below.

These results have provided the first support for the concepts described above and in Figure 1D, namely that the genetic pathway of AM symbiosis is shared in part by other similar root-microbe symbioses. However there has been no report of an enhanced resistance to pests or pathogens in legume or other mycorrhizal defective mutants, or conversely, of pest or pathogen resistance genes shown to confer mycorrhizal resistance. On the contrary, two reports on distinct *Lycopersicon esculentum* mutants have specified that the AM symbiosis mutation concerned has not altered the resistance of the host to other microbes tested (Barker et al., 1998a; Schwartz-David et al., 2001). Another report concerning several *Hordeum vulgare* mutations leading to increased susceptibility to the biotrophic pathogen *Erisyphe graminis*, has demonstrated that these mutations are correlated with increased resistance to AM symbiosis (Ruiz-Lozano et al., 1999). The tomato mutation reported by Schwartz-David et al. (2001) is only resistant to *G. intraradices* colonisation in the absence of a colonised wild type host. Using a dual compartment system, extra-radicle hyphal growth from a colonised host was demonstrated to be capable of normal colonisation of the mutant. Their results suggest either that extra-radicle hyphae have different entry signal requirements from germinating spore hyphae, or that the wild type host root exudate complemented the mutant rhizosphere to enable fungal hyphal penetration, or both. This phenotype is reminiscent of the capacity for AMF to colonise non-mycorrhizal hosts in the presence of an AM host (Koide and Schreiner, 1992).

Genetic variation in EcM symbiosis

For several decades, it has been assumed that genotypic variations within EcM tree and fungal species can influence the development of mycorrhiza (Lundeberg, 1968). Most of the experiments performed until the 1980s concerning this issue compared different seed sources and/or different fungal isolates for their ability to form EcMs. The objective was to select high performance fungal genotypes for controlled inoculation in nurseries and forest plantations (Trappe, 1977). However, comparison of isolates that usually were not parentally related, did not give any clues about the genetic basis for the ability of EcM fungi to form mycorrhizas. Optimisation of spore harvesting and *in vitro* spore germination opened the door to formal genetic studies by allowing the isolation of monocaryotic (haploid) strains and by crossing them, to obtain reconstructed dicaryotic (diploid) strains. Kropp et al. (1987) studied the relative ability of sibling monocaryotic and reconstituted mycelia of *Laccaria bicolor* to form a mycorrhiza. Interestingly, they found differences between monocaryotic strains; some were mycorrhizal whereas others lost their ability to form EcMs with *Pinus banksiana*. This indicates that the mycorrhizal phenotype segregates during meiosis and suggests that a genetic basis could be responsible at least in part for the capacity of *L. bicolor* to establish the symbiosis with its host.

Genetic studies have also been applied to specific traits supposed to be related to EcM formation (hormone production, nitrogen metabolism, extramatrical phase) (Gay and Debaud, 1987; Lamhamedi and Fortin, 1991; Wagner et al., 1988, 1989). For instance, Gay and Debaud (1987) examined the intraspecific variability within monocaryotic and dicaryotic strains of *Hebeloma cylindrosporum* for auxin production. The data showed a genetic variation between mono and dicaryotic strains and demonstrated that auxin production was a complex heritable trait. These monocaryotic strains retained their ability to form EcMs with *Pinus pinaster* (Debaud et al., 1988) but formed fewer mycorrhizas than dicaryons. This observation was true for other EcM fungal species (Lamhamedi et al., 1990). Thus, the dicaryotic state seems to be required for the full expression of the EcM ability. This suggests complex interactions between nuclei of the same cell, with possible 'cross-talk' between them in order to regulate gene expression and to optimise the cellular functions necessary for the expression of the EcM phenotype. This situation differs from the mechanism regulating infectivity of some plant pathogenic fungi such as *Ustilago maydis*, where the switch to virulence is qualitative, requires the formation of a dicaryon and is regulated by dicaryotic 'cross-talk' at a single locus (Kämper et al., 1995). A more recent study performed on *Pisolithus tinctorius* discriminated between additive and non-additive genetic components of the ability for the mycelium to form EcM (Rosado et al., 1994b). The presence of additive genetic traits suggests that improvement of fungal genotypes could be envisaged by crossing efficient monocaryotic strains. However, to our knowledge, no molecular genetics have yet been performed on the genotypic variation of fungal EcM species and establishment of genetic maps is urgently needed (Doudrick et al., 1995).

The hypothesis that susceptibility of tree roots to infection by EcM fungi is under genetic control was made in the mid 1960s. Marx and Bryan (1971) were among the first to demonstrate that host genotype from control-pollinated half-sib progenies of *Pinus elliotti* was involved in the development of EcM. Several studies have subsequently analysed the effect of natural seed source on the ability to form EcM (Cline and Reid, 1982; Dixon et al., 1987; Thomson et al., 1990). A first study of the heritability of the EcM trait was performed in *Pinus elliottii* (Rosado et al., 1994a). A very strong narrow sense heritability was found (0.81) for the percentage of EcM formed. This indicates that (i) in tightly controlled conditions, environmental effects account for less than 20% of variation in the EcM phenotype, (ii) genetic variability exists between half-sib progenies and (iii) there is significant potential for genetic gain in selection programs if these outcomes extend to field conditions. A similar experiment was performed recently on poplar (Tagu et al., 2001) and showed similar results to those with *Pinus elliottii*, but with a lower heritability value (0.49). Interestingly, some progenies had a higher percentage of EcM rootlets than the two parents, giving an indication of the possibility for improvement of this trait in a tree breeding program.

Populus was chosen in this second study because of its emergence as a model system in plant biology (Bradshaw et al., 2000). Several genetic maps are available (Bradshaw et al., 1994) and several QTLs for plant – pathogen interactions have already been published (Cervera et al., 1996; Newcombe and Bradshaw, 1996; Newcombe et al., 1996; Villar et al., 1996). In a larger scale experiment performed on 150 progenies of an inter-specific *Populus* family (Tagu et al., unpublished), one quantitative trait (QT) locus was positioned on the genetic map of the parental genotype *Populus trichocarpa*. This QT locus is localised to a linkage group previously shown to be involved in plant–microbe interactions between *Populus* leaves and the pathogenic fungus *Melampsora larici-populina*, responsible for rust infection (Faivre-Rampant et al., 1999). This is the first genetic evidence that the EcM trait is associated with loci that might be common to the interaction with mutualistic and pathogenic fungi. This observation highlights a risk for tree breeders: by selecting progenies resistant to rust infection, one might counter-select genotypes with poor mycorrhizal efficiency and possible negative consequences for nutritional capacity. It is reminiscent of a similar observation made on *Triticum aestivum*, where selected modern cultivars have lost their responsiveness to mycorrhizal infection compared to ancestral types (Hetrick et al., 1993).

Putting genetic loci on the map

A genetic map is a linear representation of the organism's genome, with the physical location of different loci identified by name. Ideally, the linear components of the map (called linkage groups) are equal in number to the number of different chromosomes in the organism's genome. The nuclear genome is usually considered separately from the mitochondrial (and chloroplast) genome(s). However, functionally the different genetic components of cells are highly coordinately regulated. Genetic maps are scaled in terms of recombination units (map units or centi-Morgans). Loci can only be positioned on a genetic map if more than one form (allele) of the locus exists in interbreeding individuals, meaning that a polymorphism in phenotype and genotype exists between these individuals. The capacity to map loci has been extended by molecular biological tools that allow determination of the physical location on a chromosome of a specific gene sequence, based on direct cross-hybridisation of a probe. In that way, as gene sequences become available for genetically characterised loci, the genetic map can become integrated with a physical chromosomal map.

Precise genetic mapping of a particular process or trait requires a 'mapping population' to be generated. This is a population of progeny plants, often the F2 generation of a cross between two parental lines that have a distinct genetic make-up with respect (at least) to the trait of interest (compatible or incompatible to mycorrhizal colonisation). However, for species with long generation times like trees, where parents are usually highly heterozygous due to outbreeding, a segregating F1 population can be used (Wu et al., 2000). The number of loci involved in the trait, for which the two parents have alleles with differing functions, can be determined by assaying every individual in the mapping population for its phenotype with respect to the trait of interest. Mendelian laws governing inheritance mean that the proportions of individuals in the population that can be grouped into different classes with respect to phenotype is directly predictive of the number of loci involved, their contribution to the trait, and their interaction with each other. Where only one locus differs between the parents, a classical 3:1 or 1:2:1 Mendelian ratio of inheritance of phenotype classes is observed in the F2 population, which can be as small as 20 or 30 individuals, although a larger population brings more statistical validity to the outcome. Pearson's chi-squared goodness of fit statistical test is sufficient to ascertain the validity of a simple ratio such as this one. Where additional loci differ and there may also be genetic interactions, the ratios become more complex and in practice, the populations being tested must be larger (100s or even 1000s) to obtain sufficient progeny in each class to work out the gene interactions and numbers involved. Software programs such as Mapmanager QT (Manly and Cudmore, 1997) and QGENE (Nelson, 1997) have been developed to assist in analysis of these more complicated situations.

Model molecular mycorrhizal host plants, for which detailed molecular genetic maps exist, include *Lycopersicon esculentum*, *Medicago truncatula*, *Lotus japonicus*, *Eucalyptus*, *Pinus* and *Populus* (see Barker and Larkan, this volume; Wu et al., 2000). The phenotype score for mycorrhizal status of each member of the mapping population is integrated with the available molecular genetic map information for the same population individuals, to assign probable map location and likely contribution to the trait (i.e. positive or negative effect; component of total phenotype score for a quantitative trait) for each identified locus. The most serious cause of errors in this process is mistaken phenotype scores for population members. Consequently, it is important to have replication in the population scoring method, as well as careful bio-assay trial design to avoid environmental influences on the bio-assay outcome.

Very little information is available for the genetic map location of mycorrhiza-associated genes, but some candidate genes have been mapped. In the case of AM, these genes were all identified by mutant screening. This work is proceeding rapidly, but has not yet been published, perhaps because the map information goes hand in hand with gene identification, which is in progress (below). The anticipation in the field is equivalent to that felt by molecular plant pathologists as the race to clone the first plant disease resistance genes was on (Staskawicz, 2001).

For EcM, one eucalypt α-tubulin and one auxin-regulated glutathione-S-transferase (*EgHypar*) cDNA, corresponding to two ectomycorrhizal regulated genes (Carnero-Diaz et al., 1996; Nehls et al., 1998) were located on *Eucalyptus urophylla* and *E. grandis* linkage maps. These two sequences belong to two different linkage groups and *EgHypar* is in the same linkage group as genes involved in lignification (e.g. caffeic acid 3-O-methyltransferase; Gion et al., 2000). Mutants of EcM trees are not available as these species are diploid, heterozygote and have a long generation time, making chemical or irradiation mutagenesis an unattractive research proposition compared to work with the fungus. However, one can think of using activation tagging (Walden et al., 1994) to select – by using genetic transformation – mutants with improved capacity of forming mycorrhiza (gain of function). Activation tagging allows the abnormal transcriptional activation of genes. Activation tagging could also be applied to ectomycorrhizal fungi which are dicaryotic and heterozygous (Figure 2). The optimisation of genetic transformation of ectomycorrhizal fungi may increase the success of this technique (Pardo et al., 2002). However, this approach and other 'reverse genetics' techniques (targeted creation of mutations to investigate gene function), such as are actively under way with various of the model AM hosts (Marsh and Schultze, 2001; Stougaard, 2001), still generate only a small percentage of mutants, thus requiring large scale experiments, which for tree host species could be handled only by a consortium.

From the map to the gene, and then. . . ?

Efforts at map-based cloning of genes identified in the mutant AM host screens are under way in various laboratories around the world and results for the first genes are anticipated shortly. These projects have benefited greatly from the direction of resources towards model genome initiatives spawned from the Human Genome project. However, such projects no longer need to be restricted to model hosts. The extent of chromosome sequence relationships (orthology) between different dicot species means that genome maps of model species such as *Arabidopsis thaliana* and *Medicago truncatula* can be used to assist map-based cloning projects in other related species, such as *Pisum sativum* (pea) (e.g. for Solanaceae, Grube et al., 2000). Once a map location for the gene is known, the closest molecular markers from that genome can be used to identify genomic DNA clones of that region in the model species. These can be used to directly

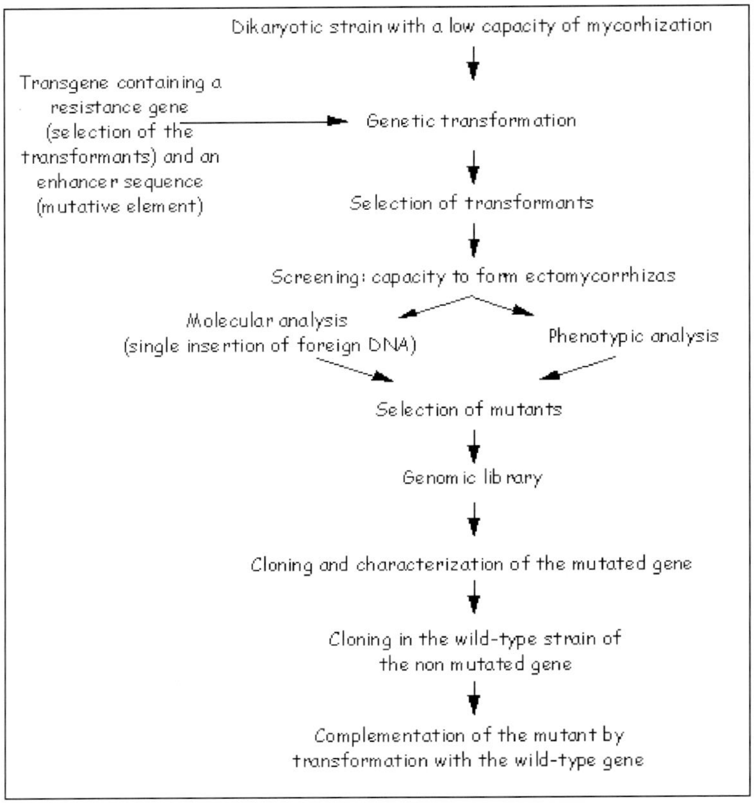

Figure 2. Activation tagging for gain of function mutations in an ectomycorrhizal fungus. The schematic illustrates the experimental steps by which a novel EcM fungal strain with enhanced symbiotic function is generated by random insertional mutagenesis of the ectomycorrhizal mycelium. Subsequently, the activated gene can be cloned using the transgene as an identification tag. This scheme indicates that the activation tagging has the additional benefit of enabling identification of fungal genes that are important for the symbiosis (or plant genes, if the same process were to be performed with a host plant).

assess potential candidate genes in the region of the mutation. Alternatively the map information from that region of the model species genome can be used to generate a greater range of DNA markers across the region of the mutation in the host under study and, using these markers, to obtain cloned genes. Whatever the method, candidate genes thus obtained from normal plants can then be tested for their functional complementation of the mutation by genetic transformation of the mutant host. Cloning the genes will enable identification of potential functions for their encoded products by sequence analysis and other methods of molecular biological analysis.

That this cloning effort will be of value in improving our understanding of how mycorrhizal symbioses are established is clear from comparison with the research outcomes in the related field of molecular plant pathology (Staskawicz, 2001). There, map based cloning of several plant disease resistance genes showed that significant commonalities in the perception mechanism leading to resistance responses existed, despite the diversity of pests and pathogens concerned (viruses, bacteria, fungi, nematodes, aphids; Ellis et al., 2000). Receptors that are linked to or associated with protein kinases form the early part of a network of signal transduction pathways which determine the plants defence response against the undesired contact (Feys and Parker, 2000; Shirasu and Schulze-Lefert, 2000).

Ultrastructural observations of the interaction between AM host mutants that do not form a mycorrhiza are suggestive of plant disease resistance responses. Observations include cell wall thickening, cytoplasmic emptying, or increased plant cell wall autofluorescence localised to the site of attempted penetration by the AM fungus (Bonfante et al., 2000; Gao et al., 2001; Peterson and Guinel, 2000; Senoo et al., 2000). Given this biochemical relationship of the mutant host responses to disease resistance responses, it might be expected that the outcome stems from an equivalent defence response signal transduction pathway. However, the nature of the biological interaction is the converse of that with plant disease resistance genes. That is, normally the AM fungus is able to enter the root and the early symbiosis host mutations that have been studied are uniformly recessive, suggesting loss of function is the cause of loss of host mycorrhizal capacity. In contrast, loss of function of a receptor-kinase pathway disease resistance gene enables the pathogen to grow in its host. Therefore, if receptor-kinase type function is found to operate as the host perception mechanism in the early stages of AM (and perhaps other such) symbioses, then triggering of the defence response pathway either does not occur, or is repressed by the action of other plant and or fungal genes. An interesting report that indirectly supports the concept that biotrophic, compatible growth might be enabled by receptor kinase activity is the identification of a pathogenicity determinant of *Colletotrichum gloeosporioides*, that appears to be a secreted protein product with a weak sequence relationship to a plant cell wall-associated receptor kinase. The gene is required to prevent a necrotic host cell response at the site of infection (Stephenson et al., 2000). The gene product might mimic the structure of a positively activated 'symbiosis' receptor-kinase thus enabling the fungus to grow inside the plant by deception.

Studies on evolution of sequenced receptor-kinase genes and database homologues indicate an ancient lineage with a modern branch restricted to dicots and a second branch found in both Angiosperm clades (Meyers et al., 1999). If a soon-to-be announced AM/nodulation mutation gene is indeed a receptor kinase, it will be interesting to see whether it can be placed as an ancestrally related out-group of the disease resistance gene family. Work could then progress to determine if the 'transient defence response' reported to occur during AM colonisation by various research groups over the last decade (Barker et al., 1998b) is in fact the elicitation by the ancestral signalling pathway, of mycorrhizal-specific members of gene families that also contain defence response members (e.g. chitinase gene family, Salzer et al., 2000). This speculative model therefore proposes that establishment of the receptor-kinase disease resistance pathway post-dated the evolution of symbiosis and involved evolutionary recruitment with requisite functional modification (i.e. duplication then divergence) not just of the signal perception genes, but also, of downstream genes involved in the symbiosis signalling pathway (as outlined in Figure 1B). This model would explain how the biochemical consequence of signal perception became reversed from compatibility (symbiosis pathway) to incompatibility (disease resistance pathway) during subsequent evolution.

However, checks and balances to intraradicle mycorrhizal fungus growth are not confined to the epidermal cell layer (Barker et al., 1998b) and the fact that plant roots are not constantly overwhelmed by pests and pathogens indicates that the host-symbiont recognition mechanism is not genetically simple to mimic. Most of the AM host mutations so far identified by prescreening for nodulation defects operate early in the mycorrhizal symbiosis. Now that a number of mutant genes that block the early stages of the symbiosis are being targeted for characterisation, it will be worth making a concerted effort to obtain more mutants without a nodulation prescreen (Barker et al., 1998a; David-Schwartz et al., 2001). Alternatively, tagging techniques that modulate but do not disrupt gene function, might enable genetic characterisation of the post-entry signalling mechanisms of endosymbiotic growth.

Acknowledgements

Francis Martin (INRA-Nancy), Sally Smith (Adelaide University) and Martin Parniske (John Innes Institute, Norwich) are acknowledged for their helpful discussions. SJB is a member of the Centre for Plant Root Symbioses, Adelaide University.

References

Barker S J, Stummer B, Gao L, Dispain I, O'Connor P J and Smith S E 1998a A mutant in *Lycopersicon esculentum* Mill. with highly reduced VA mycorrhizal colonization: isolation and preliminary characterisation. Plant J. 15, 791–797.

Barker S J, Tagu D and Delp G 1998b Regulation of root and fungal morphogenesis in mycorrhizal symbioses. Plant Physiol. 116, 1201–1207.

Bonfante P, Genre A, Faccio A, Martini I, Schauser L, Stougaard J, Webb J and Parniske M 2000 The *Lotus japonicus LjSym4* gene is required for the successful symbiotic infection of root epidermal cells. Mol. Plant Micr. Int. 13, 1109–1120.

Bradshaw H D Jr, Ceulemans R, Davis J and Stettler R 2000 Emerging model systems in plant biology: poplar (*Populus*) as a model forest tree. J. Plant Growth Regul. 19, 306–313.

Bradshaw H D Jr, Villar M, Watson B D, Otto K G, Stewart S and Stettler R 1994 Molecular genetics of growth and development of *Populus*. III. A genetic linkage map of a hybrid poplar composed of RFLP, STS and RAPD markers. Theor. Appl. Genet. 89, 167–178.

Carnero-Diaz E, Martin F and Tagu D 1996 Eucalypt α-tubulin: cDNA cloning and increased level of transcripts in ectomycorrhizal root system. Plant Mol. Biol. 31, 905–910.

Cavagnaro T R, Gao L-L, Smith F A and Smith S E 2001 Morphology of arbuscular mycorrhizal is influenced by fungal identity. New Phytol. 151, 469–475.

Cervera M T, Gusmão J, Steenackers M, Peleman J, Storme V, Vanden Broeck A, Van Montagu M and Boerjan W 1996 Identification of AFLP molecular markers for resistance against *Melampsora larici-populina* in *Populus*. Theor. Appl. Genet. 93, 733–737.

Cline M L and Reid C P P 1982 Seed sources and mycorrhizal fungus effects on growth of containerized *Pinus contorta* and *Pinus ponderosa* seedlings. Forest Sci. 28, 237–250.

Debaud J C, Gay G, Prevost A, Lei J and Dexheimer J 1988 Ectomycorrhizal ability of genetically different homokaryotic and dikaryotic mycelia of *Hebeloma cylindrosporum*. New Phytol. 108, 323–328.

Dixon R K, Garrett H E and Stelzer H E 1987 Growth and ectomycorrhizal development of loblolly pine progenies inoculated with three isolates of *Pisolithus tinctorius*. Silvae Genetica 36, 240–245.

Doudrick R L, Raffle V L, Nelson C D and Furnier G R 1995 Genetic analysis of homokaryons from a basidiome of *Laccaria bicolor* using random amplified polymorphic DNA (RAPD) markers. Mycol. Res. 99, 1361–1366.

Douds D D Jr and Millner P A 1999 Biodiversity of arbuscular mycorrhizal fungi in agroecosystems. Agric. Ecosys. Env. 74, 77–93.

Duc G, Trouvelot A, Gianinazzi-Pearson V and Gianinazzi S 1989 First report of non-mycorrhizal plant mutants (Myc-) obtained in pea (*Pisum sativum* L.) and fababean (*Vicia faba* L.). Plant Sci. 60, 215–222.

Eason W R, Webb K J, Michaelson-Yeates T P T, Abberton M T, Griffith G W, Culshaw C M, Hooker J E and Dhanoa M S 2001 Effect of genotype of *Trifolium repens* on mycorrhizal symbiosis with *Glomus mosseae*. J. Agric. Sci. 137, 27–36.

Ellis J, Dodds P and Pryor T 2000 The generation of plant disease resistance gene specificities. Trends Plant Sci. 5, 373–379.

Faivre-Rampant P, Bastien C, Augustin S, Breton V, Delpalnque A, Mourier M C, Kertadikara A, Laurans F, Lefèvre F, Lesage M C, Menard M, Pinon J, Saintagne C, Valadon A and Villar M 1999 Locating Genomic Regions Involved in Pest Resistance in Poplars. Proceedings of the International Poplar Symposium II (IPSII), IUFRO, Orléans, France, p 31.

Feys B J and Parker J E 2000 Interplay of signaling pathways in plant disease resistance. Trends Genet. 16, 449–455.

Gao L-L, Delp G and Smith S E 2001 Colonization patterns in a mycorrhizal-defective mutant tomato vary with different arbuscular-mycorrhizal fungi. New Phytol. 151, 477–491.

Gay G and Debaud J C 1987 Genetic study on indole-3-acetic acid production by ectomycorrhizal *Hebeloma* species: inter- and intraspecific variability in homo- and dikaryotic mycelia. Appl. Microbiol. Biotechnol. 26, 141–146.

Gianinazzi-Pearson V 1984 Host-fungus specificity, recognition and compatibility in mycorrhizae. *In* Genes Involved in Microbe–Plant Interactions. Eds. D P S Verma and Th Hohn. pp 225–253. Springer-Verlag, Wien.

Gianinazzi-Pearson V 1996 Plant cell responses to arbuscular mycorrhizal fungi: getting to the roots of the symbiosis. Plant Cell 8, 1871–1883.

Gion J M, Rech P, Grima-Pettenati J, Verhaegen D and Plomion C 2000 Mapping candidate genes in *Eucalytus* with emphasis on lignification genes. Molec. Breeding 6, 441–449.

Grube RC, Radwanski E R and Jahn M 2000 Comparative genetics of disease resistance within the Solanaceae. Genetics 155, 873–887.

Hetrick B A D, Wilson G W T and Cox T S 1993 Mycorrhizal dependence of modern wheat cultivars and ancestors: a synthesis. Can. J. Bot. 71, 512–518.

Hetrick B A D, Wilson G W T, Gill B S and Cox T S 1995 Chromosome location of mycorrhizal responsive genes in wheat. Can. J. Bot. 73, 891–897.

Hibbett D S, Gilbert L B and Donoghue M J 2000 Evolutionary instability of ectomycorrhizal symbioses in basidiomycetes. Nature 407, 506–508.

Kämper J, Reichmann M, Romeis T, Bölker M and Kahmann R 1995 Multiallelic recognition: nonself-dependent dimerization of the bE and bW homeodomain proteins in *Ustilago maydis*. Cell 81, 73–83.

Koide R T and Schreiner R P 1992 Regulation of the vesicular-arbuscular mycorrhizal symbiosis. Annu. Rev. Plant Physiol. Plant Mol. Biol. 43, 557–581.

Kropp B R, McAfee B J and Fortin A 1987 Variable loss of ectomycorrhizal ability in monokaryotic and dikaryotic cultures of *Laccaria bicolor*. Can. J. Bot. 65, 500–504.

Lamhamedi M S and Fortin A 1991 Genetic variations of ectomycorrhizal fungi: extramatrical phase of *Pisolithus* sp. Can. J. Bot. 69, 1927–1934.

Lamhamedi M S, Fortin A, Kope H H and Kropp B R 1990 Genetic variation in ectomycorrhiza formation by *Pisolithus arhizus* on *Pinus pinaster* and *Pinus banksiana*. New Phytol. 115, 689–697.

Langridge P J, Karakousis A, Collins N, Kretschmer J and Manning S 1995 A consensus linkage map of barley. Mol. Breeding 1, 389–395

Lundeberg G 1968 The formation of mycorrhizae in different provenances of pine (*Pinus silvestris* L.). Svensk Bot Tidskr 62, 249–255.

Manly K F and Cudmore R H Jr 1997 Map manager QT, Software for mapping quantitative trait loci. Abstracts of the 11th International Mouse Genome Conference, St Petersburg, FL.

Marmeisse R, Gay G and Debaud J C 1995 Genetics of ectomycorrhizal fungi and their transformation. *In* Biotechnology of Ectomycorrhizae. Molecular Approaches. Eds. V Stocchi, P Bonfante and M Nuti. pp. 99–114. Plenum Press, New York.

Marsh J F and Schultze M 2001 Analysis of arbuscular mycorrhizas using symbiosis-defective plant mutants. New Phytol. 150, 525–532.

Marx D H and Bryan W G 1971 Formation of ectomycorrhizae on half-sib progenies of slash pine in aseptic culture. Forest Sci. 17, 488–492.

Meyers B C, Dickerman A W, Michelmore R W, Sivaramakrishnan S, Sobral B W and Young N D 1999 Plant disease resistance genes encode members of an ancient and diverse protein family within the nucleotide-binding superfamily. Plant J. 20, 317–332.

Nehls U, Béguiristain T, Ditengou F, Lapeyrie F and Martin F 1998 The expression of a symbiosis-regulated gene in eucalypt roots is regulated by auxins and hypaphorine, the tryptophan betaine of the ectomycorrhizal basidiomycete *Pisolithus tinctorius* Planta 207, 296–302.

Nelson J C 1997 QGENE: software for marker-based genomic analysis and breeding. Mol. Breeding 3, 239–245.

Newcombe G and Bradshaw H D Jr 1996 Quantitative,trait loci conferring resistance in hybrid poplar to leaf spot caused by *Septoria populicata*. Can. J. Forest Res. 26, 1943–1950.

Newcombe G, Bradshaw H D Jr, Chastagner G A and Stettler R F 1996 A major gene for resistance to *Melampsora medusce* f. sp. *deltoidae* in hybrid poplar pedigree. Phytopath 86, 87–94.

Pardo A, Hanif M, Raudaskoski M and Gorfer M 2002 Genetic transformation of ectomycorrhizal fungi mediated by *Agrobacterium tumefaciens*. Mycol. Res. 106, 132–137.

Pearson J N, Abbott L K and Jasper D A 1993. Mediation of competition between two colonizing VA mycorrhizal fungi by the host plant. New Phytol. 123, 93–98.

Peterson R L and Guinel F C 2000 The use of plant mutants to study regulation of colonization by AM fungi. *In* Arbuscular Mycorrhizas: Physiology and Function. Eds. Y Kapulnik and D D Douds Jr. pp 147–171. Kluwer Academic Publishers, Dordrecht, The Netherlands.

Rausch C, Daram P, Brunner S, Jansa J, Lalol M, Leggewie G, Amrhein N and Bucher M 2001 A phosphate transporter expressed in arbuscule-containing cells in potato. Nature 414, 462–466.

Rosado S C S, Kropp B R and Piché Y 1994a Genetics of ectomycorrhizal symbiosis. I. Host plant variability and heritability of ectomycorrhizal and root traits. New Phytol. 126, 105–110.

Rosado S C S, Kropp B R and Piché Y 1994b Genetics of ectomycorrhizal symbiosis. II. Fungal variability and heritability of ectomycorrhizal traits. New Phytol. 126, 111–117.

Ruiz-Lozano J, Gianinazzi S and Gianinazzi-Pearson V 1999 Genes involved in resistance to powdery mildew in barley differentially modulate root colonization by the mycorrhizal fungus *Glomus mosseae*. Mycorrhiza 9, 237–240.

Salzer P, Bonanomi A, Beyer K, Vogeli-Lange R, Aeschbacher R A, Lange J, Wiemken A, Kim D, Cook D R and Boller T 2000 Differential expression of eight chitinase genes in Medicago truncatula roots during mycorrhiza formation, nodulation, and pathogen infection. Mol. Plant Micr. Int. 13, 763–777.

Sanders I R, Clapp J P and Wiemken A 1996 The genetic diversity of arbuscular mycorrhizal fungi in natural ecosystems – a key to understanding the ecology and functioning of the mycorrhizal symbiosis. New Phytol. 133, 123–134.

Schwartz-David R, Badani H, Smadar W, Levy A A, Galili G and Kapulnik Y 2001 Identification of a novel genetically controlled step in mycorrhizal colonization: plant resistance to infection by fungal spores but not extraradicle hyphae. Plant J. 27, 561–569.

Selosse M A and Le Tacon F 1998 The land flora: a phototroph-fungus partnership ? Trends Ecol. Evol. 13, 15–20.

Senoo K, Solaiman M Z, Kawaguchi M, Imaizumi-Anraku H, Akao S, Tanaka A and Obata H 2000 Isolation of two different phenotypes of mycorrhizal mutants in the model legume plant *Lotus japonicus* after EMS-treatment. Plant Cell Physiol. 41, 726–732.

Shirasu K and Schulze-Lefert P 2000 Regulators of cell death in disease resistance. Plant Mol. Biol. 44, 371–385.

Smith F A and Smith S E 1996 Mutualism and parasitism: diversity in function and structure in the 'arbuscular' (VA) mycorrhizal symbiosis. Adv. Bot. Res. 22, 1–43.

Smith S E and Read D J 1997 Mycorrhizal Symbiosis. Ed. 2. Academic Press, London.

Staskawicz B J 2001 Genetics of plant–pathogen interactions specifying plant disease resistance. Plant Physiol. 125, 73–76.

Stephenson S-A, Hatfield J, Rusu A G, Maclean D J and Manners J M 2000 cgDN3: an essential pathogenicity gene of *Colletotrichum gloeosporioides* necessary to avert a hypersensitive-like response in the host *Stylosanthes guianensis*. Mol. Plant Micr. Int. 13, 929–941.

Stougaard J 2001 Genetics and genomics of root symbiosis. Curr. Op. Plant Biol. 4, 328–335.

Tagu D, Faivre-Rampant P, Lapeyrie F, Frey-Klett Pþ, Vion P and Villar M 2001 Variation in the ability to form ectomycorrhizas in the F1 progeny of an interspecific poplar (*Populus* spp.) cross. Mycorrhiza 10, 237–240.

Thomson J, Matthes-Sears U and Peterson R L 1990 Effect of seed provenance and fungal species on bead formation in roots of *Picea mariana* seedlings. Can J. For. Res. 20, 1746–1752.

Tommerup I C and Sivasithamparam K 1990 Zygospores and asexual spores of *Gigaspora decipiens*, an arbuscular mycorrhizal fungus. Mycol. Res. 94, 897–900.

Trappe J 1977 Selection of fungi for ectomycorrhizal inoculation in nurseries. Ann. Rev. Phytopathol. 15, 203–22.

Villar M, Fefèvre F, Bradshaw H D Jr and Teissier du Cros E 1996 Molecular genetics of rust resistance in poplar (*Melampsora larici-populina Kleb./Populus* sp.) by bulked segregant analysis in a 2×2 factorial mating design. Genetics 143, 531–536.

Wagner F, Gay G and Debaud J C 1988 Genetical variability of glutamate hydrogenase activity in monokaryotic and dikaryotic mycelia of the ectomycorrhizal fungus *Hebeloma cylindrosporum*. Appl. Microbiol. Biotechnol. 28, 566–571.

Wagner F, Gay G and Debaud J C 1989 Genetical variation of nitrate reductase activity in mono- and dikaryotic populations of ectomycorrhizal fungus, *Hebeloma cylindrosporum* Romagnési. New Phytol. 113, 259–264.

Walden R, Fritze K, Hayashi H, Miklashevichs E, Harling H and Schell J 1994 Activation tagging: a means of isolating genes implicated as playing a role in plant growth and development. Plant Mol. Biol. 26, 1521–1528.

Wu R, Zeng Z B, McKeand S E and O'Malley D 2000 The case for molecular mapping in forest tree breeding. Plant Breeding Rev. 19, 41–68.

Plant and Soil **244:** 97–105, 2002.

The ectomycorrhizal symbiosis: genetics and development

Denis Tagu[1], Frédéric Lapeyrie & Francis Martin
INRA, UMR INRA/UHP 1136 "Interactions Arbres/Micro-Organismes", 54280 Champenoux, France
[1]*Corresponding author**

Received 21 August 2001. Accepted in revised form 12 December 2001

Key words: cell-wall, ectomycorrhiza, fungus, genetic, symbiosis, tree

Abstract

Ectomycorrhiza represents a symbiotic structure made between tree roots and filamentous hyphae. This new organ results from a favourable interaction between plant and microbes, taking place in the soil at the vicinity of the root. Diversity is extremely important in the rhizosphere, with large numbers of bacterial, fungal, nematode and invertebrate species. Therefore, partners of the mycorrhiza have to recognize each other and they do it by using diffusible rhizospheric molecules. This recognition leads to cellular interactions between root and fungal cells, driven by changes in gene and protein expression. The aim of this review is to describe the cellular, genetic and molecular events leading to the formation of the ectomycorrhizal tissues with an emphasis on gene expression and cell-to-cell communication.

Introduction: Communication is harmony

Living in symbiosis sounds like each partner being in harmony. However, communication in a couple is essential for a stable and enriching relationship. This seems to be the case for the mutualistic ectomycorrhizal symbiosis, a well described association which allow trees to adapt and support adverse conditions. The key to this equilibrium is probably due to a finely tuned balance between (i) genetic determinism of the ability to form a symbiosis, and (ii) environmental conditions which may enhance or repress the establishment of the relationship. Cell sensing of the environnement and cell-to-cell communications within the ectomycorrhizal tissues likely play a major role in this interchange. Any breakdown of the relationship leads to a disruption of the symbiosis. This may happen in winter for temperate climates, or during root and hyphal senescence. This may also have happened several times through evolution, since some mycorrhizal species appear to have reverted to a free-living condition (Hibbett et al., 2000).

Ectomycorrhizas are abundant mainly under temperate and boreal climates, between tree species (Angiosperms and Gymnosperms) and soil-borne filamentous fungi (ascomycetes and basidiomycetes essentially) (Smith and Read, 1997). Ectomycorrhizal mycelium growing into the substrate connects to host roots and forms a mutualistic symbiosis where partners are reciprocally benefited. Photosynthetic carbohydrates are translocated from the plant to the heterotrophic fungus which needs this rich carbon source to develop. Hyphae growing in the substrate explore the soil and absorb minerals and water which are in part tranferred to the root (Smith and Read, 1997). This exchange of metabolites is essential for the persistence of both tree and mycelium, mainly in disturbed environmental conditions. This symbiosis plays an essential role in the stability of forest ecosystems, and fulfills a role in the protection of root systems from pathogenic attacks, and from adverse abiotic soil conditions like water-stress (Smith and Read, 1997).

To allow or maintain communication and integrity in symbiotic tissues, fungal and root cells have probably developed sensing molecules (cell-anchor receptors and mobile signal-ligands similar to adhesins) (Corrêa et al., 1996) to (i) rapidly adapt to changes in their close cellular environment and (ii) send to the nucleus intra-cellular signals through transduction pathways. These cellular and molecular adaptations

* FAX No: 33383394069. E-mail: tagu@nancy.inra.fr

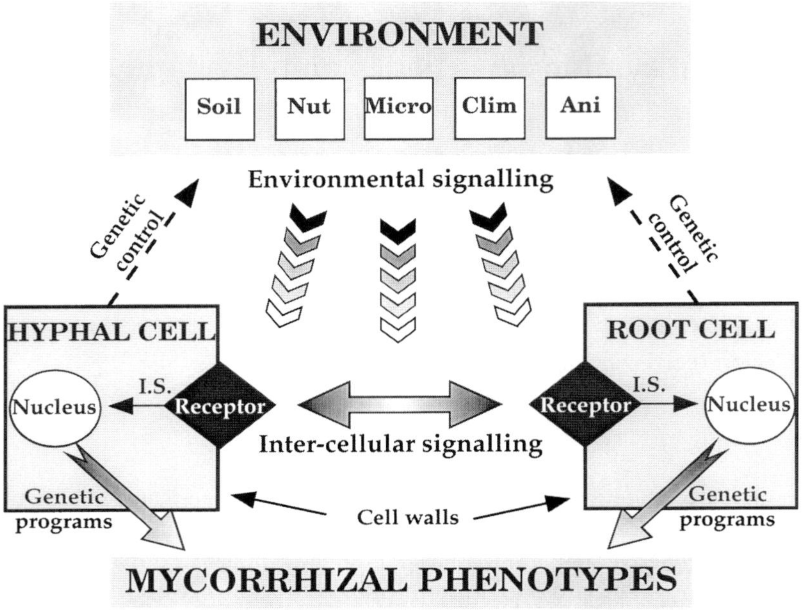

Figure 1. Inter- and intra-cellular communications between fungal hyphae and root cells in the ectomycorrhiza symbiosis. Changes in environmental conditions may produce signals sensed by both partner cells which probably transduce this information to their nuclei and provoke modifications in gene expression and consequently in phenotypes. I.S. – Intra-cellular signalling. Nut – nutrients. Micro – microorganisms. Clim – climatic factors. Ani – animals (human included).

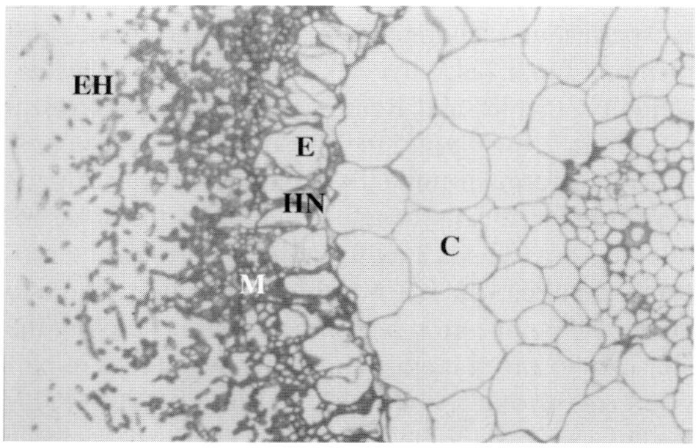

Figure 2. Cross-section of *Eucalyptus globulus - Pisolithus* ectomycorrhiza showing the mantle (M) ensheathing the epidermal root cells (E), the Hartig net (HN) made of fungal hyphae penetrating between radially elongated epidermal cells. EH – extra-matrical hyphae. Photographs by E. Carnero-Diaz, D. Tagu and F. Martin. Magnification: 700×.

are dependent on cell wall structure, organization and plasticity since fungal and root cells communicate through their apoplasts. In this review – after describing the ectomycorrhizal system – we present the actual knowledge on signalling and cell surfaces and their role in ectomycorrhiza formation (Figure 1).

Ectomycorrhizas: A sustainable interaction ?

Structure and function

Despite the large number of plant and fungal species able to form ectomycorrhizas, the general organization of the symbiotic tissues is conserved, from which variations in color, density, size and forms could appear. Ectomycorrhiza is defined as a modified lateral root. Colonizer fungal hyphae surround the root and form a sheath which isolates the root surface from its substrate (Figure 2). This ectomycorrhizal mantle is connected to a highly extended network of hyphae exploring the soil. Fruitbodies and fungal mats are formed from the differentiation of these extra-radical hyphae. Hyphae in the mantle are also connected to the root by an intra-radicular structure called the Hartig net, which is the structural and functional interface between fungal and root cells. It is made by the colonization of epidermal or cortical root cells by hyphae. In the ectomycorrhiza, the contact remains exclusively through the apoplastic compartments of both partners.

This basal organization corresponds to compartmentalization of functions. Extra-radical hyphae are responsible for mineral nutrition and water uptake. Minerals and water are taken up by hyphal tips and their distribution within the other part of the mycelium – including symbiotic hyphae – is driven by long distance trafficking of solutes involving vesicles, vacuoles and tubules (Cole et al., 1998). Extra-radical hyphae can aggregate to form multi-hyphae structures (rhizomorphs, cords) particularly adapted to long distance transport of solutes (Cairney, 1992). The ectomycorrhizal mantle is a storage compartment (Kottke et al., 1995), a buffer zone between the 'sampling' hyphae (extra-radical net) and the 'exchanging' hyphae (intra-radical net). The Hartig net is the privileged place for metabolite exchange between root cells and fungal cells. It is made of specialized hyphae presenting extended and invaginated plasma membranes that amplify the contact zone with root cells. Cell walls are altered in composition and structure compared to free living conditions (Bonfante, 2001) and their plasticity reflects their role in intra- and inter-cellular communication.

The final symbiotic tissue results from a highly physical integration of root and fungal cells. Such a cellular 'understanding' between the two partners is driven and finely regulated in time and space by interactions between plant and fungal genetic programs involved in sensing the environment, cell-to-cell communication, and constructing new cellular structures.

Genetic versus *environment*

Some ectomycorrhizal fungal species are specific to certain host plants while others are not (Kropp and Anderson, 1994) and several genetic studies indicate that the ability to form ectomycorrhiza is a heritable trait and probably polygenically controlled in fungi (e.g. Debaud et al., 1988; Lamhamedi et al., 1990). Establishment of genetic maps for ectomycorrhizal fungi are urgently needed for further understandings of the fungal genetic components of this symbiosis (Doudrick et al., 1995). Genetic maps for ectomycorrhizal trees have been available for several years, but very little data on the inheritance of the ability to form ectomycorrhiza is described. In recent studies, we have characterized the variability to form ectomycorrhizas generated by an interspecific cross in poplar, comparing the full-sib progeny to the parental genotypes. In a pilot experiment (comparing 18 progenies), we have demonstrated that 3 months after controlled inoculation with *Laccaria bicolor*, the ability to form ectomycorrhiza was genetically determined as a dominant and male inherited trait (Tagu et al., 2001b). A broad sense heritability value of 0.49 was proposed, indicating a moderate but still important genetic control of the trait. In a larger scale experiment performed on 150 progenies of the same poplar family (Tagu et al., unpublished), a QTL (quantitative trait loci) was localized on the genetic map of the parental genotype *Populus trichocarpa*. This QTL is localized to a linkage group already characterized as involved in plant-microbe-interactions between poplar leaves and the pathogenic fungus *Melampsora larici-populina* responsible for rust infection. This is the first genetic evidence that the ectomycorrhizal trait is associated to genetic bases which might be partly common to mutualistic and pathogenic fungi.

Development

Morphological and ultrastructural changes

The formation of an ectomycorrhiza is a process which follows a timely regulated succession of morphological modifications during root colonization by hyphae. In the pre-infection step, hyphae sense the presence of a host root. Colonizer hyphae can be connected to a germinated spore and have low reserves of metabolites and energy. Colonizer hyphae can also be connected to fungal cells already involved in mycorrhizas. Colonization of new root tips from a mycorrhized short root occurs before the emergence of the new roots (Tranvan et al., 2000). Cell-to-cell communication processes occurring during mycorrhiza formation from these two types of hyphae might be different since the nature and the concentration of signalling molecules may differ.

After the pre-infection step, hyphae reach the root surface. This rapidly induces large modifications in the morphology of the fungal cells. Hyphae branch and form several new tips. Fungal cells in contact with the root cell wall swell and adhere firmly to the root. Hyphal branching is probably triggered by root exudates (see below), whereas swelling might be stimulated by topographic sensing of the root surface. Colonization of the root surface is accompanied by the inhibition of pre-existing root hairs. Fungal cells excrete a fibrillar material towards the root surface which probably plays a role in anchoring hyphae. A similar fibrillar material is also deposited between two adjacent hyphae forming the mantle. Mantle formation induces important changes in extra-matrical material and involves appearance of new interfaces between fungal and root cells (Bonfante, 2001). Hyphae forming the Hartig net progress within the apoplastic compartment of the root: Cell walls and plasma membranes are highy invaginated and cells become multinucleated and rich in mitochondria and endoplasmic reticulum, indicating a high metabolic activity within these cells. During progression and penetration of hyphae, colonized root cells of some angiosperm species change their orientation of growth and these modifications are probably correlated to cytoskeleton changes (Carnero Diaz et al., 1996; Gorfer et al., 2001; Niini et al., 1996; Tarkka et al., 2000). Root apoplast is thus drastically altered by the fungal ingress. The majority of these morphological modifications are accompanied by changes in gene expression and protein accumulation.

Changes in transcript and protein patterns

It has been known for more than 10 years that the formation of ectomycorrhizas is correlated with changes in polypeptide accumulation and synthesis in both partners (e.g. Hilbert et al., 1991; Simoneau et al., 1993; Tarkka et al., 1998). Depending on the plant or fungal species, and on the experimental system, the proportion of Symbiosis-Regulated (SR) proteins varies. Only a very few of these proteins have been characterized and their functions are still unknown (Martin et al., 1999; Tagu and Martin 1996).

Analyses of transcript populations accumulated in ectomycorrhizas allowed the identification of SR-genes (e.g. Kim et al., 1999; Tagu et al., 1993). However, the development of new approaches such as transcript profiling and differential hybridization of cDNA arrays allowed the recent characterization of the major plant and fungal genetic programs regulated by the formation of ectomycorrhizas. From the initial work developed on *Eucalyptus – Pisolithus* ectomycorrhizal association, it has been shown that more than 15% of the transcripts correspond to SR-genes, but no ectomycorrhiza-specific genes were detected in this study. This suggests that the contact between *Eucalyptus* and *Pisolithus* does not trigger one specific genetic program, but rather stimulates or represses pre-existing root and fungal genetic programs (Voiblet et al., 2001). Nevertheless, Voiblet et al. (2001) demonstrated that root defense- or stress-reactions are enhanced by hyphae colonization; fungal cell wall proteins are more abundant in regions in contact with the root cells, and intra-cellular signal transduction pathways are prevalent both in *Eucalyptus* roots and *Pisolithus* hyphae during the interaction. These data clearly suggest that the most reactive root and fungal cell responses concern (i) cell-to-cell communication and (ii) cell wall plasticity.

Signalling: How to pass the border

Inter-cellular signalling

Diversity is extremely important in the rhizosphere, with billions of bacterial, fungal, nematode and invertebrate species. Therefore, partners of the mycorrhiza have to recognize each other and they do it by using diffusible rhizospheric molecules. In the case of colonizer hyphae originating from spore germination, sensing the presence of a root system might be done by sensing modifications in nutrient components of the rhizosphere compared to bulk soil. Plant roots absorb nitrogen at their vicinity, creating a depleted zone in nitrogen, and excrete sugars, locally enriching the rhizosphere in carbohydrates. This N depletion and C enrichment could serve as trophic signals by soil-borne fungi for the presence of a root system, as already demonstrated for the foliar pathogenic fungus *Cladiosporium fulvum* (Pérez–García et al., 2001). However, these trophic changes are probably not specific, and could be used as well by saprotrophic, pathogenic or symbiotic fungi to recognize the presence of a living plant. Root and fungal exudates are probably the source of more specific signals, even though they have not been characterized in ectomycorrhiza yet. Root exudates are known to stimulate spore germination and hyphal growth of several ectomycorrhizal fungi (reviewed in Duplessis et al., 2001b). Flavonoids – which are abundant molecules of the secondary metabolism excreted by the root exudates – are involved in numerous plant – microbe interactions (Broughton et al., 2000; Paiva, 2000). Recently, Lagrange et al. (2001) identified a flavonol from *Eucalyptus* root exudates – the rutin – which stimulated growth of *Pisolithus* hyphae at very low concentrations (1 pM). The action of rutin was not specific to *Pisolithus* or even to ectomycorrhizal fungi. Zeatin – a plant cytokinin – was detected in eucalypt root exudates and was able to regulate *in vitro* branching angles of *Pisolithus* hyphae (Lagrande and Lapeyrie, 2001). Thus, rutin and zeatin play complementary roles in the signalling towards *Pisolithus* hyphae. More systematic screening of active root exudates molecules are needed for their better qualitative and quantitative characterization.

Fungal hyphae also excrete rhizospheric signals. It has been known for a long time that ectomycorrhizal species synthesize and excrete auxins in the substrate (reviewed by Barker and Tagu, 2000). Auxins serve as very potent morphogenetic signals towards root systems. At low concentrations, they increase root growth and stimulate the formation of new meristems and lateral roots. These lateral roots are new targets for colonization by ectomycorrhizal hyphae. It has been clearly demonstrated that mutant fungal hyphae which hyper-produced auxins formed 3 – 5 times more mycorrhizas than a wild type mycelium (Gay et al., 1994). The use of auxin transport inhibitors indicated that ectomycorrhizal morphogenetic effects driven by auxins were dependent on the existence of concentration gradients (Kaska et al., 1999; Karabaghli–Degron et al., 1998). Once the fungal hyphae have surrounded the lateral root, subsequent division of root meristematic cells is blocked: It is hypothesized that root cells completely ensheated by ectomycorrhizal hyphae perceive a high concentration of auxins, which inhibits cell division (Barker and Tagu, 2000). These hormonal gradients are probably regulated by pools of conjugated auxins since the active auxins are known to be very unstable. Béguiristain et al. (1995) have identified an abundant indolic compound excreted by *Pisolithus* hyphae. This molecule – hypaphorine – is a tryptophane betaine and is highly accumulated in *Pisolithus* hyphal cells in contact with *Eucalyptus* roots. Hypaphorine is able to stimulate the expression of auxin-regulated genes in *Eucalyptus* and it was hypothesized that hypaphorine might act as an auxin-like compound (Nehls et al., 1998). However, recent data clearly indicate that hypaphorine is not an auxin-like molecule but, more surprisingly, counteracts auxin action, for example, on tap root elongation or root hair elongation (Ditengou and Lapeyrie, 2000; Ditengou et al., 2000). It is probable that hypaphorine competes with auxins for putative auxin-binding proteins and receptors (Kawano et al., 2001).

Intra-cellular signalling

Once signals have been released and sensed at the periphery of living cells, the information has to be transmitted to the cell nucleus to prevent or induce expression of specific genes responsible for the phenotype linked to the external signal. This transmission is performed by a cascade of biochemical regulations, called the transduction pathways. Nearly nothing is known about transduction pathways in ectomycorrhizas. Nevertheless, they are probably fundamental for ectomycorrhiza formation since in the case of pathogenic fungi, disruption of signal transduction genes often leads to the loss of pathogenicity (e. g. Mitchell and Dean, 1995; Regenfelder et al., 1997). Voiblet et al., (2001) demonstrated that genes involved in transduction pathways are among the most abundant (13% of the mRNAs) in ectomycorrhizal tissues. They correspond to sub-units of G-proteins, protein kinases and calcium binding proteins. A fungal α subunit of a G-protein was particularly stimulated in *Eucalyptus* - *Pisolithus* ectomycorrhizas, as well as a fungal Ras protein and a calmodulin gene from eucalyptus. A *ras* cDNA has also been described in the ectomycorrhizal fungus *Laccaria bicolor* and may be involved in the control of cell growth and proliferation of hyphae (Sundaram et al., 2001). Intracellular signalling in plant cells is also induced by fungal elicitors and triggered by G-proteins (Hebe et al., 1999). In turn, these plant G-proteins trigger efflux of anions and cations, extracellular alkalinization, phosphorylation or desphophorylation of unknown target proteins and finally, synthesis of H_2O_2 (Salzer and Boller, 2000).

Cell wall: A porous barrier

Cell walls and extra-cellular matrices must play several pivotal roles in mycorrhiza formation. They mainly control growth and cell extension of plant and fungal cells. They are the first cellular structures in the contact between the two partners: Apoplastic spaces are the obligate passage for signalling molecules and metabolites. Cell walls are connected to plasma membranes where receptors are anchored, and cell walls are responsible – together with the cytoskeleton – for cell structure and integrity. Mycorrhiza formation is accompanied by drastic changes in composition and structure of plant and fungal cell walls and extra-cellular matrices. The most demonstrative modification is the creation of new interfaces between fungal and root cells, where cell walls and plasma membranes are juxtaposed (Bonfante, 2001).

Chitin and ß-1,3 glucans of fungal origin are the main components of these interfaces (Bonfante, 2001; Martin et al., 1999) and their organization may change during ectomycorrhiza formation. The outer layer of fungal cell walls is connected to a fibrillar material which strongly reacts to a labelling specific to polysaccharides. These fungal fimbriae are secreted towards the root surface in the early stages of ectomycorrhiza development (Lei et al., 1990). The distribution of cystein rich components in the cell walls is completely changed between hyphae from the outer and the inner mantle (Paris et al., 1993). Biochemical and molecular analysis of *Pisolithus* – *Eucalyptus* ectomycorrhizas identified cell wall proteins regulated during the symbiosis (Martin et al., 1999; Tagu and Martin, 1996). These are cell wall mannoproteins of unknown structure and function (Tagu and Martin, 1996), alanin-rich acid polypeptides (the so-called SRAP-32 for Symbiosis Regulated Acidic Polypeptides of 32 kDa) and the hydrophobins. *Pisolithus* cell wall mannoproteins gp95 and gp72 are down regulated during the interactions with *Eucalyptus* roots. Concommitantly, SRAP-32 are highly synthesized.

SRAP-32 were first detected in two dimensionnal electrophoresis and are composed of at least 9 polypeptides differing by their apparent pI (from 4.5 to 5.5) and molecular weights (from 30 to 32 kDa). They correspond to a multigene family (Table 1). They are located in the cell walls of *Pisolithus* hyphae and are highly accumulated at fungal/root interfaces, and in penetrating hyphae forming the Hartig net (Laurent et al., 1999). Their sequences are unique to *Pisolithus* and no similarities to other proteins were found. Some of the SRAP-32 polypeptides have a signal peptide and may correspond to cell wall isoforms. Other SRAP-32 have no signal peptide and may correspond to cytosolic or vacuolar forms (Sorin and Tagu, unpublished) or secreted forms by an unknown mechanism similar to fungal galectins (Boulianne et al., 2000). Some members of the SRAP-32 family have an Arg-Gly-Asp (RGD) motif, found in many animal and microbial adhesins. This adhesion motif is known to be implicated in interactions with integrin receptors (Hynes, 1992). The role of the RGD motif found in SRAP-32 in adhesion is still speculative. It has been demonstrated that RGD motifs might also be involved in adhesion of fungal pathogens to host plasma membranes or cell walls by regulating the expression of plant defense responses during fungal penetration (Kiba et al., 1998; Mellersh and Heath, 2001). RNAs of SRAP-32 are among the most abundant in symbiotic hyphae. These genes are highly regulated during the formation of the mycorrhiza, mainly when mantle is formed and when Hartig net progresses (Martin et al., 1999). Altogether, these observations suggest that SRAP-32 could be involved in cell wall composition of hyphae during ectomycorrhiza formation.

Hydrophobins are cell wall proteins that have been well described in numerous filamentous fungi (Kershaw and Talbot, 1998). Their roles have been well studied in the saprotrophic fungus *Schizophyllum commune* for which it has been demonstrated that hydrophobins are very potent surfactant proteins (Wösten et al., 1999). Hydrophobins are able to self-assemble at any hydrophilic/hydrophobic interface. When hyphae are growing into the air, they are lined with a layer of hydrophobins: They direct their hydrophobic domains towards the air, thus rendering the surface of the hyphae very hydrophobic. This hydrophobin layer can also be involved in colonization by hyphae of any other hydrophobic surfaces like insect or plant cuticles, wood fragments or other hyphae (Wessels, 1997). Hydrophobin cDNAs have been found in *Pisolithus* – *Eucalyptus* ectomycorrhizas as 4 different sequences (Duplessis et al., 2001a; Tagu et al., 1996; Voiblet et al., 2001) (Table 1). The HYDPt-1 polypeptide has been localized at the surface of fungal cell walls (Tagu et al., 2001a). The HYDPt-1 polypeptide is able *in vitro* to decrease water surface tension (Bilewicz et al., 2001). *Pisolithus* hydrophobin mRNAs are – together with *Srap-32* transcripts – among the most abundant messengers in *Pisolithus* ectomycorrhiza and are strongly up-regulated in the early stages of symbiosis formation. They are probably involved in root colonization by hyphae and/or in agregation of hyphae during early mantle formation. Furthermore, the hydrophobicity of *Pisolithus* hyphae may reflect the importance of hydrophobins in allowing symbiotic hyphae to develop in air pockets often associated to *Pisolithus* ectomycorrhizas (Vesk et al., 2000). However, their precise roles are still ambigous and only targeted gene disruption experiments should provide further insights into these protein functions.

Both *Srap-32* and *HydPt* genes are highly up-regulated by ectomycorrhiza formation. However, the stimuli responsible for the up-regulation are not known. As suggested in Figure 1, trophic conditions could be one of this triggering. But in our conditions, no up-regulation of SRAP-32 and hydrophobin genes have been observed in different concentrations of N and C sources (Duplessis et al., 2001a, Sorin and Tagu, unpublished). The effect of root exudates have also been tested, but no regulation of these genes were observed upon contact of *Pisolithus* hyphae with eucalypt root exudates (Duplessis, Lagrange, Lapeyrie and Martin, unpublished). Thus, these highly abundant fungal cell wall proteins are probably regulated by other biotic or abiotic factors, or may represent co-regulated genes responsive to intrinsic developmental signals during ectomycorrhiza formation.

Conclusions

Ectomycorrhizal symbiosis is a complex interaction between two eucaryotic species. Despite the recent

Table 1. Pisolithus cell wall proteins regulated during ectomycorrhiza formation

	Number of ESTs[a]	Gene copies per haploid genome	Sub-classes[b]	Sequence characteristics	Properties	Cellular localization	mRNA accumulation
Hydrophobins	25	■ At least 4.	4	■ 100 amino acids in average. ■ 8 conserved cysteins. ■ Signal peptide. ■ Neutral pI.	■ Partial purification of the polypeptides. ■ Recombinant protein in *E. coli*. ■ Self-assembly of monomers. ■ Reduce water surface tension[c].	■ Cell wall located at the surface of saprotrophic and symbiotic hyphae[c].	■ Up-regulated (x 7) by mycorrhiza formation. ■ No regulation by N or C source availability[c]. ■ No regulation by root exudates[d].
Srap 32	27	■ At least 3.	3	■ 250 amino acids in average. ■ Presence of Arg-Gly-Asp motif for one sub-class. ■ Signal peptide for one sub-class. ■ Rich in Ala or Leu or Ser. ■ Acid pI (< 5,0).	■ Identification and characterization by 2D-PAGE[c].	■ Cell wall located at the surface of saprotrophic and symbiotic hyphae[c].	■Up-regulated (x 5) by mycorrhiza formation. ■ No regulation by N or C source availability[d]. ■ No regulation by root exudates[d].

[a]From 850 *E. globulus – P. tinctorius* ectomycorrhizas ESTs (Voiblet et al., 2001).
[b]A sub-class is defined when amino acid sequences diverge from more than 50% of similarity.
[c]Tagu et al. (1996); Laurent et al. (1999); Bilewicz et al. (2001); Duplessis et al. (2001a); Tagu et al. (2001a).
[d]Unpublished data.

development of cellular, genetic and molecular studies, our knowledge is still sparse since only very few signalling molecules, genes and proteins have been identified. A comprehensive view of cell-to-cell communications taking place in ectomycorrhizas requires further identification of nearly all the components involved in these processes. New technical approaches are now available whose aims are to make exhaustive lists of molecules in a given tissue. This may lead to indices of genes (genome sequencing) (International Human Genome Sequencing Consortium, 2001), of mRNA (transcriptome) (Velculescu et al., 1999), of polypeptides (proteome) (Pandey and Mann, 2000) or metabolites (metabolome) (Raamsdonk et al., 2001). These approaches, if applied to ectomycorrhizal partners, will represent departure points for further examination of the roles of these putative signalling molecules, SR-genes or SR-proteins. Gene disruption analyses should help in deciphering the function of plant or fungal SR-genes in the symbiosis. Generation of fungal and plant mutants, by gene tagging, mutated in the ectomycorrhizal phenotype would be very useful to meet this aim. For the first time, these comprehensive analyses seem possible within the next 10 years and will hopefully give a near complete image of the functioning of an ectomycorrhiza.

Acknowledgements

The authors would like to thank all their colleagues from the UMR 'IaM' for their helpful discussions and for having given unpublished data: Drs Sébastien Duplessis, Franck Ditengou, Annegret Kohler, Hubert Lagrange, Tom Kawano and Anne Jambois. Supported by INSU-PNSE.

References

Barker S J and Tagu D 2000 The roles of auxins and cytokinins in mycorrhizal symbioses. J. Plant Growth Regul. 19, 144–154.

Béguiristain T, Côté R, Rubini, P, Jay–Allemand C and Lapeyrie F 1995 Hypaphorine accumulation in hyphae of the ectomycorrhizal fungus, *Pisolithus tinctorius*. Phytochemistry 40, 1089–1091.

Bilewicz R, Witomski J, Van der Heyden A, Tagu D, Palin B and Rogalska E 2001 Modification of electrodes with self-assembled hydrophobin layers. J. Phys. Chem. B 105, 9772–9777.

Bonfante P 2001 At the interface between mycorrhizal fungi and plants: The structural organization of cell wall, plasma membrane and cytoskeleton. *In* The Mycota IX Fungal Associations. Ed. B Hock. pp 45–61. Springer Verlag, Berlin, Germany.

Boulianne R P, Liu Y, Aebi M, Lu B C and Kües U 2000 Fruiting body development in *Coprinus cinereus*: Regulated expression of two galectins secreted by a non-classical pathway. Microbiology 146, 1841–1853.

Broughton W J, Jabbouri S and Perret X 2000 Keys to symbiotic harmony. J. Bacteriol. 182, 5641–5652.

Cairney J W G 1992 Translocation of solutes in ectomycorrhizal and saprotrophic rhizomorphs. Mycol. Res. 96, 135–141.

Carnero Diaz E, Martin F and Tagu D 1996 Eucalypt α-tubulin: cDNA cloning and increased level of transcripts in ectomycorrhizal root system. Plant Mol. Biol. 31, 905–910.

Cole L, Orlovich D A and Ashford AE 1998 Structure, function, and motility of vacuoles in filamentous fungi. Fungal Genet. Biol. 24, 86–100.

Corrêa A, Staples R C and Hoch H C 1996 Inhibition of thigmostimulated cell differentiation with RGD-peptides in *Uromyces* germlings. Protoplasma 194, 91–102.

Debaud J C, Gay G, Prevost A, Lei J and Dexheimer J 1988 Ectomycorrhizal ability of genetically different homokaryotic and dikaryotic mycelia of *Hebeloma cylindrosporum*. New Phytol. 108, 323–328.

Ditengou F A and Lapeyrie F 2000 Hypaphorine from the ectomycorrhizal fungus *Pisolithus tinctorius* counteracts activities of indole-3-acetic acid and ethylene but not synthetic auxins in eucalypt seedlings. Mol. Plant-Microbe Interact. 13, 151–158.

Ditengou F A, Béguiristain T and Lapeyrie F 2000 Root hair elongation is inhibited by hypaphorine, the indole alkaloid from the ectomycorrhizal fungus *Pisolithus tinctorius*, and restored by IAA. Planta 211, 722–728.

Doudrick R L, Raffle V L, Nelson C D and Furnier G R 1995 Genetic analysis of homokaryons from a basidiome of *Laccaria bicolor* using random amplified polymorphic DNA (RAPD) markers. Mycol. Res. 99, 1361–1366.

Duplessis S, Sorin C, Voiblet C, Palin B, Martin F and Tagu D 2001a Cloning and expression analysis of a new hydrophobin cDNA from the ectomycorrhizal basidiomycete *Pisolithus*. Curr. Genet. 39, 335–339.

Duplessis S, Tagu D and Martin F 2001b Living together underground: A molecular glimpse of the ectomycorrhizal symbiosis. *In* Molecular Biology of Fungal Development: Mycological Series. Ed. H Osiewacz. Dekker & Dekker, New York, USA. In press.

Gay G, Normand L, Marmeisse R, Sotta B and Debaud J C 1994 Auxin overproducer mutants of *Hebeloma cylindrosporum* Romagnési have increased mycorrhizal activity. New Phytol. 128, 645–657.

Gorfer M, Tarkka M T, Hanif M, Pardo A G, Laitiainen E and Raudaskoski M 2001 Characterization of small GTPases Cdc42 and Rac and the relationship between Cdc42 and actin cytoskeleton in vegetative and ectomycorrhizal hyphae of *Suillus bovinus*. Mol. Plant–Microbe Interact. 14, 135–144.

Hebe G, Hager A and Salzer P 1999 Initial signalling processes induced by elicitors of ectomycorrhiza-forming fungi in spruce cells can also be triggered by G-protein-activating mastoparan and protein phosphatase-inhibiting cantharidin. Planta 207, 418–425.

Hibbett D S, Gilbert L B and Donoghue M J 2000 Evolutionary instability of ectomycorrhizal symbioses in basidiomycetes. Nature 407, 506–508.

Hilbert J L, Costa G and Martin F 1991 Ectomycorrhizin synthesis and polypeptide changes during the early stage of eucalypt mycorrhiza development. Plant Physiol. 97, 977–984.

Hynes R O 1992 Integrins: Versatility, modulation, and signaling in cell adhesion. Cell 69, 11–25.

International Human Genome Sequencing Consortium 2001 Initial sequencing and analysis of the human genome. Nature 409, 860–921.

Karabaghli–Degron C, Sotta B, Bonnet M, Gay G and Le Tacon F 1998 The auxin transport inhibitor 2,3,5-triiodobenzoic acid (TIBA) inhibits the stimulation of *in vitro* lateral root formation and the colonization of the tap-root cortex of Norway spruce (*Picea abies*) seedlings by the ectomycorrhizal fungus *Laccaria bicolor*. New Phytol. 140, 723–733.

Kaska D D, Myllylä R and Cooper J B 1999 Auxin transport inhibitors act through ethylene to regulate dichotomous branching of lateral root meristems in pine. New Phytol. 142, 49–58.

Kawano T, Kawano N and Lapeyrie F 2001 Fungal auxin antagonist hypaphorine competitively inhibits indole-3-acetic acid-dependent superoxide generation by horseradish peroxidase. Biochem. Biophys. Res. Com. 288, 546–551.

Kershaw K J and Talbot N J 1998 Hydrophobins and repellents: Proteins with fundamental roles in fungal morphogenesis. Fungal Genet. Biol. 23, 18–33.

Kiba A, Sugimoto M, Toyoda K, Ichinose Y, Yamada T and Shiraishi T 1998 Interaction between cell wall and plasma membrane via RGD motif is implicated in plant defense responses. Plant Cell Physiol. 39, 1245–1249.

Kim S J, Bernreuther D, Thumm M and Podila G K 1999 LB-AUT7, a novel symbiosis-regulated gene from an ectomycorrhizal fungus, *Laccaria bicolor*, is functionally related to vesicular transport and autophagocytosis. J. Bacteriol. 181, 1963–1967.

Kottke I, Pargney J C, Qian X M and Le Disquet I 1995 Passage and deposition of solutes in the hyphal sheath of ectomycorrhizas – The soil–root interface. *In* EUROSILVA-Contribution to Forest Tree Physiology. Eds. H Sandermann and M Bonnet–Masimbert. pp 255–271. INRA Editions, Paris, France.

Kropp B R and Anderson A J 1994 Molecular and genetic approaches to understanding variability in mycorrhizal formation and functioning. *In* Mycorrhizae and Plant Health. Eds. F L Pfleger and R G Linderman. pp 309–336. APS Press. St. Paul, Minnesota, USA.

Lagrange H and Lapeyrie F 2001 Zeatin, a cytokinin from *Eucalyptus* seedling root exudates, controls hyphal branching and hypaphorine accumulation by the ectomycorrhizal fungus *Pisolitus* sp. Submitted.

Lagrange H, Jay–Allemand C and Lapeyrie F 2001 Rutin, the phenolglycoside from eucalyptus root exudates, stimulates *Pisolithus* hyphal growth at picomolar concentrations. New Phytol. 149, 349–355.

Lamhamedi M S, Fortin A J, Kope H H and Kropp B R 1990 Genetic variation in ectomycorrhiza formation by *Pisolithus arhizus* on *Pinus pinaster* and *Pinus banksiana*. New Phytol. 115, 689–697.

Laurent P, Voiblet C, Tagu D, de Carvalho D, Nehls U, De Bellis R, Balestrini R, Bauw G, Bonfante P and Martin F 1999 A novel class of ectomycorrhiza-regulated cell wall polypeptides in *Pisolithus tinctorius*. Mol. Plant–Microbe Interact. 12, 862–871.

Lei J, Lapeyrie F, Malajczuk N and Dexheimer J 1990 Infectivity of pine and eucalypt isolates of *Pisolithus tinctorius* (Pers.) Coker & Couch on roots of *Eucalyptus urophylla* S. T. Blake *in vitro*. II. Ultrastructural and biochemical changes at the early stage of mycorrhiza formation. New Phytol. 116, 115–122.

Martin F, Laurent P, de Carvalho D, Voiblet C, Balestrini R, Bonfante P and Tagu D 1999 Cell wall proteins of the ectomycorrhizal basidiomycete *Pisolithus tinctorius*: Identification, function, and expression in symbiosis. Fungal Genet. Biol. 27, 161–174.

Mellersh D G and Heath M C 2001 Plasma membrane-cell wall adhesion is required for expression of plant defense responses during fungal penetration. Plant Cell 13, 413–124.

Mitchell T K and Dean R A 1995 The cAMP-dependent protein kinase catalytic subunit is required for appressorium formation and pathogenesis by the rice blast pathogen *Magnaporthe grisea*. Plant Cell 7, 1869–1878.

Nehls U, Béguiristain T, Ditengou F, Lapeyrie F and Martin F 1998 The expression of a symbiosis-regulated gene in eucalypt roots is regulated by auxins and hypaphorine, the tryptophan betaine of the ectomycorrhizal basidiomycete *Pisolithus tinctorius*. Planta 207, 296–302.

Niini S S, Tarkka T and Raudaskoski M 1996 Tubulin and actin protein patterns in Scots pine (*Pinus sylvestris*) roots and developing ectomycorrhiza with *Suillus bovinus*. Physiol. Plant. 96, 186–192.

Paiva N L 2000 An introduction to the biosynthesis of chemicals used in plant–microbe communication. J. Plant Growth Regul. 19, 131–143.

Pandey A and Mann M 2000 Proteomics to study genes and genomes. Nature 405, 837–846.

Paris F, Dexheimer J and Lapeyrie F 1993 Cytochemical evidence of a fungal cell wall alteration during infection of *Eucalyptus* roots by the ectomycorrhizal fungus *Cenococcum geophilum*. Arch. Microbiol. 159, 526–529.

Pérez–García A, Snoeijers S S , Joosten M H A J, Goosen T and de Wit P J G M 2001 Expression of the avirulence gene *Avr9* of the fungal tomato pathogen *Cladosporium fulvum* is regulated by the global nitrogen response factor NRF1. Mol. Plant–Microbe Interact. 14 316–325.

Raamsdonk L M, Teusink B, Broadhurst D, Zhang N, Hayes A, Walsh M C, Berden J A, Brindle K M, Kell D B, Rowland J J, Westerhoff H V, Van Dam K and Oliver S G 2001 A functional genomics strategy that uses metabolome data to reveal the phenotype of silent mutations. Nature Biotech. 19, 45–50.

Regenfelder E, Spellig T, Hartmann A, Lauenstein S, Bolker M and Kahmann R 1997 G proteins in *Ustilago maydis*: Transmission of multiple signals? EMBO J. 16, 1934–1942.

Salzer P and Boller T 2000 Elicitor induced reactions in mycorrhizae and their suppression. *In* Current Advances in Mycorrhizae Research. Eds. G K Podila and D D Douds Jr. pp 1–10. APS Press, St Paul, Minesota, USA.

Simoneau P, Viemont J D, Moreau J C and Strullu D G 1993 Symbiosis-related polypeptides associated with the early stages of ectomycorrhiza organogenesis in birch (*Betula pendula* Roth). New Phytol. 124, 495–504.

Smith S E and Read D J 1997 Mycorrhizal Symbiosis, 2nd edn. Academic Press, San Diego, California.

Sundaram S, Kim S J, Suzuki H, Mcquattie C J, Hiremath S T and Podila G K 2001 Isolation and characterization of a symbiosis-regulated ras from the ectomycorrhizal fungus *Laccaria bicolor*. Mol. Plant–Microbe Interact. 14, 618–628.

Tagu D and Martin F 1996 Molecular analysis of cell wall proteins expressed during the early steps of ectomycorrhizal development. New Phytol. 133, 73–85.

Tagu D, De Bellis R, Balestrini R, de Vries O M H, Piccoli G, Stocchi V, Bonfante P and Martin F 2001a Immunolocalization of hydrophobin HYDPt-1 from the ectomycorrhizal basidiomycete *Pisolithus tinctorius* during colonization of *Eucalyptus globulus* roots. New Phytol. 149, 127–135.

Tagu D, Faivre Rampant P, Lapeyrie F, Frey–Klett P, Vion P and Villar M 2001b Variation in the ability to form ectomycorrhizas in the F1 progeny of an interspecific poplar (*Populus* spp.) cross. Mycorrhiza 10, 237–240.

Tagu D, Nasse B and Martin F 1996 Cloning and characterization of hydrophobin-encoding cDNAs from the ectomycorrhizal basidiomycete *Pisolithus tinctorius*. Gene 168, 93–97.

Tagu D, Python M, Crétin C and Martin F 1993 Cloning symbiosis-related cDNAs from eucalypt ectomycorrhiza by PCR-assisted differential screening. New Phytol. 125, 339–343.

Tarkka M, Niini S S and Raudaskoski M 1998 Developmentally regulated proteins during differentiation of root system and ectomycorrhiza in Scots pine (*Pinus sylvestris*) with *Suillus bovinus*. Physiol. Plant. 104, 449–455.

Tarkka M T, Vasara R, Gorfer M and Raudaskoski M 2000 Molecular characterization of actin genes from homobasidiomycetes: two different actin genes from *Schizophyllum commune* and *Suillus bovinus*. Gene 251, 27–35.

Tranvan H, Habricot Y, Jeannette E, Gay G and Sotta B 2000 Dynamics of symbiotic establishment between an IAA-overproducing mutant of the ectomycorrhizal fungus *Hebeloma cylindrosporum* and *Pinus pinaster*. Tree Physiol. 20, 123–129.

Velculescu V E, Madden S L, Zhang L, Lash A E, Yu J, Rago C, Lal A, Wang C J, Beaudry G A, Ciriello K M, Cook B P, Dufault M R, Ferguson A T, Gao Y H, He T C, Hermeking H, Hiraldo S K, Hwang P M, Lopez M A, Luderer H F, Mathews B, Petroziello J M, Polyak K, Zawel L, Zhang W, Zhang X M, Zhou W, Haluska F G, Jen J, Sukumar S, Landes G M, Riggins G J, Vogelstein B and Kinzler K W 1999 Analysis of human transcriptomes. Nature Genet. 23, 387–388.

Vesk P A, Ashford A E, Markovina A L and Allaway W G 2000 Apoplasmic barriers and their significance in the exodermis and sheath of *Eucalyptus pilularis* - *Pisolithus tinctorius* ectomycorrhizas. New Phytol. 145, 333–346.

Voiblet C, Duplessis S, Encelot N and Martin F 2001 Identification of symbiosis-regulated genes in *Eucalyptus globulus* - *Pisolithus tinctorius* ectomycorrhiza by differential hybridization of arrayed cDNAs. Plant J. 25, 181–192.

Wessels J G H 1997 Hydrophobins: Proteins that change the nature of the fungal surface. Adv. Microbial Physiol. 38, 1–45.

Wösten H A B, van Wetter M A, Lugones L G, van der Mei H C, Busscher H J and Wessels J G H 1999 How a fungus escapes the water to grow into air. Curr. Biol. 9, 85–88.

Plant and Soil **244:** 107–116, 2002.

Molecular approaches to understanding mycorrhizal symbioses

Susan J. Barker[1] & Nicholas J. Larkan
Plant Biology, Faculty of Natural and Agricultural Sciences, The University of Western Australia, 35 Stirling Highway, Crawley, WA 6009, Australia. [1]*Corresponding author**

Key words: arbuscular mycorrhiza, gene expression, genetic variation, genomics, molecular biology, mycorrhizal mutation

Abstract

Molecular analyses of plant–microbe interactions have become common place in the last two decades. Although there are philosophical considerations about the application of a reductionist approach to some areas of research, the collaborative interface (e.g. molecular ecology) can provide specialised insight to the generalist, whilst adding broader relevance to the research of the specialist. However, the expense of this discipline has tended to restrict research to work on model host–microbe interactions. Molecular techniques were embraced early on by a few pioneers from the field of mycorrhizal research. Despite some high profile research, the number of molecular mycorrhizal publications has only recently begun to escalate. However the extent of literature now has exceeded the capacity for a comprehensive short review. In this paper we will briefly discuss the use of model species for molecular research and explore the range of questions that are being addressed using molecular techniques, whilst minimising use of specific jargon, to maximise the usefulness of this review to a non specialist audience. Our primary focus is on arbuscular mycorrhizal symbiosis, to complement the papers by Tagu et al., Podila et al. and Chalot et al. (all this volume), who have addressed aspects of research on ectomycorrhizal symbioses. Here we include specific citations from research groups around the world, along with reference to more detailed reviews, to provide a taste of the current excitement in this fundamental and rapidly evolving research field.

Abbreviations: AM – arbuscular mycorrhiza; AMF – arbuscular mycorrhizal fungi; EcM – ectomycorrhiza; ErM – ericoid mycorrhiza

Introduction

In 1988, Smith and Gianinazzi-Pearson noted that molecular research would provide an avenue to explore different aspects of the arbuscular mycorrhizal symbiosis. At that time molecular research on mycorrhizal symbiosis was restricted to the ectomycorrhizae, where genes and genetic variation analyses were underway. Four years later, Professor Jeff Schell closed the 1992 International Conference on Molecular Plant Microbe Interactions by predicting the next breakthroughs in that field would include AM research. In 1996 at the First International Conference on Mycorrhizas, more than one-third of the Gene Expression poster presenters could be invited to speak at the matching workshop (Tagu and Barker 1997); there were similar numbers of posters in the areas of Molecular Systematics and Molecular Ecology. The key words arbuscular and molecular identified almost 500 citations from Biological Abstracts, and about 350 citations were retrieved by the key words molecular and ectomycorrhiza, in a search of contents from 1993 to June 2001. Although molecular is a keyword that also identifies chemical studies, the situation for AM molecular biological research clearly has altered dramatically from that in 1993, when the mycorrhizal research literature was surveyed by Klironomos and Kendrick, and a broad range of reviews and commentaries have highlighted the coming of age of this research field (e.g. Barker and Tagu, 2000; Bonfante and Perotto, 1995; Gianinazzi-Pearson, 1996; Harrier, 2001; Harrison 1999a; Martin, 2001; Stougaard

* FAX No: +61-8-9380-1108. E-mail sbarker@agric.uwa.edu.au

2001). However, the situation for research on ErM and orchid mycorrhizas has not changed from that described by Klironomos and Kendrick (1993). Less than 50 papers on ErM and only one on orchid mycorrhizas (were identified in the same keyword search of Biological Abstracts in 2001, and this is an area where future research might fruitfully be directed.

Molecular research on arbuscular mycorrhizal (AM) symbiosis in our laboratory is based on the concept that the expression of sets of genes underlies the phenotype and behaviour of organisms and that evolution involves the recruitment and shuffling of different combinations of genes for new purposes (see Cronk, 2001, for a recent review of current ideas on the evolution of development). AM symbiosis may have co-evolved with land plants; certainly fossilised structures suggest the symbiosis was there before root structure had fully evolved and that the Glomalean fungi were present on land well before the evolution of the seed plants (Redecker et al., 2000; Remy et al. 1994). Although families of plants that are not mycorrhizal have evolved (Fitter and Moyersoen, 1996), and even though mycorrhizal status can be lost with a single host gene mutation, these cases have proven the exception rather than the rule (Smith and Read, 1997). This suggests that AM symbiosis (or a newer mycorrhizal symbiosis such as ErM or EcM) is highly beneficial for the survival of plants in the natural ecosystem (Selosse and Le Tacon, 1998) and, therefore, that AM symbiosis is an underlying genetic state, fundamental to root biology. A concept based on this evolutionary theory, that has been gaining strength in the literature as research evidence mounts in its support, is that there are overlaps in the gene sets expressed in different symbioses (Baron and Zambryski, 1995). There is strong evidence for overlap between nodulation and arbuscular mycorrhizal symbiosis (see below, and Barker et al., this volume) and some for other overlaps, such as nematode giant feeding cells and/or nodulation and AM symbiosis in tomato and *Medicago truncatula*, (Tahiri-Alaouhi and Antoniw, 1996; Koltai et al., 2001); powdery mildew resistance and AM symbiosis in barley (Ruiz-Lozano et al., 1999a). A prediction arising from this theory concerns how plants deal with root pathogens; the genetic routes available to plants to prevent growth of root pathogens most likely have been limited by selection that favours maintenance of a mycorrhizal status. A more comprehensive defense strategy would compromise the ability to accommodate mycorrhizal fungi. Therefore, research that aims to prevent root pathogen susceptibility should benefit from an understanding of the genetic basis for the predominant root symbiosis that has evolved in that family.

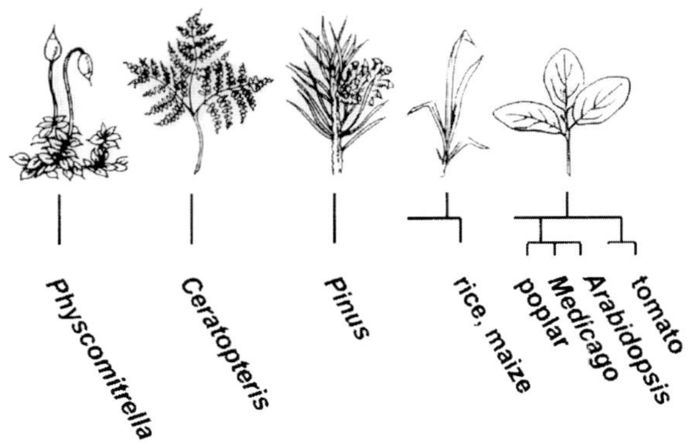

Figure 1. Model molecular plant species (adapted from Mandoli and Olmstead 2000; The Angiosperm Phylogeny Group, 1998). A range of species that have been developed as models for molecular-genetic research and that are relevant to mycorrhizal research interests is illustrated. Branched lines indicate relationships between model species within the monocots and dicots. Untipped branches indicate the existence of additional lineages within these groupings for which no models have been developed.

This focus on the fundamental genetics of mycorrhization is not the only reason for taking a molecular research approach. Molecular techniques analysing gene expression at the level of a single cell and localising gene expression to sub-cellular structures can be important in answering physiological questions about nutrient exchange (e.g. Burleigh, 2000; Ferrol et al., 2000; Harrison 1999b), as well as developmental questions about signaling between host and fungus (Barker and Tagu, 2000; Bonfante and Perotto 1995; Voiblet et al., 2001). DNA analysis has assumed a large role in assessment of mycorrhizal fungal systematics and biodiversity where often there are no tools besides analysis of DNA sequences to determine the extent of variation in populations, let alone the species identity of intra-radicle fungi (e.g. van Tuinen et al., 1998; McLean et al., 1999; Redecker 2000; and see papers in this volume). It will be important to define the correlation between functional diversity and genetic variation both in taxonomic investigations and in ecology (e.g., Buscot et al., 2000; Cairney, 1999; Sanders et al., 1996).

Requirement for model species

From the perspective of a molecular biologist, model species are those that have particular intrinsic attributes that render them easy to analyse by molecular techniques. These include small genome size, diploidy, genome sequence simplicity (little repetitive DNA), and capacity for genetic transformation. It also helps if there has been significant research tool development, for example, a genome sequencing project, molecular genetic map, stocks of well defined monogenic mutants, and an international research community devoted to that species and with genuine collaborative tendencies. Unfortunately, the prime model molecular plant, namely *Arabidopsis thaliana*, does not have the other primary required characteristic from our perspective; that it be a functionally mycorrhizal host. The Adelaide group's choice of tomato and barley as 'model' AM hosts, and the more or less recent choices of consortiums of other molecular researchers to focus on *Medicago truncatula* and *Lotus japonicus*, have all been motivated by experience of these benefits. However, it is important to point out that taxonomic classification places the majority of these model molecular hosts in very close relationship to each other (Figure 1; Mandoli and Olmstead, 2000). Nearly two-thirds of gene expression studies of AMs have been carried out in legumes. Poplar and *Eucalyptus*, which have been the subjects of molecular and genetic EcM work (D. Tagu, pers. commun; e.g., Voiblet et al., 2001), are also species that are quite close relatives of the legumes (The Angiosperm Phylogeny Group, 1998). In the development of broad genetic principles about establishment of a mycorrhizal symbiosis, it will probably be wise to develop a broader span of molecular research across the major land plant clades (e.g. Read et al., 2000).

Development of molecular models clearly has not had the same level of research input for the fungal side of the interaction as has occurred for plants. The culturability of some EcM fungi leading to a laboratory research focus on those species has been convenient, but should not be taken to mean that they are typical symbionts without appropriate expansion to other species. AM research may have suffered somewhat from the assumption that broad host range equates with functional equivalence — at any rate it has been common for molecular labs to work with only one or two fungal species; usually a *Glomus* species (see Dodd et al., 2000). Below we describe the initial approaches taken in direct assessment of fungal molecular research, and we highlight recent work with a tomato host mutation where characterisation with multiple AMF species has identified clear evidence of variation in colonisation potential.

Categories of molecular research

There are three main categories of molecular mycorrhizal research; namely, genes and their expression, genomics and genetic variation. We present some examples of the types of discoveries to be made in the first of these areas, followed by a brief discussion of the emerging field of genomics research. Finally, a brief overview of research into genetic variation will be presented. This latter area is well served by other contribution(s) to this volume.

Genes and their expression

Typically molecular researchers have either targeted particular genes or gene functions, or have carried out non-targeted searches that do not assume involvement of particular genes. Research by the Adelaide group has primarily taken the non-targeted approach (Delp et al., 1998), which also has been successfully applied by other researchers in this field (see below). Work in Adelaide began with establishment of AM colonisation procedures for our model hosts, tomato and barley, with particular emphasis on establishment of a rapid, synchronous and abundant colonisation procedure to assist in detection of genes that are altered in their expression early in the symbiosis (Rosewarne et al., 1997). We investigated gene expression using differential mRNA screening procedures; either differential screening of cDNA libraries (Murphy et al., 1997; Rosewarne, 1998) or differential display PCR (Delp et al., 2000). These techniques identify genes that are involved in mycorrhizal symbiosis by their changed transcript abundance when comparing different stages of colonisation or equivalently grown mycorrhizal and non mycorrhizal plants. A major interpretation issue for these kinds of experiments is the quantification of living fungal material in colonised root samples. Various methods may be used, including biochemical approaches (Bothe et al., 1994) and quantitative PCR amplification of plant and fungal controls (Burleigh, 2001). Contributors to a workshop on this issue at ICOM3 that was run by Dr. G Rosewarne concluded that various 'constitutive' control genes might be of use, but the key issue was to include more than one control in the experiment.

In research on plant symbiotic genes, Rosewarne (1998) found five tomato genes that were differentially expressed during establishment of the symbiosis with *Glomus intraradices*. Four of these, *LerGi* 1, 2, 4 and

Table 1. Range of plant and fungal species combinations for which AM-associated gene expression has been examined

Phylogenetic grouping	Host	AMF partner	example Reference
Monocotyledons grasses	*Hordeum vulgare* (barley)	*G. intraradices*	Murphy et al., 1997
	Oryza sativa (rice)	*G. mosseae*	Blilou et al., 2000
Dicotyledons rosids - legumes	*Vicia faba* (faba bean)	*G. fasciculatum*	Frühling et al., 1997
	Lotus japonicus	*Glomus* R-10	Senoo et al., 2000
	Medicago truncatula	*G. versiforme*	Harrison, 1996
		G. intraradices	Salzer et al., 2000
	Pisum sativum (pea)	*G. mosseae*	Ruiz-Lozano et al., 1996b
		G. versiforme	Balestrini et al., 1999
	Phaseolus vulgaris (common bean)	*G. mosseae*	Lange et al., 1999
	Glycine max (soybean)	*G. intraradices*	Lambais and Mehdy, 1996
Dicotyledons asterids	*Lycopersicon esculentum* (tomato)	*G. mosseae*	Tahiri-Alaoui and Antoniw, 1996
		G. intraradices	Rosewarne et al., 1999
	Nicotiana tabacum (tobacco)	*G. intraradices*	Shaul et al., 1999
	Solanum tuberosum (potato)	*G. intraradices*	Eausch et al., 2001

5, were increased in transcript abundance as a consequence of mycorrhizal colonisation, while the fifth gene, DB297, was decreased in abundance. *LerGi* 2 and DB297 were of particular interest as these had previously been identified as specific to root knot nematode giant feeding cells (Bird and Wilson, 1994) and *LerGi*2 was also identified independently by Tahiri-Alaoui and Antoniw (1996). To put these results into perspective, the range of host plant genes identified or examined in AM symbioses can be considered. The phylogenetic range of host species that has been examined and the fungal partners are shown in Table 1. When this list is compared with Figure 1 the narrow range of plant taxa examined is apparent. Research on model hosts has occurred but does not form the bulk of information. Plant defence response genes, particularly chitinases and β-1,3-glucanase, have formed a major focus in this field as researchers have attempted to determine whether the fungus is detected by its host and if so, how a pathogen response (PR) protein attack is avoided (e.g. review by Gianinazzi-Pearson et al., 1996). Other plant functions examined or identified include: nutrient exchange (see Harrison, 1999b, for review); cell structure and organisation proteins (Harrison, 1999a); and nodulation genes (see below and Barker et al., this volume).

A particularly thorough project recently published (Salzer et al., 2000) has highlighted the need to be careful about analysis of genes like those coding for many PR proteins, where a number of related family members exist, and especially where there are multiple functions of the family members in development. For example, ubiquitous expression of one family member can mask responses by another family member where the probe used to detect the gene transcripts hybridises to both. This research on *Medicago truncatula* chitinases (Salzer et al., 2000) shows that two of eight family members respond specifically to the symbiosis whilst others respond to pathogenic infection. This example is not unique; another recent paper (Rausch et al., 2001) has illustrated the same principle with phosphate transporter gene expression in potato. These examples support the concepts of plant evolution described above.

AMF gene research has also become more common in the last decade. In the Adelaide research efforts, Delp et al. (2000) using differential display PCR with colonised and control barley roots, identified three *G. intraradices* genes with putative regulatory functions. *Ginmyc2* was expressed both in extraradicle hyphae and in colonised roots, whilst expression of *Ginmyc 1* and *GinHB1* was only detectable in external hyphae (Delp, Smith and Barker, unpublished results). Earlier publications from other groups examined presymbiotic and symbiotic phases of fungal growth. It was established that spores contain little stored mRNA but that mRNA accumulation precedes germination in *Gigaspora rosea* (Franken et al., 1997). Research on nutritional aspects of the symbiosis resulted in identification of a phosphate transporter expressed in external hyphae of the AM fungus *Glomus versiforme* (Harrison and van Buuren, 1995). Other genes identified include chitin synthase, β-tubulin and metabolic enzymes (reviewed by Harrier, 2001).

Although gene cloning from AMF species has proven successful, there has not been an equivalent understanding about genome size and structure. Nuclear DNA content has been measured, for example; *Glomus versiforme* (0.25–0.27 pg) and *Gigaspora margarita* (0.74–0.77 pg) (Bianciotto and Bonfante, 1992). The former is comparable to the genome size of *Arabidopsis thaliana* (0.2 pg) and the latter to tomato (0.75 pg) (Grierson and Covey, 1988). However the number of genome copies present in AMF spore nuclei, and the number of chromosomes per genome, have not been determined. Other unanswered questions abound. For example, the fascinating video images of fungal nuclei moving bidirectionally along AMF hyphae, presented by Dr. Uetake at ICOM3, beg the question whether there is functional heterogeneity between nuclei. There is great scope for molecular research on AMF.

A further fascination of AMF biology has been the identification and gene characterisation of 'bacteria-like organisms' (BLOs) present throughout the life-cycle of some AMF isolates as obligate endosymbiotic bacteria, belonging to the genus *Burkholderia* (pseudomonad) (Bianciotti et al., 1996; and see Minerdi et al., this volume). These bacteria present a complication in the analysis of fungal DNA from spores as well as making an unknown contribution to gene expression and the functional contribution of AMF to the symbiosis. The presence of nitrogen fixation genes, for example, raises the question of whether BLOs might play an active role in the plant nutrition capability of the fungus (Minerdi et al., 2001).

The route to understanding functional relevance of any gene from the research described above is through genetic transformation (a form of 'reverse genetics') where the isolated sequence is used to generate a specific mutation and the consequent phenotype is identified. Genetic transformation of many of the plants used as hosts in these studies is routine, and the consequences of altered expression in some defense response genes on AM symbioses have been examined (e.g., vacuolar chitinase, which does not affect colonisation of tobacco by *Glomus mosseae*; Vierheilig et al., 1993). However the fungal partner is another story. Although successful transformation was first reported for an EcM fungus in 1990, to date there has been only one report of transient expression of a transgene in AMF (Forbes et al., 1998). Given lack of understanding of nuclear 'ploidy' and the heterokaryotic biology of these fungi, as well as possible functional heterogeneity of nuclei, as noted above, gene disruption through an RNA interference technique might prove a more reliable method of investigating gene function. This method of gene silencing is most effective when double-stranded RNA homologous to the gene of interest is introduced to the organism by injection, virus infection or transformation of an appropriately designed gene construct. It relies on an innate cellular pathway, which is probably a virus protection mechanism, and has been demonstrated in organisms wide spread through the evolutionary tree (Marx, 2000) and so is also potentially present in the Zygomycetes.

The alternative approach to investigation of genes and their expression in mycorrhizas has been the use of mutant hosts. Mutant analysis identifies key loci that are essential for development of the symbiosis but that can be disrupted without lethal consequence to the host. However until the mutated genes are cloned and sequenced their precise contribution is unknown. This contrasts with the molecular biological studies, where genes of known function have been identified as being symbiosis-regulated, but have unknown functional importance to the symbiosis. Targeted research in legume host species, involving a prescreen for nodulation mutations, has been the primary approach taken by AM symbiosis researchers. There has been particular interest in the possibility that nodulation is established by a genetic pathway that is similar to and overlapping with establishment of AM symbiosis (e.g., Hirsch and Kapulnik, 1998), following the discovery that some but not all nodulation defective mutants are also altered in their mycorrhizal phenotype (Duc et al., 1989). This has become a major focus in molecular genetic analysis of AM symbiosis and is one area where there has been a significant move to the investigation of model species (*Lotus japonicus* and *Medicago truncatula*). There have been recent thorough reviews of this research area (Marsh and Schultze, 2001; Peterson and Guinel, 2000; Stougaard, 2001; and see Barker et al., this volume).

In this area of research the Adelaide group has chosen the less travelled path, namely a non-targeted screen for mycorrhizal mutations in the non-legume hosts; barley and tomato. Our best characterised mutant is in tomato and is called *rmc* (reduced mycorrhizal colonisation; Barker et al., 1998a). This recessive mutation has no detectable phenotype beyond colonisation potential. We are currently working to determine the genetic map location of the mutation, which should enable map-based cloning of the gene. Of particular interest is that different fungal species colonise the mutant to different extents (Gao et al., 2001) and these differences also extend to defense gene expression patterns (L-L Gao, pers. comm.). These results are the first direct demonstration that fungal partners differ in their host recognition requirements, and reinforce the increasing physiological data that functional variation in AMF exists and should be taken into account in molecular research.

Genomics

Research in molecular biology has completed one turn of a spiral from the days prior to easy cloning and sequencing, when genome structure and function were investigated by 'C_0t curve' and 'R_0t curve' analyses. Those techniques identified fundamental aspects of eucaryotic genomes, including that genome size in plants varied in large part according to the contribution of repeated sequences (Flavell, 1982, cited in Grierson and Covey, 1988), and that there are an estimated 25 000 genes with varying degrees of overlap expressed in each plant organ (Goldberg, 1986). The 'new' genomics technologies essentially perform the same type of holistic analysis, but with the added benefit of enabling a focus on sequence information as well. Terms such as proteome and transcriptome have been coined to describe the complete set of proteins and transcripts from a genome, respectively. Colebatch et al., (2001) describe the tools and discuss the paradigm shift of this new science.

Array technology is particularly powerful and involves the synthesis of small filters (macro- or mini-arrays) or glass slides (micro-arrays) on which a comprehensive set of genes involved in specific metabolic functions (for example), or gene (genome) or transcript (transcriptome) clones has been spotted. Although preparation of the arrays is expensive, subsequent processing of these arrays requires considerably less materials and time than a traditional library screen and provides information about considerably larger numbers of genes at one time. Micro-array technology is still evolving. Issues for which solutions are being sought include: high cost of preparation of the materials to go onto the array; technical difficulties with preparation of accurately spotted arrays; limited options for probe labelling so that only pairwise comparisons of gene expression patterns are possible; and the requirement for biometrical trial design and interpretation of results, which has not been a common element in molecular biological research in the past.

As discussed above and by Barker et al., (this volume), the molecular mechanisms of evolution that are starting to be uncovered, along with the molecular and genetic overlaps in nodulation and AM symbioses, and other symbioses, suggest the possibility that symbiosis gene sets may have some significant overlaps. To date, only one mycorrhizal project has been published that has approached the level of detail made possible by arrays (Voiblet et al., 2001). This research examined gene expression during the early EcM colonisation of *Eucalyptus globulus* with *Pisolithus tinctorius*. The existence of a *Eucalyptus–Pisolithus* macro-array provides an exciting opportunity to begin to directly compare the EcM and AM symbiotic gene sets. The overlaps between these two symbioses are most likely to exist in the nutrient exchange mechanisms (Barker and Tagu, 2000). However, overlaps between symbioses where internal feeding sites are established in roots, such as AM symbiosis, nodulation, nematode colonisation, aphid colonisation, phylloxera infestation of grapes and/or parasitic plants, may prove to be significant and certainly can be tested using array technology and suitable plants which host several of the organisms of interest.

Another form of genomics research, namely proteomics, involves analysis of protein sets using high resolution two-dimensional polyacrylamide gel electrophoresis (2D-PAGE) followed by microsequencing. Different extracts can be analysed, from whole organs to subcellular fractions (e.g., Benabdellah et al., 2000; Vallorani et al., 2000). In fact, proteomics minus the protein sequence analysis has been one of the most long standing AM research approaches, with quite detailed investigation of protein profile changes in AM symbiosis published by Dumas-Gaudot et al. in 1994, for example. Now that protein sequencers with very small protein mass requirements are becoming increasingly available around the globe, and the genomes of model plants are being sequenced in their entireties, the identities of the 'mycorrhizins' and other differentially accumulated proteins in mycorrhizal symbiosis can be expected to be revealed.

Genetic variation

The other main areas of molecular research on mycorrhizas can be grouped as investigations of genetic variation in mycorrhizal fungi, although their purposes vary from systematics and taxonomy through to development of tools for field sample identification. The virtually complete focus of these types of research on EcM and a little on ErM associations that existed in the 1980s and early 1990s, has now broadened to being the most comprehensive range of any of the molecular research areas. Papers including orchid mycorrhizas (e.g., Sen et al., 1999) as well as the AMs (e.g., Simon, 1996) indicate the extensive interest in molecular ecology of mycorrhizas, as well as an improved understanding of their evolution and current species relationships. Revision of AMF phylogeny for example, is an ongoing process (Redecker, this volume; Schüssler, this volume).

This broad research area almost entirely relies on PCR amplification of ribosomal DNA (rDNA) sequences, with nuclear encoded small subunit (17S/18S rDNA) being the most common but mitochondrial small subunit rDNA (16S), nuclear encoded large subunit (25S rDNA) and nuclear encoded rDNA internal transcribed sequences (ITS) also playing a significant role. Random amplified polymorphic DNAs (RAPDs), and microsatellites as molecular markers have been far less popular. A paper describing amplified fragment length polymorphism (AFLP^{TM}) analysis in AMF single spore DNA extracts (Rosendahl and Taylor, 1997) may spark some interest as it provides a robust method for investigation of a broader range of loci from the genome than the single gene focus of the rDNA work. As indicated above and elsewhere in this volume, a major question remaining in this field of molecular mycorrhizal research is the significance of genetic variation within AM species (not to mention intrasporal genetic variation), and particularly the relationship between genetic variation and functional variation (e.g., Sanders et al., 1996).

Future opportunities

This brief review of trends in molecular research of mycorhizal symbioses indicates that molecular techniques have become firmly established and an increasing proportion of publications utilising these approaches can be anticipated over the coming decade. Genomics research will provide the tools to investigate thoroughly the concept that AM symbiosis gene sets have been 'reused', at least in part, in evolution of newer symbioses. The early pathway of events in each symbiosis will be thoroughly characterised and the mechanism(s) by which the host controls the symbiotic fungi will be ascertained. For example, the genetic determination of *Arum* or *Paris* forms of the AM symbiosis may be able to be established using tomato as a host, because different fungal species form *Arum* — or *Paris* — AM with this plant (Cavagnaro et al., 2001). We are particularly curious to know whether AM fungal dependency on the host for completion of its life cycle has evolved through movement of key genes and/or their regulation to the control of the plant host nucleus. Certainly, images of developing arbuscules wrapping the host cell nucleus are highly evocative of some kind of genetic signalling or exchange (Barker et al., 1998b).

Other areas of opportunity include assessment of the extent and nature of gene pathway overlaps between the range of plant–microbe symbioses and in addition, other evolutionary pathways with plant nutritional consequences, such as the similarities between EcM short root development and the developmental pathway of proteoid roots. Finally, a great deal more molecular work can be done with ErM and orchid mycorrhizas including, for the orchid association, an investigation of the mechanism by which the symbiosis is regulated to enable the development and germination of the basal zygote. Comparison of these symbioses with the AMs and EcMs will provide a very comprehensive understanding of the ways in which plant roots respond to mutualistic associations.

Acknowledgements

N. L. is funded by Australian Research Council Large Grant A19943108 to S E Smith and S J B. Thanks to Sally Smith and lab members, Leanne Forsyth and Lyn Abbott and lab members for helpful discussions, Denis Tagu for discussion of unpublished work and Ben Murphey for assistance with preparation of the manuscript. Two reviewers provided helpful comments concerning the focus of the paper.

References

Balestrini R, Perotto S, Gasverde E, Dahiya P, Guldmann L L, Brewin N J and Bonfante P 1999 Transcription of a gene encoding a lectin-like glycoprotein is induced in root cells harboring arbuscular mycorrhizal fungi in *Pisum sativum*. Mol. Plant Microbe Interact. 12, 785–791.

Barker S J and Tagu D 2000 The roles of auxins and cytokinins in mycorrhizal symbioses. J. Plant Growth Reg. 19, 144–154.

Barker S J, Stummer B, Gao L, Dispain I, O'Connor P J, and Smith S E 1998a A mutant in *Lycopersicon esculentum* Mill. with highly reduced VA mycorrhizal colonization: isolation and preliminary characterisation. Plant J. 15, 791–797.

Barker S J, Tagu D and Delp G 1998b Regulation of root and fungal morphogenesis in mycorrhizal symbioses. Plant Physiol. 116, 1201–1207.

Baron C and Zambryski P C 1995 The plant response in pathogenesis, symbiosis, and wounding: variations on a common theme? Annu. Rev. Genet. 29, 107–129.

Benabdellah K, Azcón-Aguilar C and Ferrol N 2000 Alterations in the plasma membrane polypeptide pattern of tomato roots (*Lycopersicon esculentum*) during the development of arbuscular mycorrhiza. J. Exp. Bot. 51, 747–754.

Bianciotto V and Bonfante P 1992 Quantification of the nuclear DNA content of two arbuscular mycorrhizal fungi. Mycol. Res. 96, 1071–1076.

Bianciotto V, Bandi C, Minerdi D, Sironi M, Tichy H V and Bonfante P 1996 An obligately endosymbiotic mycorrhizal fungus itself harbors obligately intracellular bacteria. Appl. Env. Microbiol. 62, 3005–3010.

Bird, D M and Wilson M A 1994 DNA sequence and expression analysis of root-knot nematode elicited giant cell transcripts. Mol. Plant Microbe Interact. 7, 419–424.

Blilou I, Ocampo J A and Garcia-Garrido J M 2000 induction of Ltp (lipid transfer protein) and Pal (phenylalanine ammonia-lyase) gene expression in rice roots colonized by the arbuscular mycorrhizal fungus Glomus mosseae. J. Exp. Bot. 51, 1969–1977.

Bonfante P and Perotto S 1995 Strategies of arbuscular mycorrhizal fungi when infecting host plants. New Phytol. 130, 3–21.

Bothe H, Klingner A, Kaldorf M, Schmitz O, Esch H, Hundeshagen B and Kernebeck H 1994 Biochemical approaches to the study of plant-fungal interactions in arbuscular mycorrhiza. Experientia 50, 919–925.

Burleigh S 2000 Cloning arbuscule-related genes from mycorrhizas. Plant Soil 226, 287–292.

Burleigh S H 2001 Relative quantitative RT-PCR to study the expression of plant nutrient transporters in arbuscular mycorrhizas. Plant Sci. 160, 899–904.

Buscot F, Munch J C, Charcosset J Y, Gardes M, Nehls U and Hampp R 2000 Recent advances in exploring physiology and biodiversity of ectomycorrhizas highlight the functioning of these symbioses in ecosystems. FEMS Microbiol. Rev. 24, 601–614.

Cairney J W G 1999 Intraspecific physiological variation: implications for understanding functional diversity in ectomycorrhizal fungi. Mycorrhiza 9, 125–135.

Cavagnaro T R, Gao L-L, Smith F A and Smith S E 2001 Morphology of arbuscular mycorrhizas is influenced by fungal identity. New Phytol. 151, 469–475.

Colebatch G, Trevaskis B and Udvardi M 2001 Functional genomics: tools of the trade. New Phytol. 153: 27–36.

Cronk Q C B 2001 Plant evolution and development in a post-genomic context. Nat. Rev. Gen. 2, 607–619.

Delp G, Rosewarne G and Barker S J 1998 The molecular route to understanding VAM symbiosis. *In* Mycorrhizae: A Modern Laboratory Manual. Ed. A Varma. pp 327–352. Springer, Berlin.

Delp G, Smith S E and Barker S J 2000 Isolation by differential display of three cDNAs coding for proteins from the VA mycorrhizal fungus *G. intraradices*, Mycol. Res. 104, 293–300.

Dodd J C, Boddington C L, Rodriguez A, Gonzalez-Chavez C and Mansur I 2000 Mycelium of arbuscular mycorrhizal fungi (AMF) from different genera: form, function and detection. Plant Soil 226, 131–151.

Duc G, Trouvelot A, Gianinazzi-Pearson V and Gianinazzi S 1989 First report of non-mycorrhizal plant mutants (Myc-) obtained in pea (*Pisum sativum* L.) and fababean (*Vicia faba* L.). Plant Sci. 60, 215–222.

Dumas-Gaudot E, Guillaume P, Tahiri-Alaoui A, Gianinazzi-Pearson V and Ganinazzi S 1994 Changes in polypeptide patterns in tobacco roots colonized by 2 *Glomus* species. Mycorrhiza 4, 215–221.

Ferrol N, Barea J M and Azcón-Aguilar C 2000 Molecular approaches to study plasma membrane H+-ATPases in arbuscular mycorrhizas. Plant Soil 226, 219–225.

Fitter A H and Moyersoen B 1996 Evolutionary trends in root-microbe symbioses. Phil. Trans. R. Soc. London B 351, 1367–1375.

Forbes P J, Millam S, Hooker J E and Harrier L A 1998 Transformation of the arbuscular mycorrhiza *Gigaspora rosea* by particle bombardment. Mycol. Res. 102, 497–501.

Franken P, Lapopin L, Meyer-Gauen G and Gianinazzi-Pearson V 1997 RNA accumulation and genes expressed in spores of the arbuscular mycorrhizal fungus, *Gigaspora rosea*. Mycologia 89, 293–297.

Frühling M, Roussel H, Gianinazzi-Pearson V, Puhler A and Perlick A M 1997 The *Vicia faba* leghemoglobin gene VFLB29 is induced in root nodules and in roots colonized by the arbuscular mycorrhizal fungus *Glomus fasciculatum*. Mol. Plant Microbe Interact. 10, 124–131.

Gao L-L, Delp G and Smith S E 2001 Colonization patterns in a mycorrhizal-defective mutant tomato vary with different arbuscular-mycorrhizal fungi. New Phytol. 151; 477–491.

Gianinazzi-Pearson V 1996 Plant cell responses to arbuscular mycorrhizal fungi: getting to the roots of the symbiosis. Plant Cell 8, 1871–1883.

Gianiniazzi-Pearson V, Dumas-Gaudot E, Gollotte A, Tahiri-Alaoui A and Gianinazzi S 1996. Cellular and molecular defence-related responses to invasion by arbuscular mycorrhizal fungi. New Phytol. 133: 45–57.

Goldberg R B 1986 Regulation of plant gene expression. Phil. Trans. R. Soc. London B 314, 343–353.

Grierson D and Covey S N 1988 Plant Molecular Biology. 2nd Ed. Chapman and Hall, Glasgow, UK Ch 1, pp 1–21.

Harrier L A 2001 The arbuscular mycorrhizal symbiosis: a molecular review of the fungal dimension. J. Exp. Bot. 52, 469–478.

Harrison M J 1996 A sugar transporter from *Medicago truncatula* — altered expression pattern in roots during vesicular–arbuscular (VA) mycorrhizal associations. Plant J. 9, 491–503.

Harrison M J 1999a Molecular and cellular aspects of the arbuscular mycorrhizal symbiosis. Annu. Rev. Plant Physiol. Plant Mol. Biol. 50, 361–389.

Harrison M J 1999b Biotrophic interfaces and nutrient transport in plant/fungal symbioses. J. Exp. Bot. 50, 1013–1022.

Harrison M J and van Buuren M L 1995 A phosphate transporter from the mycorrhizal fungus *Glomus versiforme*. Nature 378, 626–629.

Hirsch A M and Kapulnik Y 1998 Signal transduction pathways in mycorrhizal associations: Comparisons with the *Rhizobium*–legume symbiosis. Fungal Genet. Biol. 23, 205–212.

Klironomos J N and Kendrick W B 1993 Research on mycorrhizas — Trends in the past 40 years as expressed in the MYCOLIT database. New Phytol. 125, 595–600.

Koltai H, Dhandaydham M, Opperman C, Thomas J and Bird D 2001 Overlapping plant signal transduction pathways induced by a parasitic nematode and a rhizobial endosymbiont. Mol. Plant Micribol. Interact. 14, 1168–1177.

Lambais M R and Mehdy M C 1996 Soybean roots infected by *Glomus intraradices* strains differing in infectivity exhibit differential chitinase and beta-1,3-glucanase expression. New Phytol. 134, 531–538.

Lange J, Xie Z-P, Broughton W J, Vogeli-Lange R and Boller T 1999 A gene encoding a receptor-like protein kinase in the roots of common bean is differentially regulated in response to pathogens, symbionts and nodulation factors. Plant Sci. 142, 133–145.

Mandoli D F and Olmstead R 2000 The importance of emerging model systems in plant biology. J. Plant Growth Regul. 19, 249–252.

Marsh J F and Schultze M 2001 Analysis of arbuscular mycorrhizas using symbiosis-defective plant mutants. New Phytol. 150, 525–532.

Martin F 2001 Frontiers in molecular mycorrhizal research — genes, loci, dots and spins. New Phytol. 150, 499–507.

Marx J 2000 Interfering with gene expression. Science 288, 1370–1372.

McLean C B, Cunnington J H and Lawrie A C 1999 Molecular diversity within and between ericoid endophytes from the Ericaceae and Epacridaceae. New Phytol. 144, 351–358.

Minerdi D, Fani R, Gallo R, Boarino A and Bonfante P 2001 Nitrogen fixation genes in an endosymbiotic *Burkholderia* strain. Appl. Env. Microbiol. 67, 725–732.

Murphy P J, Langridge P and Smith S E 1997 Cloning plant genes differentially expressed during colonisation of roots of *Hordeum vulgare* by the vesicular–arbuscular mycorrhizal fungu *Glomus intraradices*. New Phytol. 135, 291–301.

Peterson R L and Guinel F C 2000 The use of plant mutants to study regulation of colonization by AM fungi. *In* Arbuscular Mycorrhizas: Physiology and Function. Eds. Y Kapulnik and D D Douds Jr. pp 147–171. Kluwer Academic Publishers, Dordrecht.

Rausch C, Daram P, Brunner S, Jansa J, Lalol M, Leggewie G, Amrhein N and Bucher M 2001 A phosphate transporter expressed in arbuscule-containing cells in potato. Nature 414, 462–466.

Read D J, Duckett J G, Francis R, Ligrone R and Russell A 2000 Symbiotic fungal associations in 'lower' land plants. Phil. Trans. R. Soc. London B 355, 815–831.

Redecker D 2000 Specific PCR primers to identify arbuscular mycorrhizal fungi within colonized roots. Mycorrhiza 10, 73–80.

Redecker D, Kodner R and Graham L E 2000 Glomalean fungi from the Ordovician. Science 289, 1920–1921.

Remy W, Taylor TN, Hass H and Kerp H 1994 Four hundred-million-year-old vesicular arbuscular mycorrhizas. Proc. Natl. Acad. Sci. USA 91, 11841–11843.

Rosendahl S and Taylor JW 1997 Development of multiple genetic markers for studies of genetic variation in arbuscular mycorrhizal fungi using AFLPþ. Mol. Ecol. 6, 821–829.

Rosewarne G M 1998 Molecular characterisation of the vesicular-arbuscular mycorrhizal symbiosis in *Lycopersicon esculentum* Mill. Ph.D. Thesis. The University of Adelaide, South Australia.

Rosewarne G, Barker S J and Smith S E 1997 Production of near synchronous fungal colonisation in tomato for developmental and molecular analyses of mycorrhiza. Mycol. Res. 101, 966–970.

Rosewarne G M, Barker S J, Smith S E, Smith F A and Schachtman D P 1999 A *Lycopersicon esculentum* phosphate transporter (*LePT1*) involved in phosphorus uptake from a vesicular–arbuscular mycorrhizal fungus. New Phytol. 144, 507–516.

Ruiz-Lozano J M, Gianinazzi S and Gianinazzi-Pearson V 1999a Genes involved in resistance to powdery mildew in barley differentially modulate root colonization by the mycorrhizal fungus *Glomus mosseae*. Mycorrhiza 9, 237–240.

Ruiz-Lozano J M, Roussel H, Gianinazzi S and Gianinazzi-Pearson V 1999b Defense genes are differentially induced by a mycorrhizal fungus and *Rhizobium* sp in wild-type and symbiosis-defective pea genotypes. Mol. Plant Microbe Interact. 12, 976–984.

Salzer P, Bonanomi A, Beyer K, Vogeli-Lange R, Aeschbacher R A, Lange J, Wiemken A, Kim D, Cook D R and Boller T 2000 Differential expression of eight chitinase genes in *Medicago truncatula* roots during mycorrhiza formation, nodulation, and pathogen infection. Mol. Plant Microb. Interact. 13, 763–777.

Sanders I R, Clapp J P and Wiemken A 1996 The genetic diversity of arbuscular mycorrhizal fungi in natural ecosystems — a key to understanding the ecology and functioning of the mycorrhizal symbiosis. New Phytol. 133, 123–134.

Selosse M A and Le Tacon F 1998 The land flora: a phototroph-fungus partnership? Trends Ecol. Evol. 13, 15–20.

Sen R, Hietala A M and Zelmer C D 1999 Common anastomosis and internal transcribed spacer RFLP groupings in binucleate *Rhizoctonia* isolates representing root endophytes of *Pinus sylvestris*, *Ceratorhiza* spp. from orchid mycorrhizas and a phytopathogenic anastomosis group. New Phytol. 144, 331–341.

Senoo K, Solaiman M Z, Kawaguchi M, Imaizumi-Anraku H, Akao S, Tanaka A and Obata H 2000 Isolation of two different phenotypes of mycorrhizal mutants in the model legume plant *Lotus japonicus* after EMS-treatment. Plant Cell Physiol. 41, 726–732.

Shaul O, Galili S, Volpin H, Ginzberg I, Elad Y, Chet I and Kapulnik Y 1999 Mycorrhiza-induced changes in disease severity and PR protein expression in tobacco leaves. Mol. Plant Microbe Interact. 12, 1000–1007.

Simon L 1996 Phylogeny of the Glomales: deciphering the past to understand the present. New Phytol. 133, 95–101.

Smith S E and Gianinazzi-Pearson P 1988 Physiological interactions between symbionts in vesicular–arbuscular mycorrhizal plants. Annu. Rev. Plant Physiol. Plant Mol. Biol. 39, 221–244.

Smith S E and Read D J 1997 Mycorrhizal Symbiosis. 2nd Edn, Academic Press, London. Stougaard J 2001 Genetics and genomics of root symbiosis. Curr. Opin. Plant Biol. 4, 328–335.

Tagu D and Barker S J 1997 At the root of mycorrhizal symbioses. Trends Plant Sci. 2, 2-3. The Angiosperm Phylogeny Group 1998 An ordinal classification for the families of flowering plants. Ann. Missouri Bot. Gard. 85, 531–553.

Tahiri-Alaoui A and Antoniw J F 1996 Cloning of genes associated with the colonization of tomato roots by the arbuscular mycorrhizal fungus *Glomus mosseae*. Agronomie 16, 699–707.

Vallorani L, Bernardini F, Sacconi C, Pierleoni R, Pieretti B, Piccoli G, Buffalini M and Stocchi V 2000 Identification of *Tuber borchii* Vittad. mycelium proteins separated by two-dimensional polyacrylamide gel electrophoresis using amino acid analysis and sequence tagging. Electrophoresis 21, 3710–3716.

van Tuinen D, Jacquot E, Zhao B, Gollotte A and Gianinazzi-Pearson V 1998 Characterization of root colonization profiles by a microcosm community of arbuscular mycorrhizal fungi using 25S rDNA-targeted nested PCR. Mol. Ecol. 7, 879–887.

Vierheilig H, Alt M, Neuhaus J-M, Boller T and Wiemken A 1993 Colonization of transgenic *Nicotiana sylvestris* plants, expressing different forms of *Nicotiana tabacum* chitinase, by the root pathogen *Rhizoctonia solani* and by the mycorrhizal symbiont *Glomus mosseae*. Mol. Plant Microb. Int. 6, 261–264.

Voiblet C, Duplessis S, Encelot N and Martin F 2001 Identification of symbiosis-regulated genes in *Eucalyptus globulus–Pisolithus tinctorius* ectomycorrhiza by differential hybridisation of arrayed cDNAs. Plant J. 25, 181–191.

Plant and Soil **244:** 117–128, 2002.

Fungal gene expression in early symbiotic interactions between *Laccaria bicolor* and red pine

G.K. Podila[1], J. Zheng S. Balasubramanian, S. Sundaram, S. Hiremath, J.H. Brand & M.J. Hymes
Department of Biological Sciences, Michigan Tech University, 1400 Townsend Drive, Houghton, MI 49931, USA
[1]*Corresponding author**

Received 21 August 2001. Accepted in revised form 18 January 2002

Key words: Ectomycorrhiza interaction specific genes, *Laccaria bicolor*, *Pinus resinosa*, suppression subtractive hybridization screening

Abstract

Ectomycorrhizas are mutualistic symbiotic organs formed by interaction between plant roots and fungi. Mycorrhizal initiation, development and functional maintenance involve morphological changes that are mediated by activation and suppression of several fungal and plant genes. During the pre-infection stage, a harmonized cross-talk takes place between the symbionts, to determine their compatibility. Upon mutual recognition, the symbionts initiate further physiological and morphological changes essential for the formation of the symbiotic organ. In order to understand the molecular mechanisms underlying these events, we developed an interaction-specific cDNA library from *Laccaria bicolor* that represents fungal genes regulated by its interaction with *Pinus resinosa* roots. Membrane array analyses of these cDNAs suggested that a wide variety of genes are involved in the pre-infection stage processes.

Introduction

Ectomycorrhizas are formed from symbiotic associations between plant roots and certain soil fungi known as ectomycorrhizal fungi. They are complex organs consisting of both plant root and fungal mycelial tissues. A vast majority of vascular plants, including a variety of conifers, exhibit such symbiotic relationships with ectomycorrhizal fungi that colonize their roots (Allen, 1992; Harley and Smith, 1983). Extensive nutrient exchange takes place between the plant and fungal partners in the ectomycorrhizal organ (Smith and Read, 1997). The ectomycorrhizal phenomenon, essential for health and survival of many conifers and hardwood species, has been studied widely from ecological and physiological perspectives. However, to optimize the benefits of this natural phenomenon, it is essential to gain a complete understanding of its formation and functioning at the molecular level.

The formation of the ectomycorrhizal organ can be broadly divided into four stages: pre-infection, initiation, differentiation, and functioning (Martin et al., 1997). The amount of time required to form functional ecomycorrhiza varies. In the *Laccaria bicolor – Pinus resinosa* interaction, this process takes approximately eight weeks from the time of initial exposure, under sterile conditions. In the case of the Petri-dish technique of raising *Pisolithus tinctorius – Eucalyptus globus* ectomycorrhiza, active mycorrhizae are formed within two weeks (Burgess and Dell, 1996). In the pre-infection stage, the plant root and a potential fungal partner exchange signals in the soil leading to the determination of compatibility for symbiosis. Pre-infection stage exchanges are mediated by a variety of molecular, biochemical and morphogenetic events and play a critical role towards the final establishment of the functional mycorrhizas (Martin et al., 2001;

* FAX No: 1-906-487-3167. E-mail: gkpodila@mtu.edu

Sundaram et al., 2001). Once 'mutually-recognized', the root and the mycelial tissues start growing towards each other and finally establish physical contact with each other (initiation). Thereafter, the hyphae gradually grow and differentiate within the root cortex (differentiation) forming the functional mycorrhiza (functioning) (Hampp et al., 1995; Martin and Botton, 1993; Martin et al., 1997). The dynamic events during the pre-infection stage, before the physical contact is established, therefore, are crucial in determining the direction and ultimate success of the formation of a symbiotic association.

The formation of the symbiotic organ involves a myriad of molecular events that form the basis for a coordinated regulation of morphological, biochemical and physiological changes in both the partners (Martin et al., 2001; Tagu et al., 2000; Voiblet et al., 2001). While many elegant molecular genetic studies have been conducted on the later stages of ectomycorrhiza development (Burgess et al., 1995; Hilbert et al., 1991; Laurent et al., 1999; Martin et al., 1997, 1999; Nehls et al., 1999; Voiblet et al., 2001), only limited information is available about the very early interaction events. Although it is apparent that signaling genes are involved in the early stages of interaction, it is not known what other types of genes may be regulated during the pre-infection stage. We have been pursuing studies focused on understanding a variety of the critical events taking place during the pre-infection stage, especially from the fungal point of view. We chose the fungus *L. bicolor* because of its aggressive nature in forming ectomycorrhizal symbiotic relationship with a wide variety of tree species in North America (Harley and Smith, 1983; Richter and Bruhn, 1989). The ectomycorrhiza formation of this fungus has been shown to be highly beneficial for the survival and health of its hosts, which include many conifer and hardwood tree species (Cripps and Miller, 1993; Kropp, 1990; Richter and Bruhn, 1989).

Investigation of temporal regulation of gene activity during ectomycorrhiza development necessitates a model system where it is possible to determine fungal gene expression in different environments. We have developed a model *in vitro* system to isolate and analyze genes from the ectomycorrhizal fungus *Laccaria bicolor* during its symbiotic interactions with *Pinus resinosa* (red pine) (Kim et al., 1998, 1999a, b; Sundaram et al., 2001). Our *in vitro* system (Kim et al., 1998) facilitates this, where gene expression in the ectomycorrhizal fungus can be studied during (i) the free-living stage, (ii) the development of ectomycorrhiza, and (iii) in established ectomycorrhiza. In the past, we used this system and classical methods such as differential display reverse transcription-PCR (DDRT-PCR) (Liang and Pardee, 1992) to identify and characterize several interaction specific genes in *L. bicolor* (Kim et al., 1998; Sundaram et al., 2001). However, a highly complex system like ectomycorrhiza requires a broad-based analysis tool in order to identify and monitor a large number of genes that are apparently involved in the process of formation and functioning of the ectomycorrhiza. Microarray technology and bioinformatics have demonstrated that they have the potential and capability for characterizing such complex systems (Gyorgyey et al., 2000; Schena et al., 1996; Wodicka et al., 1997).

The aim of these experiments is to obtain a broad overview on various genes involved in the pre-infection stages of the ectomycorrhiza formation, in the fungal partner. We report here identification and regulation of several genes from *L. bicolor* that potentially play a role in the pre-infection stage during its interaction with red pine. An interaction-specific (IS) cDNA library was generated using subtractive hybridization, and differential expression of these cDNAs was investigated by membrane array analyses. Our ultimate goal is to apply the microarray technology to understand the molecular mechanisms of recognition and development processes taking place during interactions between *L. bicolor* and pine roots. However, as a preliminary step towards this goal, we used the membrane array analysis to examine global changes happening in the fungal gene expression due to its interaction with plant roots.

Materials and methods

Media, cultures and Laccaria bicolor – Pinus resinosa *interactions*

The basidiocarp isolate *Laccaria bicolor* (dikaryotic) (Maire Orton) DR170 (obtained from the roots of red pine trees from Michigan's Upper Peninsula and provided by Dr. Dana Richter, School of Forestry, Michigan Technological University, Houghton, MI) was grown and maintained on MMN medium (Bills et al., 1995). *In vitro* interactions and harvesting of interacted fungal tissue have been described before (Kim et al., 1998, 1999b; Wong et al., 1990).

cDNA library construction

The cDNA library construction in lambda Screen phagemid vector (Novagen, WI) using the RNA from interaction tissue was as described before (Sundaram et al., 2001). PCR analysis of inserts of 50 random clones revealed that the average size of cDNA inserts was about 1.2 kbp; some clones had inserts larger than 4 kbp. Sequence analysis revealed that 96% of the clones contained the poly(A/T) tail.

Probe from suppression subtractive hybridization and screening

The SMART-PCR cDNA Synthesis Kit (Clontech, CA) was used to obtain doubled-stranded cDNAs representing mRNAs expressed in *L. bicolor* during (i) its interaction with red pine for 6–72 h (tester), and (ii) free-living condition (driver). Following the manufacturer's instructions (PCR-Select cDNA Subtraction Kit, Clontech) based on the procedure described by Diatchenko et al. (1996, 1999), the tester cDNA pool was subtracted twice by hybridization to an excess of the driver cDNA. The resulting cDNAs, enriched in IS sequences, were labeled with ^{32}P-dCTP by random priming (Rediprime kit, Amersham Corp, NJ) and used as a probe to screen the interaction cDNA library described above (Ausubel et al., 1998). Positive clones were plaque-purified and converted into plasmid clones through automatic cre-lox excision (Novagen, WI). Plasmid DNA from each of these clones was prepared (QIAvac 24 kit; Qiagen, CA) and used to transform DH5α *E.coli* cells for further use.

Sequencing and analysis

Plasmid DNA purified using the Qiagen QIAvac24 Miniprep kit (Qiagen, CA) was subjected to nucleotide sequencing using the Big Dye Terminator Cycle Sequencing Ready Reaction kit (PE Applied Biosystems, CA) and the ABI PRISM 310 genetic Analyzer (PE Applied Biosystems, CA). The sequence output, after scanning visually to confirm all the base calls, was edited using the SEQUENCE ANALYSIS (version 3.3) (PE Biosystems) and the MacDNAsis software version 3.5 (Hitachi Corp., CA). Leading and trailing vector and polylinker sequences as well as sequences having more than 3% ambiguous base calls were removed by filters. Sequences shorter than 200 bp were not included in further analyses. Homology comparisons with the GenBank database (non-redundant nucleic acid sequences) at the National Center for Biotechnology Information (NCBI, NIH, USA) were performed by batch processing using the BLASTX. In most instances, amino acid sequences with E values $\leq 10^{-5}$ were considered to specify sufficient homology and to represent *L. bicolor* homologues of genes known in other systems. However, the significance of the homology was manually determined by scrutinizing the BLASTX matches between the query and the candidate sequences. The edited EST sequences were deposited in dbEST at NCBI (accession #s BI 094571- BI 094677). Functional categories were assigned according to the classification scheme utilized by the Expressed Gene Anatomy Database available at WWW.TIGR.ORG.

Membrane array analyses

Preparation of membrane arrays

The cDNA insert from each of the IS library clones was amplified through standard PCR using the flanking T7 and SP6 primers. The products were cleaned using the QIAquick PCR Purification Kit (Qiagen, CA) and quantitated using the DNA Quick Test system (Geno-Tech, St. Louis, MO). After preparing DNA samples for spotting onto the nylon membrane as described (Schummer et al., 1997), they were bound to a Hybond N+ nylon membrane (Pharmacia, NJ) in a grid pattern using the Vicki Ultra High Density Array system (VP-Scientific, CA). Duplicate pattern of cDNA clones were spotted to confirm that the hybridization signals were consistent for each clone. Control DNAs including polylinker from the lambda screen vector and positive controls, such as the symbiosis-regulated *L. bicolor* cDNAs PF6.2 and LB-AUT7 (Kim et al., 1998, 1999a, b), were included to determine the efficiency of hybridization as well as to calibrate the membrane arrays.

Preparation of exponential cDNA probes from RNA and hybridization were performed according to Gonzalez et al. (1999). Briefly, 0.5 μg RNA was reverse transcribed (RT) using the SMART PCR cDNA synthesis system (Clontech, CA) and Superscript II RNaseH^{-} Reverse Transcriptase (Life Technologies, MD). Two μl of sscDNA from the RT reactions was used for amplification using the Qiagen HotStar Taq polymerase (Qiagen, CA). The dscDNAs obtained were purified and labeled with ^{32}P-dATP and ^{32}P-dCTP using the Ready-to-Go system (Amersham, IL). The probes, after purification from the reaction mixtures using the QIAquick Nucleotide Removal Kit

(Qiagen, CA), were checked for their specific activity using a scintillation counter.

In order to reduce background problems during hybridization, the buffer contained the poly(A)–poly(T) linker, DNA from the pScreen vector and salmon sperm DNA. However, washing conditions were kept to a moderate stringency level (two times in 2X SSC, 1% SDS at 65 °C/ 20 min each) to accommodate variation in target size, and to retain all the hybrids formed at individual sequence's DNA COT values. After washing, the membranes were subjected to autoradiography, and the autoradiograms were scanned using a high-resolution scanner at 1200×1200 dpi, with each spot normalized and quantitated using the ImageTool 2.0 software (UTHS, San Antonio, TX). Clones were selected as true positives when their signals from duplicate membranes matched (Jordan 1998). Signal intensities for each clone were determined by subtracting background from the signal value. The expression values are normalized based on three replicate blots, and checking the expression signals against internal controls such as PF6.2 which we have already well characterized. After normalization of the signals, the data from the ImageTool software was used to create a scatter plot in Microsoft Excel (Microsoft, WA). The plot compared normalized values from free-living fungus versus those from the interaction signal.

Northern analysis

The total RNA was isolated from *L. bicolor* before (free-living state) and after interaction with red pine roots for various time points as described (Kim et al., 1998, 1999a). The cDNA inserts of selected IS clones (LbSSH # 00007, BiP; LbSSH # 00013, PF6.2) were purified by agarose gel electrophoresis followed by extraction of DNA using the QIAEX gel extraction kit (Qiagen Inc., CA). Approximately 25 ng of cDNA was labeled using the REDIPRIME labeling kit and ^{32}P-dCTP (3000 Ci/mmol, Amersham Corp, NJ) and employed as a probe. Prehybridization, hybridization and washings were done as described before (Kim et al., 1998).

Results

Isolation and characterization of the IS library

Our aim is to understand the changes in IS gene expression. We enriched for IS genes, first by using interacted fungal mycelium (6 – 72 h) to generate the cDNA library and then using SSH to generate IS probe as described in 'Materials and methods'. This method identified 121 library clones of which only 98 appeared potentially significant for further analysis. The remaining clones either had ambiguities and other sequencing errors, or were very short with no identifiable ORFs of significant size. A list of these 98 clones and the degree of their sequence similarity to known sequences in the Genbank is given in Table 1. The IS clones with E values of $\leq 10^{-5}$at the protein level were adjudged to specify adequate identity to known entries in the GenBank, since this level of homology is routinely considered significant for various EST projects. However, every BLASTX match was manually scrutinized to confirm that the matching regions spanned significant stretches of amino acids. These 98 clones were designated as the IS library.

Based on the BLAST analysis (Altschul et al., 1997), 33% of the 98 clones showed no significant similarity to any sequences in the NCBI database. The sequences of the remaining 67%, however, suggested that they were homologues of genes previously identified in other systems. These were classified into groups based on their probable function. About 13% were related to signal transduction, 15% to metabolism, 10% to cellular protein synthesis/processing and turnover, 9% to transport and movement of ions/peptides and amino acids, 7% to structural proteins, 5% to RNA/DNA processing, 6% to transcriptional regulation, and about 10% to hypothetical proteins with no known function. In addition, one clone sequence suggested it was related to apoptosis.

Identity of these sequences was based on matches with sequences in the database, the degree of which varied among different sequences. About 25% of them showed strong similarity (up to 80%) while 55% of them exhibited moderate similarity (from 30 to 45%). For the remaining, only small patches of similarities were observed.

As expected, the sequence characterization showed that the IS library also contained sequences corresponding to *L. bicolor* genes that were previously identified and characterized in detail. Sequences of five clones were identical to that of the PF6.2 gene (Kim et al., 1998, 1999a), three matched the *ras* related proteins (Kim et al., 1999a; Sundaram et al., 2001), and two matched with the peroxisomal β-keto thiolase (Zheng et al., unpublished). In addition, there was one member each related to the peroxisomal malate synthase (Balasubramanian et al., 2002), the Chitin

Table 1. The *L. bicolor* interaction specific (IS) library clones selected by screening with subtraction hybridization derived probes. Their nucleotide sequences were determined and deposited in the GenBank (accession numbers given). Searches using the BLASTX indicated varying degrees of similarity (*E* values) to genes known in other systems. The figures to the right indicate the increase in their expression (*E* ratio) in response to interaction with the red pine roots. The *E* ratio was calculated by comparing the expression levels in the interacted tissue with those in the free-living fungus (see 'Results and discussion').

L. bicolor clone ID	GenBank Accession no.	*E* value	Database match	Expression ratio
LbSSH00002	BI094571		No match	6.01
LbSSH00003	BI094572	2e-6	a1A Voltage-dependent Calcium channel (*H. sapiens*)	3.21
LbSSH00005	BI094574		No match	3.20
LbSSH00006	BI094575	1e-6	Profilaggrin, human	2.12
LbSSH00007	BI094576	3e-69	BiP protein (*Aspergillus nidulans*)	4.10
LbSSH00008	BI094577		No match	2.60
LbSSH00009	BI094578	1e-5	Unknown protein, Human (Image 3627860)	1.59
LbSSH00010	BI094579	1e-09	3-GPD (*Agaricus bisporus*)	3.23
LbSSH00012	BI094581	2e-6	*Gallus gallus* ssDNA binding protein	1.34
LbSSH00013	BI094582	2e-69	PF6.2.1 (*Laccaria bicolor*)	8.00
LbSSH00014	BI094583	2e-48	α-tubulin (*Ustilago maydis*)	4.51
LbSSH00015	BI094584		No match	3.02
LbSSH00016	BI094585	9e-13	40S ribosomal protein (*Saccharomyces cerevisiae*)	4.81
LbSSH00017	BI094586		No match	2.53
LbSSH00018	BI094587	1e-10	Homeo box genes Hox-2.6 (*Mus musculus*)	3.30
LbSSH00019	BI094588	1e-12	Glutathione peroxidase (*Arabidopsis thaliana*)	1.62
LbSSH00021	BI094590	7e-11	RAS (*Aspergillus fumigatus*)	2.91
LbSSH00022	BI094591	1e-10	Unknown protein (*Drosophila melanogaster* CG2962)	1.24
LbSSH00023	BI094592	2e-10	PEP carboxykinae (*Mus musculus*)	4.33
LbSSH00024	BI094593		No match	1.01
LbSSH00025	BI094594		No match	2.42
LbSSH00026	BI094595		No match	1.21
LbSSH00028	BI094597	1e-10	p300/CBP (*Xenopus laevis*)	2.86
LbSSH00029	BI094598		No match	1.11
LbSSH00031	BI094599	1e-06	Receptor kinase like protein (*A. thaliana*)	1.06
LbSSH00033	BI094601	2e-63	LB-AUT7 (*L. bicolor*)	3.73
LbSSH00035	BI094603		No match	1.10
LbSSH00036	BI094604	1e-6	Mg2+ chelatase (*Glycine max*)	2.21
LbSSH00037	BI094605	4e-10	TBP like CYP450 (*Nicotiana tabacum*)	2.62
LbSSH00038	BI094606	3e-09	β-importin (*Schizosaccharomyces pombe*)	4.43
LbSSH00039	BI094607	7e-16	Ferrodoxin (S. *cerevisiae*)	1.81
LbSSH00042	BI094610		No match	1.02
LbSSH00043	BI094611	1e-08	AcylCoA oxidase (*Hordeum vulgare*)	3.61
LbSSH00044	BI094612	2e-35	Malate synthase (*L. bicolor*, *Pichia pastoris*)	3.82
LbSSH00045	BI094613	1e-06	DnaJ protein (peanut witches broom Phytoplasma)	2.91
LbSSH00046	BI094614		No match	1.22
LbSSH00047	BI094615	9e-36	TEF (EF1a) (*Schizophyllum commune*)	3.61
LbSSH00051	BI094619	1e-06	IRS 1-like protein (*Xenopus laevis*)	3.13
LbSSH00053	BI094621	1e-31	Ras related protein (*L. bicolor*)	3.81
LbSSH00054	BI094622	9e-25	Aryl-alcohol dehydrogenase (*Phanerochaete chrysosporium*)	2.21
LbSSH00055	BI094623	1e-06	LZK protein kinase (*Homo sapiens*)	2.90
LbSSH00057	BI094625		No match	2.51
LbSSH00058	BI094626	3e-14	Glutathione S-transferase III (*Naegleria fowleri*)	2.61
LbSSH00059	BI094627	1e-08	Deoxyhypusine synthase (*Neurospora crassa*)	4.83
LbSSH00060	BI094628	1e-08	Brain seleno protein (*H. sapiens*)	2.61

Continued on p. 122

Table 1. contd.

L. bicolor clone ID	GenBank Accession no.	*E* value	Database match	Expression ratio
LbSSH00061	BI094629	7e-13	Lactonohydrolase (*Fusarium oxysporum*)	3.91
LbSSH00062	BI094630	1e-06	Homeobox Hox-b4 gene (*Mus musculus*)	2.14
LbSSH00063	BI094631		No match	1.94
LbSSH00064	BI094632	9e-07	E-MAP-115 (*H. sapiens*)	4.21
LbSSH00065	BI094633	5e-32	Disulfide isomerase (*Humicola insolens*)	3.82
LbSSH00066	BI094634	1e-07	S-receptor kinase (*A. thaliana*)	2.41
LbSSH00067	BI094635	1e-26	SUG1 26S protease regulatory subunit 8 (*S. cerevisiae*)	4.61
LbSSH00068	BI094636	1e-06	PCAR apoptotic-related (*H. sapiens*)	1.36
LbSSH00069	BI094637		No match	1.32
LbSSH00072	BI094638	2e-06	Teashirt patterning gene (*D. melanogaster*)	2.51
LbSSH00073	BI094639	1e-06	Septin Spn3 (*S. pombe*)	3.92
LbSSH00074	BI094640		No match	2.52
LbSSH00079	BI094641	1e-06	Probable membrane protein YOL063c (*S. cerevisiae*)	1.81
LbSSH00080	BI094642		No match	1.21
LbSSH00081	BI094643		acid sensitive K^+channel (*Cavia porcellus*)	2.32
LbSSH00082	BI094644		No match	1.64
LbSSH00083	BI094645		No match	1.06
LbSSH00085	BI094646		No match	1.13
LbSSH00086	BI094647	7e-07	17.5 KD hypothetical protein (*S. cerevisiae*)	3.11
LbSSH00092	BI094649	1e-07	Peptide transport protein (*A. thaliana*)	2.80
LbSSH00093	BI094650		No match	2.25
LbSSH00094	BI094651		No match	1.71
LbSSH00095	BI094652		No match	1.03
LbSSH00096	BI094653	4e-18	Ras-related protein (*L. bicolor*)	3.21
LbSSH00097	BI094654		No match	1.22
LbSSH00098	BI094655		No match	1.01
LbSSH00099	BI094656	1e-09	Squalene monooxygenase (*Mus musculus*)	2.81
LbSSH00100	BI094657	1e-06	Clathrin adapter protein (*A. thaliana*)	2.08
LbSSH00101	BI094658	5e-06	Glutamate transporter (*Caenorhabditis elegans*)	2.13
LbSSH00102	BI094659		No match	2.92
LbSSH00103	BI094660	6e-21	AcetylCoA acetyltransferase (*L. bicolor*)	4.25
LbSSH00104	BI094661	1e-06	Zn transporter (*Mus musculus*)	2.31
LbSSH00105	BI094662	1e-06	Ran interacting Giant nulceopore protein (*H. sapiens*)	2.93
LbSSH00106	BI094663		No match	1.07
LbSSH00107	BI094664	1e-06	You-too homeotic gene (*Danio rerio*)	2.01
LbSSH00108	BI094665		No match	1.22
LbSSH00109	BI094666	3e-14	β-keto thiolase-1 (*L. bicolor*)	4.26
LbSSH00110	BI094667	1e-06	β-transducin (*S. pombe*)	4.11
LbSSH00111	BI094668	1e-06	Putative Ca^{2+}transporting ATPase (*Allomyces capsulatus*)	3.28
LbSSH00112	BI094669		No match	1.81
LbSSH00113	BI094670		No match	4.40
LbSSH00114	BI094671	1e-07	Unknown protein F27G20-7 (*A. thaliana*)	1.42
LbSSH00116	BI094672	2e-10	G3PD (*D. melanogaster*)	3.06
LbSSH00118	BI094674	1e-08	Cystathione beta synthase (*H. sapiens*)	2.65
LbSSH00119	BI094675		No match	2.22
LbSSH00120	BI094676	2e-37	Chitin synthase I (*U. maydis*)	3.71
LbSSH00121	BI094677		No match	2.04

synthase (unpublished), the BiP protein (Kim et al., 1999a), and an autophagosis-related protein (Kim et al., 1999b). All the *L. bicolor* genes mentioned above were shown to be differentially expressed either by direct northern analyses or by DDRT-PCR (Kim et al., 1999a, b).

Membrane array and northern analyses

We examined quantitative changes in the expression of the IS library genes using the membrane array technique. An *E* ratio, that indicates the relative increase in the expression of each gene in the interaction over the free-living state, is used to quantitate differential expression. There is an overall increase in levels of expression of clones tested. A third of the clones tested show a marginal difference in expression levels. The highest differential expression (*E* ratio 8) was observed for the PF6.2 (LbSSH00013) (Table 1). This was further confirmed by northern analysis where RNA samples from tissues that had interacted for 6–72 h were probed with cloned PF6.2 sequences. (Figure 2, panel PF6.2). The second highest increase in transcript level (6-fold increase) was for the clone LbSSH0001, the identity of which is not known since its sequence did not correspond to any known sequences in the Genbank. Similarly, there was a significant increase (4-fold) in the expression of the BiP protein (LbSSH0007), previously identified as an early interaction regulated gene (Kim et al., 1999a). Further analysis for this gene by northern analysis (Figure 2, panel BiP) showed increased expression during the first 6–24 h after interaction, and thereafter its levels went down gradually. Levels of expression of other genes are listed in Table 1. The scatter plot of the normalized data from signal analysis of the membranes is presented in Figure 1 which shows global changes in the expression of IS genes.

Discussion

The ectomycorrhizal symbiosis is a complex and modular process during which a new complex organ is formed by a well-coordinated interaction between the plant roots and compatible fungi (Martin and Tagu, 1999; Smith and Read, 1997). The pre-infection stage, during which the symbionts mutually recognize each other through a 'cross-talk', plays a crucial role in the successful formation of this organ. This stage therefore, is very important in determining the direction each of the partners will pursue. The 'cross-talk' is mediated by drastic changes in genetic activity among the partners in response to each other.

Isolation and characterization of the IS library

We used subtractive hybridization to generate the IS probe as described in 'Materials and methods'. The use of large excess of driver cDNA in the SSH significantly reduced the pool of common sequences, such as those of housekeeping genes, in the tester cDNA. The resultant cDNA preparation, used for preparing the probe, was enriched in the IS sequences, which included transcripts of fungal genes *de novo* induced due to interaction as well as genes whose activity was increased substantially after interaction. However, the subtractive hybridization technique cannot remove all the transcripts common to both phases of fungal growth. This is especially true in the case of genes that are expressed in high abundance in the interaction phase. Therefore, some common sequences were still represented in the IS probes. Also, genes that are down regulated by the interaction, escape detection by this technique. This latter group may include genes that are involved in eliciting defense responses in the plant host.

BLAST analysis (Altschul et al., 1997) suggested that 33% of the 98 clones showed no significant similarity to any sequences in the NCBI database, indicating that expression of these genes or the genes themselves might be unique to *L. bicolor*. It is possible that these represent very rare transcripts that have not been identified and/or characterized previously. To date, very few genes associated with mycorrhizal symbiosis have been isolated and characterized. The remaining 67% however, were identified in other systems. We discuss these genes by categories such as signal transduction, metabolism, protein synthesis and processing, developmental regulation and transport.

One of the largest group of genes in the *L. bicolor* IS library was related to signal transduction proteins (13%). The PF6.2, previously identified as related to signal transduction was also included in this group (Table 1, Figure 1). Other genes that appeared to be involved in early signaling included *ras* and *ras*-related genes, beta transducin genes, protein kinases, kinase receptors (LbSSH00021, LbSSH00031, LbSSH00051, LbSSH00053, LbSSH00066, LbSSH00096) and receptor substrate proteins (LbSSH00110) (Figure 1) (Liu et al., 1995). Certain genes that are in-

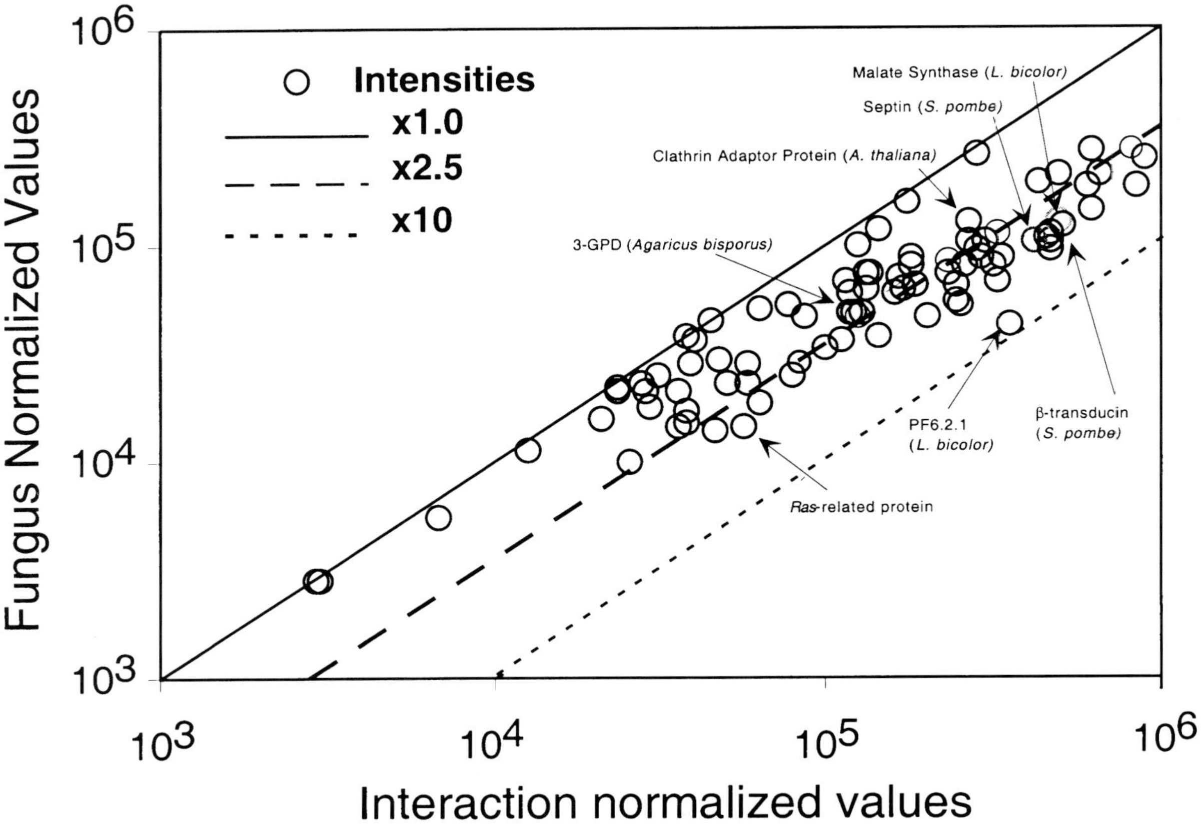

Figure 1. Expression profiling of interaction-specific genes between free-living *Laccaria bicolor* and *L. bicolor* interacted with *Pinus resinosa* seedling roots. For each gene, transcript levels were calculated for the free-living mycelium and the mycelium that has interacted with red pine seedling root signals. Solid lines indicate an expression level of 1 fold or above the free-living mycelium, dashed line 2.5–6 fold increase and dotted line 8 fold or higher levels of expression. Only non-redundant clones are represented in the plot. Selected clones that showed significant differential expression are highlighted.

volved in regulating ion fluxes during signaling were also present. These included genes related to the voltage dependant Ca^{2+} channel (LbSSH0003) and K^{+}channel protein (LbSSH00081) and Ca2+ ATPase (LbSSH00111). Presence of a large number of signaling genes in the IS library was expected since the 'cross-talk' between symbionts is apparently mediated by these genes.

The PF6.2 clone (LbSSH00013) was represented five times in the IS library. This was probably due to the relatively higher expression level of this gene as compared to others (see below 'Membrane array analysis'). The PF6.2 expression in *L. bicolor* is initiated within 6 h in response to signals from red pine roots and it continues to be expressed in the fully established ectomycorrhiza (Kim et al., 1998, 1999a). Because of its abundance and proven dependence on interaction for its expression, we chose this gene as a positive internal control for calibrating results from membrane array analyses.

The largest number of IS genes (15%) were related to the fungal metabolism. These included several genes involved in the peroxisomal beta-oxidation of lipids and gluconeogenesis such as acetyl CoA oxidase, acetylCoA acetyl transferase (b-keto thiolase), malate synthase (Figure 2) and PEP carboxykinase (LbSSH00024, LbSSH00043, LbSSH00044, LbSSH00103, LbSSH000109). We have recently cloned and characterized the malate synthase and acetylCoA acetyl transferase from *L. bicolor* (Balasubramanian et al., 2002; Zheng et al., 2000, unpublished). The data suggest that gluconeogenesis may play an important role in the pre-infection stage processes. In addition, two different clones corresponding to glyceraldehyde-3-P dehydrogenase (LbSSH00010, LbSSH00116), which are known to have functions in

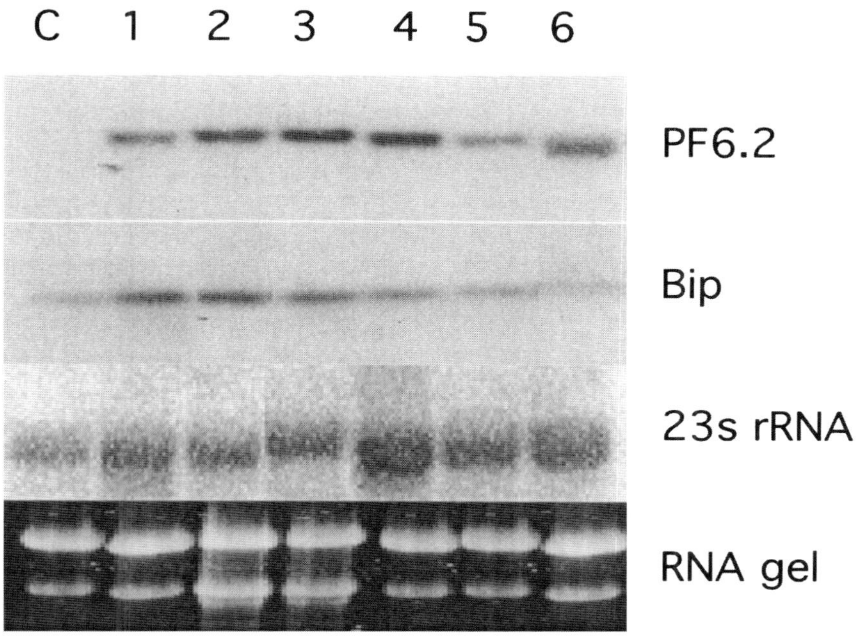

Figure 2. Pre-infection regulated expression of selected interaction-specific clones of *L. bicolor*. RNA was extracted from *L. bicolor* interacted with red pine roots for specified amounts of time. RNA from free-living *L. bicolor* was used as control. Total RNA (10 μg) was separated on a denaturing agarose gel and transferred to a nylon membrane. The blot was first hybridized to ^{32}P-labeled PF6.2 (LbSSH00013) or BiP sequence (LbSSH00007). After hybridization followed by autoradiography, the membrane was stripped and rehybridized to a *L. bicolor* 23S rRNA probe in order to correctly estimate the relative amounts of total RNA loaded in each lane. Lane C represents RNA from the free-living fungus used as a control; lanes 1–6 have RNA from *L. bicolor* that had interacted with red pine roots for 6, 12, 24, 36, 48, 72 h, respectively. The bottom panel shows ethidium bromide stained picture of the gel before transferring RNA to the nylon membrane.

primary metabolism, were also identified. Other prominent members of this group were the aryl-alcohol dehydrogenase (AAD) (LbSSH00054) (Reiser et al., 1994), the lactonohydrolase (LH) (LbSSH00061), the glutathione peroxidase (LbSSH00019), and the Glutathione S-transferase (LbSSH00058). The last four have not been identified from mycorrhizal fungi previously. It is possible that AAD may have a role in processing root exudates containing plant elicitors that might include aromatic aldehydes. The LH has been shown to be involved in hydrolyzing lactone-containing aldehydes and aromatics, which serve as signals or hormones in the microbial world (Kobayashi et al., 1998). The increased expression of LH during pre-infection suggests that it may play a role by maintaining the levels of lactone-containing aromatics released by the red pine roots. Further studies on temporal and spatial expression of AAD and LH would be helpful in determining their role, if any, in the signaling process during the pre-infection stage.

It was not surprising that several metabolic genes were represented in the IS library. Interaction with roots activates the relatively dormant fungus to undergo drastic changes that lead to its growth and proliferation. During this phase, the fungus needs to reorient its metabolic processes to those that will facilitate formation and functioning of the ectomycorrhiza. In the ectomycorrhizal organ, unlike in the free-living state, the fungus acquires carbon from the host plant and utilizes it for synthesizing other metabolites necessary for its growth and function. In addition, it initiates processes that are necessary to sequester phosphorus and nitrogen from the soil for transferring them to the host plant. These new symbiosis-related functions, therefore, require changes in the expression of metabolic genes.

We also observed significant modulation in the expression of: (i) several homeotic genes involved in the transcriptional regulation of development and patterning (Hox and the CBP genes; LbSSH00018, LbSSH00028, LbSSH00062, LbSSH00072, LbSSH00107), (ii) genes involved in the transport of proteins across the cell wall (LbAut7; LbSSH00033) and the nuclear membrane

(β-importins, Ran/GTP nucleoporins; LbSSH00038, LbSSH00106), (iii) several genes involved in protein synthesis, processing and transport, such as the elongation factor 1a (eIF1a)(LbSSH00047), deoxyhypusine synthase (DSH) (activates eIF5a; LbSSH00059), the BiP protein involved in protein folding (LbSSH0007), 40 S ribosomal proteins (LbSSH00016), the disulfide isomerase (LbSSH00065), and amino acid/ peptide transport and processing proteins (LbSSH00095, LbSSH00101, LbSSH00118). Increased expression of the protein synthesis/processing genes is expected since development of the ectomycorrhiza necessitates it. The homeotic genes identified were related to the transcriptional regulation of genes involved in pattern formation and development (Caubit et al., 2000; Graham et al., 1988; Karlstrom et al., 1999). This is consistent with the metamorphic changes that take place during the development of ectomycorrhiza, where a directed growth of the fungus towards the roots occurs in response to plant signals.

The structural genes identified in the IS library were related to the cytoskeleton, cell division, and cell wall biogenesis. These included the α-tubulin and E-MAP 115 (LbSSH0003, LbSSH00064), Septin (LbSSH00073) (involved in yeast cell separation; Shafaatian et al., 1996) and Chitin synthase I (LbSSH00120). These genes are directly related to cell growth and proliferation, which happens in the development of the ectomycorrhiza.

As expected, the sequence characterization showed that the IS library contained sequences corresponding to *L. bicolor* genes that were previously identified and characterized in detail. Sequences of five clones were identical to that of the PF6.2 gene (Kim et al., 1998, 1999a), three matched the *ras* related proteins (Kim et al., 1999a; Sundaram et al., 2001), and two matched the peroxisomal β-keto thiolase (Zheng et al., unpublished). In addition, there was one member each related to the peroxisomal malate synthase (Balasubramanian et al., 2002), the Chitin synthase (unpublished), the BiP protein (Kim et al., 1999a), and an autophagosis-related protein (Kim et al., 1999b). All the *L. bicolor* genes mentioned above were shown to be differentially expressed either by direct northern analyses or by DDRT-PCR (Kim et al., 1999a, b). These results further validate the effectiveness of the interaction system we used here to isolate and characterize the pre-infection stage-related genes (Table 1, Figure 1).

Membrane array analyses

As expected, there was a general increase in the expression levels of the clones tested (Table 1). However, the degree of increase varied among different clones. Expression went up by a factor of 2 for 33% of the genes and by a factor of 3 or higher for another 33% of them. However, there was only a marginal difference in the remaining third of the clones. The latter, presumably, represent housekeeping and other genes common to the fungus during its free-living and the interactive stages. It is to be noted here that the IS probe for these analyses is derived after subtractive hybridization. While this technique is meant to eliminate the free living state transcripts, it is possible that the free-living levels of transcript become saturated and the excess becomes included in the probe. Thus, we would have to consider that an E ratio value of 1 represents an increase in expression after elimination of transcript levels from free-living mycelium. Apparently, the subtractive hybridization could not remove all the common sequences. Nevertheless, the analyses clearly showed that the pre-infection stage of conifer ectomycorrhizal symbiosis involves differential expression of genes belonging to many functional groups.

A differential expression ratio value of 8 is the highest for PF6.2 (LbSSH00013) (Table 1, Figure 2, panel PF6.2). The second highest (6-fold increase) was for the clone LbSSH0001, the identity of which is not known since its sequence did not correspond to any known sequences in the Genbank. A 4-fold increase in the expression of the BiP protein (LbSSH0007), previously identified as an early interaction regulated gene (Kim et al., 1999a) was observed. Northern analysis of this gene (Figure 2, panel BiP) suggests a tight temporal regulation of this gene. This stringency in regulation possibly extends to several others during the pre-infection stage. A more critical evaluation of expression of these genes will be necessary, where expression levels at shorter intervals are measured. The deoxyhypusine synthase (DHS), associated with activation of the protein synthesis factor eIF5a, also showed a 4-fold increase in its expression. The activities of DHS and eIF5a have been demonstrated to be correlated with cell proliferation (Tao and Chen, 1995). Also worth noting were the increases in the expression of metabolic enzymes peroxisomal acetyl CoA oxidase (3.6-fold); β-keto thiolase (4.2-fold); malate synthase (3.8-fold); PEP carboxykinase (4-fold), as well as two different alleles of glyceraldehyde-3 phosphate dehydrogenase (3–4-fold).

The data here identified several putative regulated genes and provided insights into gene activities involved in the pre-infection stage of the symbiosis. Our results demonstrate the utility of isolating IS libraries through subtractive hybridization-generated probes to study pre-infection stage-related fungal genes. This technique significantly minimizes the number of samples that need to be analyzed. Thus, it would be possible to use a simple membrane array analysis to obtain useful information regarding some of the most important genes involved in the pre-infection stage of the ectomycorrhizal development. Furthermore, as an initial step, we used a probe corresponding to a combined 6–72 h interaction point. Although, it did not provide information on temporal changes in the very early periods, as in the case of the BiP (Figure 2), the analysis helped us to get a quick look at the pre-infection stage gene expression identifying possible roles played by different genes. These results will form the basis for a more precise molecular dissection of the complex genetic networks that control symbiosis development and function. In order to investigate temporal regulation of key genes, it would be necessary to perform membrane array analyses using probes representing shorter interaction periods. However, the revelation that major metabolic shifts leading to changes in the carbon flow are probably occurring within the cell during the pre-infection stage is a significant departure from other studies, which could be a feature of conifer ectomycorrhiza that requires further attention. The identified symbiosis-regulated genes might be especially interesting targets for functional analysis of symbiosis, as these genes predetermine the next steps in the development of ectomycorrhiza. More temporal gene expression studies of selected target genes are necessary to fully understand the dynamics of the pre-infection stage of the ectomycorrhizal symbiosis.

Acknowledgement

This work was supported by USDA NRICGP Grant 99-35107-7810 to G. K. Podila and S. Hiremath.

References

Allen M F 1992 Mycorrhizal Functioning: An Integrative Plant-fungal Process. Chapman & Hall, Inc. New York. 530 pp.

Altschul S F, Madden T L, Schäffer A L, Zhang J, Zhang Z, Miller W and Lipman D J 1997 'Gapped BLAST and PSI-BLAST: a new generation of protein database search programs'. Nucl. Acids Res. 25, 3389–3402.

Ausubel F M, Brent R, Kingston R E, Moore D, Siedman J G, Smith J A and Struhl K 1998 Current Protocols in Molecular Biology. 13.7.1–13.7.10. John Wiley & Sons, USA.

Balasubramanian S, Kim S J and Podila G K 2002 Differential expression of a malate synthase gene during the pre-infection stage of symbiosis in the ectomycorrhizal fungus *Laccaria bicolor*. New Phytol. (In press)

Bills SN, Richter D L and Podila G K 1995 Genetic transformation of the ectomycorrhizal fungus *Paxillus involutus* by particle bombardment. Mycol. Res. 99, 557–561.

Burgess P and Dell B1996 Changes in protein biosynthesis during the differentiation of *Pisolithus* – *Eucalyptus grandis* ectomycorrhiza. Can. J. Bot. 74, 553–560

Burgess T, Laurent P, Dell B, Malajczuk N and Martin F 1995 Effect of the fungal isolate aggressivity on the biosynthesis of symbiosis-related polypeptides in differentiating eucalypt ectomycorrhiza. Planta 195, 408–417.

Caubit X, Core N, Boned A, Kerridge S, Djabali M and Fasano L 2000 Vertebrate orthologues of the Drosophila region-specific patterning gene teashirt. Mech. Dev. 91, 445–448.

Cripps C L and Miller O K, Jr. 1993 ectomycorrhizal fungi associated with aspen on three sites in the north-central rocky mountains. Can. J. Bot. 71, 1414–1420.

Diatchenko L, Lau YF, Campbell A P, Chenchik A, Moqadam F, Huang B, Lukyanov S, Lukyanov K, Gurskaya N, Sverdlov E D and Siebert P D 1996 Suppression subtractive hybridization: a method for generating differentially regulated or tissue-specific cDNA probes and libraries. Proc. Natl. Acad. Sci. USA 93, 6025–6030.

Diatchenko L, Lukyanov S, Lau Y F and Siebert P D 1999 Suppression subtractive hybridization: a versatile method for identifying differentially expressed genes. Methods Enzymol. 303, 349–380.

Gonzalez P, Zigler S, Epstein D L and Borras T 1999 Identification and isolation of differentially expressed genes from very small tissue samples. Biotechniques 26, 884–892.

Gyorgyey J, Vaubert D, Jimenez-Zurdo J I, Charon C, Troussard L, Kondorosi A and Kondorosi E. 2000 Analysis of *Medicago truncatula* nodule expressed sequence tags. Mol. Plant-Microb. Interact. 13, 62–71.

Graham A, Papalopulu N, Lorimer J, McVey J H, Tuddenham E G and Krumlauf R 1988 Characterization of a murine homeo box gene, Hox-2.6, related to the *Drosophila* deformed gene. Genes Dev. 2, 1424–1438.

Hampp R, Schaffer C, Wallenda T, Stülten C, Johann P and Einig W 1995 Changes in carbon partitioning or allocation due to mycorrhiza formation: biochemical evidence. Can J. Bot. 73, S548–S556.

Harley J L and Smith S E 1983 Mycorrhizal Symbiosis. Academic Press, London.

Hilbert J L, Costa G and Martin F 1991 Ectomycorrhizin synthesis and polypeptide changes during the early stage of eucalypt mycorrhiza development. Plant Physiol. 97, 977–984.

Jordan B R 1998 Large-scale expression measurement by hybridization methods: from high-density membranes to 'DNA chips'. J. Biochem. 124, 251–258.

Karlstrom R O, Talbot W S and Schier A F 1999 Comparative synteny cloning of zebrafish you-too: mutations in the Hedgehog target gli2 affect ventral forebrain patterning. Gene Dev. 13, 388–393.

Kim S J, Bernreuther D, Thumm M and Podila G K 1999b LB-AUT7, a novel symbiosis-regulated gene from an ectomycorrhizal fungus, *Laccaria bicolor*, is functionally related to vesicular transport and autophagocytosis. J. Bacteriol. 181, 1963–1967.

Kim S J, Hiremath S T and Podila G K 1999a Cloning and characterization of symbiosis-regulated genes from the ectomycorrhizal *Laccaria bicolor*. Mycol. Res. 103, 168–172.

Kim S J, Zheng J, Hiremath S T and Podila G K 1998 Cloning and characterization of a symbiosis-related gene from an ectomycorrhizal fungus *Laccaria bicolor*. Gene 222, 203–212.

Kobayashi M, Shinohara M, Sakoh C, Kataoka M and Shimizu S 1998 Lactone-ring-cleaving enzyme: genetic analysis, novel RNA editing, and evolutionary implications. Proc. Natl. Acad. Sci USA 27, 12787–12792.

Kropp B R 1990 Variable interactions between non-mycorrhizal ectomycorrhizal strains of the basidiomycete *Laccaria bicolor*. Mycol. Res. 94, 412–415.

Laurent P, Voiblet C, Tagu D, de Carvalho D, Nehls U, De Bellis R, Balestrini R, Bauw G, Bonfante P and Martin F 1999 A novel class of ectomycorrhiza-regulated cell wall polypeptides in *Pisolithus tinctorius*. Mol. Plant-Microb. Interact. 12, 862–871.

Liang P and Pardee A B 1992 Differential display of eukaryotic messenger RNA by means of the polymerase chain reaction. Science 257, 967–971.

Liu X J, Sorisky A, Zhu L and Pawson T 1995 Molecular cloning of an amphibian insulin receptor substrate 1-like cDNA and involvement of phosphatidylinositol 3-kinase in insulin- induced *Xenopus* oocyte maturation. Mol. Cell. Biol. 15, 3563–3570.

Martin F and Botton B1993 Nitrogen metabolism of ectomycorrhizal fungi and ectomycorrhizas. *In* Advances in Plant Pathology, Vol. 9, pp 83–102. Academic Press.

Martin F, Duplessis S, Ditengou F, Lagrange H, Voiblet C and Lapeyrie F 2001 Development of cross talking in the ectomycorrhizal symbiosis: Signals and communication genes. New Phytol. 151, 145–154.

Martin F and Hilbert J L 1991 Morphological, biochemical and molecular changes during ectomycorrhiza development. Experientia 47, 321–331.

Martin F, Lapeyrie F and Tagu D 1997 Altered Gene Expression during Ectomycorrhiza Development in the Mycota, Vol. VI. Plant Relationships Eds. P. Lemke and G. Caroll. pp. 223–242. Springer-Verlag, Berlin.

Martin F, Laurent P, De Carvalho D, Voiblet C, Balestrini R, Bonfante P and Tagu D 1999 Cell wall proteins of the ectomycorrhizal basidiomycete *Pisolithus tinctorius*: identification, function, and expression in symbiosis. Fungal Genet. Biol. 27, 161–174.

Martin F and Tagu D 1999 Developmental biology of a plant-fungus symbiosis: the ectomycorrhiza. In Mycorrhiza: Structure, Function, Molecular Biology and Biotechnology Eds. AK Varma and B Hock. pp. 51–73. Springer-Verlag, Berlin.

Massicotte HB, Peterson R L, Ackerley C A and Ashford A E 1987 Ontogeny of *Eucalyptus pilularis – Pisolithus tinctorius* ectomycorrhizae. II. Transmission electron microscopy. Can. J. Bot. 65, 1940–1947.

Nehls U, Mikolajewski S, Ecke M and Hampp R 1999 Identification and expression-analysis of two fungal cDNAs regulated by ectomycorrhiza and fruit body formation. New Phytol. 144, 195–202.

Reiser J Muheim A Hardegger M Frank G and Fiechter A 1994 Aryl-alcohol dehydrogenase from the white-rot fungus *Phanerochaete chrysosporium*. Gene cloning, sequence analysis, expression, and purification of the recombinant enzyme. J. Biol. Chem. 269, 28152–28159.

Richter D L and Bruhn J N 1989 Field survival of containerized red and jack pine seedlings inoculated with mycelial slurries of ectomycorrhizal fungi. New Forests 3, 247–258.

Richter D L and Bruhn J N 1990 *Pinus resinosa* ectomycorrhizae:seven host-fungus combinations synthesized in pure culture. Symbiosis 7, 211–228.

Schena M, Shalon D, Heller R, Chai A, Brown P O and Davis R W 1996 Parallel human genome analysis: microarray-based expression monitoring of 1000 genes. Proc. Natl Acad. Sci. USA 93, 10614–10619.

Schummer M, Ng W L, Nelson P S, Bumgarner R B and Hood L 1997 A simple high-performance DNA arraying device for comparative expression analysis of a large number of genes. BioTechniques 23, 1087–1092

Smith S E, Gianinazzi-P-Pearson V, Koide R and Cairney J W G 1994. Nutrient transport in mycorrhizas: structure, physiology and consequences for efficiency of the symbiosis. Plant Soil 159, 103–113.

Smith S E and Read D J 1997 Mycorrhizal Symbiosis, 2nd edn. Academic Press, London.

Sundaram S, Kim S J, Suzuki H, Mcquattie C J, Hiremath S T and Podila G K 2001 Isolation and characterization of a symbiosis-regulated *ras* from the ectomycorrhizal fungus *Laccaria bicolor*. Mol. Plant Microbe. Int. 14, 618–628.

Tagu D, Lapeyrie F, Ditengou F, Lagrange H, Laurent P, Missoum N, Nehls U and Martin F 2000 Molecular aspects of ectomycorrhiza development. *In* Current Advances in Myorrhiza Research. Eds. G K Podila and D D Douds Jr pp. 69–89. American Phytopathological Society Press, St. Paul, MN, USA.

Tao Y and Chen K Y 1995 Molecular cloning and functional expression of Neurospora deoxyhypusine synthase cDNA and identification of yeast deoxyhypusine synthase cDNA. J. Biol. Chem. 270, 23984–23987.

Voiblet C, Duplessis D, Encelot N and Martin F 2001 Identification of symbiosi-reglated gene in *Eucalyptus globulus–Pisolithus* ectomycorrhiza by differential hybridization of arrayed cDNAs. Plant J 25, 181–191.

Wendland J and Kothe E 1997 The translation elongation factor 1a (EF1a) of *Schizophyllum commune*. Mycol. Res. 10, 798–798.

Wodicka L, Dong, H L, Mittmann M, Ho M H and Lockhart D J 1997 Genome-wide expression monitoring in *Saccharomyces cerevisiae*. Nature Biotech. 15, 1359–1367.

Wong K K, Piche Y and Fortin J A 1990 Differential development of root colonization among four closely related genotypes of ectomycorrhizal *Laccaria bicolor*. Mycol. Res. 94, 876–884.

Zheng J, Balasubramanian S, Podila G K and Hiremath S T 2000 Expression of Acetyl-CoA acetyltransferase in an ectomycorrhizal fungus *Laccaria bicolor* during symbiosis. Phytopathology 90, S88.

Plant and Soil **244:** 129–139, 2002.

Early developmentally regulated genes in the arbuscular mycorrhizal fungus *Glomus mosseae*: identification of GmGIN1, a novel gene with homology to the C-terminus of metazoan hedgehog proteins

Natalia Requena[1,3], Petra Mann[2], Rüdiger Hampp[1] & Philipp Franken[2]
[1]*Botanical Institute, Department of Physiological Ecology of Plants, University of Tübingen, Auf der Morgenstelle 1, D-72076 Germany.* [2]*Max-Planck-Institute for terrestrial Microbiology and Philipps-University of Marburg, Dept. of Microbiology, Karl-von-Frisch-Str., D-35043 Marburg.* [3]*Corresponding author**

Received 21 August 2001. Accepted in revised form 10 March 2002

Key words: arbuscular mycorrhiza, development, gene regulation, *Glomus mosseae*, hedgehog proteins

Abstract

The life cycle of the obligate biotrophic arbuscular mycorrhizal fungi comprises several well-defined developmental stages whose genetic determinants are still unknown. With the aim of understanding the molecular processes governing the early developmental phase of the AM fungal life cycle, a subtractive cDNA library was constructed using a suppressive subtractive hybridization technique. The library contains more than 600 clones with an average size of 500 bp. The isolated cDNAs correspond to genes up-regulated during the early development of the AM fungus *Glomus mosseae* versus genes expressed in extraradical hyphae. The expression of several of the isolated genes was further confirmed by RT-PCR analysis. Among the isolated clones, a novel gene named *GmGIN*1 only expressed during early development in *G. mosseae* was found. The full-length *GmGIN*1 cDNA codes for a protein of 429 amino acids. The most interesting feature of the deduced protein is its two-domain structure with a putative self-splicing activity. The N-terminal domain shares sequence similarity with a novel family of GTP binding proteins while the C-terminus has a striking homology to the C-terminal part of the hedgehog protein family from metazoa. The C-terminal part of hedgehog proteins is known to participate in the covalent modification of the N-terminus by cholesterol, and in the self-splicing activity which renders the active form of the protein with signalling function. We speculate that the N-terminal part of GmGIN1, activated through a similar mechanism to the hedgehog proteins, has GTP-binding activity and participates in the signalling events prior to symbiosis formation.

Abbreviations: AM – arbuscular mycorrhizal; SSH – suppressive subtractive hybridization; sscDNA – single stranded cDNA; RT-PCR – reverse transcription-polymerase chain reaction; hh – hedgehog

Introduction

The mycorrhizal association between the Glomalean fungi and the roots of vascular plants represent for both partners a whole set of morphological and metabolic adaptations to the life in symbiosis. However, it is the fungal partner which depends absolutely on the plant to complete its life cycle. In the absence of it, the arbuscular mycorrhizal spores can only germinate and have a very restricted saprotrophic growth, mainly upon its own carbon expenses (Bago et al., 1999). This growth is very limited in extension but not in time, since the AM fungi have the ability to re-germinate several times. For that, the hyphae expanding out of the spore arrest growth after a certain time, about 2–3 weeks for *G. mosseae*, and suffer apical septation with retrieval of protoplasm towards the spore (Mosse, 1959). This rare mechanism is understood as a survival strategy of the AM fungi under conditions where

* FAX No: 7071-2973083.
E-mail: natalia.requena@uni-tuebingen.de

to find a root partner becomes difficult or impossible (Logi et al., 1998). These spores and growing hyphae have, therefore, a very different role from the extraradical symbiotic hyphae coming out of the root. The first ones are using their stored energy in form of triacylglyerides, glycerol and trehalose to grow enough to find a compatible root, and if not, to arrest growth and suffer a process similar to apoptosis (Bago et al., 1998; Logi et al., 1998). In contrast, the extraradical hyphae have as a main mission to scavenge mineral nutrients from the soil (mainly phosphate) for nursing its symbiotic partner and, eventually, also to colonize other roots. Our goal is to study gene regulation between these two different developmental stages.

The method employed here to identify genes specific from the early developmental stage, the suppressive subtractive hybridization, is meant to isolate transcripts of medium or rare abundance. That is, genes which are not very highly expressed, and that are regulated at that developmental stage. The SSH technique allows then, to identify key regulators of the morphological and metabolic adaptations that the cell suffer to accommodate a specific developmental change. This is one of the first times that this SSH technique has been applied to fungi (Watson et al., 2000). It was originally described for mammalian cells (Diatchenko et al., 1996), and it has been very recently employed for plant cells (Xiong et al., 2001). Other methods, such as the differential RNA display (DDRT-PCR) also described for the isolation of specific and rare transcripts (Lapopin et al., 1999; Martin-Laurent et al., 1997; Requena et al., 1999), or the construction of subtractive libraries (van Buuren et al., 1999) have been previously employed for the isolation of mycorrhizal specific genes. However, DDRT-PCR has been proven to be very time consuming because of the high amount of false positives rendered (Requena et al., 2000), and subtractive libraries allow in principle only the isolation of very highly expressed genes (for review see Franken and Requena, 2001a,b). In the case of fungal genes from arbuscular mycorrhizas, the main handicap is the small amounts of pure fungal material available due to its unculturability. This made SSH our method of choice since it allows starting from amounts of RNA (i.e. tissue) as little as 50 ng.

Materials and methods

Isolation of fungal pre-symbiotic material

Sporocarps from the AM fungus *G. mosseae* (Nicol & Gerd.) Gerdemann & Trappe (BEG12) were purchased from Biorize (France) and surface-sterilized according to Budi et al. (1999). Sporocarps were germinated in water-agar plates as described elsewhere (Requena et al., 1999) and grown in an incubator at 25 °C in the dark. After 12 days, the germinated sporocarps were harvested using fine forceps to recover all hyphae and immediately frozen in liquid nitrogen.

Isolation of fungal symbiotic material

Parsley (*Petroselinum crispum*) plants were inoculated with *G. mosseae* and grown in pots divided into compartments separated by a thin mesh (50 μm) which allows the growth of extraradical hyphae but not of roots (Redecker et al., 1995). The hyphal compartment contained 2 mm glass beads from where the hyphae could be easily recovered. Plants were grown for three months in a phytochamber with controlled conditions of light (100 μE m^{-2} s^{-1}), temperature (20 °C) and photoperiod (14 h light). After that time, extraradical hyphae were isolated from the glass bead compartment, washed several times with chilled distilled water and frozen in liquid nitrogen.

RNA extraction and Subtractive Library Construction

Total RNA was extracted using the RNAeasy kit from Qiagen (Heidelberg) from both the pre-symbiotic material and the extraradical hyphae. RNA quality and quantity was assessed by Northern-blot using a DIG probe containing the ribosomal genes from *G. mosseae* (Franken and Gianinazzi-Pearson, 1996). Equal amounts from both RNA populations (500ng) were used to synthesise cDNA using the SMART protocol from Clontech (Palo Alto, CA). The cDNA populations were then subtracted using the PCR-Select^{TM} cDNA Subtraction kit from Clontech according to manufacturer recommendations, where pre-symbiotic material represented the tester and the hyphal material the driver. Subtracted cDNA fragments were cloned into the TA TOPO vector from Invitrogen (Groningen).

Reverse Northern-blot screening

One hundred clones randomly selected were PCR-amplified using the primers Nested PCR primer 1 and Nested PCR primer 2R (www.clontech.com) that flank the borders of every insert from the subtractive library. PCR was performed in 50 μl using the standard Gibco Taq polymerase, in a Biometra thermal-cycler. Five μl aliquots were separated in a 1.5% agarose gel and Southern blotted in parallel into two nylon membranes according to Sambrook et al. (1989). A radioactive probe was prepared using as template SMART cDNA from each population (pre-symbiotic mycelium and extraradical hyphae) by incorporation of $\alpha-^{32}$P dATP in a PCR reaction using as primers the SMART PCR primer from the SMART kit. The membranes were then hybridized using the standard DIG buffer (Boehringer) at 65 °C overnight. Membranes were then washed twice using 2X SSC; 0.1% SDS buffer at room temperature, and one time with 0.5XSSC; 0.1% SDS buffer at 65 °C. Membranes were then exposed for 24–72 h to an X-ray film.

Sequencing and homology searches

Clones which were found to be either up or down regulated in the Reverse Northern-blots were sequenced and their sequence compared to the non-redundant NCBI database using the programme BlastX 2.0.11 (Altschul et al., 1997). The parameters used for the sequence comparison were an expected value of 10, Word size 3, using the Blosum62 matrix with Gap cost: Existence 11, Extension 1. Only sequences showing an E value lower than 10 and with an overall homology higher than 40% were considered as significant homologies and are listed in Table 1. Sequences listed in Table 1 were submitted to the EMBL database and Accession numbers are given.

RT-PCR expression analyses

To confirm the expression pattern observed in the Reverse Northern-blot, the expression of ten of the sequenced cDNA fragments was further studied by RT-PCR. Single stranded cDNA was synthesised from 50 ng of total RNA using an oligo (dT) primer. cDNA was synthesised from each population using the MMLV reverse transcriptase from Promega at 37 °C for 1 h. Specific primers designed for each of the genes analyzed were used (Table 2). One microliter of a dilution 1:5 of the sscDNA was used as template for a 30 cycles PCR reaction. The analysis was repeated twice.

Isolation of the full length cDNA of the gene GmGIN1

A modified RACE (rapid amplification of cDNA ends) technique was used to amplify the 5′ and 3′ ends of the cDNA encoding the gene *GmGIN*1. A nested PCR reaction was performed on the SMART cDNA using two pairs of primers based on the sequence of the cDNA fragment in combination with either the PCR primer (SMART kit) for the 5′ end or the CDS primer (SMART kit) for the 3′ end. The PCR reactions were performed using the proof-reading enzyme Advantage 2 (Clontech). The resulting fragments were cloned into the pCR2.1 TOPO vector (Invitrogen) and sequenced (MWG Biotech, Germany).

PCR amplification of GmGIN1 in genomic DNA

Genomic DNA from *Glomus mosseae* was isolated from germinated sporocarps (100) with the Qiagen DNeasy Plant Mini kit. Two PCR reactions were performed. In the first one, the primers used (Gm79F2/R2) amplified the region encoding the C-terminus of *GmGIN*1. In the second one, a combination of primers (Gm79F1/R2) was used that amplify a larger region that covers part of the N-terminus as well as part of the C-terminus of GmGIN1. Both PCRs used as template150pg of genomic DNA in a 25 μl reaction. Primers are listed in Table 2.

Expression of GmGIN1 during different stages of the life cycle

Germinated sporocarps, as well as extraradical hyphae and colonized roots were used as template to study the expression of *GmGIN*1. The cDNA was synthesised with hexamers primers using the Superscript II reverse transcriptase from GIBCO (Life Technologies, UK) according to manufacturer recommendations. Equal amounts of total RNA (70 ng) were used for cDNA synthesis except for the colonized root were five times more RNA belonging to plant and fungus together were employed. cDNA amounts were calibrated by PCR using the ribosomal primers VANS1 and VA-GLO (Simon et al., 1992, 1993). Equal fungal cDNA amounts were then used for the expression analysis.

Table 1. List of isolated cDNA clones from *G. mosseae* expressed during pre-symbiosis with significant homologies to proteins from the NCBI database

Clone	Acc. number	Size (bp)	Position	Definition	Organism	Acc. number	E value	Homol.
1	AJ315702	374	polyA	spindle pole body component Alp6	*Schizosaccharomyces pombe*	BAA94097	0.002	41%
9	AJ315703	265	?	TrihydrophobinCpth	*Claviceps fusiformis*	CAB61236	0.66	43%
11	AJ315704	455	3′ end	proteasome subunit, beta type, 2	*Rattus norvegicus*	NP_058980	1e-15	71%
14	AJ315705	807	5′	F-box protein Fb12	*Homo sapiens*	AAF04510	1.5	45%
16	AJ315706	194	5′	60S ribosomal protein L15	*S. pombe*	T41421	6.6	93%
18	AJ315707	546	mid.	P-type H+ATPase	*Pneumocystis carinii*	AAB06958	1e-29	59%
69	AJ315708	427	mid.	Sec14	*S. pombe*	Q10137	9e-39	72%
79	AJ315709	1132	mid.	N-terminus: AIG1 protein	*Arabidopsis thaliana*	AAG52215	1e-12	56
				C-terminus:Desert hedghegog	*Xenopus laevis*	Q91611	2e-12%	51%
80	AJ315710	362	mid.	28S ribosomal RNA gene	*Lobaria pulmonaria*	AF183934	3e-22	94%
85	AJ315711	465	mid.	DNAJ tetratricopeptide	*S. pombe*	CAC05244	7e-21	54%
89	AJ315712	293	mid.	cytosolic leucine aminopeptidase	*Rickettsia prowazekii*	P27888	2e-17	67%
92	AJ315713	348	3′	purine nucleotide binding protein Fet5	*S. pombe*	T38803	1e-09	78%
101	AJ315714	425	5′	hypothetical protein YNR020p	*Saccharomyces cerevisiae*	NP_014417	2e-06	55%
111	AJ315715	250	3′	CG3060 gene product	*Drosophila melanogaster*	AAF47123	1e-07	60%
118	AJ315716	425	midd.	hypothetical protein T08B2.5	*Caenorhabditis elegans*	T28754	0.046	57%
135	AJ315717	500	poly A	Glucose repression regulatory protein RGR1	*S. cerevisiae*	P19263	2.4	50%
142	AJ315718	308	poly A	Polyubiquitin	*Neurospora crassa*	UQNC	3e-04	60%
149	AJ315719	405	3′	alpha NAC	*D. melanogaster*	AAB97513	4e-38	78%
512	AJ315720	942	midd.	Inositol polyphosphate 5-phosphatase	*H. sapiens*	Q01968	8e-53	57%
513	AJ315721	586	midd.	ubiquitin carboxyl-terminal hydrolase	*A. thaliana*	AAK25908	0.005	53%
519	AJ315722	507	5′	phosphatidylglycerol/phosphatidylinositol of transfer protein	*Aspergillus oryzae*	AAD16095	2.0	48%
520	AJ315723	342	5′	Vacuolar ATP synthase subunit E	*N. crassa*	Q01278	7e-23	82%
1000	AJ315724	1045	5′	CGI-15 protein similar to phosphate/phosphoenol piruvate translocator	*H. sapiens*	XP_009673	6e-27	56%
1001	AJ315725	594	5′	CDC48	*Dyctiostelium discoideum*	U83085	4e-51	73%
1002	AJ315726	292	5′	Putative 3-Ketoacyl-CoA Thiolase	*Aspergillus oryzae*	Q9HG46	6e-16	70%
1003	AJ315727	578	midd.	Heat shock protein 70	*P. carinii*	AAD00455	4e-57	91%
1004	AJ315728	791	5′	Splicing factor, arginine/serine-rich 4	*H. sapiens*	XP_001807	7e-21	75%

Results

Establishment of a suppressive subtractive library

SSH was employed as described in the 'Material and methods' section to obtain a library of cDNA clones up-regulated during early development. A total of 625 colonies containing cDNA fragments from genes expressed during early development in *G. mosseae* were isolated. From those, about 9% of the clones contained either no insert or a very small insert to be seen on an agarose gel. One hundred clones were PCR amplified using primers Nested PCR primer 1 and Nested PCR primer 2R to analyse their expression by reverse-northern blot. Most of the clones gave a single amplification signal as expected but some of them showed more than one band. This can be the result of artefacts during cloning in which more than one cDNA fragment are ligated into the vector (Figure 1). From those 100 clones, 70 were sequence analysed; none of the sequences was redundant, and only two or three of them could potentially belong to different fragments from the same gene. Only one ribosomal sequence corresponding to the large ribosomal subunit and one 60S ribosomal protein were found. The average size of the library clones is around 500 bp.Twenty seven percent of the clones corresponded to the poly A

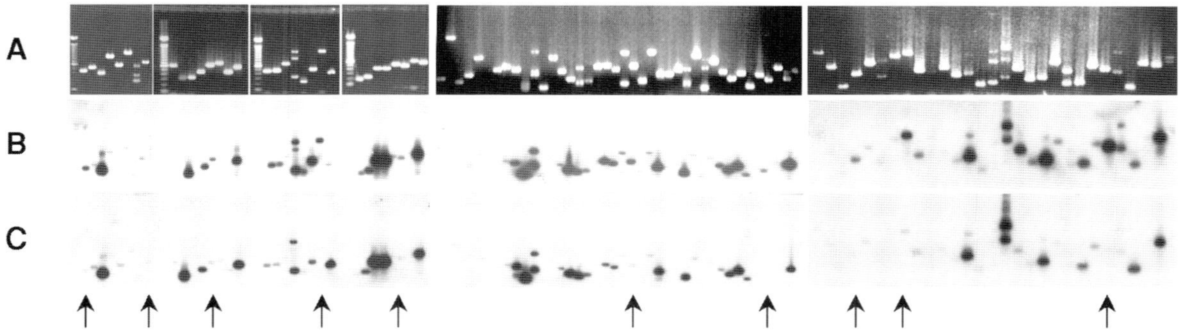

Figure 1. Reverse-northern blot screening for identification of regulated genes between pre-symbiotic and symbiotic hyphae. Panel A shows the PCR amplification of 100 clones from the cDNA library using the primers Nested1 and Nested2R resolved in an agarose gel and stained with ethidium bromide. Panel B and C show the differential hybridization of the nylon membranes, where the clones were blotted, with radioactive labelled cDNA synthesised either from pre-symbiotic material (B) or from extraradical hyphae (C). Clones further analysed by RT-PCR are indicated by arrows.

Table 2. List of the primers used in this paper

Primer name	Sequence
Gm18F3	ccagtcaagaaagttcccaag
Gm18R1	cacgttcagctgctgatatcg
Gm79F2	ggagactacgtatgttgtgg
Gm79R2	aacaagagcagagaacctcg
Gm79F1	gatgtctcggattcagctgc
Gm85F1	gtcaggaagcattgcgttgc
Gm85R1	catatcactcaatgcctccg
Gm1000F1	ggtagtaacgatagttcaccacc
Gm1000R1	gcggatcatctgttcatacttcc
Gm111F1	ttggaatttgaagtctcgcc
Gm11R1	tcttcatttgatacgaaccc
Gm520F1	atgtcgaatctacgtgc
Gm520R1	agaagcagcttgaatgg
Gm1001F1	ctacatacctttacctgatgagc
Gm1001R1	ctcttgtgatttgtggaactgg
Gm1003F1	agcgaccaagaagttcaatccg
Gm1003R1	gacatcaaaaactccaccaccc
Gm69F1	cgtgaagaattcaaggaaggagg
Gm69R1	caggcaggtaatctgacgtcagc
Gm512F1	atccga ttgtgtcggtgtcg
Gm512R1	gatccaatcatccctcgcac

tail while only 15% to the 5′ end of the corresponding genes. Only 27 of the sequenced clones (38%) had a significant homology (Table 1) and from those only four clones were located at the poly A tail. Most of the sequences had homology to fungal proteins (60%) while the rest showed homology to proteins from different eukaryotes, and only one showed significant homology to a bacterial sequence. The GC content of the analysed sequences was always lower than 50% which seems to be characteristic of arbuscular mycorrhizal fungi (Hosny et al., 1997). In several cases where the cDNA clone corresponded to the 5′ region of the gene, a large poly T region could be observed in the 5′ untranslated region (data not shown).

Expression analyses

The Reverse-Northern blot experiment (Figure 1) showed that as expected, the membranes hybridized with the pre-symbiotic labelled cDNA gave a more intense hybridization signal than those hybridized with labelled hyphal cDNA. There was one exception, clone Gm80 (AJ315710), which belonged to the ribosomal large subunit gene. Some clones, such as clone Gm79 (AJ315709), with similarity to GTPases and hedgehog proteins, showed high level of regulation in this screening with a high expression during pre-symbiosis in contrast to no signal in extraradical mycelium.

The RT-PCR analysis of the 10 analysed clones revealed that most of them were indeed up-regulated during pre-symbiosis versus extraradical hyphae. Half of the clones showed at least more than three times up-regulation while the other half were only slightly up-regulated. Only one of the analysed clones (Gm69) showed no changes in gene expression between pre-symbiosis and extraradical hyphae, while in another one (Gm512) the level of expression was higher in hyphae, than in spores.

```
ACCCACTATAGGGCGAATTGGGCCCTCTAGATGCATGCTCGAGCGGCCGCCAGTGTGATGGATATCTGCAGAATTCGCCC 80
                                            M  H  A  R  A  A  A  S  V  M  D  I  C  R  I  R  P
TTAAGCAGTGGTAACAACGCAGGTACGCGGGGAAGTTGATACAATGTCAAATTGCCCATCTATCCTATTGATAGGAAA 160
  .                                                   M  S  N  C  P  S  I  L  L  I  G  K
AACTGGCGTTGGTAAAAGTACTTTGGGAAATTTACTTTTGGGTCGTGATGTATTTGATGTCTCGGATTCAGCTGCATCTT 240
  T  G  V  G  K  S  T  L  G  N  L  L  L  G  R  D  V  F  D  V  S  D  S  A  A  S
TAACTCAAGGATATCAAACTGCCCCGATCGAAATAAACGAGAAAACGTTCAATGTTGTTGATACACACGGATTTTTTGAC 320
L  T  Q  G  Y  Q  T  A  P  I  E  I  N  E  K  T  F  N  V  V  D  T  H  G  F  F  D
ACAAACAGAACGAATCAAGAAATATTGAAGGAGGTTACCCAAGAAATATTACAATGCGAAAATGGAATTCAGGCTTTTGT 400
 T  N  R  T  N  Q  E  I  L  K  E  V  T  Q  E  I  L  Q  C  E  N  G  I  Q  A  F  V
ATTTGTCATAGAGGCTACACGTTTTACCAAAGAACAGAGGGACACTATTAATCAAATTATCAACTTCCTTGGCGAAGACT 480
  F  V  I  E  A  T  R  F  T  K  E  Q  R  D  T  I  N  Q  I  I  N  F  L  G  E  D
CATTAAATAATATGATCGCGGTATTTTCAAAATGCAGAAAGGCTCCGACAATTAACCCTGACCGGCTCTTCAATTCATTT 560
S  L  N  N  M  I  A  V  F  S  K  C  R  K  A  P  T  I  N  P  D  R  L  F  N  S  F
TCTCAGGAGGAAAAGGACTTTCTCAACCGTATCGGGAACAGATTTACTATCTCACCTAATCTCGAGATTTTTGACGAACC 640
 S  Q  E  E  K  D  F  L  N  R  I  G  N  R  F  T  I  S  P  N  L  E  I  F  D  E  P
AAACGACCCAATTGTCGTGCGCCATATGACGAAACTTAAAGAATACATTGTCAATTTTCCAGACTTATATACAACAGCTG 720
  N  D  P  I  V  V  R  H  M  T  K  L  K  E  Y  I  V  N  F  P  D  L  Y  T  T  A
TATTTGAAAAAGTTTTAATGGCTCGAGAAAATGAATACAAAAGAAAACGTGTAAATGCTGGGCTAAAAGGATTCCTTACT 800
V  F  E  K  V  L  M  A  R  E  N  E  Y  K  R  K  R  V  N  A  G  L  K  G  F  L  T
CGTAAACCTGTTGAATTAATTAATGATTTCGAAGGATGTTTTGCCGCAGATTCAAAAGTTATTTTGAAAAATGGAAAAGT 880
 R  K  P  V  E  L  I  N  D  F  E [G  C  F] A  A  D  S  K  V  I  L  K  N  G  K  V
CACCAAAATTTCAGAACTTGTTATTGGAGACTACGTATGTTGTGGATTTGAGGACGGAAAGCAAGTTTATAGTGAAGTGT 960
  T  K  I  S  E  L  V  I  G  D  Y  V  C  C  G  F  E  D  G  K  Q  V  Y  S  E  V
TTCTAATGATCCATGCCGACCCAAATGCAGTGACGAAATTTCAATCAATTGACTTTGTGAAACAAGACGGATCACAAGGA 1040
F  L  M  I  H  A  D  P  N  A  V  T  K  F  Q  S  I  D  F  V  K  Q  D  G  S  Q  G
AATCTTCATATTACCCCCAAACATCACATCTTCGTGAACAACGGTGAGACTGATTTCGCAAATAATGTTACAACAAACAC 1120
 N  L  H  I [T] P  K [H] H  I  F  V  N  N  G  E  T  D  F  A  N  N  V  T  T  N  T
TAAACTTTTTGTTTCTGATGGAGAGAAATTTGTTACCGTGCTTCCAATTCGTGTAACAAAAGAACGAAGAAAAGGTTATT 1200
  K  L  F  V  S [D] G  E  K  F  V  T  V  L  P  I  R  V  T  K  E  R  R  K  G  Y
ATAGTCCCTTGACTCGAAGCGGCACCATTCTTGTTGACGAGGTTCTCTGCTCTTGTTATGCTTCTGCACCGCCTTATCAA 1280
Y  S  P  L  T  R  S  G  T  I  L  V  D  E  V  L  C  S  C  Y  A  S  A  P  P  Y  Q
GCCTTGTTAAACTTTGTCCTCGTGCCATTAAGAATGTACACGAAAATTTTTCCCTCTAATTACTTAGACAAAGAGATCCA 1360
 A  L  L  N  F  V  L  V  P  L  R  M  Y  T  K  I  F  P  S  N  Y  L  D  K  E  I  H
TCCCTATGTAAAATTCTTATACAAGGGTCGCTGGATCATGGGGTGCCTTTAAATAAAAAGATACTTCTCGGAAAGATTTG 1440
  P  Y  V  K  F  L  Y  K  G  R  W  I  M  G  C  L  .
TTTTTTTAAAACTTTTGATAGAAAATAACTAAAGAATTTATCTGAATTCAAACTTCAAAATATGGCTACTTACTATTGTA 1520
ATAAAATCGAATTTGTCGTACCAAAAAAAAAAAAAAAAAAAAAAAAAAAAAAA 1571
```

Figure 2. Full length cDNA from *GmGIN*1 showing the large open reading frame encoding the 429 aa GmGIN1 protein. The GTP CDC motif (PF 00735) which extends through most of the amino terminus of the protein is highlighted in dark grey. Inside of that region the nucleotide binding site (ATP/GTP P-loop, PS00017) is indicated in bold. The carboxy-terminus of the protein contains the Hint domain (PF01079) characteristic of hedgehog proteins (highlighted in light grey). Conserved amino acids involved in the autoprocessing of the protein or in cholesterol binding are boxed.

Analysis of GmGIN1

The amplification of 5′ and 3′ cDNA ends from the clone Gm79 rendered a 1571bp fragment with a large open reading frame (ORF) of 1287 bp and the corresponding gene was called *GmGIN1* (Figure 2). The deduced protein, GmGIN1, contains 428 aa and has a calculated molecular mass of 48 kDa. A small ORF (17 aa) was found 96 bp upstream of the first codon. A profile search of the deduced protein using the Pfam-A and Prosite data banks (www.isrec.isb-sib.ch/cgi-bin/PFSCAN) showed three significant matches. The first 209 aa matched to the GTP CDC motif (PF 00735) characteristic of proteins involved in cell division, shadowed in dark grey in Figure 2, with an ATP/GTP binding domain (PS00017) located between amino acids 11–18 (in bold in Figure

```
1    M S N - - - - - - - - - - - - - - - - - - - - - - - - - - - - - - - - - - - - -        GmGinN
1    M K G L R K F Q Q M A N D Q K N S E S F P A K E D H K K D D A A A P A E V D H K        AtAIG

4    - - - - - - - - - - C P S  I L L I G K T G  V G K S  T L G N  L L L G R D  V F D V S   GmGinN
41   D E F S A S Q P H P V E N  I V L V G R T G  N G K S  A T G N  S I V R S K  V F K S K   AtAIG

34   D S A A S  L T Q G Y Q T  A - P I E I N E K T F  N V V D T  H G F F D T N R T N Q E   GmGinN
81   T K S S G  V T M E C H A  V K A V T P E G P I L  N V I D T  P G L F D L S V S A E F   AtAIG

73   I L K E V  T Q E I L Q C E N G I Q A F V F V I E A T R F T K E Q R  D T I N Q I I    GmGinN
121  I G K E I  V K C L T L A D G G L H A V L L V L S V R T R I S Q E E  E M V L S T L    AtAIG

113  N F L G E D S L N N M I A V F S K C R K  A P T I N P D R  L F N S F S Q E E K  D F  GmGinN
161  Q V L F G S K I V D Y L I V V F T G G D  V L E D D G M T  L E D Y L G D N M P  D F  AtAIG

153  L N R I G  N R F T I S P N  L E I F D E P N  D P I V V R H M T K L K E Y I V N F P  GmGinN
201  L K R V L  I L C G Q R M I  L F D N K T K D  D E K K T K Q V H E L L K L I D L V R  AtAIG

193  D L Y T T  A V F E K V L M A R E N  E Y K R K R V N A G L K G F L T R K P V E L I    GmGinN
241  K Q N N N  I P Y T D E M Y H M I K  E E N E R H K K E Q E E L E S K G H S E E Q L    AtAIG

233  N D F E                                                                             GmGinN
281  A A L M                                                                             AtAIG


237                                                                                 G C  GmGinC
161                                                                                 G C  XlDHH2

239  F A A D S K V I L K N  G K V T K  I S E L  V I G D  Y V C C G F  E D G K Q V Y S E V  GmGinC
201  F P G T A M V M M E T  G K K K P  L S E L  K L G D  T V F T T D  E T G L L I H S V V  XlDHH2

279  F L M I H  A D P N A V T K  F Q S I D F  V K Q D G  S Q G N  L H I T P K H H I F V N  GmGinC
241  L L F L H  R D P Y K T A T  F V - - - L  I E A E G  H P T K  L L V T P N H L L F I K  XlDHH2

319  N G E T D - - - - - - -  F A N N V T T N T K  L F V  S D G E K F V  T V L P  I R V T  GmGinC
278  S S S S T G F Q P T F A  Y R V Q I G D L I Q  I Y V  - N G T Q V Q  S S K V  V R V S  XlDHH2

352  K E R R K  G Y Y S P L T R S  G T I L V D  E V L C S C Y A S A  P P Y Q A  L L N F V  GmGinC
317  V D E Q T  G V Y A P M T E H  G T L L V D  G V L T S C Y A T V  E S - H T  L A H A S  XlDHH2

392  L V P L R  M Y - - - T K I F  P S N Y L D K E  I H P Y V K F L Y K G - - - - - R  W  GmGinC
356  L A P L R  L F Q G I A S M L  P D L H T S D G  V H W Y C H I L Y V L A K Y V L W  W  XlDHH2

424  I M G C L                                                                            GmGinC
396  D M P                                                                                XlDHH2
```

Figure 3. Alignment of the amino and carboxy terminus from the predicted GmGIN1 protein with their closest homologues, *Arabidopsis thaliana* AIG1 (At AIG, AAG52215) and *Xenopus laevis* Desert hedgehog (Xl DHH2, Acc. Q91611) respectively. Conserved amino acids are highlighted.

2). A Hint module (PF01079) characteristic of hedgehog proteins and inteins (proteins that suffer splicing as post-translational modification) was found between residues 227 and 426 (in light grey in Figure 2).

The similarity search of the large protein in the NCBI database using the programme BlastP showed that the C-terminal part of the gene has high homology (50%) to the C-terminal domain of the hedgehog protein family from vertebrates. This conserved domain is not found outside the metazoa and indeed, no significant homology to any fungal protein was found when the GmGIN C-terminus was blasted against fungal databases. Several conserved amino acids essential for the autoproteolysis activity catalyzed by the C-terminus of hedgehog proteins were found. Thus, a conserved cystein residue (Cys-236 in GmGIN1) in the conserved tripeptide GlyCysPhe (GCF), is known to be essential for the initial nucleophilic attack as well as for cholesterol binding, while Thr-310 and His-313 are essential for auto-cleavage (boxed in Figure 2). An aspartic residue involved in cholesterol binding was also found (Asp-285).

The N-terminal part of GmGIN1 showed significant homology (46%) to several proteins with GTPase

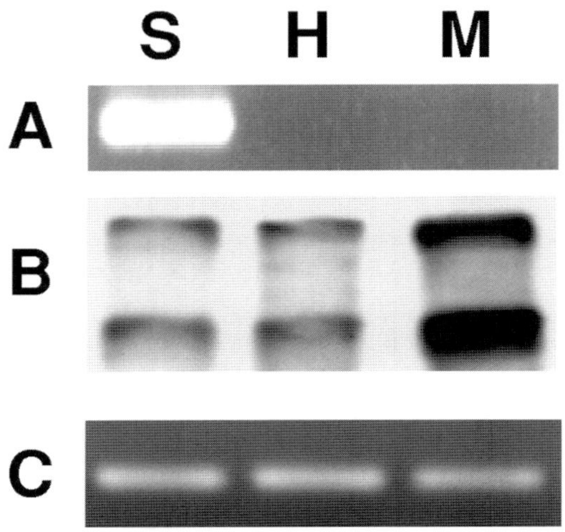

Figure 4. Differential expression of GmGIN1 during distinct developmental stages of the life cycle of *G. mosseae*, S (early development), H (extraradical hyphae), and M (mycorrhizal roots). Panel A shows the expression of the gene studied by RT-PCR and analysed on a 1.5% agarose gel and stained with ethidium bromide. Panel B shows the northern blot calibration with the amount of RNA used from each sample to synthesise the cDNA. Panel C shows the cDNA calibration from each sample by PCR using the primers VANS1 and VAGLO specific for *G. mosseae* and analysed on agarose gel and ethidium bromide staining. One microliter of a 1:10 000 dilution of the synthesized cDNA was used as template in a 28, 30 and 32 cycles PCR. Panel C shows the result after 30 cycles.

activity, and particularly to a novel family of GTP-binding proteins found in plants and more recently in animals. Weak similarities were found between the amino terminus of GmGIN1 and the septin family of fungal proteins probably due to the presence of a CDC type GTP binding domain. Alignments of the N- and the C-terminus from GmGIN1 with their respective putative homologues are presented in Figure 3. Because GmGIN1 showed similarity to different proteins at both ends, we performed a PCR analysis on genomic DNA to exclude the possibility of a PCR artefact during the construction of the library. The PCR analysis performed on genomic DNA confirmed that indeed both ends of the cDNA are part of a single gene.

The analysis of *GmGIN*1 expression was studied at different developmental stages of the fungal life cycle. Thus, while the gene is highly expressed during presymbiosis it seems to be completely shut down during symbiosis, both at the intraradical structures and in the external mycelium (Figure 4).

Discussion

Early developmental genes in the life cycle of arbuscular mycorrhizal fungi must encode proteins responsible for spore germination as well as for the pre-symbiotic hyphal growth in search of a host plant. Some of those should be in charge of the plant recognition and further signaling events that lead to the formation of the symbiosis. In contrast, other genes have to regulate the growth arrest and cytoplasm retraction that takes place in the absence of signals from a compatible host root. Our aim when constructing the library described in this paper was to isolated genes related to the signaling events taking place in the fungus prior to symbiosis formation.

The suppressive-subtractive library constructed in this paper contains 625 clones, about the same number of other SSH libraries described (Xiong et al., 2001). The average size of the clones produced, 500 bp, were demonstrated to be appropriate to find significant homologies in the databases with a good degree of certainty. The reverse-northern blot screening served as a pre-selection for sequencing of highly regulated genes. However, this method underestimates the number of regulated genes because many of the clones showed very weak or no signal even after several days of exposure. They probably belong to genes expressed at very low levels that might be then underscored. A detailed analysis by RT-PCR for those clones would probably disclose their expression pattern. The sequence similarity analysis of the sequenced clones to known proteins from the NCBI database revealed that only a small percentage corresponds to highly conserved proteins whose identity could be predicted. The majority of the sequences, however, belonged to evolutionarily not so highly conserved proteins or to untranslated cDNA regions. The RT-PCR analyses of some of the sequenced clones confirmed their expression pattern and showed that indeed the SSH method allowed identification of specifically regulated genes during early development.

Among those regulated genes, *GmGIN*1 appears to be a novel gene encoding a protein with two different domains. The amino-terminus of GmGIN1, GmGIN1-N has homology with a series of proteins all containing a nucleotide-binding site. In fact, a characteristic ATP/GTP binding site A (P-loop) motif was found between amino acids 11 and 18. A significant match with the GTP CDC cell division motif was also found between amino acids 7 and 209 which constitute most of the N-terminal part of the protein. This motif is present in proteins related to control of cell cycle, i.e. septins, and they all contain a GTP binding site. However, the similarity search in the databases showed the highest similarity with a new family of GTP-binding proteins first described in plants, and now recently also in mammals, which are pathogen-induced and could be related to the control of an apoptosis response (Poirier et al., 1999; Reuber and Ausubel, 1996). It seems that GmGIN1-N could be a new member of this class of proteins and probably also possess GTPase activity.

The C-terminal domain of GmGIN1, GmGIN1-C has a striking similarity to the C-terminal domain of hedgehog (hh) proteins. The carboxy-terminus domain from hh proteins functions as a cholesterol transferase. It catalyzes the autoprocessing of the protein that renders an amino-terminus covalently modified by a cholesterol moiety (Hall et al., 1997; Porter et al., 1996a, b). Such modification produces the active form of the protein as well as determining its tissue specific spatial localization at the cell membrane (Ingham, 2000; Mann and Beachy, 2000; Porter et al., 1996a). The mechanism by which such a process takes place is a nucleophilic attack by the thiol group of a very conserved cystein (in our case Cys-236) to the carbonyl group of the preceding amino acid residue (a glycine) to form a thioester linkage in place of the peptide bond. In a second step, the thioester suffers also nucleophilic attack from the 3β hydroxl group of a cholesterol molecule, releasing the N-terminal part of the protein covalently attached to cholesterol. The attachment of cholesterol increases dramatically the hydrophobicity of the protein that is then targeted to the cell membrane (Porter et al., 1996a). The mechanism has similarities to the protein splicing mechanism of inteins (Paulus, 2000), and therefore proteins having such a conserved hh carboxy-terminus are recognized in the database to contain the so-called Hint (*H*edgehog *int*ein) motif.

The whole autoprocessing mechanism in hh proteins seems to be independent of the N-terminus (Hall et al., 1997; Porter et al., 1996a, b). Interestingly, a number of unrelated protein families from the worm *Caenorhabditis elegans* have been recently found to possess similar carboxy-terminal regions albeit a totally non-related amino terminus. These proteins have been found to be cleaved in a similar manner to hedgehog proteins (Burglin, 1996; Porter et al., 1996a), although the participation of sterols other than cholesterol is presumed (Aspöck et al., 1999). Only members from the metazoa have been found, so far, to belong to this family of hh C-terminus homologues. GmGIN1 represents, therefore, the first member of this family from lower eukaryotes. A detailed analysis of active sites in the C-terminal domain of hedgehog proteins showed that, besides the above mentioned cystein, two other conserved downstream amino acids, histidine and threonine had an essential role in the autoprocessing (Hall et al., 1997). When mutated, the formation of the internal thioester and the autocleaving activity is severely hindered. In GmGIN1 they are also conserved and correspond to His-310 and Thr-313, indicating a possible similar mechanism of action. The question is whether cholesterol catalyses also the second nucleophilic attack in *G. mosseae*. An important amino acid, an aspartic residue about 50 amino acids downstream of the nucleophilic cystein was found to be dispensable for the first nucleophilic attack but required for the cholesterol transfer (Hall et al., 1996). This residue found in hedgehog proteins is not apparent in inteins sequences, where cholesterol is not the responsible of the processing activity. In GmGIN1, an aspartic residue (Asp-285) is located 47 amino acids downstream the Cys-236. It raises the possibility that in *G. mosseae* cholesterol, or another sterol as in the *C. elegans* proteins, could be responsible for the nucleophilic attack to the thioester formed between the cystein and the preceding glycine. AM fungi are known to contain large amounts of lipids that serve as the main carbon source during pre-symbiotic growth. It would not be surprising that they would also be involved in the signaling processes taking place at this stage.

Our current hyphothesis is that GmGIN1 suffers autoprocessing in a similar manner to hh proteins (see Figure 5). A sterol transfer activity located in the carboxy-terminus would provoke the cleavage of the protein between the amino acids 236 and 237 releasing an amino terminal domain attached to a cholesterol or another sterol moiety. This would cause the activation and proper localization of the N-terminal part of the protein with GTPase activity at the plasma membrane. It remains to be determined which precise function this GTPase would have. Because the expression data show that the gene is mostly expressed in the pre-symbiotic stage prior to contact with the plant, this GTPase activity could be involved in the signaling cascade controlling growth arrest and further programmed cell death in the absence of a signal from the host plant.

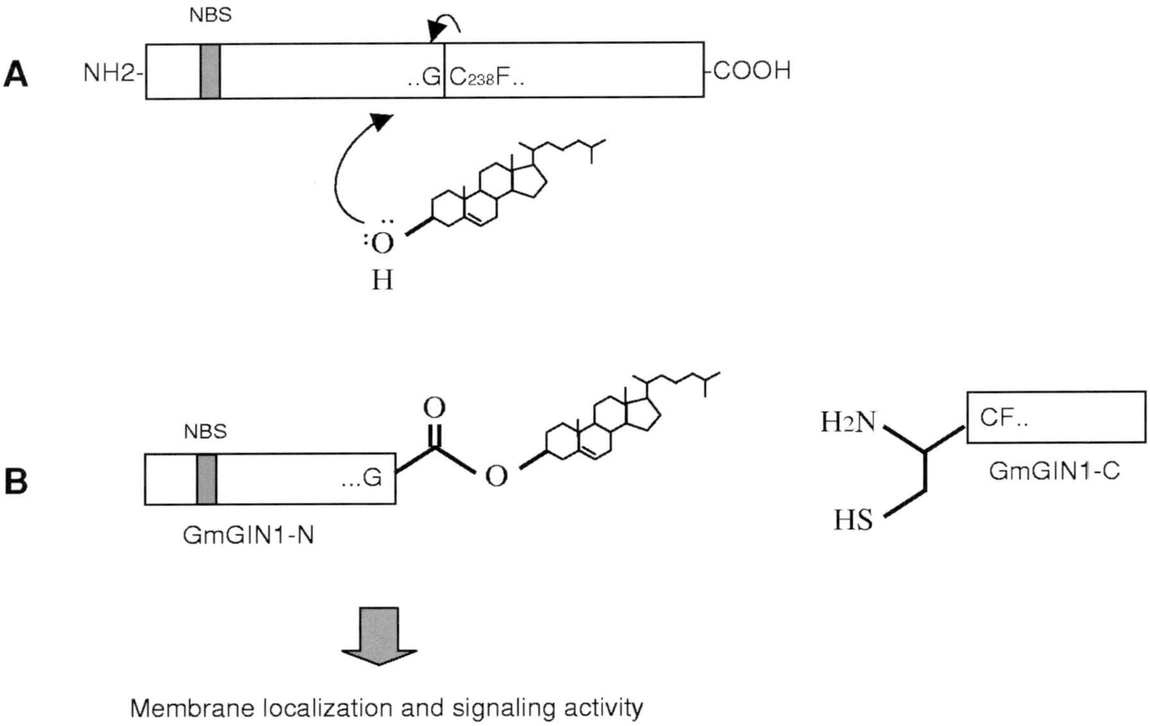

Figure 5. Diagram of the proposed autoprocessing reaction of GmGIN1 according to the model proposed for hedgehog proteins (Hall et al., 1997). A shows the protein structure of GmGIN1 with the nucleotide binding site (NBS) in the N-terminus and the GCF tripeptide containing the cysteine-238 in the middle of the protein. The Cys-238 initiates the nuclephilic attack on the carbonyl carbon to the precedent amino acid residue. This attack forms an intermediate thioester linkage between Gly-237 and Cys-238 which is resolved with the second nucleophilic attack performed by the electrons from the 3β hydroxyl group of a cholesterol molecule. As a result, the protein is spliced in two and the N-terminus results covalently attached to a cholesterol moiety (B). This modification increases considerably the hydrophobicity of the N-terminal part of the protein that may then localize to the membrane and exert a signaling function through its GTPase activity.

Acknowledgements

We thank A. Ocón-Garrido, V. Karidaki and B. Achatz for their help with the expression studies. Special thank to Drs. R. Fischer and M. Takka for critical reading of the manuscript and helpful discussions.

References

Altschul, S F, Madden T L, Schäffer A A, Zhang J, Zhang Z, Miller W and Lipman D J 1997 Gapped BLAST and PSI BLAST: A new generation of protein database search programs. Nucleic Acids Res. 25, 3389–3402.

Aspöck Gq, Kagoshima H, Niklaus G and Bürglin T R 1999 *Caenorhabditis elegans* has scores of hedgehog-related genes: Sequence and expression analysis. Genome Research 9, 909–923.

Bago B, Zipfel W, Williams R M, Chamberland H, Lafontaine J G, Webb W W and Piché Y 1998 In vivo studies of the nuclear behaviour of the arbuscular mycorrhizal fungus *Gigaspora rosea* grown under axenic conditions. Protoplasma 203, 1–15.

Bago B, Pfeffer P E, Douds D D Jr, Brouillette J, Becard G and Shachar-Hill Y 1999 Carbon metabolism in spores of the arbuscular mycorrhizal fungus *Glomus intraradices* as revealed by nuclear magnetic resonance spectroscopy. Plant Physiol 121, 263–72.

Budi S W, Blal B and Gianinazzi S 1999 Surface-sterilization of *Glomus mosseae* sporocarps for studying endomycorrhization *in vitro*. Mycorrhiza 9, 65–68.

Burglin T R 1996 Warthog and Groundhog, novel families related to Hedgehog. Curr. Biol. 15, 2294–2303.

Diatchenko L, Lau Y F, Campbell A P, Chenchik A, Moqadam F, Huang B, Lukyanov S, Lukyanov K, Gurskaya N, Sverdlov E D and Siebert P D 1996 Suppression subtractive hybridization: A method for generating differentially regulated or tissue-specific cDNA probes and libraries. Proc. Natl. Acad. Sci. USA 93, 6025–6030.

Franken P and Gianinazzi-Pearson V 1996 Phage cloning of ribosomal RNA genes from the arbuscular mycorrhizal fungi *Glomus mosseae* and *Scutellospora castanea*. Mycorrhiza 6, 167–173.

Franken P and Requena N 2001a Molecular approaches to arbuscular mycorrhiza functioning. *In* The Mycota IX. Fungal Associations. Ed. B Hock. pp 19–28. Springer-Verlag, Berlin.

Franken P and Requena N 2001b Analysis of gene expression in arbuscular mycorrhizas: New approaches and challenges. New Phytol. 150, 517–523.

Hall T M, Porter J A, Young K E, Koonin E V, Beachy P A and Leathy D J 1997 Crystal structure of a hedgehog autoprocessing domain: homology between hedgehog and self-splicing proteins. Cell 91, 85–97.

Hosny M, Pais de Barros J P, Gianinazzi-Pearson V and Dulieu H 1997 Base composition of DNA from glomalean fungi: high amounts of methylated cytosine. Fungal Genet. Biol. 22, 103–11.

Ingham P W 2000 How cholesterol modulates the signal. Curr. Biol. 10, R180–R183.

Lapopin L, Gianinazzi-Pearson V and Franken P 1999 Comparative differential RNA display analysis of arbuscular mycorrhiza in *Pisum sativum* wild type and a mutant defective in late stage development. Plant Mol. Biol. 41, 669–77.

Logi C, Sbrana C and Giovannetti M 1998 Cellular events involved in survival of individual arbuscular mycorrhizal symbionts growing in the absence of the host. Appl Environ Microbiol 64, 3473–3479.

Mann R K and Beachy P A 2000 Cholesterol modification of proteins. Biochim. Biophys. Acta1529, 188–202.

Martin-Laurent F, van Tuinen D, Dumas-Gaudot E, Gianinazzi-Pearson V, Gianinazzi S and Franken P 1997 Differential display analysis of RNA accumulation in arbuscular mycorrhiza of pea and isolation of a novel symbiosis-regulated plant gene. Mol. Gen. Genet. 256, 37–44.

Mosse B 1959 The regular germination of resting spores and some observations on the growth requirements of an *Endogone* sp. causing vesicular-arbuscular mycorrhiza. Trans. Br. Mycol. Soc. 42, 273–286.

Paulus H 2000 Protein splicing and related forms of protein autoprocessing. Ann. Rev. Biochem. 69, 447–496.

Poirier G M C, Anderson G, Huvar A, Wagaman P C, Shuttleworth J, Jenkinson E, Jackson M R, Peterson P A and Erlander M G 1999 Immune-associated nucleotide-1 (IAN-1) is a thymic selection marker and defines a novel gene family conserved in plants. J. Immunol. 163, 4960–4969.

Porter J A, Ekker S C, Park W-J, von Kessler D P, Young K E, Chen C H, Ma Y, Woods A S, Cotter R J, Koonin E V and Beachy P A 1996a Hedgehog patterning activity: Role of a lipophilic modification mediated by the carboxy-terminal autoprocessing domain. Cell 86, 21–34.

Porter J A, Young K E and Beachy P A 1996b Cholesterol modification of Hedgehog signaling proteins in animal development. Science 274, 255–297.

Redecker D, Thierfelder H and Werner D 1995 A new cultivation system for arbuscular mycorrhizal fungi on glass beads. Angewandte Botanik 69, 189–191.

Requena N, Fuller P and Franken P 1999 Molecular characterization of GmFOX2, an evolutionarily highly conserved gene from the mycorrhizal fungus *Glomus mosseae*, down-regulated during interaction with rhizobacteria. Mol. Plant-Microbe Interact.12, 934–42.

Requena N, Mann P and Franken P 2000 A homologue of the cell cycle checkpoint TOR2 from *Saccharomyces cerevisiae* exists in the arbuscular mycorrhizal fungus *Glomus mosseae*. Protoplasma 212, 89–98.

Reuber T L and Ausubel F M 1996 Isolation of *Arabidopsis* genes that differentiate between resistance responses mediated by the RPS2 and RPM1 disease resistance genes. Plant Cell 8, 241–249.

Sambrook J, Fritsch E F and Maniatis T A 1989 Molecular cloning: a laboratory manual. 2nd edn. Cold Spring Harbor Laboratory, Cold Spring Harbor, NY.

Simon L, Lalonde M and Bruns T D 1992 Specific amplification of 18S fungal ribosomal genes from vesicular-arbuscular endomycorrhizal fungi colonizing roots. Appl. Environ. Microbiol. 58, 291–295.

Simon L, Lévesque R C and Lalonde M 1993 Identification of endomycorrhizal fungi colonizing roots by fluorescent single-strand conformation polymorphism-polymerase chain reaction. Appl. Environ. Microbiol. 59, 4211–4215.

van Buuren M L, Maldonado-Mendoza I E, Trieu A T, Blaylock L A and Harrison M J 1999 Novel genes induced during an arbuscular mycorrhizal (AM) symbiosis formed between *Medicago truncatula* and *Glomus versiforme*. Mol. Plant Microbe-Interact. 123, 171–81.

Watson, A J, Worley J, Elliott R M, Jeenes D J and Archer D B 2000 Cloning stress-induced genes from *Aspergillus niger* using polymerase chain reaction-augmented subtractive hybridization. Anal Biochem 277, 162–165.

Xiong L, Lee M-W, Qi M and Yang Y 2001 Identification of defense-related rice genes by suppression subtractive hybridization and differential screening. Mol. Plant Microb. Interact. 14, 685–692.

Plant and Soil **244:** 141–148, 2002.

Expression in an arbuscular mycorrhizal fungus of genes putatively involved in metabolism, transport, the cytoskeleton and the cell cycle

Jeongwon Jun[1], Jehad Abubaker[1], Charles Rehrer[1], Philip E Pfeffer[2], Yair Shachar-Hill[1,3] & Peter J Lammers[1]

[1]*Department of Chemistry and Biochemistry, New Mexico State University, Las Cruces, NM 88001, USA.* [2]*USDA-ARS, ERRC, 600 East Mermaid Lane, Wyndmoor, PA 19038, USA.* [3]*Corresponding author**

Received 21 August 2001. Accepted in revised form 16 May 2002

Key words: Arbuscular mycorrhiza, cell cycle, cytoskeleton, meiosis, gene expression, metabolism, transport

Abstract

Arbuscular mycorrhizal (AM) fungi are multinucleate, coenocytic, obligate symbionts with no known sexual stages and very wide host and habitat ranges. While contributing vitally to the growth of land plants they face unique challenges in metabolism, transport, growth and development. To provide clues to the strategies that AM fungi have adopted, random sequencing of cDNA's from *Glomus intraradices* was undertaken. Putative genes for enzymes, transporters, structural proteins and cell-cycle regulatory factors were discovered. Among the EST's of particular interest are sequences with homology to known trehalase, arsenite transporter, cysteine synthase, tubulins, actin, dynein, cell cycle regulatory proteins, and three meiosis-related proteins. The significance of these sequences is discussed in the context of what is known about AM metabolism, transport, growth and phylogeny.

Abbreviations: AM – Arbuscular Mycorrhizal; copy DNA; EST – Expressed sequence tag; NMR – Nuclear Magnetic Resonance; GBA – GenBank Accession

Introduction

Arbuscular mycorrhizal (AM) fungi lead a biologically unique life style. As multinucleate, coenocytic, obligate symbionts with no known sexual stages and extraordinary host and habitat ranges they face unusual biological challenges. They therefore present a fascinating opportunity to examine fundamental biological mechanisms. Three areas of cellular and physiological biology invite particular scrutiny because they are very probably uniquely adapted in the AM fungi. These are metabolism and transport, cell cycle regulation and signaling, and structure and morphogenesis.

The study of the arbuscular mycorrhizal symbiosis has been hindered in the past by the small amounts of fungal tissue that are generally available, by the fact that AM fungi are obligate symbionts and by the confounding influence of other microorganisms present in field or greenhouse-grown mycorrhizal systems. The emergence of *in vitro* cultures (Becard and Fortin, 1988; St-Arnaud et al., 1996) and the application of molecular biological tools have greatly expanded the range of analyses that are possible (see Harrison, 1999; and Harrier, 2001 for reviews). Here we present some of our recent findings of putatively expressed genes that are related to the three areas of interest identified above.

Studies of AM metabolism have included analytical determination of metabolites, enzymatic assays of individual metabolic steps, microscopy and labeling with radioactive and stable isotopes (for review see Bago et al., 2000). From the literature and our own NMR based studies we have proposed working models of carbon and nitrogen metabolism (Bago et al., 2000; Bago et al., 2002). To test and extend this model, and

* FAX No.: +1-240-352-8021. E-mail: yairhill@nmsu.edu

to provide tools for studying regulation, gene identification and verification of expression are necessary. The expression of putative genes for several enzymes involved in the fluxes of carbon have been reported previously. These include 3-phosphoglycerate kinase (Harrier et al., 1998), glycerhaldehyde 3-phosphate dehydrogenase (Franken et al., 1997), isocitrate lyase and malate synthase (Lammers et al., 2001), acylcoA dehydrogenase (see accompanying paper by Bago et al. (2002)), and glycogen synthase (Abubaker et al., unpublished results).

Transport processes in the AM symbiosis have also been studied by a combination of isotopic labeling and molecular biological techniques. Of particular importance from a 'phytocentric' perspective is the movement of phosphate, and a phosphate uptake transporter has been identified from an AM fungus and its characteristics studied (Harrison and Van Buuren, 1995). A number of proton ATP'ase genes have also been identified in AM fungi (Ferrol et al., 2000). Further identification of the organic and mineral nutrients taken up and transferred by the fungal partner and of the transporters involved is needed if we are to understand the nutritional basis of the symbiosis. A different type of transport system is the intracellular trafficking of organelles (Bago et al., 1999b) vacuoles and lipid bodies (see accompanying paper by Bago et al. (2002)) that can occur over long distances within AM fungal hyphae. Such systems require structural and motility proteins and the identification of these and their organization (Butehorn et al., 1999; Timonen et al., 2001) is important for studying both transport and growth.

The cell cycle is not well understood in AM fungi, with little being known about basic questions like the ploidy level of nuclei, the identity of signals for mitosis, whether AM fungal nuclei undergo meiosis, and whether in the absence of known sexual reproduction there is evidence for lateral gene transfer. In this context, cytological studies are important and genomic level analyses are very desirable. In the meantime, clues obtained from identifying putative homologues to cell cycle genes in other organisms (Requena et al., 1998) can generate hypotheses and guide investigations of cell cycle regulation in the AM fungi.

Materials and methods

Tissue culture

Extraradical mycelium and spores used for labeling experiments as well as for cDNA library construction were obtained from *in vitro* cultures of Ri-T DNA transformed roots of carrot (*Daucus carota* L.) colonized by *Glomus intraradices* (DAOM 197198 Biosystematics Research Center, Ottawa, Canada). Conditions for this monoxenic culture were as previously described (St. Arnaud et al., 1996; Pfeffer et al., 1999). Briefly, transformed carrot roots were grown together with *G. intraradices* inoculum from a previous monoxenic culture on one side of divided Petri plates containing solidified M medium (Chabot et al., 1992). The fungal extraradical hyphae, but not the roots were permitted to grow over the dividing barrier into the other half of the plate which contained M medium without sucrose, where they proliferated and sporulated (St. Arnaud et al., 1996). To obtain axenic cultures of *G. intraradices*, spores collected from such plates were germinated in liquid M medium without sucrose for 11–14 days (1% CO_2, 32 °).

EST library construction

A cDNA library from 11-day-old germinating *G. intraradices* spores was constructed using the SMART kit (Clontech). One μg of total RNA served for first strand cDNA synthesis, and the resulting single stranded cDNA was amplified by PCR, and ligated into the λTriplEx2 vector (Lammers et al., 2001)

cDNA sequencing and analyses

λTriplEx2 clones were converted to pTriplEx2 plasmid using *E. coli* BM25.8 as the bacterial host. Plasmid DNA was prepared using QIAprep spin miniprep kit (Qiagen). Nucleotide sequences of the cDNA inserts were determined with the dideoxy chain-termination method using SequiTherm Excel II kit (Epicentre). Sequencing data was obtained from Li-Cor 4200 IR^2 automated sequencers using 66-cm plates with 4% acrylamide, 7M urea gels or an ABI 3100 instrument equipped with 50 cm capillaries.

All sequences in the *G.i.* database were compared against the GenBank NonRedundant database (National Center for Biotechnology Information, NCBI, Bethesda, MD) using BLASTX and an expectation value (E-value) cutoff 1e-5 or less. The sequences are available from NCBI with GenBank numbers

BE603746-BE604036, BI246179-BI246194, and also at http://praseo.nmsu.edu/glomus/.

Results and discussion

Metabolism and transport

Table 1 lists seven EST sequences that are homologous to known enzymes and two sequences homologous to transporters from other organisms. The length of each sequence, its expectation value (see 'Methods') and the GenBank Accession (GBA) number are also shown for each sequence.

The net breakdown of trehalose during germination of AM spores has previously been directly observed (Becard et al., 1991), and trehalose turnover has been inferred from isotopic labeling patterns in intraradical and extraradical mycelium (Lammers et al., 2001; Shachar-Hill et al., 1995). These observations are consistent with finding a sequence (GBA BI246186) that is 58% identical to a neutral trehalase from *Schizosaccharomyces pombe*, which has been characterized by Cansado et al. (1998). The sequence is also homologous to parts of the trehalase from *Neurospora crassa* (58% identical over 51 amino acids) and to the *Emericella nidulans* neutral trehalase (treB) gene (80% identical over 26 amino acids). Trehalase activity and gene expression has been demonstrated in the tissues of many plants and may be involved in carbohydrate partitioning (Muller et al., 2001). Once confirmed as a functional trehalase this AM fungal sequence should permit the regulation of plant and fungal activities to be investigated together in mycorrhizal roots.

Ectomycorrhizal fungi make polyamines apparently from *S* adenosyl methionine (Zarb and Walters, 1994) and we (Pfeffer et al., 1999) have reported the labeling of an unidentified trimethylamine that implies flux through nonphotosynthetic one carbon metabolism. Three EST sequences identified here strongly resemble the sequences of genes involved in non-photosynthetic one carbon metabolism.

(1) GBA BI246181 is a sequence with 73% identity to part of the purH gene of Neurospora. PurH includes phosphoribosyl-aminoimidazolecarboxamide formyltransferase and inosine monophosphate cyclohydrolase activities and is involved both in the transfer of formyl units and a cyclization reaction required for purine biosynthesis. The presence of a purH homolog suggests that nucleotides may be formed during spore germination that may be required for DNA replication and/or RNA biosynthesis, both of which have been demonstrated in germinating spores (Becard and Pfeffer, 1993; Bianciotto and Bonfante, 1993; Bielby, 1983). The expression of this putative gene suggests that DNA replication and transcription may involve the biosynthesis of new nucleotides. (2) GBA BI246182 is a sequence having high homology (67% identity over 180 amino acids) to the eth-1 gene of *Neurospora crassa* that encodes an *S*-adenosylmethionine synthetase (Mautino et al., 1996). *S*-adenosylmethionine is the methyl donor for a large number of biosynthetic reactions. (3) GBA BE603835 is 82% identical to a *Schizosaccharomyces pombe* *S*-adenosyl homocycsteine hydrolase gene over 98 amino acid residues. The hydrolase activity putatively encoded by this sequence is responsible for recycling the *S*-adenosyl homocycsteine generated when *S*-adenosylmethionine acts as a methyl donor. The activated methyl cycle in which this enzyme functions makes methionine and allows the transfer of methyl groups from tetrahydrofolate to a spectrum of biomolecules. The random identification of three putative sequences for non-photynthetic on carbon metabolism in germinating spores is consistent with active transcription of the genes of the different sections of this metabolic network. This is consistent with the fact that biosynthesis, signaling and other functions expected to be active at this stage of the AM fungal life cycle depend on fluxes through non-photosynthetic one carbon metabolism.

The acquisition and use of exogenous nutrients would be required to allow presymbiotic growth to continue after internal stores are consumed. Although carbon does not appear to limit presymbiotic growth and sugars are only taken up to a limited extent (Bago et al., 1999a), this may not be true for other nutrients. The sequence GBA BI246184 is extremely similar (82% identical over 185 amino acids) to the cysteine synthase of *Aspergillus nidulans*. This enzyme produces cysteine from *O*-acetylserine is part of an enzyme complex that is central to sulfur assimilation in plants and fungi. If confirmed in its putyative function this sequence will be a useful tool for studying sulfur nutrition in the AM symbiosis, and area about which there appears to be no information at present.

GBA BI246187 bears high homology to the arsenite transporter (arsenite-translocating ATP'ase) from *Schizosaccharomyces pombe* (67% identity over 170 amino acids). These transporters confer arsenate res-

Table 1. cDNAs encoding putative enzymes and transporters

Predicted function based on homology		Expectation value	EST sequence length (bp)	Genbank accession number
Neutral Trehalase		4e-26	793	BI246186
Non-photosynthetic one carbon metabolism	*S*-adenosyl homocysteinase	2e-39	489	BE603835
	S-adenosyl methionine synthetase	4e-60	570	BI246182
	Bi-functional purine biosynthesis enzyme	2e-79	823	BI246181
Lipid metabolism	Sterol *O*-acyl transferase	7e-22	776	BE604005
	Linoleate diol synthase	2e-38	730	BI246194
Cysteine synthase		2e-86	817	BI246184
Arsenite translocating ATP'ase		4e-61	605	BI246187
Glyoxal oxidase		4e-20	793	BI246189

istance by pumping arsenic out of cells in the form of arsenite. The uptake of phosphate by mycorrhizal fungi is central to mycorrhizal symbioses and AM fungi have an active phosphate transporter (Harrison and Van Buuren, 1995). Because of its chemical similarity to phosphate, arsenate can enter cells, and this is expected to be a particular problem for mycorrhizal fungi when arsenate is present at significant levels. At least one Ericoid mycorrhizal fungus is known to be arsenate resistant and it has been shown that an enhanced efflux of arsenite is present in resistant compared to susceptible varieties (Sharples et al., 2000) An arsenite exporting pump would be of advantage to AM fungi in allowing them to operate active phosphate uptake systems in the presence of soil arsenate.

GBA BI246194 is similar (39% identical over 222 amino acids) to a known linoleate diol synthase from the fungus *Gaeumannomyces graminis*. This enzyme produces dihydroxylinoleate and is a heme lipoxygenase related to prostaglandin producing enzymes in animals but its role in fungal metabolism is undetermined. Linoleate is a C18 dienoic fatty acid and at least one such fatty acid is present in the storage lipid of *G. intraradices* spores (Jabaji-Hare, 1988).

BE604005 is homologous to acyl-CoA:cholesterol acyltransferases (ACATs) from various fungi and in particular it is most similar in sequence to an ACAT-related gene in *S. cerevisiae*. ACATs esterify cholesterol to fatty acids and this esterification is involved in the incorporation of sterol into lipid droplets. Sterols are known to be present in the AM fungal spores (Grandmougin-Ferjani et al., 1999) and it is possible that this putative ACAT or ACAT-related sequence is involved in sterol handling during germination. An ACAT related protein has also been reported as being involved in the synthesis of triglyceride (TAG) in yeast (Sandager et al., 2000). ^{13}C labeling experiments show that there is no detectable *de novo* formation of TAG in germinating spores (Bago et al., 1999a) but there is incorporation of exogenously supplied glycerol into TAG, presumably by exchange of the glyceryl moieties (Bago et al., unpublished results). It is possible that this homolog of ACAT-related protein may be involved in the turnover of triglycerides.

As well as biosynthetic functions, AM fungi in the presymbiotic phase have catabolic needs for using endogenous storage metabolites. It seems likely that extracellular catabolic enzymes may be involved in establishment of the symbiosis and perhaps also in nutrient acquisition and/or defense. GBA BI246189 is similar to Glyoxal oxidase from the basidiomycete *Phanerochaete chrysosporium*. This enzyme is involved in lignin degradation of wood rotting fungi (Whittaker et al., 1999). The sequence also resembles a putative *Arabidopsis* pectinase. The homology of this sequence to enzymes of cell wall breakdown is interesting in a presymbiotic fungal stage since breakdown of host cell walls is probably important in penetration of the host root and the establishment of the symbiosis (Garcia-Garrido et al., 2000).

Cytoskeleton/Filaments

The cytoskeleton of AM fungi is known to contain microtubles and microfilaments (Astrom et al., 1994) and Beta tubulin has previously been shown to be expressed in germinating AM spores (Franken et al., 1997). Tubulin has also been visualized in AM fungal mitotic spindles (Requena et al., 2000) and forming a network connecting the extraradical and extraradical fungal mycelium (Timonen et al., 2001). Sequences putatively encoding several other cytoskeletal proteins were discovered in this sequencing work.

GBA BI246191 is 88% identical to alpha tubulin from *Drosophila melanogaster* and is also very similar to alpha tubulins from a range of other plant and animal species. Two independent EST sequences overlap to form a contiguous sequence, BI246179, which is very similar (close to 90% percent identical) to beta tubulin sequences from several fungi. Beta tubulin has been shown to be expressed at substantial levels at different stages of the AM fungal life cycle (Butehorn et al., 1999). GBA BI246188 has over 90% identity at the amino acid level to Beta Actin sequences from the basidiomycete *Schizophyllum commune* and other fungi as well as other organisms. GBA BI246192 is a sequence with homology (54% identity over 83 amino acids) to human dynein. Dynein functions with microtubules in the trafficking of organelles and in motility. It is known to be involved in nuclear movement and spacing in filamentous fungi (Alberti-Segui et al., 2001) and AM fungi are known to move their nuclei along hyphae and to locate them at regular intervals (Bago et al., 1999b) The trafficking of lipid bodies is an active pathway of carbon movement within the fungus (see accompanying paper) so that dynein may also be involved in this important nutritional flux, as well as in the movement of polyphosphate-containing vacuoles.

Nuclear migration and distribution also involves other genetic loci and an EST sequence, BI246180, displays a high level of homology (73% percent identity over 145 amino acids) to actin-related protein **Arp** I from *Aspergillus nidulans* that is known to be required for normal nuclear distribution along hyphae (Xiang et al., 1999). **Arp** I is a component of the dynactin complex and its presence in germinating spores is therefore likely to be related to the expression of the putative dynein sequence. The actin-related protein (**Arp**) subfamily has been identified based on sequence homology to conventional actin, and although some Arps are clearly involved in cytoskeletal functions, both actin and/or Arps have been found as stoichiometric subunits of several nuclear chromatin-remodeling enzymes (Boyer and Peterson, 2000).

Cell cycle

The life style of AM fungi dictate that controls on chromosome replication and nuclear division are unlinked to cytokinesis. The genetic basis for the asexual lifestyle of AM fungi is also unknown. Our gene identification efforts in *G. intraradices* via EST sequencing have revealed putative homologs to cell cycle and meiosis-specific genes from other fungi as shown in Table 3. These examples suggest that known signaling pathways may be present in AM fungi but in the absence of almost any other information in this area (Requena et al., 2000), caution is important when interpreting these findings.

A cDNA sequence from *G. intraradices* germinating spores was found to be similar to a MAT-2 like gene product from *Fusarium oxysporum* (BE603853). Pairwise alignments show 49/122 amino acid identities for the 126 residue protein. MAT gene products have been shown to be sufficient to determine mating type in many heterothallic and homothallic fungi (Yun et al., 1999). The best match with our sequence was from *F. oxysporum*, an asexual fungus that nevertheless carries MAT1 and MAT2 genes that are transcribed and processed to mature mRNAs (Yun et al., 2000).

A composite made from two overlapping independent cDNA clones (BE603782, BE603969) was found to have significant homology to STE12-like transcription factors that are normally components of fungal mating-type specific pheromone response signaling pathways. The *G. intraradices* cDNA best aligns with the N-terminal region of a *Neurospora* STE12-like sequence (86/100 amino acid identities). The region of similarity encompasses the DNA binding domain. The STE12 like proteins are transcription factors that cooperate with other factors in activating genes that promote mating and subsequent fusion of complementary haploid cells (Laroche et al., 1995). STE12 appears to be important for pseudohyphal differentiation in *S. cerevisieae,* and in *Cryptococcus neoformans* (Chang et al., 2000), so the *Glomus* STE12 like sequence could be associated with either morphological differentiation or perhaps with the known anastamosis of hyphae from different spores of the same species (Giovannetti et al., 1999).

Another *G. intraradices* sequence was identified with similarity to a meiosis-specific gene called NDT80 that is more difficult to rationalize as meiosis-independent. The closest database match to this sequence was with a N. *crassa* NDT80 protein that has

Table 2. cDNAs encoding putative cytoskeleton proteins

Predicted function based on BLASTX matches	Expectation value	EST sequence length (bp)	Genbank accession number
Alpha Tubulin	1e-51	550	BI246191
Beta Tubulin	1e-102	810	BI246179
Beta Actin	6e-91	746	BI246188
Dynein Intermediate Chain	6e-23	536	BI246192
Actin Related Protein (Actin II)	8e-95	768	BI246180

Table 3. cDNAs encoding putative cell cycle and meiosis-related proteins

Predicted function based on BLASTX matches	Expectation value	EST sequence length (bp)	Genbank Accession Number
Similar to MAT-2 (Mating Type) Protein – *Fusarium*	3e-15	992	BE603853
Similar to STE12α – *Filobasidiella*	6e-38	729	BE603782, BE603969
Related to Meiosis-Specific Protein, NDT80	2e-18	757	BI246190
Similar to myeloblastosis (myb-B) oncogene	8e-23	663	BI246185
rRNA intron-encoded homing endonuclease	5e-15	785	BI246193
Adenylate cyclase regulatory protein, also known as CDC25 in *Saccharomyces cerevisiae*	7e-32	760	BI246183

not been well studied, however second-best match was to the yeast NDT80 protein which activates middle sporulation genes and is required for exit from the pachytene phase of meiosis (Xu et al., 1995). Specifically, NDT80 functions to activate transcription of a set of genes required for both meiotic division (e.g., B-type cyclins) and gamete formation (Chu and Herskowitz, 1998) and it is expressed only during the middle phase of sporulation (Chu et al., 1998). A sequence similar to the myeloblastosis (myb) oncogenes was also identified in the EST survey. The myb genes have been implicated in both meiotic differentiation and somatic cell proliferation in mammalian systems (Latham et al., 1996). These *G. intraradices* sequences provide an avenue for studying the regulation of chromosome replication within multinucleate hyphae.

GBA BI246193 shows a strong similarity to a homing endonuclease associated with lateral transfer of an rDNA intron. Evidence presented by Goddard and Burt (1999) suggests that this selfish gene has undergone repeated lateral transfer in the yeasts. Such lateral gene transfer might occur in the AM fungi though nuclear exchange that has been observed to occur through anastamoses formed between the hyphae of different germinating spores of AM fungi (Giovennetti et al., 1999). Further studies of this rDNA intron may well provide useful insights about the evolution of AM fungi and their potential for lateral gene transfer.

Finally, a homolog of a cell cycle regulatory protein known as CDC25 from yeast was identified in *G. intraradices* cDNA. The CDC25 proteins are thought to act as phosphatases on the phosphorylated form of adenylate cyclase, involved in the RAS signaling pathway (Broek et al., 1987), while in *Schizosaccharomyces Pombe* CDC25 is thought to play a regulatory role in cAMP-dependent cell size regulation (Kishimoto and Yamashita, 2000).

Conclusion

The identification of putative homologues of known genes in other organisms by random sequence analysis and comparative functional genomics is an effective way to learn more about hard-to-study organisms like AM fungi (Sawaki and Saito, 2001; Stommel et al., 2001). Homologous sequences provide support for the existence of basic mechanisms for processes that are known to take place in AM fungi – for example motility of nuclei and other organelles. Such sequences also provide independent evidence for processes that have been deduced from functional studies, such as the activity of particular metabolic pathways, fluxes through which had been deduced from enzymological or isotopic labeling data. The cDNA's also provide probes for studying the regulation of different processes at the level of transcription – this should provide powerful tools for studying signaling events and for analyzing responses to environmental or developmental changes. Finally, the existence of homologues for genes involves in processes that are either little studied or unknown in AM fungi suggests interesting hypotheses that may be tested using those sequences or other means. However the limitations of the random sequencing approach must always be borne in mind. The existence of sequence homology does not prove function since proteins of similar sequence may perform different functions in different species. No methods currently exist for making knockouts or transgenics in AM fungi, so it is difficult to unambiguously confirm such functions. Heterologous expression, complementation and/or other methods can be applied to support functional assignments for some but not all of the putative assignments made here.

References

Alberti-Segui C, Dietrich F, Altmann-Johl R, Hoepfner D and Philippsen P 2001 Cytoplasmic dynein is required to oppose the force that moves nuclei towards the hyphal tip in the filamentous ascomycete *Ashbya gossypii*. J. Cell Sci. 114, 975–986.

Astrom H, Giovannetti M and Raudaskoski M 1994 Cytoskeletal components in the arbuscular mycorrhizal fungus *Glomus mosseae*. Mol. Plant-Microbe Interact. 7, 309–312.

Bago B, Pfeffer P E, Douds D D, Brouillette J, Becard G and Shachar-Hill Y 1999a Carbon metabolism in spores of the arbuscular mycorrhizal fungus *Glomus intraradices* as revealed by nuclear magnetic resonance spectroscopy. Plant Physiol. 121, 263–271.

Bago B, Pfeffer P E and Shachar-Hill Y 2000 Carbon metabolism and transport in arbuscular mycorrhizas. Plant Physiol. 124, 949–957.

Bago B, Pfeffer P E and Shachar-Hill Y 2001 Could the urea cycle be translocating nitrogen in the arbuscular mycorrhizal symbiosis? New Phytol. 149, 4–8.

Bago B, Zipfel W, Williams R M and Piche Y 1999b Nuclei of symbiotic arbuscular mycorrhizal fungi as revealed by *in vivo* two-photon microscopy. Protoplasma 209, 77–89.

Bago B, Pfeffer P E, Zipfel W, Lammers P and Schachar-Hill Y 2002 Trading metabolism and imaging transport in arbuscular mycorrhizal fungi. Plant Soil 244, 191–199.

Becard G, Doner LW, Rolin D B, Douds D D and Pfeffer P E 1991 Identification and quantification of trehalose in vesicular-arbuscular mycorrhizal fungi by *in vivo* C-13 NMR and HPLC analyses. New Phytol. 118, 547–552.

Becard G and Fortin J A 1988 Early events of vesicular arbuscular mycorrhiza formation on ri T-DNA transformed roots. New Phytol. 108, 211–218.

Becard G and Pfeffer P E 1993 Status of nuclear division in arbuscular mycorrhizal fungi during *in-vitro* development. Protoplasma 174, 62–68.

Beilby J P 1983 Effects of inhibitors on early protein, RNA, and lipid synthesis in germinating vesicular-arbuscular mycorrhizal fungal spores of *Glomus caledonium*. Can. J. Bot. 29, 596–601.

Bianciotto V and Bonfante P 1993 Evidence of DNA-replication in an arbuscular mycorrhizal fungus in the absence of the host-plant. Protoplasma 176, 100–105.

Boyer L A and Peterson C L 2000 Actin-related proteins (Arps): conformational switches for chromatin-remodeling machines? Bioessays 22, 666–672.

Broek D, Toda T, Michaeli T, Levin L, Birchmeier C, Zoller M, Powers S and Wigler M 1987 The S. cerevisiae CDC25 gene product regulates the RAS/adenylate cyclase pathway. Cell 48, 789–799.

Butehorn B, Gianinazzi-Pearson V and Franken P 1999 Quantification of beta-tubulin RNA expression during asymbiotic and symbiotic development of the arbuscular mycorrhizal fungus *Glomus mosseae*. Mycol. Res. 103, 360–364.

Cansado J, Soto T, Fernandez J, Vicente-Soler J and Gacto M 1998 Characterization of mutants devoid of neutral trehalase activity in the fission yeast *Schizosaccharomyces pombe*: partial protection from heat shock and high-salt stress. J. Bacteriol. 180, 1342–1345.

Chabot S, Becard G and Piche Y 1992 Life-cycle of Glomus intraradix in root organ culture. Mycologia 84, 315–321.

Chang Y C, Wickes B L, Miller G F, Penoyer L A and Kwon-Chung K J 2000 *Cryptococcus neoformans* STE12 alpha regulates virulence but is not essential for mating. J. Exp. Med. 191, 871–881.

Chu S and Herskowitz I 1998 Gametogenesis in yeast is regulated by a transcriptional cascade dependent on NDT80 Mol. Cell 1, 685–696.

Chu S, DeRisi J, Eisen M, Mulholland J, Botstein D, Brown PO and Herskowitz I 1998 The transcriptional program of sporulation in budding yeast. Science 282, 699–705.

Ferrol N, Barea J M and Azcon-Aguilar C 2000 The plasma membrane H+-ATPase gene family in the arbuscular mycorrhizal fungus Glomus mosseae. Current Genetics 37, 112–118.

Franken P, Lapopin L, MeyerGauen G and Gianinazzi-Pearson V 1997 RNA accumulation and genes expressed in spores of the arbuscular mycorrhizal fungus, *Gigaspora rosea*. Mycologia 89, 293–297.

Garcia-Garrido J M, Tribak M, Rejon-Palomares A, Ocampo J A and Garcia-Romera I 2000 Hydrolytic enzymes and ability of arbuscular mycorrhizal fungi to colonize roots. J. Exp. Bot. 51, 1443–1448.

Giovannetti M, Azzolini D and Citernesi A S 1999 Anastomosis formation and nuclear and protoplasmic exchange in arbuscular mycorrhizal fungi. Appl. Env. Microbiol. 65, 5571–5575.

Grandmougin-Ferjani A, Dalpe Y, Hartmann M A, Laruelle F and Sancholle M 1999 Sterol distribution in arbuscular mycorrhizal fungi. Phytochemistry 50, 1027–1031.

Goddard M R and Burt A 1999 Recurrent invasion and extinction of a selfish gene. Proc. Natl. Acad. Sci. USA 96, 13880–13885.

Harrier L A 2001 The arbuscular mycorrhizal symbiosis: a molecular review of the fungal dimension. J. Exp. Bot. 52, 469–478.

Harrier L A, Wright F and Hooker 1998 JE Isolation of the 3-phosphoglycerate kinase gene of the arbuscular mycorrhizal fungus *Glomus mosseae* (Nicol. & Gerd.) Gerdemann & Trappe. Current Genetics 34, 386–392.

Harrison M J 1999 Molecular and cellular aspects of the arbuscular mycorrhizal symbiosis. Ann. Rev. Plant Physiol. Plant Mol. Biol. 50, 361–389.

Harrison M J and Van Buuren M L 1995 A phosphate transporter from the mycorrhizal fungus glomus versiforme. Nature 378, 626–629.

Jabaji-Hare S 1988 Lipid and fatty-acid profiles of some vesicular-arbuscular mycorrhizal fungi – contribution to taxonomy. Mycologia 80, 622–629.

Kishimoto N and Yamashita I 2000 Cyclic AMP regulates cell size of *Schizosaccharomyces pombe* through Cdc25 mitotic inducer. YEAST 16(#6) pp 523–529.

Lammers P J, Jun J, Abubaker J, Arreola R, Gopalan A, Bago B, Hernandez-Sebastia C, Allen J W, Douds D D, Pfeffer PE and Shachar-Hill Y 2001 The glyoxylate cycle in an arbuscular mycorrhizal fungus: carbon flux and gene expression. Plant Physiology, in press.

Laroche S D, Shafer B K and Strathern JN 1995 A STE12-allele having a differential effect on a versus alpha-cells. Mol. Gen. Genet. 246, 80–90.

Latham K E, Litvin J, Orth J M, Patel B, Mettus R and Reddy E P 1996 Temporal patterns of A-myb and B-myb gene expression during testis development. Oncogene 13, 1161–1168.

Mautino M R, Barra J L and Rosa A L 1996 ETH-1, the *Neurospora crassa* locus encoding *S*-adenosylmethionine synthetase: molecular cloning, sequence analysis and *in vivo* overexpression. Genetics 142, 789–800.

Muller J, Aeschbacher R A, Wingler A, Boller T and Wiemken 2001 A Trehalose and trehalase in *Arabidopsis*. Plant Physiol. 125, 1086–1093.

Pfeffer P E, Douds D D, Becard G, Shachar-Hill Y 1999 Carbon uptake and the metabolism and transport of lipids in an arbuscular mycorrhiza. Plant Physiol. 120, 587–598.

Redecker D, Kodner R and Graham L E 2000 Glomalean fungi from the Ordovician. Science 289, 1920–192.

Requena N, Mann P and Franken P 2000 A homologue of the cell cycle check point TOR2 from *Saccharomyces cerevisiae* exists in the arbuscular mycorrrhizal fungus *Glomus mosseae*. Protoplasma 212, 89–98.

Sandager L, Dahlqvist A, Banas A, Stahl U, Lenman M, Gustavsson M and Stymne S 2000 An acyl-CoA:cholesterol acyltransferase (ACAT)-related gene is involved in the accumulation of triacylglycerols in *Saccharomyces cerevisiae*. Biochem. Soc. Trans. 28, 700–702.

Sawaki H and Saito M 2001 Expressed genes in the extraradical hyphae of an arbuscular mycorrhizal fungus, *Glomus intraradices*, in the symbiotic phase. FEMS Microbiol Lett. 195, 109–113.

Shachar-Hill Y, Pfeffer P E, Douds D, Osman S F, Doner L W and Ratcliffe R G 1995 Partitioning of intermediary carbon metabolism in vesicular-arbuscular mycorrhizal leek. Plant Physiol. 108, 7–15.

Sharples J M, Meharg A A, Chambers S M and Cairney J W G 2000 Mechanism of arsenate resistance in the ericoid mycorrhizal fungus *Hymenoscyphus ericae*. Plant Phsyiol. 124, 1327–1334.

Simon L 1996 Phylogeny of the Glomales: Deciphering the past to understand the present. New Phytol. 133, 95–101.

Simon L, Bousquet J, Levesque R C and Lalonde M 1993 Origin and diversification of endomycorrhizal fungi and coincidence with vascular land plants. Nature 363, 67–69.

St-Arnaud M, Hamel C, Vimard B, Caron M and Fortin J A 1996 Enhanced hyphal growth and spore production of the arbuscular mycorrhizal fungus *Glomus intraradices* in an *in vitro* system in the absence of host roots. Mycol. Res. 100, 328–332.

Stommel M, Mann P and Franken P 2001 EST-library construction using spore RNA of the arbuscular mycorrhizal fungus *Gigaspora rosea*. Mycorrhiza 10, 281–285.

Timonen S, Smith F A and Smith S E 2001 Microtubules of the mycorrhizal fungus *Glomus intraradices* in symbiosis with tomato roots. Can. J. Bot. 79, 307–313.

Whittaker M M, Kersten P J, Cullen D and Whittaker J W 1999 Identification of catalytic residues in glyoxal oxidase by targeted mutagenesis. J. Biol. Chem. 274, 36226–36232.

Xiang X, Zuo W Q, Efimov V P and Morris N R 1999 Isolation of a new set of *Aspergillus nidulans* mutants defective in nuclear migration. Curr. Genet. 35, 626–630.

Xu L Z, Ajimura M, Padmore R, Klein C and Kleckner N 1995 NDT80, a meiosis-specific gene required for exit from pachytene in *Saccharomyces cerevisiae* Mol. Cell. Biol. 15, 6572–6581.

Yun S H, Berbee M L, Yoder O C and Turgeon B G 1999 Proc. Natl. Acad. Sci. USA 96, 5592–5597.

Yun S W, Arie T, Kaneko I, Yoder O C and Turgion B G 2000 Molecular organization of mating type loci in heterothallic, homothallic, and asexual gibberella/fusarium species. Fungal Genet. Biol. 31, 2–20.

Zarb J and Walters D R 1994 The formation of cadaverine, aminopropylcadaverine and n,n bis (3-aminopropyl) cadaverine in mycorrhizal and phytopathogenic fungi. Lett. Appl. Microbiol. 19, 277–280.

Plant and Soil **244:** 149–163, 2002.

Recent developments in the study of orchid mycorrhiza

Hanne N. Rasmussen
*Danish Forest & Landscape Research Institute, 11 Hoersholm Kongevej, DK 2970 Hoersholm, Denmark**

Received 21 August 2001. Accepted in revised form 12 December 2001

Key words: basidiomycetes, mycoheterotrophy, Orchidaceae, plant-fungal relationships, specificity, symbiosis

Abstract

Orchids are mycoheterotrophic during their seedling stage and in many species the dependency on fungi as a carbohydrate source is prolonged into adulthood. The mycobionts in orchid mycorrhiza belong in at least 5 major taxonomic groups of basidiomycetes. Traditional records have mainly focused on saprotrophic mycobionts but the participation of both ectomycorrhizal and parasitic fungi in orchid mycorrhiza has been corroborated. There is an increasing evidence of specific relationships between orchids and fungi, though usually not on a species-to-species level. Physiological compatibility demonstrated under artificial conditions, as *in vitro*, may be much broader, however. Recent development of field sowing techniques has improved the possibilities of evaluating orchid-fungal relations in an ecological context. Although the general nutrient flow in orchid mycorrhiza is well known, some questions remain regarding breakdown processes of fungi within orchid tissues, especially the ptyophagic syndrome that has recently been illustrated at the ultrastructural level for the first time.

Energy sources for orchid mycorrhiza in the field

Fungi associated with orchid mycorrhiza (OM) have traditionally been mostly regarded as saprotrophs, dead organic material thus being the energy source for the symbiosis. This is supported by well documented cases where orchids have developed with fungi cultured on organic additives *in vitro*, or on organic debris. A water agar with the addition of ground woodchips can sustain some orchid-fungus symbioses from seed to small plant (Whigham et al., MS). Many of the orchid endophytes that are referred to *Rhizoctonia* DC (*sensu lato*) are saprophytes (Roberts, 1999). *Thanatephorus ochraceus* (Massee)P.Roberts (Table 1) for example, grows on decaying wood and dead fern fronds (Roberts, 1998). Rhizoctonias associated with orchids produce a range of carbohydrate-degrading and other enzymes enabling the breakdown of plant debris (a summary in Rasmussen, 1995). *Rhizoctonia* strains referred to *Ceratorhiza* R.T. Moore produce polyphenoloxidases, which are active in the breakdown of lignin (Zelmer et al., 1996). Species of *Mycena* that associate with species of *Cymbidium* and *Gastrodia* (Fan et al., 1996; Lan et al., 1996), are acknowledged saprotrophs. *Lentinus edodes* Berk., the shiitake mushroom, that is a white-rot saprotroph, can support the development of a chlorophyll-deficient orchid, *Erythrorchis ochobiensis* (Hayata) Garay, as can a range of other wood rotting fungi (Umata, 1998a). The natural symbiont of *E. ochobiensis*, as well as *Galeola altissima* (Bl.) Bl., another chlorophyll-deficient liana, is *Erythromyces crocicreas* (Berk.&Br.) Hjortst. & Ryv., the causal agent for white pocket rot in wood (Hjortstam and Tellería, 1990; Umata, 1995).

Some anastomosis groups of *Rhizoctonia solani* Kühn, a well known plant pathogen, are found in constant association with orchids (Carling et al., 1999; Perkins and McGee, 1995); similar associations have been described previously but usually as unstable relationships (survey in Rasmussen, 1995). The pathogenic rhizoctonias are generally necrotrophic parasites, first killing their host and subsequently living =saprotrophically from it (Roberts, 1999). This life form thus requires a considerable saprotrophic capacity (Garrett, 1962). Hypovirulent strains of *R. solani* exist and it is not known whether the energy

* FAX No: 45-763233. E-mail: hnr@fsl.dk

Table 1. The diversity of fungi in mycorrhizal relationship with orchids involve saprotrophs, ectomyccorhizal and parasitic fungi and 5 major basidiomycete groups: **Heterobasidiomycetes**: *Ceratobasidium, Oliveonia, Thanatephorus, Tulasnella, Sebacina*, **'Hericianae'**: *Russula* , **'Hymenocaetanae'**: *Erythromyces*, **'Thelephoranae'**: *Thelephora-Tomentella*, and '**Agaricanae**': *Armillaria, Mycena* (higher basidiomycete; systematics according to informal system by Henrik Petersen, http://www.mycokey.com/aau/systematics/systematicsbasifuturistic.html.). Species combinations from well documented symbioses are listed, but this is not meant to imply that these species are exclusive to other symbionts. The list is not intended to be exhaustive but mainly comprises recent developments. ∗ Source not seen. ∗[2]Not evident from reference; found repeatedly in tissues of C.s. but germination test failed; yet germination test succeeded with *Gastrodia elata* and *Dendrobium* spp. ∗[3]According to Henrik Petersen *Oliveonia* belongs in Ceratobasidiales. ∗[4] Source not seen.

Teleomorph fungus	Anamorph fungus	Species of orchid	Synomyms of fungus	Fungal group	Orchid lifeform	Fungal lifeform	Source to fungal identity and synonymy	Source to relationship
Armillaria jezoensis Cha and Igarashi		*Galeola septentrionalis* Reichb.f.		Tricholomatales	terrestrial, chlorophyll deficient	parasite		Cha and Igarishi, 1996
Armillaria mellea (Vahl.Fr.) Karst.		*Gastrodia elata* Bl.		Tricholomatales	terrestrial, chlorophyll deficient, climber	parasite		Lan et al., 1994
Ceratobasidium bicorne J. Erikss. and Ryvarden		*Prasophyllum macrostachyum* R. Br.	*Ypsilonidium anomalum* Warcup and Talbot =*Thanatephorus anomala*	Ceratobasidiales	terrestrial geophyte	saprotroph, parasite	Roberts, 1999	Warcup and Talbot, 1980
Ceratobasidium cornigerum (Bourdot) Rogers		*Acianthus reniformis* (R.Br.) Schltr.	*C. obscurum* Aut.	Ceratobasidiales	terrestrial geophyte	saprotroph, parasite	Roberts, 1999	Warcup and Talbot, 1971
Ceratobasidium cornigerum (Bourdot) Rogers		Canadian orchids	*C. obscurum* Aut.	Ceratobasidiales		saprotroph, parasite	Roberts, 1999	Currah et al, 1987
Ceratobasidium cornigerum (Bourdot) Rogers		*Goodyera repens* (L.) R.Br.		Ceratobasidiales	terrestrial rhizomatous evergreen	saprotroph, parasite	Roberts,1999	
Ceratobasidium cornigerum (Bourdot) Rogers	*Ceratorhiza goodyerae -repentis* (Costantin & Dufour) Moore	*Platanthera obtusata* (Banks ex Pursh) Lindl.		Ceratobasidiales	terrestrial geophyte	saprotroph	Roberts, 1999	Currah et al., 1990
Ceratobasidium cornigerum (Bourdot) Rogers		*Pterostylis* sp., *Prasophyllum* sp.		Ceratobasidiales	terrestrial geophyte	saprotroph, parasite	Roberts, 1999	
Ceratobasidium cornigerum (Bourdot) Rogers		*Sarcochilus* sp.	*C. papillatum* Warcup & Talbot	Ceratobasidiales	epiphytic, monopodial	saprotroph, parasite	Roberts, 1999	
Ceratobasidium cornigerum (Bourdot) Rogers		*Thrixspermum congestum* (F. M. Bail) Dockr.	*C. papillatum* Warcup & Talbot	Ceratobasidiales	epiphytic, monopodial	saprotroph, parasite	Roberts, 1999	
Ceratobasidium globisporum Warcup & Talbot		*Calanthe triplicata* (Willem.) Ames		Ceratobasidiales	terrestrial, pseudobulbous	only known as orchid symbiont	Roberts, 1999	
Ceratobasidium globisporum Warcup & Talbot		*Trichoglottis australiensis* Dockr.		Ceratobasidiales	epiphytic, monopodial	only known as orchid symbiont	Roberts, 1999	
Ceratobasidium pseudocornigerum M.P.Christ.		*Pterostylis mutica* R.Br. Warcup & Talbot	*C. angustisporum*	Ceratobasidiales	terrestrial geophyte	saprotroph	Roberts, 1999	Warcup & Talbot, 1980
Ceratobasidium sp.		*Taeniophyllum obtusum* Bl.		Ceratobasidiales	epiphytic, leafless, photosynthetic roots	saprotroph		Irawati, 1993*
Ceratobasidium sphaerosporum Warcup & Talbot		members of Sarcanthinae		Ceratobasidiales	epiphytic, monopodial	only known as orchid symbiont	Roberts, 2000	
Ceratobasidium sphaerosporum Warcup & Talbot		*Pomatocalpa macphersonii* (F. Muell.) Hunt		Ceratobasidiales	epiphytic, monopodial	only known as orchid symbiont	Roberts, 1999	Warcup & Talbot, 1971
Ceratobasidium sphaerosporum Warcup & Talbot		*Robiquetia wassellii* Dockrill		Ceratobasidiales	epiphytic, monopodial	only known as orchid symbiont	Roberts, 1999	Warcup & Talbot, 1971

Table 1. Continued

Erythromyces crocicreas (Berk. & Br.) Hjortst.&Ryv.	*Galeola altissima* (Bl.) Bl.	*Hymenochaete crocicreas* Berk. & Br.	Aphyllophorales	terrestrial, chlorophyll deficient, climber	saprotroph		Umata, 1995
Erythromyces crocicreas (Berk. & Br.) Hjortst.& Ryv.	*Erythrorchis ochobiense* (Hayata) Garay	*Hymenochaete crocicreas* Berk. & Br.	Hymenochaetales	terrestrial, chlorophyll deficient, climber	saprotroph		Umata, 1998
Moniliopsis anomala Currah	*Coeloglossum viride* (L.) Hartm.		Ceratobasidiales	terrestrial geophyte	only known as orchid symbiont		Currah et al., 1990
Moniliopsis anomala Currah	*Platanthera hyperborea* (L.) Lindl.		Ceratobasidiales	terrestrial geophyte	only known as orchid symbiont		Currah et al., 1990
Mycena orchidicola Fan et Guo	*Cymbidium sinense* (Andr.) Willd.		Tricholomatales	terrestrial rhizomatous	saprotroph		Fan et al., 1996*[2]
Mycena osmundicola Lange	*Gastrodia elata* Bl.		Tricholomatales	terrestrial, chlorophyll deficient, climber	saprotroph		Lan et al., 1996
Oliveonia pauxilla FK(H.S.Jacks.) Donk	?		Exidiales*[3]		saprotroph	Roberts, 1999	Warcup, 1975
Russulaceae spp.	*Corallorhiza maculata* (Rafin.) Rafin.		"Gloecystidiales"	terrestrial, chlorophyll deficient	ectomycorrhizal		Taylor and Bruns, 1999
Russulaceae spp.	*Corallorhiza mertensiana* Bongard		"Gloecystidiales"	terrestrial, chlorophyll deficient	ectomycorrhizal		Taylor and Bruns, 1999
Serendipita vermifera (Oberw.) P. Roberts	*Acianthus* spp.	*Sebacina vermifera* Oberw.	Exidiales	terrestrial geophyte	saprotroph, ectomycorrhizal	Roberts, 1999	
Serendipita vermifera (Oberw.) P. Roberts	*Caladenia* spp.	*Sebacina vermifera* Oberw.	Exidiales	terrestrial geophyte	saprotroph, ectomycorrhizal	Roberts, 1999	
Serendipita vermifera (Oberw.) P. Roberts	*Elythranthera* spp.	*Sebacina vermifera* Oberw.	Exidiales	terrestrial geophyte	saprotroph, ectomycorrhizal	Roberts, 1999	
Serendipita vermifera (Oberw.) P. Roberts	*Eriochilus* spp.	*Sebacina vermifera* Oberw.	Exidiales	terrestrial geophyte	saprotroph, ectomycorrhizal	Roberts, 1999	
Serendipita vermifera (Oberw.) P. Roberts	*Glossodia major* R.Br.	*Sebacina vermifera* Oberw.	Exidiales	terrestrial geophyte	saprotroph, ectomycorrhizal	Roberts, 1999	
Serendipita vermifera (Oberw.) P. Roberts	*Microtis* spp.	*Sebacina vermifera* Oberw.	Exidiales	terrestrial geophyte	saprotroph, ectomycorrhizal	Roberts, 1999	
Serendipita vermifera (Oberw.) P. Roberts	*Prasophyllum* spp.	*Sebacina vermifera* Oberw.	Exidiales	terrestrial geophyte	saprotroph, ectomycorrhizal	Roberts, 1999	
Sistotrema sp.	*Piperia unalascensis* (Spreng.) Rydb.			terrestrial geophyte	saprotroph		Currah et al., 1990
Sistotrema sp.	*Platanhera obtusata*(Banks ex Pursh) Lindl.			terrestrial geophyte	saprotroph		Currah et al., 1990
Thanatephorus cucumeris (A.B.Frank) Donk	*Prasophyllum odoratum* Rogers		Ceratobasidiales	terrestrial geophyte	parasite	Roberts, 1999	
Thanatephorus cucumeris (A.B.Frank) Donk	*Pterostylis foliata* Hook.f.		Ceratobasidiales	terrestrial geophyte	parasite	Roberts, 1999	
Thanatephorus gardneri Warcup nom dub.	*Rhizanthella gardneri* Rogers		Ceratobasidiales	subterranean, chlorophyll-deficient	orchid symbiont, ectomycorrhizal		Warcup, 1991
Thanatephorus ochraceus (Massee) P. Roberts	*Calypso bulbosa* (L) Oakes	*T. orchidicola* Warcup & Talbot, *T. pennatus* Currah	Ceratobasidiales	terrestrial geophyte	saprotroph	Roberts, 1998	Currah, 1987
Thanatephorus ochraceus (Massee) P.Roberts	*Coeloglossum viride* (L.) Hartm.	*T. orchidicola* Warcup & Talbot, *T. pennatus* Currah	Ceratobasidiales	terrestrial geophyte	saprotroph	Roberts, 1998	Warcup & Talbot 1967

Table 1. Continued

Teleomorph fungus	Anamorph fungus	Species of orchid	Synomyms of fungus	Fungal group	Orchid lifeform	Fungal lifeform	Source to fungal identity and synonymy	Source to relationship
Thanatephorus ochraceus (Massee) P. Roberts		*Orchis mascula* (L.) L.	*T. orchidicola* Warcup & Talbot, *T. pennatus* Currah	Ceratobasidiales	terrestrial geophyte	saprotroph	Roberts, 1998	Warcup & Talbot 1967
Thanatephorus sp		*Acampe praemorsa* (Roxb.) Blatter & McCann		Ceratobasidiales	epiphytic, monopodial	parasite or saprotroph		Senthilkumar, Vengadeswari and Krishnamurthy 2000
Thanatephorus sterigmaticus (Bourdot) Talbot		*Thelymitra antennifera* Hook.f.		Ceratobasidiales	terrestrial geophyte	saprotroph	Roberts, 1999	Warcup & Talbot 1967
Thelephoraceae sp.		*Cephalanthera austinae*(A.Gray) Heller		Thelephorales	terrestrial, chlorophyll deficient	ectomycorrhizal		Taylor & Bruns, 1997
Tulasnella cruciata Warcup & Talbot		*Acianthus caudatus* R. Br.		Tulasnellales	terrestrial geophyte	only known as orchid symbiont	Roberts, 1999	Warcup & Talbot, 1971
Tulasnella cruciata Warcup & Talbot		*Thelymitra* spp.		Tulasnellales	terrestrial geophyte	only known as orchid symbiont	Roberts, 1999	Warcup & Talbot, 1971
Tulasnella deliquescens (Juel) Juel	*Epulorhiza repens* (N. Bernard) R.T. Moore	*Dactylorhiza purpurella* (T & T.A. Stephenson) Soo	*T. calospora* aut.	Tulasnellales	terrestrial geophyte	saprophyte	Roberts, 1999	
Tulasnella deliquescens (Juel) Juel	*Epulorhiza repens* (N. Bernard) R.T. Moore	*Dendrobium* sp.	*T. calospora* aut.	Tulasnellales	epiphyte pseudobulbous	saprophyte	Roberts, 1999	
Tulasnella deliquescens (Juel) Juel	*Epulorhiza repens* (N. Bernard) R.T. Moore	*Diuris, Acianthus, Caladenia, Thelymitra, Lyperanthes* spp.	*T. calospora* aut.	Tulasnellales	terrestrial geophyte	saprophyte	Roberts, 1999	
Tulasnella irregularis Warcup & Talbot		*Dendrobium dicuphum* F. Muell.		Tulasnellales	epiphyte pseudobulbous	only known as orchid symbiont	Roberts, 1999	Warcup & Talbot 1980
Tulasnella pinicola Bres.		*Dendrobium* sp.	*T. assymmetrica* Warcup & Talbot	Tulasnellales	epiphyte pseudobulbous	saprotroph	Roberts, 1999	Warcup & Talbot, 1967 eller 1972
Tulasnella pinicola Bres.		*Thelymitra luteocilium* Fitzg.	*T. assymmetrica* Warcup & Talbot	Tulasnellales	terrestrial geophyte	saprotroph	Roberts, 1999	Warcup & Talbot, 1967 eller 1971
Tulasnella violea (Quél.) Bourd. & Galz.		*Thelymitra aristata* Lindl.		Tulasnellales	terrestrial geophyte	saprotroph	Roberts, 1999	Warcup & Talbot 1971
	Rhizoctonia repens N. Bernard	*Cymbidium goeringii* Reichb.f.		Fungi imperfecti	terrestrial rhizomatous	saprophyte		Lee et al., 1998*[4]
	Epulorhiza calendulina Zelmer & Currah	*Amerorchis rotundifolia* (Banks) Hulten		Fungi imperfecti	terrestrial geophyte	only known as orchid symbiont		Zelmer and Currah, 1995b
	Epulorhiza anaticula(Currah) Currah	*Coeloglossum viride* (L.) Hartm.		Fungi imperfecti	terrestrial geophyte	only known as orchid symbiont		Currah et al., 1990
	Epulorhiza repens (N. Bernard) R.T. Moore	*Microtis parviflora R. Br.*		Fungi imperfecti	terrestrial geophyte	saprophytes		Perkins et al, 1995
	Epulorhiza anaticula(Currah) Currah	*Plantanthera obtusata*(Banks ex Pursh) Lindl.		Fungi imperfecti	terrestrial geophyte	only known as orchid symbiont		Currah et al., 1990
	Epulorhiza anaticula(Currah) Currah	*Platanthera hyperborea* (L.) Lindl.		Fungi imperfecti	terrestrial geophyte	only known as orchid symbiont		Currah et al., 1990
	Ceratorhiza pernacatena Zelmer & Currah	*Platanthera praeclara* Sheviak & Bowles		Fungi imperfecti	terrestrial geophyte	only known as orchid symbiont		Zelmer & Currah, 1995b

source in OMs involving pathogenic Rhizoctonias is ever a living organism. Sen et al. (1999) assumed that the *Ceratorhiza* spp. they consistently found in roots of *Goodyera repens* (L.) R.Br. lived in a mildly pathogenic association with *Pinus silvestris* L. The main substrate of *Ceratorhiza* sp. in association with Scottish *G. repens* was assumed to be pine needle litter (Downie, 1943), but parasitic strains of *Ceratobasidium cornigerum* (Bourdot) Rogers (usually the teleomorph form of *G. repens* endophytes) exist on woody and herbaceous plants (Roberts, 1998). The mycobionts in certain species of *Galeola* (Cha et al., 1996; Terashita, 1996) belong to *Armillaria*, a genus of tree parasites.

Recent findings demonstrate that ectomycorrhizal fungi (ECM) may also participate in OM which means that living trees provide photosynthates in a triple symbiosis. This type of relationship has been suspected to exist in species of *Corallorhiza* since Campbell (1970) observed rhizomorphs adjacent to the rhizomes of *C. striata* Lindl. Recently, the relationship has been studied in considerable detail. Zelmer and Currah (1995a) isolated a clamp-bearing basio- diomycete from *C. trifida* Chatelain and verified that it could form ECM with seedlings of *Pinus contorta in vitro*. In populations of *C. maculata* (Rafin.) Rafin., Taylor and Bruns (1997) found endophytes which, by DNA-analyses, were identified to Russulaceae, again strongly suggesting an ECM relationship as the carbohydrate source. Most recently, McKendrick et al. (2000a) obtained isolates from field sown seedlings of *C. trifida* and by DNA sequencing identified them to the *Thelephora-Tomentella* complex of Thelephoraceae. In microcosms, they subsequently linked the OM to roots of *Betula pendula* and *Salix repens* in an ECM relationship, and by isotope tracing demonstrated the transfer of carbon from the tree through the fungal partner to the orchid (McKendrick et al., 2000b). This would seem to settle the case, as far as *Corallorhiza* is concerned. Since a small and short-lived inflorescense with little chlorophyll is the only aboveground structure produced in species of *Corallorhiza*, the contribution from the tree probably is crucial to the survival of the orchid. The drain on the trees, on the other hand, of supporting the orchids was estimated to be very modest (McKendrick et al., 2000b). Another case where an ectomycorrhizal tree could be involved as carbohydrate source is the chlorophyll-deficient orchid *Cephalanthera austinae* (A.Gray) Heller whose mycobionts belong to Thelephoraceae and showed similarity to fungi forming ECM on adjacent tree roots (Taylor and Bruns, 1997).

The nutritional basis of OM relationships may be further complicated when more than one endophyte form pelotons at the same time in the same orchid tissue, such as found in both chlorophyll-deficient and photosynthetic orchids (Scrugli and Cogoni, 1994; Zelmer et al., 1996), because these fungi could have different energy sources and life forms. If the nutritional basis is evaluated over the life time of the orchid, several different fungi could also be involved sequentially, utilizing different substrates. One such example is *Gastrodia elata* Bl. whose seedlings develop with the saprotroph *Mycena osmundicola* Lange and whose subsequent development depends on the parasite *Armillaria mellea* (Vahl.Fr.) Karst. (Xu and Mu, 1990).

The general picture is thus complex, the orchids utilizing a great diversity of fungi with different nutritional strategies, and new research, particularly based on DNA-identification, seems to be widening the range of orchid mycobionts. Saprotrophs are possibly overrepresented in the earlier records because they usually are easy to keep in pure culture. It is still premature to say whether orchids with certain mycotrophic strategies depend on fungi with a particular trophic strategy. Ectomycorrhizal or mildly pathogenetic fungi may represent a rather stable nutrient source, provided that the host is a long-lived tree, which could explain why several of the most chlorophyll-deficient orchid species have established that type of association. This is not without exceptions, however (Table 1). Saprotrophic fungi growing on plant debris are a more transitional energy source, especially in the tropics, and hence more suitable for opportunistic orchid species with a short life span and high recruitment rate and for species with a low dependency on fungi in their adult stage. Orchids relying on saprotrophs could improve their chances of longevity by a low degree of specificity towards mycobionts.

Mycorrhiza as a factor in orchid recruitment and distribution

Orchid mycorrhiza has an impact on plant fitness from germination through seedling stage and in many cases throughout life. The great seed production in orchids suggests that the mortality of seeds and seedlings is exceedingly large. Unsuitable substrate and adverse physical conditions are challenges to any viable seed, but orchid seeds have the additional problem of locating a compatible mycobiont. Depending on the requirements of the fungus in question, the proximity of certain plants or debris of a particular kind may, by providing substrate for an appropriate fungus, be decisive for a successful establishment of seedlings. This challenge may persist in older plants that need to be reinfected every year.

The existence of fungal exudates that stimulate the rate of germination as well as the percentage, is well established *in vitro* (Rasmussen et al., 1998; review in Rasmussen, 1995). Extremely lean wood or soil based media are recommended if an indication of the natural situation for germination is to be obtained (for instance water agar amended with 0.5% ground wood, Whigham et al., unpubl.). In some cases, there is an all-or-none reaction *in vitro,* as in *Encyclia tampensis* Small, *Liparis liliifolia* A.Rich ex Lindl., or *Taeniophyllum obtusum* Bl. which only germinated in the presence of a fungus (Irawati, 1993; Rasmussen and Whigham, 1998b; Zettler et al., 1999). Recently developed techniques for field sowing *in situ* and retrieval of seeds (Masuhara and Kutsuya, 1994; Rasmussen and Whigham, 1993; van der Kinderen, 1995, Zelmer et al., 1996) have opened new possibilities for analysing germination behavior, fungal availability and natural substrates as they affect orchid recruitment. Some species, such as *Goodyera pubescens* R.Br., germinate freely in the ground without infection but the seedlings are not always successful in establishing mycorrhiza and high seedling mortality follows. Others, such as *Corallorhiza odontorhiza* Nutt. and *Liparis lilifolia*, are invariably infected when they are found in germinated condition (Rasmussen and Whigham, 1998b). Hence, the two latter species depend more on fungi during the germination process than *G. pubescens*, consistent with observations *in vitro*.

The reaction of the seed to fungi, however, is fairly unspecific and germination may be stimulated by less than optimum mycobionts and even in some cases by fungi and other microorganisms unable to participate in OM. Symbionts that are not fully compatible may result in high seedling mortality (Zettler et al., 1999). High germination percentage *in vitro* reflects physiological compatibility which does not necessarily apply in natural situations. For instance, Masuhara et al. (1993) germinated *Spiranthes sinensis* Ames with strains of *Rhizoctonia solani*, although the regular mycobiont appears to be *Tulasnella deliquescens* (Juel)Juel (Table 1) and Umata (1997 a ,b) was able to germinate *Erythorchis ochobiensis* with various fungi, with which the plant is not known to associate in nature. Further development of the seedling, however, often requires a narrower range of fungi than germination.

How fungi affect the seeds before actual invasion of the tissues remains conjectural. Ethylene and auxins are produced by some orchid mycobionts and are stimulatory to some seeds when added *in vitro* (Miyoshi and Mii, 1995; Rasmussen, 1995; Wilkinson et al., 1994). External addition of kinetin also often breaks dormancy of seeds *in vitro* (Miyoshi and Mii, 1998) but cytokinin production has not been detected in orchid fungi.

Burial of seed packets in prepared soil/wood mixtures showed that decaying wood is important to the germination and establishment of mycorrhiza in some orchids, and the species of wood and stage of decay influence the outcome (Rasmussen et al., 1998; Whigham et al., MS). The wood substrate increased the speed of germination even in asymbiotic controls, hence the seeds of these species seem to have developed the ability to react positively to a substrate that is likely to provide an appropriate mycobiont. Unsuitable substrate may be the reason why the immediate vicinity of mature plants – and their mycobionts – does not always appear to be suitable for seed germination. Masuhara and Katsuya (1994) thus found no connection between proximity of adult *Spiranthes sinensis* and success of seed packets in producing seedlings, and McKendrick et al. (2000a) made a similar observation in *Corallorhiza trifida*. Furthermore, seedlings of *Tipularia discolor* Nutt. were consistently found on decaying logs and stumps (which would eventually disappear) while plants in flowering stage were always found on the forest floor (Rasmussen and Whigham, 1998a). These examples suggest a dependency on successional vegetation for optimum seed-fungus interaction and seedling development.

Obligate associations of a species of orchid to a particular species of tree have been reported. The epiphytic species *Lepanthes caritensis* Tremblay and Ackerman was found only on *Micropholis guyanensis* in combination with thick moss cover (Tremblay et al., 1998), and the climber *Erythorchis altissima* Bl. almost exclusively occurred on the tree *Castanopsis sieboldii*, especially on dead trunks, often together with the shiitake mushroom, *Lentinus edodes* Berk. (Umata et al., 1994). The reason could either be that the tree in question is a living host, or an ectomycorrhizal partner, for the mycobiont, or that dead wood or litter of the tree offer a specialized substrate. Indirect effects of the tree on microclimate and soil composition are also possible.

Perkins and McGee (1995) found that colonies of *Rhizoctonia solani* associated with *Pterostylis acuminata* R. Br. extended up to about 0.5 m from the plants. This, as well as heterogeneity of the substrate, contribute to the patchiness of germination found in all succesfull recent field sowing experiments. Distribution of seed packets along transect lines showed that germination (and hence mycobiont presence) was positively correlated with sites of adult plants, with high content of non-decomposed organic matter, and with a low content of free nutrients (Batty et al., 2001).

Life forms, life history and phenology of orchid mycorrhiza

Previous reports that epiphytic orchids have low intensity of infection, compared to terrestrial species (Hadley and Williamson, 1972), tend to be contradicted by more recent data (Goh et al., 1992; Rivas et al., 1998; Senthilkumar et al., 1998, 2000 and references therein). These reports cover a considerable seasonal and local variation as well as variation among species. Aerial roots in epiphytic species generally are devoid of infection and roots in contact with the substrate extensively infected (Goh et al., 1992). Among temperate species, tuberous orchids appear to be more mycorrhized than rhizomatous species (Tatarenko, 1995).

Orchid mycorrhiza is present throughout the year in the tropical epiphyte *Vanda tessellata* Hook. ex G.Don with most mycophagy taking place while the plant is in active growth (Shagufta et al., 1993). Seasonal patterns of mycorrhization in temperate species mainly follows the phenology of the mycorrhized organs (Masuhara and Kutsuya, 1992; Rasmussen and Whigham, unpubl.) and is not linked to the seasonality of the photosynthetic structures.

The dependency of the plant on fungi indisputably changes over the lifetime of the orchid; in most orchids a decrease is evident from the seedling stage onwards. Some species remain heavily mycorrhized, in spite of photosynthetic capacity of the leaves (Rasmussen and Whigham, unpubl.). Zelmer et al. (1996) found that seedlings associated with a wider range of fungi than older plants did. This might indicate a change of mycobionts during development of the individual plant; however, it might also reflect a fairly unspecific germination with subsequent decimation of seedlings, leaving only those with optimum symbiosis to develop into adults. In the case of *Gastrodia elata*, however, a shift in mycobiont seems to be the rule, from *Mycena osmundicola* to *Armillaria mellea* (see above, Xu and Mu, 1990). It is known from symbiotic propagation programs that the pelotons extracted from roots of adult plants often have no positive effect on seedling development *in vitro* which suggests that it is common for adult orchids to contain more fortuitous infections. Dijk et al. (1997) suggest an adaptive advantage of a symbiont shift during the life history of an orchid. A seedling which has established with a saprotrophic mycobiont on a carbohydrate-rich substrate may find the substrate gradually changing over time. This appears to be the case in *Tipularia discolor* (see above, Rasmussen and Whigham, 1998a). While a slow growing, weak competitor with a specialized substrate requirement may be an optimum mycobiont in early life of the orchid (or in a short-lived species), a more aggressive fungus with an extensive mycelium and generalized substrate utilization might be a better support for the adult orchid in mature vegetation.

Identification of mycobionts

Pure cultures based on isolations from orchid roots, preferably from single pelotons, have up till now been the main source of information about orchid symbionts. The fungal partner in OM is always a basidiomycete and the isolates generally are sterile mycelia with very little tendency to sporulate in culture. Certain mycelial characters, particularly the ultrastructure of the septal pore (Currah and Sherburne, 1992), the size and shape of monilioid cells, formation of sclerotia, and enzyme activity are useful as distinguishing characters in sterile mycelia of *Rhizoctonia sensu lato* (Zelmer and Currah, 1995b), which comprise most of the orchid mycobionts now known. However, the mycelia generally do not yield many distinguishing characters that enable identification below generic level.

A much finer distinction between strains is now possible by means of DNA analysis of infected orchid tissue. These techniques are mostly based on PCR with fungal primers on macerated orchid tissue. This reaction will amplify selected parts of the fungal genome for further characterization of the DNA either by sequencing or fragmentation patterns. The level of distinction between strains depends on the variability of the part of the genome under study. Provided corresponding DNA from suitable identified reference fungi have been processed, a phylogenetic analysis will place the unknown strain in a taxonomic group. Hopefully, these techniques will eventually reveal the full diversity of orchid symbionts, including those whose nutritional requirements cannot easily be met in pure culture. DNA analyses of infected tissue, however, holds the same kind of drawbacks as pure cultures obtained from plating tissue fragments, i.e. that mixed infection and fortuitous contaminants can obscure results. Since peloton formation is an important criterion for an orchid symbiont, DNA-analysis based on single pelotons extracted from fresh or dried orchid tissue is a new promising approach (Kristiansen et al., 2001).

Many OM isolates are referred to the form genus *Rhizoctonia* in sterile condition. Roberts (1999) lists 15 species with *Rhizoctonia* anamorphs that are known to be orchid symbionts: 5 species of *Ceratobasidium*, 3 species of *Thanatephorus*, 1 species of *Serendipta* (*Sebacina*) and *Oliveonia*, and 5 species of *Tulasnella* (Table 1). These all belong in Heterobasidiomycetes.

Agaricanae are represented among orchid symbionts by *Armillaria* (Cha and Igarishi, 1996; Lan et al., 1994; Terashita, 1996) and *Mycena* (Fan et al., 1996; Lan et al., 1996), 'Hericianae' and 'Thelophoranae' are represented by Russulaceae and Thelephoraceae (McKendrick et al., 2000a; Taylor and Bruns, 1997), and 'Hymenochaetanae' by *Erythromyces* (Umata, 1995, 1998a). Five major taxonomic groups are thus represented among OM fungi (Table 1).

The relevance of Mycelia Radices Atrovirens (MRA) that are sometimes isolated from orchid tissue (Currah et al., 1990) is uncertain. MRA fungi have not been successfully used for germination *in vitro* and they do not appear to form pelotons. Possibly they are fortuitously present within the roots or superficial contaminants that arise in cultures produced from incubating slices of surface sterilized roots. This isolation procedure is widely used (e.g., Currah et al., 1990; Vertenyi and Bratek, 1996), but not as reliable as single-peloton isolations.

Specificity between orchid and mycobiont

It is necessary to distinguish between associations that may be viable under certain experimental conditions, as *in vitro*, and thus showing a *physiological compatibility* and those association that are possible and competitive under natural conditions, i.e. a *specificity* in ecological context. Ideally, specificity should be tested either *in situ* or under realistic climate conditions *in vitro*, with the mycobiont growing on a near-natural substrate, and with fully viable, nondormant seeds; furthermore, a full identification of the species participating is essential. Most often, however, our evidence is based on case studies quite far from that ideal.

Even so it has become clear that orchids and fungi do associate according to some compatibility barriers, though usually not on a species-to-species level. The nature of these barriers is still entirely unknown. A narrow specificity in the orchid could be a reason for rarity and vulnerability of the plant species, just as a narrow food preference would in an animal species, and so is an issue of conservational interest. One suggestive example is *Corallorhiza mertensiana* Bong., a rather rare species (Freudenstein, 1997) which was found to associate with a narrower range of mycobionts than the wide-spread relative *C. maculata* (Taylor and Bruns, 1999). However, very little is known at present about how the symbiosis with fungi affects competition, survival and distribution of orchids and we have not even begun to consider how the relationship may affect the fungal community.

Germination may be stimulated by a range of fungi, with little bearing on their compatibility or actual relationship with the plant in nature. Adult plants may also house a variety of fungi. Hence, the holomycotrophic seedling phase appears as the bottleneck stage at which tolerance towards suboptimum symbioses is lowest. It is also the life history stage at which mortality is probably the highest (Rasmussen and Whigham, 1998b). A balanced seedling development may thus be the best criterion for compatibility.

Even with unidentified mycobionts it is often possible to ascertain that orchid species under standardized (if fairly artificial) conditions differ in symbiont preference: Tomita and Konno (1998) showed that *Aorchis cyclochila* (Franch. and Sav.) T.Hashimoto, *Dactylorhiza aristata* (Fisch. ex Lindl.) So, *Gymnadenia camtschatica* Miyabe & Kudo had other preferences for fungi than *Amitostigma kinoshitae* (Ohwi & Hashimoto) Hashimoto and *Ponerorchis graminifolia* Rchb.f.; *Cypripedium macranthos* Sw. appeared to have a narrower endophyte range than the rest. Comparative studies of sympatric or closely related species are more illustrative. For instance, Zettler and Hofer (1998) noted that a strain of *Epulorhiza inquilina* Currah et al., obtained from *Platanthera clavellata* (Michx)Luer, was more efficient *in vitro* in promoting seedling development of that species than isolates from three other co-habiting species of *Platanthera*. This suggests that sympatric orchid species avoid mutual competition for food by employing different species of fungi. This pattern is also evident in *Pterostylis acuminata* compared to other surrounding orchid species (Perkins and McGee, 1995) and in comparison of two species of *Corallorhiza*. In the latter case, although all the symbionts of both orchid species were placed in Russulaceae, samplings over a wide geographic range did not show any sharing of fungal species (Taylor and Bruns, 1999).

Constant yield of the same fungus or group of fungi from repeated sampling of an orchid species over a geographic range supports the assumption of a large degree of specificity. For instance, 14 samples from *Cephalanthera austinae* over a wide geographic range all belonged to Thelephoraceae, and 18 samples from *Corallorhiza maculata* were all referred to Russulaceae (Taylor and Bruns, 1997). Isolations from *Galeola septentrionalis* Rchb.f. consistently yield a range of species belonging to *Armillaria* (Terashita, 1996). Collections of *Erythrorchis ochobiensis* always contained *Erythromyces crocicreas*; in spite of that, this orchid could germinate *in vitro* and develop mycotrophic seedlings with a diversity of fungi: *Ganoderma australe* (Fr.) Pat., *Loweporus tephroporus* (Mont.) Ryv., *Microporus affinis* (Fr.) Kunt., *Phellinus* sp., *Auricularia polytricha* (Mont.)Sacc. and *Lentinula edodes* (Umata, 1997a, 1998a,b). These physiological symbionts do not appear to be ecological symbionts. Even physiological compatibility has its limits in *Erythrorchis*, though: seeds tested with *Lyophyllum shimeji* (Kawamura) Hongo germinated but without peloton formation or normal seedling development, and seeds incubated with *Tricholoma fulvocastaneum* Hongo did not germinate at all (Umata 1997b).

Dijk and Eck (1995) found compatibility of *Anacamptis morio* (L.)Batem. et al. with *Epulorhiza* sp., but not with *Ceratorhiza* sp.; with three species of *Dactylorhiza*, *Ceratorhiza* showed greater efficiency than *Epulorhiza* but also less tolerance to high N concentrations. Within the genetically determined range of possible mycobionts, geography and habitat conditions may influence which one(s) is active in a particular population (Taylor and Bruns, 1999).

Interactions with other micro-organisms

Germination of either *Caladenia latifolia* R. Br. or *Diuris magnifica* D.L.Jones on mixed or single inoculations suggests that competition between the OM fungus and other fungi can be a significant factor in nature (Quay et al., 1995). Synergy between microorganisms in OM is also possible; bacteria associated with *Pterostylis vittata* Lindl. stimulated symbiotic germination, probably because they produced IAA or induced the plant into auxinproduction (Wilkinson et al., 1994). These bacteria belonged to *Pseudomonas putida* (but not all strains), *Zanthomonas maltophilia* and *Bacilllus cereus*, whereas other bacterial strains were ineffective.

Structure and ultrastructure

Traditionally, two types of orchid mycorrhiza have been recognized, i.e. tolypophagy, found in the great majority of species, and ptyophagy, only noted in a number of highly mycotrophic tropical orchids (Rasmussen, 1995, and references therein). In 1995 the most recent reference to ptyophagy dated from 1936, and ptyophagy remained an obscure phenomenon. However, recent studies throw new light on this type of orchid-fungal symbiosis.

In *Gastrodia elata*, the hyphae of *Armillaria mellea* extend in bundles along the roots in cortical canals (Wang et al., 1997). These canals develop from lines of 'passage cells' whose adjoining cell walls and original cell content deteriorate. The outer cortex, outside the passage canal consists of 'host cells' where the hyphae coil and apparently persist, and the inner cortex contains 'digestion cells' (Figure 1). When hyphae enter a digestion cell, an interface is formed between the receding plant plasmalemma and the hyphal wall. Electron transparent vesicles with lysosomic (or fungistatic?) properties pass the perihyphal plant plasmalemma and are released into the interface (Figure 2). Subsequently, electron-dense vesicles appear along the plant plasmalemma and elongate to form a radiating tubular system around the hyphae. These vesicles are believed to be endocytic ('endocytic tubes') and contain products from hyphal dissolution; they appear to cut off minor vesicles that migrate into the plant cytoplasm (Figures 3 and 4). Fusion between these and electron-transparent, presumably lysosomic vesicles scattered in the cytoplasm is believed to mark the final step in the breakdown of the liquid hyphal products (Figure 5). The fungal wall shatters in the interfacial space (or digestion vacuole). Invaginations from the plasmalemma, to which electron transparent vesicles and parts of ER fuse, appear to pinocytose chips of fungal wall for further breakdown (Figures 6 and 7).

According to Burgeff (1936), the distinguishing element in ptyophagy is that fungal tips of the intracellular hyphae become lysed and the hyphal cell contents thus released, whereas in tolypophagy there is an overall collapse and breakdown of hyphae. In view of the electron microscopic evidence now available this discription may need to be modified. However, there is little doubt that ptyophagy differs rather markedly from the more well-known tolypophagic pattern, both at the histological level (the formation of passage canals) and in the ultrastructure of the digestion process (endocytic tubes, pinocytosis of hyphal wall fragments). The clumps of collapsed and aggregated hyphal walls, so characteristic of tolypophagy, seems to be lacking. Some differences could be due to properties of the mycobionts in question, others to the plant. However, ptyophagy adds to the diversity of plant/fungal interactions, apparently in a rather outstanding manner and deserves further study.

The tolypophagic interaction is comparatively well investigated at the ultrastructural level, but some unsolved questions remain: The origin of the interfacial matrix, the actual transfer of fungal products to the plant cytoplasm across that barrier, the mechanisms in cells, where hyphae are present without being digested, and in cells where hyphae are rejected, and finally the exportation of mycotrophic products from digestion cells into the stele (Rasmussen, 1995).

A cytochemical localization of adenylate cyclase activity showed accumulation along the plasmalemma at the plant cell walls in both infected and non-infected cells, but not along the membrane invaginations enveloping invading hyphae. While these membranes were clearly derived from the plasmalemma, their proporties changed in contact with the hyphae (Uetake and Ishizaka, 1995). Adenylate cyclase catalyses the formation of cyclic AMP, presumed to be intracellular messenger molecule. Its functions and possible role in preventing or promoting hyphal invasion are uncertain.

The interfacial matrix was shown by Peterson et al. (1996) to contain pectins, cellulose and β1–3 glucans when the peloton hyphae are collapsing but none of these substances were detected before peloton breakdown.The cortical microtubule system (MT) disappeared in cells during infection but short MT's were observed between hyphae in colonized cells, forming a network through the peloton and connecting to the nucleus. During lysis MT's were observed between hyphae within and around the collapsing peloton (Uetake et al., 1997; Uetake and Peterson, 1998). The fact that other researchers did not observe MT's during infection (Dearnaley and McGee, 1996) could be due to a different fixation and microscopy technique by which short MT's could easily be overlooked. Also actin filaments (AT) were rearranged during infection into a network radiating from the perihyphal membrane towards the cell walls. This network remained during peloton lysis, but the cortical AT system subsequently reappeared (Uetake and Peterson, 1997). Both MT's and AT's in digestion cells seem to stabilize the peloton and possibly guide vescicles and other organelles towards and away from the interface. The AT system found in infected roothair cells, however, was oriented along the longitudinal axis of the cell (Uetake and Peterson, 1997) and could thus act as a guide leading the hyphae towards the protocorm body.

The content of digestive enzymes is much higher in infected tissues than in uninfected, and by histochemical localization Senthilkumar et al. (2000) implicated peroxidase, glutamate dehydrogenase, esterase as well as malate dehydrogenase in the lysis of pelotons.

Physiology of orchid mycorrhiza

The control of hyphal invasion in orchid tissues has traditionally been attributed to the production of phytoalexins, known from tubers and rhizomes of orchids; only recently has the phytoalexin orchinol also been demonstrated in protocorms (Beyrle et al., 1995). Fungal invasion and wounding of orchid tissues induce the production of phytoalexins (Gehlert and Kindl, 1991; Reinecke and Kindl, 1994a, b).

The balance between the symbionts is affected amongst other things by the source of nitrogen (Beyrle et al., 1995), so that a low supply coinciding with a high availability of carbohydrates gave a balanced mycorrhiza in *Anacamptis morio*, whereas high supply of nitrogen and carbohydrates resulted in rejection of the fungus. Any combinations of with low carbohydrate supply resulted in parasitism of the fungus on the plant. Shortage of external carbohydrate sources tended to increase fungal virulence. High nitrogen availability could lead to rejection of the fungi which

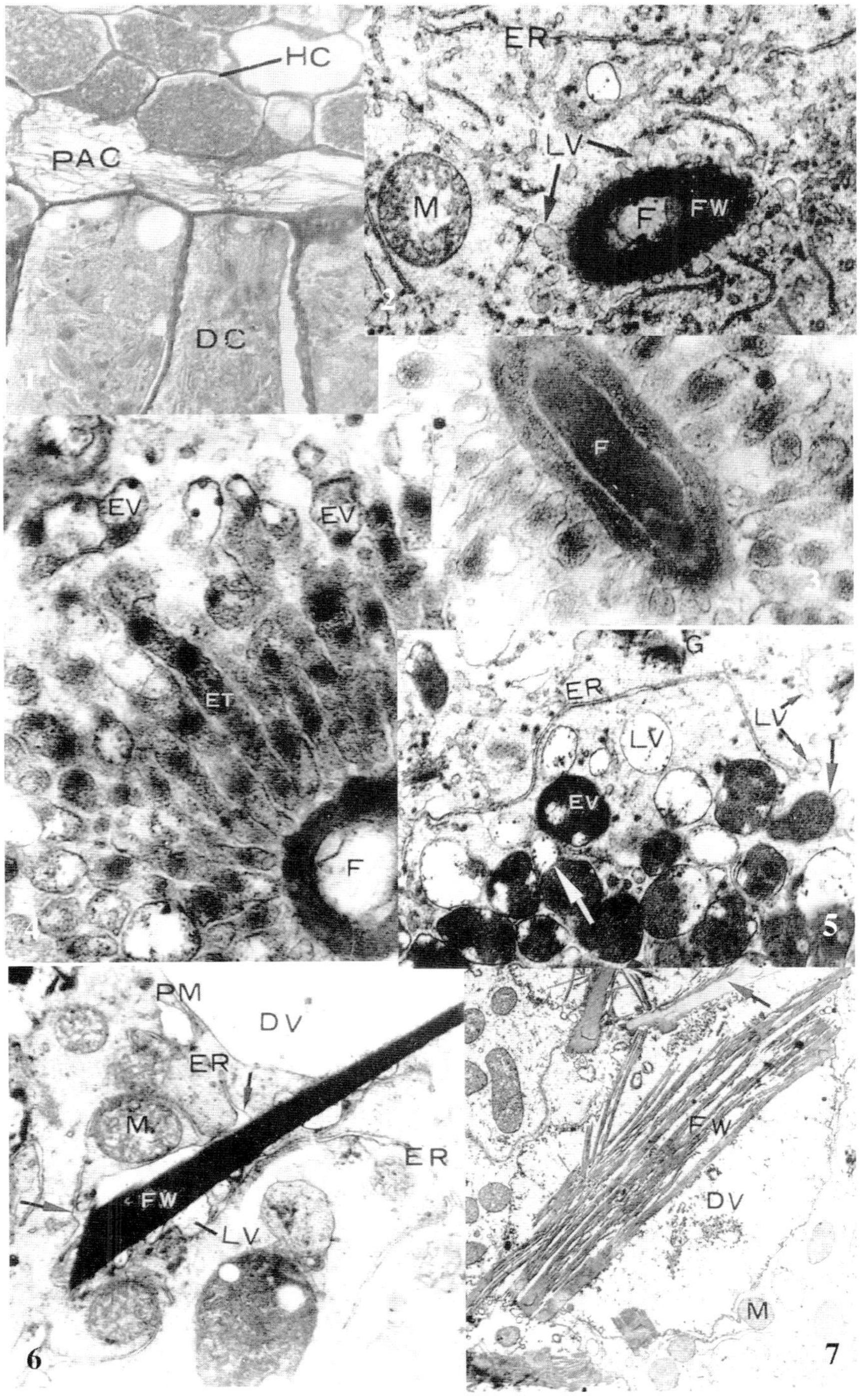

Figure 1. Longitudinal section of an infected root of *Gastrodia elata* showing passage canal (PAC), host cells (HC) and digestion cells (DC). (2) Digestion cell showing abundant ER, electron-transparent lysosomal vesicles (LV) fusing into the space around the thickened wall of invading hypha (F). (3) Endocytic vesicles forming aroung hypha (F) in early stage of fungal disintegration. (4) A radiating system of endocytic tubes (ET) forms around hypha, and cut off endocytic vesicles (EV), presumably filled with fungal products. (5) Endocytic vesicles (EV) fusing (white arrow) with ER-produced lysosomal vesicles (LV). (6) LV and ER fusing with perifungal plasmalemma to produce a large digestion vesicle around the remains of fungal wall material. Larger pieces of wall material (FW) partly enveloped by digestion vacuole membrane in the process of breakdown to smaller parts. (7) Digestion vacuole with finely chipped hyphal wall remains. Arrow points to larger piece, as in Figure 6. Magnifications: 1: × 120, 2: × 20000, 3: × 24000, 4: × 26000, 5: × 26000, 6: × 30000, 7: × ca. 15000. From Wang et al. (1997) printed with permission from Acta Botanica Sinica.

was accomplished by thickening of plant cell wall and accumulation of phenolics (Beyrle et al., 1995). Dijk and Eck (1995) noticed a negative effect of high nitrogen supply on protocorm yield *in vitro*, when the mycobiont was a representative of *Ceratorhiza* but could not detect any adverse effects when *Epulorhiza* sp. was involved. This is consistent with field observations that populations of *Dactylorhiza majalis* (Rchb.) Hunt & Summerh. responded negatively to fertilization with nitrogen, as well as phosphorus and potassium (Dijk and Olff, 1994), and not only as a result of intensified competition from the surrounding vegetation. Thus orchid mycorrhiza seems more competitive on poor soils which is consistent with many field observations. Symbiotic seedlings *in vitro* reach a higher nitrogen concentration in their tissues than asymbiotic controls which confirms that the mycobionts assist in nutrient uptake for the plants (Lee et al., 1997).

The root/shoot ratio increased, and leaf development was inhibited, in a *Cattleya* hybrid grown *in vitro*, when external carbohydrate was added (Beyrle and Smith, 1993a). In *Anacampsis morio*, greening of leaves was prevented by high carbohydrate concentration in the substrate (Beyrle and Smith, 1993b). Carbohydrate availability could thus be one of the factors determining the trade-off between photo- and mycotrophism. Another factor could be light, since exposure to light is required by some species before leaves develop from primordia (Zettler et al., 1995). The transfer of soluble carbohydrates from the mycorrhizal fungus to the heterotrophic plant has been further corroborated by a recent study: labelled glucose was traced from the mycelium of *Mycena osmundicola* into the seedlings of *Gastrodia elata* and labeling subsequently appeared in meristematic (non-infected) tissues of the plant (Lan et al., 1996). However, traces of label were observed in rhizomorphs of *Armillaria mellea* after feeding *Gastrodia elata* with labeled glucose (Lan et al., 1994). This is the first ptyophagic relationship to be studied with respect to nutrient exchange and it is possible that it differs from tolypophagy, so that a limited flow of carbohydrates towards the fungus is possible. However, the observation might also indicate a senescence phenomenon. There is little doubt that the general carbohydrate flow occurs in the direction of the clearly chlorophyll-deficient *G. elata*.

Mycorrhizal infection seems to enhance the uptake of water since seedlings of both the terrestrial species *Platanthera integrilabia* (Correll) Luer and the epiphyte *Epidendrum conopseum* R.Br. had higher water content than non-infected controls (Yoder et al., 2000). This would seem a particularly valuable feature in the epiphytic and epilithic life forms of orchids.

Ex situ applications of the mycorrhizal association

The study of orchid mycorrhiza was founded almost 100 years ago by Bernard (1904, 1909) and Burgeff (1909, 1932, 1936), but since interest in the subject was revived about 20 years ago, new insight has been accumulating and, perhaps more important, application to horticulture and conservation has begun. Several conservation projects utilize symbiotic techniques to propagate plants, to bank seeds and fungal cultures, and to evaluate natural growing sites (e.g. Dixon, 1994; Stewart, 1993). Rare and endangered species of orchids have been propagated symbiotically with the purpose of *ex situ* conservation or reintroduction (e.g. Zettler and McInnis, 1992).

Orchids are mostly grown commercially for their ornamental value but other uses exist; the symbiotic cultivation of *Gastrodia elata* has become an important enterprise based on the medicinal use of the tubers (Xu and Mu, 1990), and the edible rhizomes of *G. cunninghamii* Hook.f. are reported to be a delicious vegetable (Harris, 1997). Symbiotic cultivation techniques widen the range of species that can be grown profitably.

Maintaining a living orchid collection may present considerable difficulties; many species of orchids still defy cultivation except in seminatural conditions in botanical gardens. Most holomycotrophic taxa are deemed to be impossible to grow. The likely reason is that orchid species with a large dependency on their mycobiont have difficulties in establishing or maintaining a functional symbiosis in a garden or greenhouse environment. The survey by Goh et al. (1992) comparing collections from nature and cultivated plants growing in garden beds or pots clearly showed that mycorrhization was low in pot culture in commercial nursery potting mixtures. Rivas et al. (1998) found that although most cultivated plants under seminatural conditions were extensively colonized by mycorrhizal fungi, a few introduced species (2 out of 24) apparently were sparsely mycorrhized. Pest control by means of some fungicides of course presents a special problem to symbiotic orchid cultures (Kummuang, 1997).

Large scale symbiotic propagation may take place on complex carbohydrate substrates such as sawdust

(Umata, 1997a, 1998a), either inoculated by known orchid endophytes or amended with soil samples from natural orchid sites (Johnson, 1994). Inoculates of orchid mycobionts for larger-scale inoculation may be produced by letting the fungi invade sterilized expanded clay particle that can subsequently be dried and ground before application to soil or plant growing media (Beyrle et al., 1989). Techniques are now developing for producing seeds for commercial production or conservation encapsulated with suitable inoculum to secure an initial compatible symbiosis (Tan et al., 1998; Wood et al., 2000).

Acknowledgements

The author is grateful for support from the Carlsberg Foundation and for previous grants from the Smithsonian Institution without which the present study could not have been carried out.

References

Batty A L, Dixon K W, Brundrett M and Sivasithamparam K 2001 Constraints to symbiotic germination of terrestrial orchid seed in a mediterranean bushland. New Phytology, 152, 511–520.

Bernard N 1904 Recherches expérimentales sur les orchidées. La germination des orchidées. Revue Générale de Botanique 16, 405–451, 458–475.

Bernard N 1909 L'evolution dans la symbiose des Orchidées et leur champignons commensaux. Annales des Sciences Naturelle Paris, 9. ser. 9, 1–196.

Beyrle H, Penningsfeld F and Hock B 1989 Mycorrhizal synthesis in horticultural growth substrates and in natural soils. Orchidee 40, 173–177.

Beyrle H F and Smith S E 1993a The effect of carbohydrate on the development of a *Cattleya* hybrid in association with its mycorrhizal fungus. Mycorrhiza 3, 57–62.

Beyrle H F and Smith S E, 1993b Excessive carbon prevents greening of leaves in mycorrhizal seedlings of the terrestrial orchid *Orchis morio*. Lindleyana 8, 97–99.

Beyrle H F, Smith S E, Peterson R L and Franco C M M 1995 Colonization of *Orchis morio* protocorms by a mycorrhizal fungus: Effects of nitrogen nutrition and glyphosate in modifying the responses. Can. J. Bot. 73, 1128–1140.

Burgeff H 1909 Die Wurzelpilze der Orchideen, ihre Kultur und ihre Leben in der Pflanze. Gustav Fischer, Jena.

Burgeff H 1932 Saprophytismus und Symbiose. Gustav Fischer, Jena.

Burgeff H 1936 Samenkeimung der Orchidéen. Gustav Fischer, Jena.

Campbell E O 1970 Morphology of the fungal association in three species of *Corallorhiza* in Michigan. Michigan Bot. 9, 108–113.

Carling D E, Pope E J, Brainard K A and Carter D A 1999 Characterization of mycorrhizal isolates of *Rhizoctonia solani* from an orchid, including AG-12, a new anastomosis group. Phytopathology 89, 942–946.

Cha J Y and Igarishi T 1996 *Armillaria jezoensis*, a new symbiont of *Galeola septentrionalis* (Orchidaceae) in Hokkaido. Mycoscience 37, 21–24.

Currah R S 1987 *Thanatephorus pennatus* sp. nov. isolated from mycorrhizal roots of *Calypso bulbosa* (Orchidaceae) from Alberta. Can. J. Bot. 65, 1957–1960.

Currah R S and Sherburne R 1992 Septal ultrastructure of some fungal endophytes from boreal orchid mycorrhizas. Mycological Research 96, 583–587.

Currab R S, Siegler L and Hamilton S 1987 New records and new taxa of fungi from the mycorrhizae of terrestrial orchids of Alberta. Can. J. Bot. 65, 2473–2482.

Currah R S, Smreciu E A and Hambleton S 1990 Mycorrhizae and mycorrhizal fungi of boreal species of *Platanthera* and *Coeloglossum* (Orchidaceae). Can J. Bot. 68, 1171–1181.

Currah R S and Zelmer C 1992 A key and notes for the genera of fungi mycorrhizal with orchids and a new species of the genus *Epulorhiza*. Reports of the Tottori Mycological Institute 30, 43–59.

Dearnaley J D W and McGee P A 1996 An intact microtubule cytoskeleton is not necessary for interfacial matrix formation in orchid protocorm mycorrhizas. Mycorrhiza 6, 175–180.

Dijk E and Eck N D 1995 Effects of mycorrhizal fungi on in vitro nitrogen response of some Dutch indigenous orchid species. Can. J. Bot. 73, 1203–1211.

Dijk E, Willems J H and van Andel J 1997 Nutrient responses as a key factor to the ecology of orchid species. Acta Bot. Neerlandica 46, 339-363.

Dijk E and Olff H 1994 Effects of nitrogen, phosphorus and potassium fertilization on field performance of *Dactylorhiza majalis*. Acta Bot. Neerlandica 43, 383–392.

Dixon K W 1994 Towards integrated conservation of Australian endangered plants – the Western Australian model. Biodiversity and Conservation 3, 148–159.

Downie D G 1943 Source of the symbiont of *Goodyera repens*. Trans. Bot. Soc. Edinburgh 35, 120–125.

Fan L, Guo S X and Xiao P G 1999 Study on structure and localization of acid phosphatase of mycorrhizal root of *Cymbidium sinense* (Orchidaceae). Acta Bot. Yunnanica 21, 197–201.

Fan L, Guo S, Cao W, Xiao P, Xu J, Fan L, Guo S X, Cao W Q, Xiao P G and Xu J T 1996 Isolation, culture, identification and biological activity of *Mycena orchidicola* sp. nov. in *Cymbidium sinense* (Orchidaceae). Acta Mycol. Sin. 15, 251–255.

Freudenstein J V 1997 A monograph of *Corallorhiza* (Orchidaceae). Harvard Papers in Botany, 10: 5–51.

Garrett S D 1962 Decomposition of cellulose in soil by *Rhizoctonia solani* Kühn. Trans. British Mycol. Soc. 45, 115–120.

Gehlert R and Kindl H 1991 Induced formation of dihydrophenanthrenes and bibenzyl synthase upon destruction of orchid mycorrhiza. Phytochemistry 30, 457–460.

Goh C J, Sim A A and Lim G 1992 Mycorrhizal associations in some tropical orchids. Lindleyana 7, 13–17.

Hadley G and Williamson B 1972 Features of mycorrhizal infection in some Malayan orchids. New Phytol. 71, 1111–1118.

Harris G 1997 Huperei – the black orchid. New Zealand Garden Journal 2, 7–9, 2 pl.

Hjortstam K and Tellería M A 1990 *Columnocystis*, a synonym of *Veluticeps*. Mycotaxon 37, 53–56. (Description of *Erythromyces* on p. 55).

Irawati 1993 Orchid mycorrhiza in *Taeniophyllum obtusum* L. J. Biol. Indonesia 1, 6–16.

Johnson S R 1994 Symbiotic seed germination in relation to potential naturalization ability of *Bletilla striata* (Orchidaceae). Lindleyana 9, 99–101.

Kristiansen K A, Kjøller R, Rasmussen H N and Rosendahl S 2001 Identification of orchid mycorrhizal fungi from *Dactylorhiza majalis* (Orchidaceae) basd on PCR, SSCP and sequencing of mitochondrial ribosomal LsDNA from single pelotons. Molecular Ecol. 10: 2089–2093.

Kummuang N 1997 Effect of some fungicides on mycorrhizae in ground orchids. Kaen Kaset = Khon Kaen Agriculture Journal 25, 139–145. NB: cited from abstract.

Lan J, Xu J T and Li J S 1994 Study on symbiotic relation between *Gastrodia elata* and *Armillariella mellea* by autoradiography. Acta Mycol. Sin. 13, 219–222.

Lan J, Xu J and Li J 1996 Study on the infecting process of *Mycena osmundicola* on *Gastrodia elata* by autoradiography. Acta Mycol. Sin. 15, 197–200.

Lee S, Park S, Kim T, Paek K, Lee S S, Park S S, Kim T J and Paek K Y 1997 Effect of orchid habitat soil on growth of tissue cultured *Cymbidium kanran* and *C. goeringii*, and root infection by orchid mycorrhizal fungus. J. Korean Soc. Horticultural Sci. 38, 176–182.

Lee S, Oh C, Paek K, Lee T, Lee S S, Oh C H, Paek K Y and Lee T S 1998 Isolations of the orchid mycorrhizal fungi from the roots of the Korean native orchids and inoculations of the isolates to four different orchids. Korean J. Plant Pathol. 14, 536–542.

Masuhara G and Katsuya K 1992 Mycorrhizal differences between genuine roots and tuberous roots of adult plants of *Spiranthes sinensis* var. *amoena* (Orchidaceae) *in vitro*. Bot. Magazine Tokyo 105, 453–460.

Masuhara G and Katsuya K, 1994 *In situ* and *in vitro* specificity between *Rhizoctonia* spp. and *Spiranthes sinensis* (Persoon) Ames. var. *amoena* (M. Bieberstein) Hara (Orchidaceae). New Phyto. 127, 711–718.

Masuhara G, Katsuya K and Yamaguchi K 1993 Potential for symbiosis of *Rhizoctonia solani* and binucleate *Rhizoctonia* with seeds of *Spiranthes sinensis* var. *amoena in vitro*. Mycol. Res. 97, 746–752.

McKendrick S L, Leake J R, Taylor D L and Read D J 2000a Symbiotic germination and development of myco-heterotrophic plants in nature: ontogeny of *Corallorrhiza trifida* and characterization of its mycorrhizal fungi. New Phytol. 145, 523–537.

McKendrick S L, Leake J R and Read D J 2000b Symbiotic germination and development of myco-heterotrophic plants in nature: Transfer of carbon from ectomycorrhizal *Salix repens* and *Betula pendula* to the orchid *Corallorrhiza trifida* through shared hyphal connections. New Phytol. 145, 539–548.

Miyoshi K and Mii M 1995 Phytohormone pretreatment for the enhancement of seed germination and protocorm formation by the terrestrial orchid *Calanthe discolor* (Orchidaceae) in asymbiotic culture. Sci. Horticul. 63, 263–269.

Miyoshi K and Mii M 1998 Stimulatory effects of sodium and calcium hypochlorite, pre-chilling and cytokinins on the germination of *Cypripedium macranthos* seed *in vitro*. Physiol. Plant. 102, 481–486.

Johnson S R 1994 Symbiotic seed germination in relation to potential naturalization ability of *Bletilla striata* (Orchidaceae). Lindleyana 9, 99–101.

Perkins A J, Masuhara G and McGee P A 1995 Specificity of the associations between *Microtis parviflora* (Orchidaceae) and its mycorrhizal fungi. Aust. J. Bot. 43, 85–91.

Perkins A J and McGee P A 1995 Distribution of the orchid mycorrhizal fungus, *Rhizoctonia solani*, in relation to its host, *Pterostylis acuminata*, in the field. Aust. J. Bot. 43, 565–575.

Peterson R L, Bonfante P, Faccio A and Uetake Y 1996 The interface between fungal hyphae and orchid protocorm cells. Can. J. Bot. 74: 1861–1870.

Quay L, McComb J A and Dixon K W 1995 Methods for *ex vitro* germination of Australian terrestrial orchids. HortScience 30, 1445–1446.

Rasmussen H N 1995 Terrestrial orchids. From seed to mycotrophic plant. Cambridge University Press, Cambridge.

Rasmussen H N, Caldwell B A, Whigham D F and O'Neill J P 1998 Evidence of pregermination, preinfection signaling in the orchid-fungal symbiont-substrate relationship. Eds. U Ahonen-Jonnarth, E Danell, P Fransson, O Kårén, B Lindahl, I Rangel, and R Finlay. pp. 142–142. Uppsala, Sweden

Rasmussen H N and Whigham D F 1993 Seed ecology of dust seeds *in situ*: a new study technique and its application in terrestrial orchids. Am. J. Bot. 80, 1374–1378.

Rasmussen H N and Whigham D F 1998a Importance of woody debris in seed germination of *Tipularia discolor* (Orchidaceae). Amer. J. Bot. 85, 829–835.

Rasmussen H N and Whigham D F 1998b The underground phase: A special challenge in studies of terrestrial orchid populations. Bot. J. Linn. Soc. 126, 49–64.

Rasmussen H N and Whigham D F, MS Phenology of roots and mycorrhiza in five orchid species differing in phototrophic strategy. New Phytol., submitted.

Reinecke T and Kindl H 1994a Characterization of bibenzyl synthase catalysing the biosynthesis of phytoalexins of orchids. Phytochemistry 35, 63–66.

Reinecke T and Kindl H 1994b Inducible enzymes of the 9,10-dihydro-phenanthrene pathway. Sterile orchid plants responding to fungal infection. Molec Plant Microbe Interactions 7, 449–454.

Rivas M, Warner J and Bermudez M 1998 Presence of mycorrhizas in orchids of a neotropical botanical garden. Revista de Biologia Tropical 46, 211–216.

Roberts P 1998 *Thanatephorus ochraceus*: A saprotrophic and orchid endomycorrhizal species. Sydowia 50, 252–256.

Roberts P 1999 *Rhizoctonia*-forming fungi. A taxonomic guide. The Herbarium, Royal Botanic Gardens, Kew. ISBN 1 900347695

Scrugli A and Cogoni A 1994 Considerazioni morfologiche sugli endofiti micorrizici delle orchidee 'saprofite' delle Sardegna. Micologia Italiana 2, 105–107.

Sen R, Hietala A M and Zelmer C D 1999 Common anastomosis and internal transcribed spacer RFLP groupings in binucleate *Rhizoctonia* isolates representing root endophytes of *Pinus sylvestris*, *Ceratorhiza* spp. from orchid mycorrhizas and a phytopathogenic anastomosis group. New Phytol. 144, 331–341.

Senthilkumar S, Krishnamurthy K V and Vengadeshwari G N 1998 Studies on the mycorrhizal association of the ornamental orchid, *Papilonanthe subulata* (J. Koenig) Garay. Philippine J. Sci. 127, 189–199.

Senthilkumar S, Britto S J, Krishnamurthy K V and Hariharan C 2000 Biochemical analysis of mycorrhizal roots of *Aerides maculosum*. Phytomorphology 50, 273–279.

Shagufta S, Arun R, Siddique S and Raghuvanshi A 1993 Seasonal changes in *Vanda tessellata* mycorrhizae. J Orchid Soc India 7, 83–85.

Stewart J 1993 The Sainsbury orchid conservation project: The first ten years. Kew Magazine 10, 38–43.

Tan T K, Loon W S, Khor E and Loh C S 1998 Infection of *Spathoglottis plicata* (Orchidaceae) seeds by mycorrhizal fungus. Plant Cell Rep. 18, 14–19.

Tatarenko I V 1995 Mycorrhiza of orchids (Orchidaceae) of the Primorye territory. Botanicheskii Zhurnal 80, 64–72.

Taylor D L and Bruns T D 1997 Independent, specialized invasions of ectomycorrhizal mutualism by two nonphotosynthetic orchids. Proc. Nat. Acad. Sci. 94, 4510–4515.

Taylor D L and Bruns T D 1999 Population, habitat and genetic correlates of mycorrhizal specialization in the 'cheating' orchids *Corallorhiza maculata* and *C. mertensiana*. Mol. Ecol. 8, 1719–1732.

Terashita T 1996 Biological species of *Armillaria* symbiotic with *Galeola septentrionalis*. Nippon Kingakukai-Kaiho 37, 45–49. NB only cited on abstract.

Tomita M and Konno S 1998 A preliminary report on the symbiotic germination of nine Japanese terrestrial orchids. J. Jap. Soc. Horticultural Sci. 67, 696–698.

Tremblay R L, Zimmerman J K, Lebron L, Bayman P, Sastre I, Axelrod F and Alers G J 1998 Host specificity and low reproductive success in the rare endemic Puerto Rican orchid *Lepanthes caritensis*. Biol. Conserv. 85, 297–304.

Uetake Y, Farquhar M L and Peterson R L 1997 Changes in microtubule arrays in symbiotic orchid protocorms during fungal colonization and senescence. New Phytol. 135, 701–709.

Uetake Y and Ishizaka N 1995 Cytochemical localization of adenylate cyclase activity in the symbiotic protocorms of *Spiranthes sinensis*. Mycol. Res. 100, 105–112.

Uetake Y and Peterson R L 1997 Changes in actin filament arrays in protocorm cells of the orchid species, *Spiranthes sinensis*, induced by the symbiotic fungus *Ceratobasidium cornigerum*. Can. J. Bot. 75, 1661–1669.

Uetake Y and Peterson R L 1998 Association between microtubules and symbiotic fungal hyphae in protocorm cells of the orchid species, *Spiranthes sinensis*. New Phytol. 140, 715–722.

Umata H 1995 Seed germination of *Galeola altissima*, an achlorophyllous orchid, with aphyllophorales fungi. Mycoscience 36, 369–372.

Umata H 1997a Formation of endomycorrhizas by an achlorophyllous orchid, *Erythrorchis ochobiensis*, and *Auricularia polytricha*. Mycoscience 38, 335–339.

Umata H 1997b *In vitro* germination of *Erythrorchis ochobiensis* (Orchidaceae) in the presence of *Lyophyllum shimeji*, an ectomycorrhizal fungus. Mycoscience 38, 355–357.

Umata H 1998a *In vitro* symbiotic association of an achlorophyllous orchid, *Erythrorchis ochobiensis*, with orchid and non-orchid fungi. Memoirs of the Faculty of Agriculture, Kagoshima University 34.

Umata H 1998b A new biological function of shiitake mushroom, *Lentinula edodes*, in a myco-heterotrophic orchid, *Erythrorchis ochobiensis*. Mycoscience 39, 85–88.

Umata H 1999 Germination and growth of *Erythrorchis ochobiensis* (Orchidaceae) accelerated by monokaryons and dikaryons of *Lenzites betulinus* and *Trametes hirsuta*. Mycoscience 40, 367–371.

Umata H, Kanetani S and Mori T 1994 The habitat and habit of a chlorophyll-free orchid, *Erythrorchis altissima* (Bl.) Bl. Acta Phytotaxonomica Geobotanica 45, 131–138.

Van der Kinderen G 1995 A method for the study of field-germinated seeds of terrestrial orchids. Lindleyana 10: 68.

Vertenyi G and Bratek Z 1996 Isolation of mycorrhizal fungi of terrestrial orchids and its problems. Mikologiai Kozlemenyek 35, 31–36. NB cited from abstract.

Wang H, Wang Z, Zhang F, Liu J and He X 1997 A cytological study on the nutrient-uptake mechanism of a saprophytic orchid *Gastrodia elata*. Acta Bot. Sin. 39, 500–504.

Warcup J H 1975 Factors affection symbiotic germinaiton of orchid seed. In Endomycorrchizas. Eds. F E Sanders, B Mosse and P B Tinker. pp 88–104. Academic Press, London.

Warcup J H 1991 The *Rhizoctonia* endotypes of *Rhizanthella* (Orchidaceae). Mycol. Res. 95, 656–659.

Warcup J H and Talbot P H B 1967 Perfect states of rhizoctonias associated with orchids. New Phytol. 66, 631–641.

Warcup J H and Talbot P H B 1971 Perfect states of rhizoctonias associated with orchids. II. New Phytol. 70, 35–40.

Warcup J H and Talbot P H B 1980 Perfect states of rhizoctonias associated with orchids. III. New Phytol. 86, 267–272.

Whigham D F, Rasmussen H N, O'Neill J, Caldwell B , Smith C, McCormick M and Daniell T, MS Interactions betwen decomposing wood, mycorrhizas and terrestrial orchid seeds and protocorms. In Underlying mechanisms of trends and fluctuations in terrestrial orchid populations. Eds. P Kindlmann, J Willems and D F Whigham. In Press

Wilkinson K G, Dixon K W, Sivasithamparam K and Ghisalberti E L 1994 Effect of IAA on symbiotic germination of an Australian orchid and its production by orchid-associated bacteria. Plant Soil 159, 291–295.

Wood C B, Pritchard H W and Miller A P 2000 Simultaneous preservation of orchid seed and its fungal symbiont using encapsulation-dehydration is dependent on moisture content and storage temperature. CryoLetters 21, 125–136.

Xu J T and Mu C 1990 The relation between growth of *Gastrodia elata* protocorms and fungi. Acta Bot. Sin. 32, 26–31.

Yoder J A, Zettler L W and Stewart S L 2000 Water requirements of terrestrial and epiphytic orchid seeds and seedlings, and evidence for water uptake by means of mycotrophy. Plant Sci. Limerick 156, 145–150.

Zelmer C D and Currah R S 1995a Evidence for a fungal liaison between *Corallorhiza trifida* (Orchidaceae) and *Pinus contorta* (Pinaceae). Can. J. Bot. 73, 862–866.

Zelmer C D and Currah R S 1995b *Ceratorhiza pernacatena* and *Epulorhiza calendulina* spp nov.; mycorrhizal fungi of terrestrial orchids. Can. J. Bot. 73, 1981–1985.

Zelmer C D, Cuthbertson L and Currah R S 1996 Fungi associated with terrestrial orchid mycorrhizas, seeds and protocorms. Mycoscience 37, 439–448.

Zettler L W and McInnis T M J 1992 Propagation of *Platanthera integrilabia* (Correll) Luer, an endangered terrestrial orchid, through symbiotic seed germination. Lindleyana 7, 154–161.

Zettler L W, Barrington F V and McInnis T M J 1995 Developmental morphology of *Spiranthes odorata* seedlings in symbiotic culture. Lindleyana 10, 211–216.

Zettler L W and Hofer C J 1998 Propagation of the little club-spur orchid (*Platanthera clavellata*) by symbiotic seed germination and its ecological implications. Environ. Exp. Bot. 39, 189–195.

Zettler L W, Burkhead J C and Marshall J A 1999 Use of a mycorrhizal fungus from *Epidendrum conopseum* to germinate seed of *Encyclia tampensis in vitro*. Lindleyana 14, 102–105.

Plant and Soil **244:** 165–175, 2002.

An update on nutrient transport processes in ectomycorrhizas

Michel Chalot[1,5], Arnaud Javelle[1], Damien Blaudez[1], Raphaël Lambilliote[2], Richard Cooke[3], Hervé Sentenac[2], Daniel Wipf[4] & Bernard Botton[1]
[1]*UMR 1136 INRA/UHP IaM, Faculté des Sciences, 54506 Vandoeuvre, France.* [2] *BPMP, UMR 5004 CNRS / ENSA-Montpellier / INRA / UM2, Montpellier, France.* [3] *LPBMP, UMR 5545 CNRS, Université de Perpignan, Perpignan, France.* [4]*Zentrum für Molekularbiologie der Pflanzen, Universität Tübingen, Auf der Morgenstelle 1, D-72076 Tübingen, Germany.* [5]*Corresponding author**

Received 19 September 2001. Accepted in revised form 7 February 2002

Key words: ectomycorrhizas, *Hebeloma cylindrosporum*, hyphae translocation, nutrient mobilization, transport processes.

Abstract

Nutrient transport, namely absorption from the soil solution as well as nutrient transfer from fungus to plant and carbon movement from plant to fungus are key features of mycorrhizal symbiosis. This review summarizes our current understanding of nutrient transport processes in ectomycorrhizal fungi and ectomycorrhizas. The identification of nutrient uptake mechanisms is a key issue in understanding nutrition of ectomycorrhizal plants. With the ongoing functional analysis of nutrient transporters, identified during sequencing of fungal and tree genomes, a picture of individual transport systems should be soon available, with their molecular functions assessed by functional characterization in, e.g., yeast mutant strains or *Xenopus* oocytes. Beyond the molecular function, systematic searches for knockout mutants will allow us to obtain a full understanding of the role of the individual transporter genes in the physiology of the symbionts. The mechanisms by which fungal and plant cells obtain, process and integrate information regarding nutrient levels in the external environment and the plant demand will be analyzed.

Introduction

The favourable effects of ectomycorrhizal (ECM) fungi on plant nutrition have traditionally been attributed to the quantitative effect of the extraradical mycelium on uptake of dissolved nutrients from the soil solution (Smith and Read, 1997). This interconnected network of hyphae (or specialized aggregates, i.e. rhizomorphs) forms a supracellular compartment for the transport of nutrients from sites of nutrient capture to sites of nutrient utilization and transfer. It has been estimated that the external mycelium made by far the greatest contribution to the overall potential absorbing surface area of pine seedlings inoculated with *Pisolithus tinctorius* or *Cenococcum geophilum* (Rousseau et al., 1994). Attention has been also paid to the utilization of organic nitrogen and phosphorus forms from more complex substrates (Perez-Moreno and Read, 2000; Smith and Read, 1997), and to the direct mobilization of nutrients from minerals (for review, see Landeweert et al., 2001). Although interactions between mycorrhizal fungi and other soil microbes will not be detailed in the present review, this key feature may have a substantial role on the nutrient economy of ectomycorrhizal trees, with respect to hydrolysis of complex organic compounds (Koide and Kabir, 2001) or organic acid excretion (Landeweert et al., 2001). Translocation through the hyphal network and further transfer of nutrients from fungus to host root has also been discussed in detail (Smith and Read, 1997), but the intimate anatomical connections between fungal and root cells presents considerable technical difficulties for unambiguous experimental investigations of nutrient transfer between fungus and host. Biochemical studies using detached mycorrhizal roots or mycelia revealed the kinetic complexity of

* FAX No: +33 3-83 91 22 43.
E-mail: Michel.Chalot@scbiol.uhp-nancy.fr

nutrient transport. The interpretation of these studies was mainly hampered by the multiplicity of cell types in symbiotic tissues and the uncertainty of whether the different transport activities derived from one or more cell types (Javelle et al., 1999; Kreuzwieser et al., 2000; Van Tichelen and Colpaert, 2000). It is therefore essential to isolate individual carriers and to study their properties, function and regulation before relating them back to the whole plant/fungus level. This review will cover recent studies that have led us to consider the transport processes occurring in the ectomycorrhizal network as a key determinant in the nutrient economy of ectomycorrhizal plants. Recent papers have reviewed the actual knowledge in plant-fungus interactions on the basis of C and N availability as investigated on selected laboratory models (Buscot et al., 2000; Hampp et al., 1999; Nehls et al., 2001), and therefore this aspect will not be detailed in the present paper, except at the molecular level.

Mobilization, absorption, translocation and transfer of nutrients

Phosphorus

Although the total amount of P in the soil may be high, it is a very immobile and unavailable element for plant uptake because of its adsorption, its precipitation or its conversion to recalcitrant organic forms. It has been suggested that ECM fungi stimulate the uptake of P from poorly soluble sources in soil. Using mycorrhizal birch seedlings grown in observation chambers with the ECM fungus *Paxillus involutus*, losses of P from a range of different fermentation horizons could be measured (Perez-Moreno and Read, 2000). This ability to forage P from litter horizons was related to the capacity of the fungus to produce the necessary enzymes (Bending and Read, 1995). Wallander et al. (1997) showed that some ECM fungi were able to improve the uptake of apatite-P by ECM pine seedlings. However, weathering capacities clearly varied among ECM fungi and it has been recently found that apatite-weathering ability of ECM fungi were somehow related to their ability to produce oxalic acid (Wallander, 2000a).

Furthermore, uptake of P by mycorrhizal fungi is a key determinant of the improved P nutrition of ectomycorrhizal plants. Kinetics of net phosphate uptake by mycorrhizal *versus* non-mycorrhizal pine seedlings has been recently studied in details (Colpaert et al., 1999; Van Tichelen and Colpaert, 2000). The intact extraradical mycelium greatly increased the absorption surface area of the roots (Van Tichelen and Colpaert, 2000). *P. involutus* was found to be the most efficient mycobiont with regard to P influx, with a lower K_m value than the other pine-fungus associations tested in this study (Colpaert et al. 1999, Figure 1). Comparing the effects of arbuscular mycorrhizal (AM) and ectomycorrhizal fungi on P uptake by *Eucalyptus coccifera*, Jones et al. (1998) found that P inflow rates of ECM and AM seedlings were 3.8 times, and 2.0–2.7 times those of non-mycorrhizal seedlings. They also found that ECM and AM plants are more efficient than NM plants at acquiring P per unit of C allocated below-ground. Although numerous data clearly indicate that ECM fungi can improve P_i uptake, determination of P fluxes under various conditions also clearly demonstrate the heterogeneity of P uptake capacities among different mycobionts (Cairney and Smith, 1993). The identification of the P uptake mechanisms is therefore a key issue in understanding P nutrition of ectomycorrhizal plants.

In a recent experiment, Jentschke et al. (2001a) demonstrated that *P. involutus* hyphae were able to take up and translocate significant amounts of P to their P-deficient host plant and therefore strongly improved host P status that in turn stimulated seedling growth. There is increasing evidence that P translocation through hyphae occurs by active processes (Ashford, 1998; Timonen et al., 1996). A role of tubular vacuole system in longitudinal transport of P in *P. tinctorius* hyphae has been hypothesized (Cole et al., 1998; see Ashford and Alloway, 2002). Using X-ray spectroscopy, Bücking and Heyser (2000a) concluded that the translocation of P from fungus to host plant across the mycorrhizal interface is driven by P concentration in the cytoplasm of the Hartig net and by efflux rate into the interfacial apoplast. However, the detailed mechanism of P transfer at the symbiotic interface remains obscure.

Nitrogen

In forest soils, N is present either as NH_4^+ and/or as NO_3^- and organic compounds such as amino acids, peptides and proteins (Chalot and Brun, 1998; Marschner, 1995). The significance of mycorrhization for improved uptake of ammonium and organic N has been shown in several studies (Javelle et al., 1999; Plassard et al., 2000), whereas its influence on NO_3^-

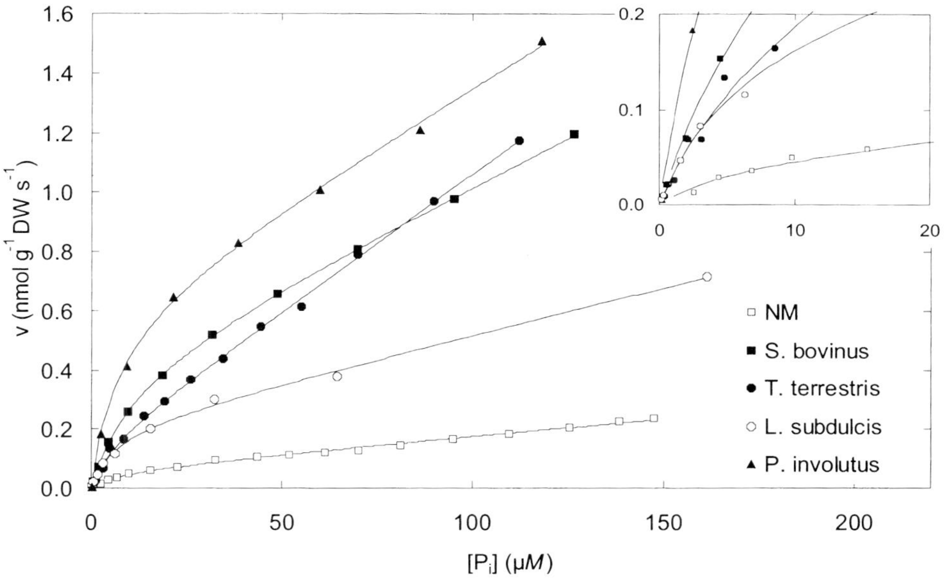

Figure 1. Michaelis–Menten plot showing the net P_i uptake in intact mycorrhizal and non-mycorrhizal *Pinus sylvestris* seedlings as a function of solution P_i concentration. The inset figure shows the relationship at the lower P_i concentrations, most representative for soil solution P_i. Measurements were performed 9 weeks after inoculation and data points were fitted to a two-phase Michaelis–Menten equation. Data obtained with excised *Fagus sylvatica* – *Lactarius subdulcis* mycorrhiza are also plotted (Colpaert et al., 1999; with permission).

uptake is still a controversial issue (Kreuzwieser et al., 2000; Plassard et al., 2000).

Ammonium mobilization by hyphae from soil sources is directly linked to hyphae uptake capacities. The kinetics and energetics of ammonium/methylammonium transport in ectomycorrhizal fungi have been characterized (Javelle et al., 1999; Jongbloed et al., 1991). Comparing the ammonium uptake capacity of the two partners separately or in symbiosis, it was found that mycelia have much higher capacities for ammonium uptake than non-mycorrhizal roots and ectomycorrhizal fungi increase ammonium uptake capacities of their host roots (Javelle et al., 1999; Plassard et al., 1997). On the contrary, recent investigations indicated that mycorrhization leads to reduced rates of NO_3^- net uptake in *Laccaria laccata* mycorrhizal beech (Kreuzwieser et al., 1997, 2000). The authors attributed this effect to a reduced influx plus an enhanced efflux of NO_3^- as compared with non-mycorrhizal beech roots. The molecular mechanisms involved in inorganic N uptake will be further discussed in the last section.

A number of observations support the view that ectomycorrhizal fungi have the ability to degrade macromolecular nitrogen and furthermore to take up and assimilate the products of hydrolytic degradation (Chalot and Brun, 1998; Näsholm and Persson, 2001; Wallenda and Read, 1999). All of the major hydrolytic enzymes involved in mobilization of nitrogen from organic compounds have been detected in ericoid and some in ectomycorrhizal fungi (Chalot and Brun, 1998). Transport of amino acids was investigated in the mycorrhizal fungi *P. involutus* (Chalot et al., 1996) and *Amanita muscaria* (Nehls et al., 1999), which demonstrated their ability to take up a variety of amino acids. Recent data indicate that birch root colonization by *Paxillus involutus* increased glutamate uptake and that mycorrhiza formation induced a profound alteration of the metabolic fate of exogenously supplied glutamate (Blaudez et al., 2001). Plassard et al. (2000) further demonstrated that *Hebeloma cylindrosporum*/*Pinus pinaster* ectomycorrhizas had higher capacities for taking up glutamate compared with non-mycorrhizal pine roots. However, the critical question is to what extent this potential to use organic N occurs under natural conditions (Näsholm and Persson, 2001). The potential utilization of isotopic signatures for the determination of resource acquisition patterns by mycorrhizal plants has been discussed in detail (Evans, 2001; Hobbie et al., 1999). However, recent data indicates that this must be considered with caution, given the large isotopic fractionation patterns during uptake and metabolism in mycorrhizal fungi (Emmerton et al., 2001).

Based on both ^{15}N labelling and mass balance data, it was demonstrated that hyphal ammonium acquisition contribute 45% to total plant N uptake under N deficiency (Jentschke et al., 2001b). In similar experiment using P deficient seedlings, they found that hyphal contribution to total plant N uptake amounted to 12% (Jentschke et al., 2001a). Disrupting the external mycelium from ectomycorrhizas greatly decreased [^{15}N]ammonium uptake by birch seedlings (Javelle et al., 1999). Therefore, external hyphae can be considered as the ammonium absorbing structure of ectomycorrhizal roots. These results confirmed the function of the extraradical mycelium in translocating N from sources to roots.

The physiological processes involved in the further transfer of N within the symbiotic tissues are still poorly understood. The exchange of nutrients between the fungus and the plant requires passage across the fungal plasmalemma, the interfacial matrix and plasmalemma of the plant. A number of hypotheses, which are derived from experiments on higher plants or bacteria, have been drawn in a previous review (Chalot and Brun, 1998). Indirect evidence from ^{15}N labeling experiments supports the view of amino acid transfer from the fungus to the host (Smith and Read, 1997). However a direct transfer of NH_4^+, as hypothesized for AM symbiosis (Bago et al., 2001), cannot be ruled out. The answer to whether amino acids or ammonium are preferentially transferred from the fungus to the host through the apoplast will probably arise from molecular approaches coupled with biochemical techniques. Direct analysis of apoplast fluid (Solomon and Oliver, 2001) together with tracer experiments for instance, will be necessary to elucidate transfer between fungal and root cells. Also detailed analysis of individual compounds using specific isotope analysis may be a key to further address these important questions (Evans, 2001).

Potassium, magnesium and calcium

Silicate minerals may provide calcium, magnesium and potassium, after chemical or biological dissolution (Landeweert et al., 2001). Potassium mobilization from apatite by mycorrhizal plants has been observed in long-term pot experiments (Wallander and Wickman, 1999; Wallander, 2000b). It was found that *Pinus sylvestris* seedlings colonized by *Suillus variegatus* enhanced K release from biotite as compared to

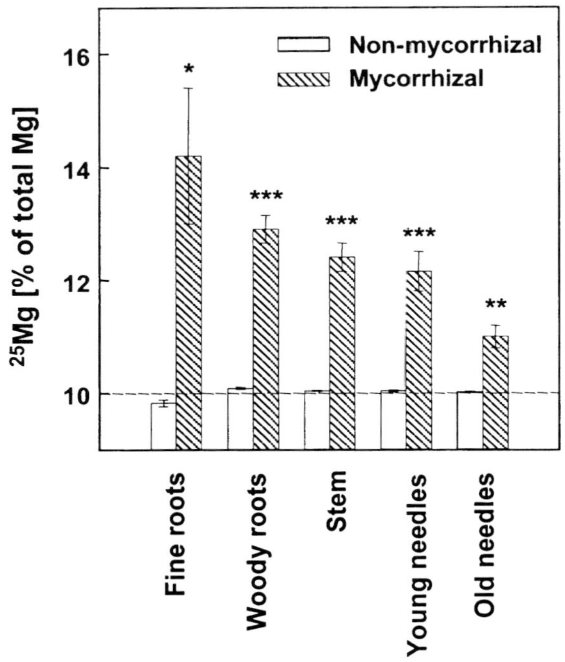

Figure 2. Concentration of the ^{25}Mg label (expressed as percentage ^{25}Mg of total Mg) in tissues of non-mycorrhizal and mycorrhizal Norway spruce seedlings after labeling the hyphal compartment in the culture system with ^{25}Mg for 6 weeks. Broken line indicates natural abundance of ^{25}Mg (10.0%). Values are mean values of 4 replicate pots. Small bars indicate standard error. Significance levels for differences between mycorrhizal and non-mycorrhizal seedlings: *, $P \leq 0.01$; **, $P \leq 0.001$; ***, $P \leq 0.0001$ (Jentschke et al., 2000, with permission).

non-mycorrhizal seedlings (Wallander and Wickman, 1999).

Further translocation of K and Mg (Figure 2) through mycorrhizal hyphae has been recently examined (Jentschke et al., 2000; Jentschke et al., 2001a). Hyphal K and Mg acquisition by mycorrhizal spruce roots was significant and contributed to a minimum of 5–6% of total plant uptake (Jentschke et al., 2001a). These data demonstrate the potential role of the ectomycorrhizal mycelium in K and Mg acquisition.

Nutrient entry in root cells colonized by mycorrhizal fungi has been recently examined using isotopic tracers (^{44}Ca, ^{25}Mg and ^{41}K) and LAMMA (laser-microprobe-mass-analyser) (Kuhn et al., 2000). Based on entry kinetics and temperature dependence, it was concluded that Ca and Mg may enter the cortex via a free apoplastic path and that the endodermis is a major barrier for the further passage of these divalent cations into the xylem. For K, although kinetics were not shown, the authors found a temperature dependence of K entry, which suggests that one component in K uptake mechanism may be carried out by an active process. This will be further discussed in the last section. However, depending on the fungal partner, variations in the intensity of apoplastic movement of nutrients could be observed (Bücking and Heyser, 2000b). For instance X-ray microanalyses indicate that in mycorrhizas of *Pinus sylvestris*/*Suillus bovinus* the apoplastic movement of cations across the fungal sheath is strongly reduced. In eucalypt mycorrhizas, the exodermal Casparian bands have a low permeability to apoplasmic tracers, suggesting that an isolated shared apoplasmic compartment exists at the interface between the partners, also not tightly sealed (Vesk et al., 2000).

Interdependence of nutrient translocation

P deficiency in spruce triggers hyphal foraging and translocation not only of P from P-rich sites, but also of N, K and Mg, which indicates potential co-translocation of a limiting element (P) with non-limiting nutrients (N, K, Mg) by *P. involutus* (Jentschke et al., 2001a). In this experiment, K translocation was of the same order of magnitude as P translocation whereas N translocation exceeded P translocation five-fold. Mg fluxes through hyphae were the lowest of all nutrient fluxes determined. In agreement with these studies, Cole et al. (1998), using X-ray mapping demonstrated that vacuoles in growing hyphae retain high levels of P and K.

Transport processes at the molecular level

Fungal transporters: a Hebeloma cylindrosporum *database*

Genomic and cDNA libraries have been prepared, and an EST (expressed sequence tags) resource of ca. 3000 sequences has been produced (Table 1). Preliminary analysis of the library reveals that ca. 60% of the tags correspond to genes that have counterparts in other organisms, and 50% of known function. Among them, 38 tags correspond to genes encoding carriers or channels that could play a role in nutrient uptake from the soil solution or from the host plant apoplast, or in nutrient secretion towards the host plant (Table 1). Of particular interest are several K^+ channels (homologous to Shaker/Shab channels identified in animals), one K^+ carrier (homologous to yeast TRK systems), 2 Pi transporters (homologous to yeast and *Glomus versiforme* Pi uptake systems), one NH_4^+ carrier, one peptide carrier, several hexose carriers, organic acid transporters, a large set of ATP binding cassette (ABC) transporters, or metal (Fe, Cu, Zn) transporters. Also, several enzymes that could play crucial roles in improving P (phosphatases) and N (aminotransferases, glutamine synthetase, glutamate synthase, urease, asparagine synthetase) nutrition have been identified.

Ammonium transporters

Ammonium transporters were first cloned from yeast (Mep1) and higher plant (Amt1) (Marini et al., 1994; Ninnemann et al., 1994). Heterologous expression of the yeast triple *mep*Δ mutant (Marini et al., 1997) has enabled the characterization of HcAmt2 and HcAmt3 from *H. cylindrosporum*, the first members of the ammonium transporter protein family (Amt) in basidiomycetes (Figure 3). HcAmt2 expressed in yeast had a higher affinity but a lower capacity than HcAmt3. Northern blot analyses in *H. cylindrosporum* showed that *HcAMT2* and *HcAMT3* were up-regulated upon nitrogen deprivation and down-regulated by ammonium (Javelle et al., 2001). A third member of this family showing a high homology to HcAmt2 was recently isolated by direct complementation from a cDNA library (Javelle, unpublished results). Results also indicate that the HcAmts could play a role in ammonium retention in *H. cylindrosporum*, as already demonstrated for the Mep proteins in *S. cerevisiae* (Marini et al., 1997). An immunological approach is currently being developed in our laboratory to elucidate the intriguing questions concerning the post-translational regulation of this protein family and its cytolocalisation in ectomycorrhizas.

Amino acid transporters

Amino acid transporters have been characterized physiologically in detail in animals, plants and yeasts (for review see Fischer et al., 1998; van Belle and André, 2001; Williams and Miller, 2001; Wipf et al., 2002). On the basis of physiological studies, the existence of a large number of transporters has been postulated differing in their substrate spectrum, in their transport mechanism, i.e. the ions used in co-transport, and in tissue specificity. Three transporter superfamilies are found in plants and yeast (the APC, amino acid-polyamine-choline transporter superfamily, the ATF, amino acid transporter superfamily, the

Table 1. Hebeloma cylindrosporum EST resource

Protein function prediction	Number of clones (Percentage of the matching ESTs)	Number of putative different proteins
Metabolism	367 (34,04)	172
Protein synthesis	256 (23,74)	131
Postranscriptional modification and proteolysis	215 (19,94)	76
Cytoskeleton	56 (5,19)	22
Signal transduction	61 (5,65)	48
Nuclear proteins	28 (2,59)	23
Transporters	38 (3,52)	29
Probable membrane proteins	28 (2,59)	19
Cell wall proteins	29 (2,69)	15
Unclassified	855	655

[a]Protein function prediction from ESTs was made with BlastX program (Altschul et al., 1990). Around 1000 sequences were absent from this table because they did not match with any protein of database (expect $> e^{-2}$). ESTs matching known proteins with BlastX expect $< e^{-15}$ have been classified into functional categories, ESTs with expect $> e^{-15}$ were placed into unclassified category together with ESTs which can not be classified into functional categories. The percentage of each individual functional class (calculated from the total number of classified proteins) is given in parenthesis. Third column indicates number of different accession numbers obtained in the first line of BlastX program.

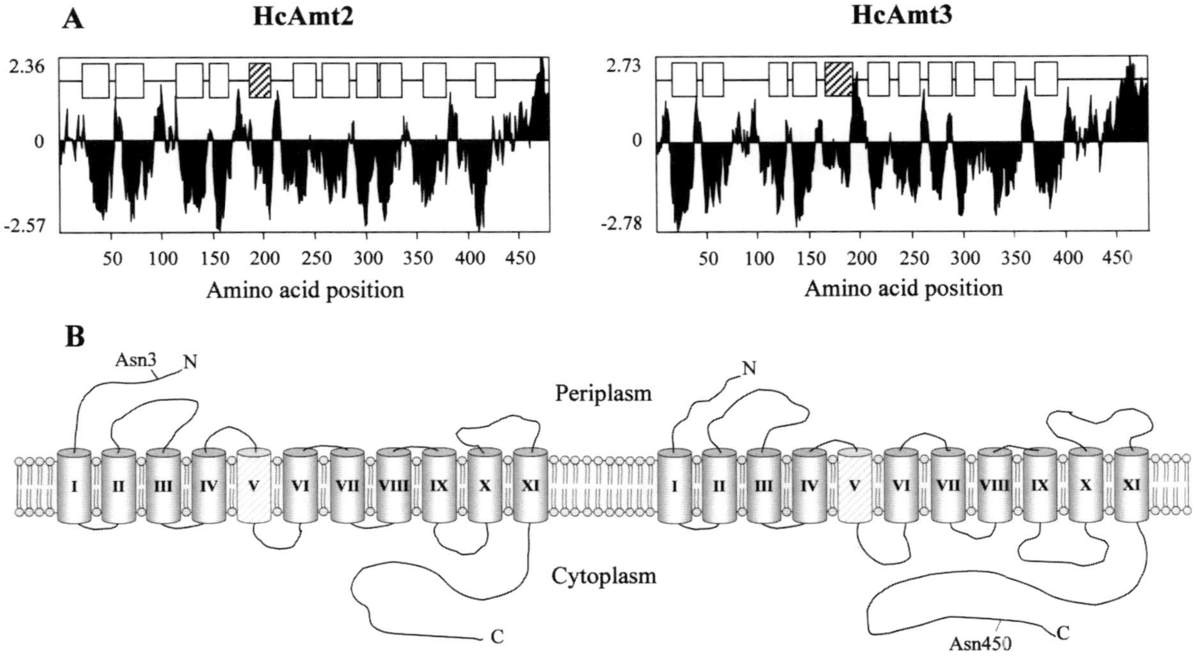

Figure 3. Topology of the ammonium transporters from *Hebeloma cylindrosporum*. (A) Hydropathy profile of the HcAmt proteins. Hydropathy analyses were performed using the Kyte and Doolittle algorithm. The upper boxes represent the relative positions of the membrane-spanning segments predicted by the algorithm TMpred, HMMTOP and TMHMM. The hatched boxes correspond to the ammonium-signature containing-TM. (B) Schematic model of the two HcAmt proteins. The predicted glycosylation sites (Asn3 for HcAmt2, Asn450 for HcAmt3) are indicated. The hatched boxes correspond to ammonium-signature containing-transmembranes (Javelle et al., 2001, with permission).

VGT, vesicular glutamate transporter family). Recently, two amino acid transporters from mycorrhizal fungi have been characterized. AmAap1 was identified from *A. muscaria* by screening a cDNA library (Nehls et al., 1999). HcBap1 was isolated from *H. cylindrosporum* by functional complementation of a yeast strain deficient in amino acid transporters (Wipf et al., unpublished results). Their expression in yeast revealed that they function as high affinity transporters. Two main functions can be expected for those genes, namely uptake of amino acid from the soil for nutrition, and the prevention of amino acid loss by hyphal leakage in the absence of a suitable N source at a low internal N status. However, given the variety of amino acid transporters described in animals, plants and yeasts, the characterization of many more fungal transporters is expected and needed.

Potassium transporters

Potassium transport has received much attention in plants and fungi (for review see Rodriguez-Navarro, 2000). It is clear that the EST resource described above now offers invaluable molecular tools for studying the molecular mechanisms of K transport in ectomycorrhizal fungi. For instance, regarding nutrient ion uptake and secretion towards the host plant, sequence analysis of the clones corresponding to the K^+ transport systems identified in the EST library provides the following working hypotheses. The K^+ carrier that displays similarities with the yeast TRK systems could be responsible for high affinity active K^+ uptake by peripheral hyphae from the soil solution. On the other hand, the K^+ channels displaying similarities with animal Shaker/Shab outward K^+ channels could play a role in K^+ secretion towards the host plant. Within these working hypotheses, a preliminary step will be to assess the molecular function of these transport systems via their expression in heterologous systems (e.g. functional complementation of yeast mutants, electrophysiological characterization in *Xenopus* oocytes).

Host tree transporters

Although a variety of ammonium, amino acid, peptide, phosphate and sugar transporters have been confirmed to be expressed in plant roots (Delrot et al., 2000; Fischer et al., 1998; von Wiren et al., 2000), data on tree roots is lacking and urgently needed. Current sequencing projects (i.e. poplar database, http://mycor.nancy.inra.fr/PoplarDB.html) may provide potential tools for detailed analysis of transporter genes in tree roots.

Tree & fungal hexose transporters

It is assumed that sucrose is delivered into the apoplast at the plant-fungus interface (Hartig net) and hydrolyzed via a plant-derived acid invertase (Nehls et al., 2001). The resulting hexoses are then taken up by fungal cells as well as by plant root cells (Nehls et al., 2000; Wright et al., 2000). A prerequisite for rapid uptake of monosaccharides is a membrane transport system. So far only one hexose transporter (AmMst1) has been identified from the ectomycorrhizal fungus *A. muscaria* (Nehls et al., 1998). The expression of AmMst1 in a *Saccharomyces cerevisiae* strain, unable to take up hexoses, demonstrated that AmMst1 was a functional monosaccharide transporter (Wiese et al., 2000). A *Picea abies* hexose transporter cDNA (*PaMST1*) that encodes an open reading frame of 513 amino acids was isolated by a RT-PCR based strategy (Nehls et al., 2000). PaMst1 is expressed in the hypocotyl and in roots of *P. abies* seedlings, but not in needles sampled at different developmental stages (cotyledons and young needles). In addition, two putative hexose transporter gene fragments were identified from birch by RT-PCR (Wright et al., 2000). While the expression of the hexose transporter gene of Norway spruce was only slightly (approx. 30%) reduced in mycorrhizas, the transcript level of both hexose transporter genes of birch was reduced by a factor of three. Even if a reduced transcript level does not mean a decreased transporter activity, and although the number of monosaccharide transporters expressed in plant roots is not yet clear, these results suggest that plants do not increase their hexose import capacity during symbiosis. In contrast, the expression of the fungal monosaccharide transporter gene increases significantly in mycorrhizas. Since, in addition, the transcript level of the fungal (*A. muscaria*) monosaccharide transporter gene is much higher than that encoding the *P. abies* hexose transporter (U. Nehls, unpublished), the fungus represents the major carbohydrate sink in infected fine roots. It could thus be assumed that the plant does not compete for hexose import at the plant-fungus interface, and that the fungal activity determines the sink strength for carbohydrates in mycorrhizas.

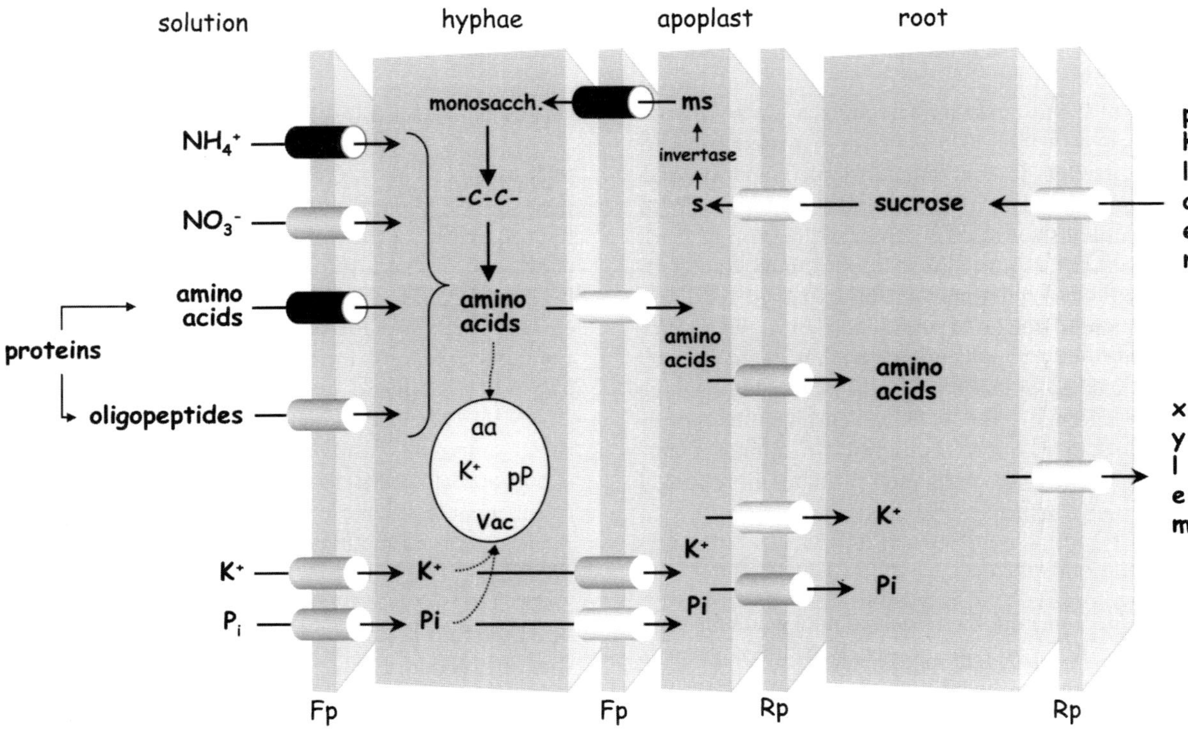

Figure 4. Summary of current understanding of transporters in ectomycorrhizal tissues. fp: fungal plasma membrane, rp: root plasma membrane, pP: polyphosphate. Black cylinder: one member of the transporter family has been fully characterized, i.e., functional complementation of a yeast deficient strain; grey cylinder: putative transporters (candidate genes available in i.e. EST resources); white cylinder: hypothetic transporters.

Transporters in ectomycorrhizal tissues: overview and perspectives

Our current understanding of transport processes in ectomycorrhizal tissues is summarized in Figure 4. Up to now, only 5 transporters from ectomycorrhizal fungi have been fully characterized, i.e., their function demonstrated by functional complementation of deficient yeasts (HcAmt2, HcAmt3, Javelle et al., 2001; HcBap1, Wipf et al., unpublished results; AmMst1, Nehls et al., 1998; AmAap1, Nehls et al., 1999). A number of other candidate genes from fungi or from host roots have been isolated either by degenerate PCR or from EST resources. However, the precise function and localization of these putative proteins in mycorrhizal tissues remain unknown. Nutrient fluxes from fungus to host largely depend upon coordination with nutrient release by the fungus into, and nutrient reuptake by the root cells from, the apoplastic interface. Unequivocal elucidation of these transport mechanisms will rely on innovative physiological and molecular approaches. Heterologous complementation of deficient yeasts with transporter genes and further expression analysis will probably help in understanding the uptake processes.

However, the most difficult mechanism to perceive will be the release of amino acids into the apoplast by the fungus. Studies indicate that amino acid release by ectomycorrhizal fungi indeed does occur (Plassard et al., 1996). Whether the mechanism involved, specifically applied to the symbiotic interface, and whether it concerns a limited range of amino acids remain intriguing questions. The lack of deficient yeasts in amino acid excretion greatly hampered the study of amino acid excretion in other organisms. Research in these directions however are in progress and should allow the characterization of amino acid exporters in mycorrhizal fungi in the near future.

Current research work on the model plant *Arabidopsis thaliana* clearly illustrates the usefulness of models allowing development of genetic, reverse genetic and transgenesis approaches (Zimmermann and Sentenac, 1999). For instance, expression of promoter-reporter gene constructs in transgenic fungi as well as immunological-based approaches will allow us to localize the expression of these various transport systems in the fungus (peripheral hyphae, Hartig network) and to study the effects of nutrient deficiency on their expression. In a recent work, the expression and the cellular localization of a proline transporter was successfully studied in *Aspergillus nidulans* using GFP fusion protein (Tavoularis et al., 2001). It is exciting to consider the value of this approach in studies of plant-fungus interactions, especially since fluorescent protein labeling patterns can be extended to the ultrastructural level by immunolocalization with GFP antibodies (Gordon et al., 2000). However, use of this technology relies on the availability of high throughput transformation system for mycorrhizal fungi. In some of the experiments described in the present paper, *H. cylindrosporum* was chosen as a model of ectomycorrhizal fungus, since this species can be grown in vitro from spore to spore (Debaud and Gay, 1987) and can be transformed (Marmeisse et al., 1992), allowing genetic and reverse genetic approaches. More recently, a technique was developed for transforming mycorrhizal fungi, including *H. cylindrosporum*, based on *Agrobacterium tumefaciens* mediated T-DNA transfer (Pardo et al., 2002). *Agrobacterium tumefaciens* mediated transformation has been successfully applied to the wheat pathogen *Mycosphaerella graminicola* for the disruption of an ABC transporter, generated disruptants with high efficiency (Zwiers and De Waard, 2001).

Ultimately, the combination of biochemical and molecular approaches will provide a platform for discovering how release and transport events at the apoplastic interface are coordinated and how controls either by the fungus or by the host are exerted.

References

Ashford A E 1998 Dynamic pleiomorphic vacuole systems: Are they endosomes and transport compartments in fungal hyphae? Adv. Bot. Res. 28, 119–159.

Ashford A E and Allaway W G, 2002 The role of a motile tubular vacuole system in mycorrhizal fungi. Plant Soil 244, 179–189 (this issue).

Altschul S F, Gish W, Miller W, Myers E W and Lipman D J 1990 Basic local alignment search tool. J. Mol. Biol. 215, 403–410.

Bago B, Pfeffer P and Shachar-Hill Y 2001 Could the urea cycle be translocating nitrogen in the arbuscular mycorrhizal symbiosis ? New Phytol. 149, 4–8.

Bending g D and Read D J 1995 The structure and function of the vegetative mycelium of ectomycorrhizal plants. VI. Activities of nutrient mobilizing enzymes in birch litter colonized by *Paxillus involutus* (Fr.) Fr. New Phytol. 130, 411–417.

Blaudez D, Botton B, Dizengremel P and Chalot M 2001 The fate of [^{14}C]glutamate and [^{14}C]malate in birch roots is strongly modified under inoculation with *Paxillus involutus*. Plant Cell Environ. 24, 449–457.

Bücking H and Heyser W 2000a Subcellular compartmentation of elements in non-mycorrhizal and mycorrhizal roots of *Pinus sylvestris*: an X-ray microanalytical study. I. The distribution of phosphate. New Phytol. 145, 311–320.

Bücking H and Heyser W 2000b Subcellular compartmentation of elements in non-mycorrhizal and mycorrhizal roots of *Pinus sylvestris*: an X-ray microanalytical study. II. The distribution of calcium, potassium and sodium. New Phytol. 145, 321–331.

Buscot A, Munch J C, Charcosset J Y, Gardes M, Nehls U and Hampp R 2000 Recent advances in exploring physiology and biodiversity of ectomycorrhizas highlight the functioning of these symbioses in ecosystems. FEMS Microbiol. Rev. 24, 601–614.

Cairney J W G and Smith S E 1993 Efflux of phosphate from the ectomycorrhizal basidiomycete *Pisolithus tinctorius*: general characteristics and the influence of intracellular phosphate. Mycol. Res. 96, 673–676.

Chalot M and Brun A 1998 Physiology of organic nitrogen acquisition by ectomycorrhizal fungi and ectomycorrhizas. FEMS Microbiol. Rev. 22, 21–44.

Chalot M, Brun A, Botton B and Söderström B 1996 Kinetics, energetics and specificity of the general amino acid transporter from the ectomycorrhizal fungus *Paxillus involutus*. Microbiology 142, 1749–1756.

Cole L, Orlovich D A and Ashford A E 1998 Structure, function, and, motility of vacuoles in filamentous fungi. Fungal Genet. Biol. 24, 86–100.

Colpaert J V, van Tichelen K K, van Assche J A and Van Laere A 1999 Short-term phosphorus uptake rates in mycorrhizal and non-mycorrhizal roots of intact *Pinus sylvestris* seedlings. New Phytol. 143, 589–597.

Debaud J C and Gay G 1987 *In vitro* fruiting under controlled conditions of the ectomycorrhizal fungus *Hebeloma cylindrosporum* associated with *Pinus pinaster*. New Phytol. 105, 429–435.

Delrot S, Atanassova R and Maurousset L 2000 Regulation of sugar, amino acid and peptide plant membrane transporters. Biochim. Biophys. Acta 1465, 281–306.

Emmerton K S, Callaghan T V, Jones H E, Leake J R, Michelsen A and Read D J 2001 Assimilation and isotopic fractionation of nitrogen by mycorrhizal fungi. New Phytol. 151, 503–512.

Evans R D 2001 Physiological mechanisms influencing plant nitrogen isotope composition. Trends Plant Sci. 6, 121–126.

Fischer W N , André B, Rentsch D, Krolkiewicz S, Tegeder M, Breitkreuz K and Frommer W B 1998 Amino acid transport in plants. Trends Plant Sci. 3, 188–195.

Gordon C L, Khalaj V, Ram A F, Archer D B, Brookman J L, Trinci A P, Jeenes D J, Doonan J H, Wells B, Punt P J, van den Hondel C A, Robson G D 2000 Glucoamylase::green fluorescent protein fusions to monitor protein secretion in *Aspergillus niger*. Microbiology 146, 415–26.

Hampp R, Wiese J, Mikolajewski S and Nehls U 1999 Biochemical and molecular aspects of C/N interaction in ectomycorrhizal plants: an update. Plant Soil 215, 103–113.

Hobbie E A, Macko S A and Shugart H H 1999 Interpretation of nitrogen isotope signatures using the NIFTE model. Oecologia 120, 405–415.

Javelle A, Chalot M, Söderström B, and Botton B 1999 Ammonium and methylamine transport by the ectomycorhizal fungus *Paxillus involutus* and ectomycorrhizas. FEMS Microbiol. Ecol. 30, 355–366.

Javelle A, Rodríguez-Pastrana B R, Jacob C, Botton B, Brun A, André B, Marini A M and Chalot M 2001 Molecular characterization of two ammonium transporters from the ectomycorrhizal fungus *Hebeloma cylindrosporum*. FEBS Lett. 505: 393–398.

Jentschke G, Brandes B, Kuhn A J, Schröder W H and Godbold D L 2001a Interdependence of phosphorus, nitrogen, potassium and magnesium translocation by the ectomycorrhizal fungus *Paxillus involutus*. New Phytol. 149, 327–338.

Jentschke G, Brandes B, Kuhn A J, Schröder W H, Becker J S and Godbold D L 2000 The mycorrhizal fungus *Paxillus involutus* transports magnesium to Norway spruce seedlings. Evidence from stable isotope labeling. Plant Soil 220, 243–246.

Jentschke G, Godbold D L and Brandes B 2001b Nitrogen limitation in mycorrhizal Norway spruce (*Picea abies*) seedlings induced mycelial foraging for ammonium: implications for Ca and Mg uptake. Plant Soil 234, 109–117.

Jongbloed R H, Clement J M A M and Borst-Pauwels G W F H 1991 Kinetics of K^+ uptake by ectomycorrhizal fungi. Effect of NH_4^+ on K^+ uptake. Physiol. Plant. 83, 427–432.

Jones M D, Durall D M and Tinker P B 1998 Comparison of arbuscular and ectomycorrhizal *Eucalyptus coccifera*: growth response, phosphorus uptake efficiency and external hyphal production. New Phytol. 140, 125–134.

Koide R T and Kabir Z 2001 Nutrient economy of red pine is affected by interactions between *Pisolithus tinctorius* and other forest-floor microbes. New Phytol. 150, 179–188.

Kreuzwieser J, Herschbach C, Stulen I, Wiersema P, Vaalburg W and Rennenberg H 1997 Interactions of NH_4^+ and L-glutamate with NO_3^- transport processes of non-mycorrhizal *Fagus sylvatica* roots. J. Exp. Bot. 48, 1431–1438.

Kreuzwieser J, Stulen I, Wiersema P, Vaalburg W and Rennenberg H 2000 Nitrate transport processes in *Fagus- Laccaria*-mycorrhizae. Plant Soil 220, 107–117.

Kuhn A J, Schröder W H and Bauch J 2000 The kinetics of calcium and magnesium entry into mycorrhizal spruce roots. Planta 210, 488–496.

Landeweert R, Hoffland E, Finlay R D, Kuyper T W and van Breemen N 2001 Linking plants to rocks: ectomycorrhizal fungi mobilize nutrients from minerals. Trends Ecol. Evol. 16, 248–254.

Marini A M, Soussi-Boudekou S, Vissers S and André B 1997 A family of ammonium transporters in *Saccharomyces cerevisiae*. Mol. Cell. Biol. 17, 4282–4293.

Marini A M, Vissers S, Urrestarazu A and André B 1994 Cloning and expression of the MEP 1 gene encoding an ammonium transporter in *Saccharomyces cerevisiae*. EMBO J. 13, 3456–3463.

Marmeisse R, Gay G, Debaud J C and Casselton L A 1992 Genetic transformation of the symbiotic basidiomycete fungus *Hebeloma cylindrosporum*. Cur. Genet. 22, 42–45.

Marschner H 1995 Mineral Nutrition of Higher Plants, Ed 2, Academic Press, London.

Näsholm T and Persson J 2001 Plant acquisition of organic nitrogen in boreal forests. Physiol. Plant. 111, 419–426.

Nehls U, Kleber R, Wiese J and Hampp R 1999 Isolation and characterization of a general amino acid permease from the ectomycorrhizal fungus *Amanita muscaria*. New Phytol. 144, 343–349.

Nehls U, Micholajewski S, Magel E and Hampp R 2001 Carbohydrate metabolism in ectomycorrhizas: gene expression, monosaccharide transport and metabolic control. New Phytol. 150, 533–541.

Nehls U, Wiese J and Hampp R 2000 Cloning of a *Picea abies* monosaccharide transporter gene and expression-analysis in plant tissues and ectomycorrhizas. Trees 14, 334–338.

Nehls U, Wiese J, Güttenberger M and Hampp R 1998 Carbon allocation in ectomycorrhizas: identification and expression analysis of an *Amanita muscaria* monosaccharide transporter. Mol. Plant Microbe Interact. 11, 167–176.

Ninnemann O, Jauniaux J C and Frommer W B 1994 Identification of a high affinity ammonium transporter from plants. EMBO J. 13, 3464–3471.

Pardo A G, Hanif M, Raudaskoski M and Gorfer M 2002 Genetic transformation of ectomycorrhizal fungi mediated by *Agrobacterium tumefaciens*. Mycol. Res. 106, in press.

Perez-Moreno J and Read D J 2000 Mobilization and transfer of nutrients from litter to tree seedlings via the vegetative mycelium of ectomycorrhizal plants. New Phytol. 145, 301–309.

Plassard C, Bonafos B and Touraine B 2000 Differential effects of mineral and organic N sources, and of ectomycorrhizal infection by *Hebeloma cylindrosporum*, on growth and N utilization in *Pinus pinaster*. Plant Cell Environ. 23, 1195–1205.

Plassard C, Chalot M, Botton, B and Martin F 1997 Le rôle des ectomycorhizes dans la nutrition azotée des arbres forestiers. Rev. For. Fr. 49, 82–98.

Plassard C, Reid R J and Tester M 1996 Amino acid release by the ectomycorrhizal fungus *Hebeloma cylindrosporum* grown *in vitro*. *In* Mycorrhizas in integrated systems, from genes to plant development. Eds. C Azcon-Aguilar and J M Barea. pp 364–367. Proceedings of the Fourth European Symposium on Mycorrhizas. European Commission, Brussels, Luxembourg.

Rodriguez-Navarro A 2000 Potassium transport in fungi and plants. Biochim. Biophys. Acta 1469, 1–30.

Rousseau J V D, Sylvia D M and Fox A J 1994 Contribution of ectomycorrhiza to the potential nutrient-absorbing surface of pine. New Phytol. 128, 639–644.

Smith S E and Read D J 1997 Mycorrhizal Symbiosis, Ed 2, Academic Press, London.

Solomon P S and Oliver R P 2001 The nitrogen content of the tomato leaf apoplast increases during infection by *Cladosporium fulvum*. Planta 213, 241–249.

Tavoularis S, Scazzocchio C, Sophianopoulou V 2001 Functional expression and cellular localization of a green fluorescent protein-tagged proline transporter in *Aspergillus nidulans*. Fung. Genet. Biol. 33, 115–25.

Timonen S, Finlay R D, Olsson S and Söderström B 1996 Dynamics of phosphorus translocation in intact ectomycorrhizal systems: Non-destructive monitoring using a beta-scanner. FEMS Microbiol. Ecol. 19, 171–180.

van Belle D and André B 2001 A genomic view of yeast membrane transporters. Cur. Opin. Cell. Biol. 13, 389–398.

van Tichelen K K and Colpaert J V 2000 Kinetics of phosphate absorption by mycorrhizal and non-mycorrhizal Scots pine seedlings. Physiol. Plant. 110, 96–103.

Vesk P A, Ashford A E, Markovina, A L and Allaway W G 2000 Apoplasmic barriers and their significance in the exodermis and sheath of *Eucalyptus pilularis–Pisolithus tinctorius* ectomycorrhizas. New Phytol. 145, 333–346.

von Wiren N, Gazzarrini S, Gojon A and Frommer W B 2000 The molecular physiology of ammonium uptake and retrieval. Cur. Opin. Plant Biol. 3, 254–261.

Wallander H 2000a Uptake of P from apatite by *Pinus sylvestris* seedlings colonised by different ectomycorrhizal fungi. Plant Soil 218, 249–256.

Wallander H 2000b Use of strontium isotopes and foliar K content to estimate weathering of biotite induced by pine seedlings colonised by ectomycorrhizal fungi from two different soils. Plant Soil 222, 215–229.

Wallander H and Wickman T 1999 Biotite and microcline as a K source in mycorrhizal and non-mycorrhizal *Pinus sylvestris* seedlings. Mycorrhiza 9, 25–32.

Wallander H, Wickman T and Jacks G 1997 Apatite as a P source in mycorrhizal and non-mycorrhizal *Pinus sylvestris* seedlings. Plant Soil 196, 123–131.

Wallenda T and Read D J 1999 Kinetics of amino uptake by ectomycorrhizal roots. Plant Cell Environ. 22, 179–187.

Wiese J, Kleber R, Hampp R and Nehls U 2000 Functional characterization of the *Amanita muscaria* monosaccharide transporter AmMST1. Plant Biol. 2, 1–5.

Williams L E and Miller A J 2001 Transporters responsible for the uptake and partitioning of nitrogenous solutes. Ann. Rev. Plant Physiol. Plant Mol. Biol. 52, 659–688.

Wipf D U Ludewig M, Tegeder D, Rentsch W, Koch, Frommer W B 2002 Amino acid/neurotransmitter transporters are highly conserved between fungi, plants and animals. Trends Biochem. Sci. 27, 139–147.

Wright D P, Scholes J D, Read D J and Rolfe S A 2000 Changes in carbon allocation and expression of carbon transporter genes in *Betula pendula* Roth. colonized by the ectomycorrhizal fungus *Paxillus involutus* (Batsch) Fr. Plant Cell Environ. 23, 39–49.

Zimmermann S and Sentenac H 1999 Plant ion channels: from molecular structures to physiological functions. Curr. Opin. Plant Biol. 2, 477–82.

Zwiers L H and De Waard M A 2001 Efficient *Agrobacterium tumefaciens*-mediated gene disruption in the phytopathogen *Mycosphaerella graminicola*. Curr Genet. 39, 388–393.

Plant and Soil **244:** 177–187, 2002.

The role of the motile tubular vacuole system in mycorrhizal fungi

Anne E Ashford[1] & William G Allaway[2,3]
[1]*School of Biological Science, The University of New South Wales, Sydney, NSW 2052, Australia.* [2]*School of Biological Sciences, The University of Sydney, NSW 2006, Australia.* [3]*Corresponding author**

Key words: cytoskeleton, motile tubular vacuole, mycorrhiza, nutrient transport, polyphosphate

Abstract

Mycorrhizal fungi, to be effective for the plant, must be able to transfer mineral nutrient elements from sites of uptake at hyphal tips across various distances to the exchange region in the mycorrhiza. Vacuoles are likely to be important in this transport, since they contain elements of nutritional significance in abundance. In tip cells of hyphae of most fungi — known to include three ectomycorrhizal basidiomycetes, an ericoid mycobiont, and two arbuscular mycorrhizal fungi — the vacuoles form a motile tubular reticulum. The vacuoles are most active in hyphal tips, but non-motile vacuoles at a distance from the tip can be induced to become motile by environmental changes. Neither the tubular vacuolar reticulum nor its contents are properly preserved by conventional fixation and embedding. Vacuolar tubules are readily shown *in vivo* with fluorescent tracers, throughout the extramatrical mycelium and in outer hyphae of the sheath in eucalypt mycorrhizas synthesised with *Pisolithus* sp., but they have proved harder to label in field-collected ectomycorrhizas and ericoid mycorrhizas. Freeze-substitution does preserve the structure of vacuoles and vacuolar tubules, and careful anhydrous techniques allow them to be microanalysed, indicating high content of K and P in vacuoles of hyphal tips, and also in sheath and Hartig net of ectomycorrhizas. Vacuoles contain polyphosphate in diffuse, non-granular form. Polyphosphate is present right up to the tip region of hyphae as well as in sheath and Hartig net: thus important mineral nutrient elements are present at both ends of the long hyphal transport pathway. Exactly what happens in between, however, remains to be elucidated.

Introduction

Ectomycorrhizal fungi benefit their plant partner by transferring nitrogen, phosphorus and other elements from distant soil sites to the fine roots. Ectomycorrhizal fungi not only envelop individual fine roots, so that they have a characteristic morphology (Figure 1), but mycelium may grow out for many centimetres from ectomycorrhizal roots into the soil (Read, 1984). This may be highly differentiated and produce linear hyphal aggregates (rhizomorphs or strands) that are thought to function as specialised conduits for transport. Even in cases where rhizomorphs are not obviously developed the mycelium extends the effective soil exploitation volume by a significant amount (Figure 2).

Hyphal tips at the growing edge of soil mycelium are thought to play a crucial role in exploiting nutrient-rich patches of organic material in the soil, where hyphae proliferate (Bending and Read, 1995a; see also Perez-Moreno and Read, 2000). They secrete enzymes which mobilise organic compounds, and take up the products of hydrolysis in a manner which appears to parallel that of saprotrophs (Bending and Read, 1995b). However, nutrients taken up by hyphal tips must be transported back across the soil mycelium to the mycorrhizal interface for exchange with the partner if they are enhance its nutrition. There is substantial evidence for transport, especially of N and P, to the plant interface (e.g., see Brandes et al., 1998 and references therein). However we understand very little about how this transport occurs, but we do know that vacuoles are an early port of call. Vacuoles are known to be the major storage organelles for phosphorus (often in the form of polyphosphate, Figure 3)

* FAX No.: +61-2-9351-4771; Tel.: +61-2-9351-2280; E-mail: allaway@bio.usyd.edu.au

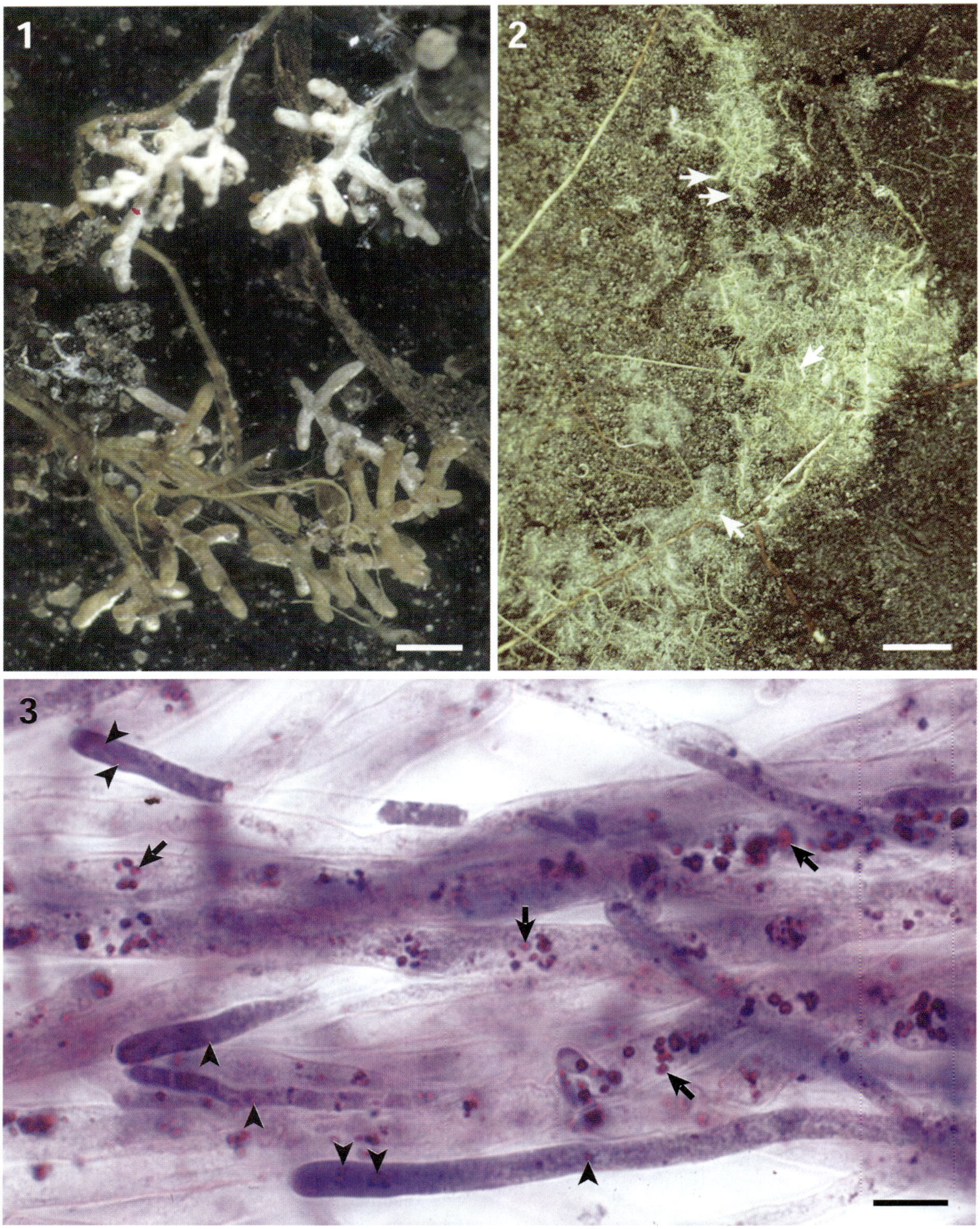

Figures 1–3.

Figure 1. ECM tips of *Eucalyptus pilularis* Smith collected near Pennant Hills, Sydney, Australia. Bar, 1 mm. *Figure 2. In situ* photograph of mycelium surrounding ECM and fine roots of *Lophostemon confertus* (R.Br.) Peter G. Wilson et Waterhouse in a suburban garden, Sydney, Australia. A stone was removed to expose the patch of roots. The only fruit bodies that were associated with this mycelium were a *Scleroderma* sp. and *Laccaria ohiensis*. Note the growing mycelial front of white hyphae, white mycorrhizas (arrows) and brownish un-infected fine roots. Bar, 10 mm. *Figure 3.* Polyphosphate in hyphal tips (arrowheads) and more mature cells (arrows) of a MMN agar-grown culture of *Pisolithus* after ethanol precipitation and staining with toluidine blue O at pH 4.4. The polyphosphate (located in vacuoles) stains bright red, while phosphate groups of RNA in the cytoplasm stain lilac/purple. Polyphosphate, although clearly present, is less abundant in the hyphal tips than in more mature regions. Note the relatively large amounts of cytoplasm in the hyphal tips, and empty regions of the hyphae caused by ethanol damage. *Pisolithus* isolate DI-15, re-isolated from a synthesised *Eucalyptus pilularis* mycorrhiza. Bar, 20 μm.

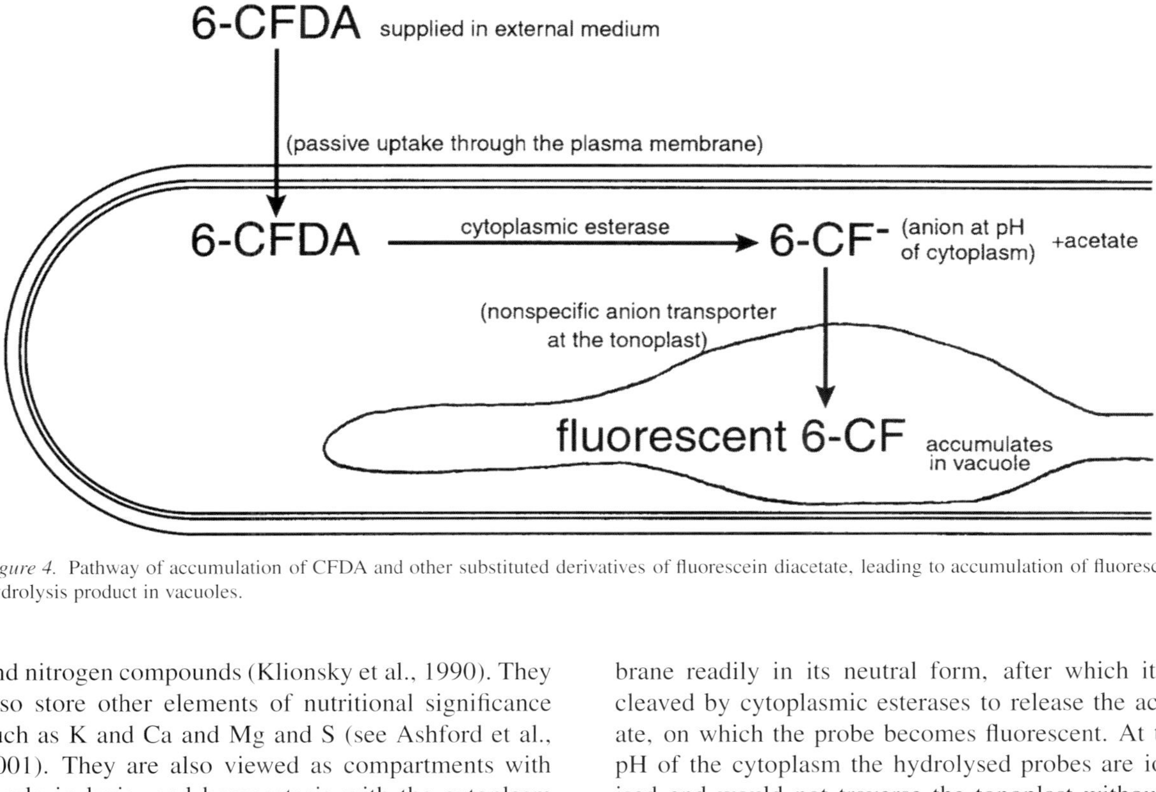

Figure 4. Pathway of accumulation of CFDA and other substituted derivatives of fluorescein diacetate, leading to accumulation of fluorescent hydrolysis product in vacuoles.

and nitrogen compounds (Klionsky et al., 1990). They also store other elements of nutritional significance such as K and Ca and Mg and S (see Ashford et al., 2001). They are also viewed as compartments with a role in lysis, and homeostasis with the cytoplasm (Ashford, 1998; Klionsky et al., 1990). Recent work with an Australian species of *Pisolithus*, a close relative of *Pisolithus tinctorius* but ectomycorrhizal with eucalypts, has added another role, that of transport.

Visualisation and occurrence

The fungal vacuole system is best visualised by loading with fluorescent probes that accumulate in the vacuole lumen. Many of these probes are derivatives of the vital dye fluorescein diacetate used to test membrane integrity and cell viability. The pathway of accumulation is shown in Figure 4. The probe which is non-fluorescent permeates the membrane readily in its neutral form, after which it is cleaved by cytoplasmic esterases to release the acetate, on which the probe becomes fluorescent. At the pH of the cytoplasm the hydrolysed probes are ionised and would not traverse the tonoplast without a specific ion channel. It is thought that they enter vacuoles via a non-specific anion transporter which probably has the role of sequestering toxic materials from the cytoplasm into the vacuoles. Evidence that this is so is based on experiments with the anion-transport inhibitor probenecid (see Steinberg et al., 1987). This drug inhibits uptake into the vacuoles and the fluorescent compound accumulates in the cytoplasm instead, but when the probenecid is removed fluorescence is rapidly (within minutes) lost from the cytoplasm and appears in the vacuoles (Cole et al., 1997). Of these probes Oregon Green 488 carboxylic acid diacetate (DFFDA) is our preference. Its fluorophore is a fluorinated analogue of fluorescein. Absorp-

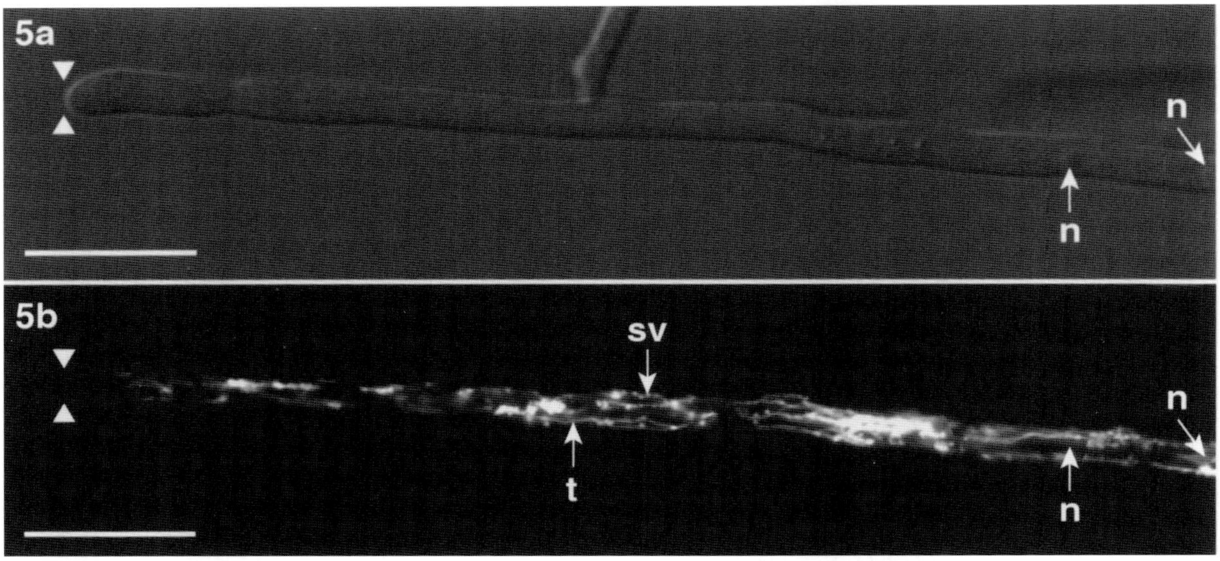

Figure 5. Tubular vacuole system of *Pisolithus* showing transition from tubules to spherical vacuoles in a hyphal tip cell labelled with DFFDA. DIC image above (a) and fluorescence image below (b). Micrographs by Rita Verma while at UNSW, Australia. Bar, 20 μm.

tion and emission spectra resemble that of fluorescein but it is less pH sensitive in the physiological range and much more photostable (Haugland, 2001), allowing longer video sequences. Other probes such as 7-amino-4-chloromethylcoumarin (CMAC) and 5-chloromethylfluorescein diacetate (CMFDA) dependent on glutathione transfer reactions, a quite different mechanism, also label the lumen of vacuoles well (Cole et al., 1997).

The organelle system that labels with Oregon Green 488 carboxylic acid is identified as the vacuole system by a number of criteria. Firstly, the pH of the large spherical components interconnected by tubules has been measured directly and can be quite acid, with a pH ranging from 4.3–7.5 (Rost et al., 1995). Secondly, the tubular components are shown to have lumen continuity with the large vacuoles by differential interference contrast (DIC) and fluorescence microscopy as well as by electron microscopy of freeze-substituted hyphae (Cole et al., 1998; Shepherd et al., 1993a). These are all part of the vacuole system. They are bounded by a single membrane and have a typical vacuolar content, including deposits of polyphosphate. Thirdly, the behavour and motility of the labelled system and its overall control by microtubules parallels that of endosomes, components of which have long been equated with the vacuoles of plants and fungi (see Ashford, 1998). Finally, Oregon Green 488 carboxylic acid labelled tubules and their associated spherical vacuoles have been differentiated from other potentially tubular organelles such as mitochondria, endoplasmic reticulum and Golgi bodies in dual labelling experiments with other organelle-specific probes (Ashford and Orlovich, 1994; Cole et al., 2000a,b). In *Pisolithus* many small vesicles, some of which interact with larger vacuolar tubules, also become labelled. If, as has been suggested, the vacuole system is equivalent to endosomal systems, then it is likely that these represent other associated structures, not normally recognised as part of the vacuole system, such as small vesicles and multivesicular bodies (Cole et al., 2002).

Tubular vacuole systems have been demonstrated in a wide range of fungi. All members of Basidiomycotina, Ascomycotina, Deuteromycotina and Zygomycotina examined have them, including *Paxillus involutus* (Batsch: Fr.) Fr. and *Suillus granulatus* (L: Fr.) O. Kuntze, both important ectomycorrhizal fungi from the Northern Hemisphere (Rees et al., 1994). The only species where we used the same loading techniques but could not demonstrate vacuoles was *Candida albicans*. Tubular vacuoles have proved hard to demonstrate in arbuscular mycorrhizal fungi. Nevertheless a tubular system, possibly vacuolar, has been observed in living arbuscular mycorrhizal hyphae and video sequences show that it has a motility rather like that of *Pisolithus* (Jock Nielsen, Risø National Research Laboratory, Denmark, pers. comm.). Further-

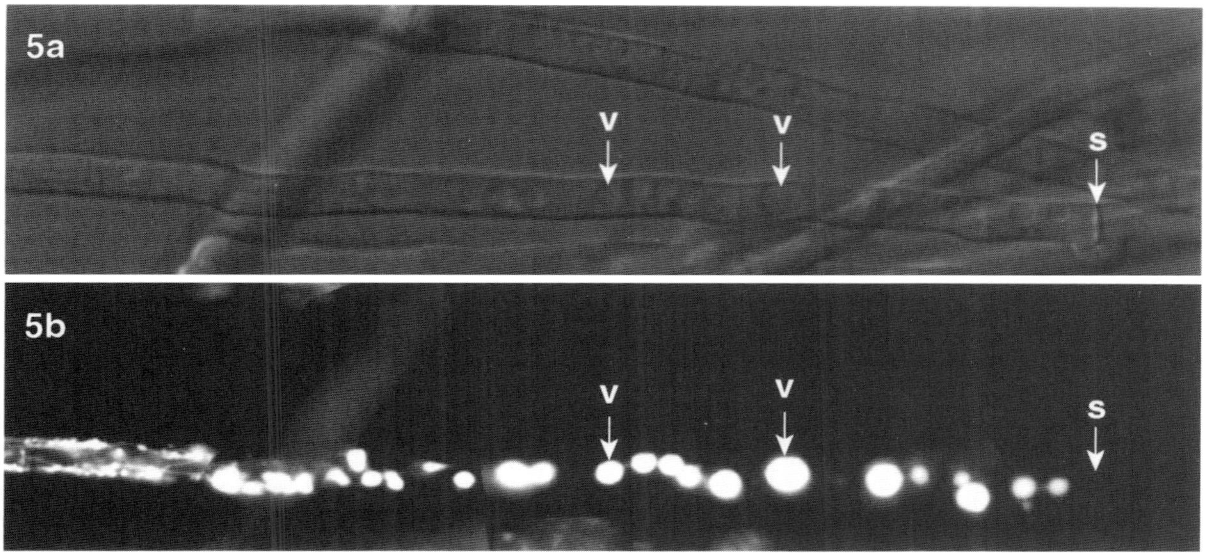

Figure 5. (Continued)

more, Uetake et al. (2001) have found motile tubular vacuoles in germ tubes as well as extra radical and intra radical hyphae of *Gigaspora margarita* mycelium mycorrhizal with *Allium*. These are demonstrated with confocal microscopy by their fluorescence after labelling with DFFDA. It is important to realise that, while cultures grown in hydrophilic conditions usually label well at least in the growing region, hyphae taken from relatively dry situations such as natural soil rarely label well, probably because the wall has low permeability and impedes entry of the fluorochrome (see Vesk et al., 2000).

Behaviour and structure

The vacuole system in *Pisolithus* is shown to be a very complex pleiomorphic organelle. It ranges in form from a complex reticulum of tubules with few spherical vacuoles, to a series of larger spherical components variously interconnected by tubules. This transition between these two extremes is typically seen in the same cell (Figure 5). The tubules are very motile. They can extend, retract, fuse and separate, and videos often show fluorochrome being 'squirted' from one large vacuole to the next, via tubular connections (see Figure 3 of Cole et al., 1998). The larger spherical components are less mobile. They are often appressed to the plasma membrane and either do not move at all during the short observation periods (video sequences are captured at 4 frames/s. for >1 min) or are seen to move slowly along this membrane (see Figure 3 of Cole et al., 1998). The connection between tonoplast, plasma membrane and wall in these appressed regions is so strong that when the plasma membrane pulls away from the wall as sometimes happens in freeze substituted material, the membranes break rather than separate, leaving a pad of both membranes on the wall surface (Figure 6 of Cole et al., 1998). This indicates that both the tonoplast and plasma membrane may in some way be anchored in the wall in these regions, perhaps by a fibronectin-like protein, but this linkage nevertheless still allows movements of these vacuoles. These regions of close contact are certainly anchor points for the vacuole system within individual cells. They may also be sites of uptake directly into the vacuole from the external solution, by-passing the cytoplasm. Tubules extend from the larger less motile vacuoles and interconnect them. They can be seen to traverse distances of 60 μm or so and they can extend in either direction. Tubule tips advance at rates of the order of 1.5 μm/s.

Tubules are not always present, and hyphae very often contain a series of apparently separate (usually large) vacuoles. This is usually the case in more mature regions of agar-grown fungal colonies. However these can be changed into a vacuolar continuum. This has been demonstrated in non-labelled as well as fluorochrome-labelled vacuoles by simply changing the external solution (Hyde and Ashford, 1997). In this case increased motility may simply be caused by rise in external O_2 levels, since motility declined

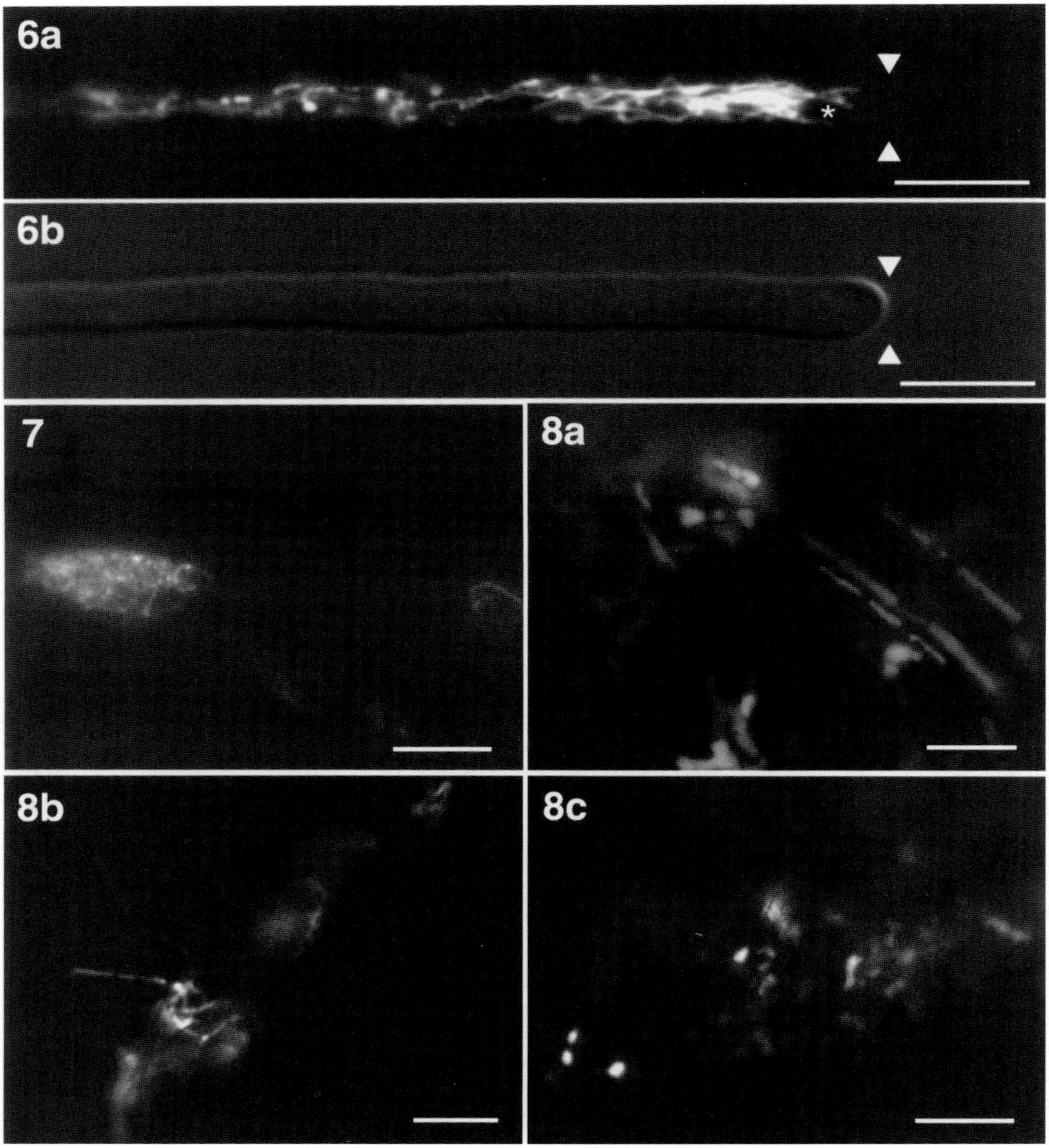

Figure 6. Effect of high sodium phosphate (100 mM) on distribution of vacuoles in hyphal tip cells. Tubular vacuoles are accumulated in the tip region as in latrunculin-treated hyphae (cf. Figure 9 of Hyde et al. 1999) though there is still a vacuole-free zone in the extreme apical 5 μm. Labelled with DFFDA. (a) Fluorescence image. (b) DIC image of the same hypha. Micrographs by Rita Verma while at UNSW, Australia. Bar, 10 μm. *Figure 7.* Vacuoles in *Calluna vulgaris* L. coils labelled with DFF following treatment of hair roots with 10 μg/ml DFFDA in distilled water. Plant material from southern Sjaelland, Denmark, provided by Mayra Gavito, Lund University, Sweden. Bar, 10 μm. *Figure 8.* Vacuoles in field-collected ectomycorrhizas, visualised by treatment with 10 mg/ml DFFDA in water. Vacuoles of varied forms including some tubular ones in superficial hyphae of ectomycorrhiza of (a) oak, *Quercus robur* L. and (b) beech, *Fagus sylvatica* L. In the densely-packed hyphae of the compact sheath in a beech mycorrhiza, large vacuoles occupy most of the irregular cell volume (c). The majority of the hyphae in these field-collected mycorrhizas showed no accumulation of fluorochrome. Collected in Skåne, southern Sweden. Bars: (a) 10 μm, (b,c) 40 μm.

progressively with time in specimens kept under the coverslip, but could be restored by drawing a fresh solution of either distilled water or fluorochrome over them. Other treatments can either suppress or promote motility and tubulation, indicating that external conditions can modulate vacuole activity. The two distinct components — large spherical vacuoles and highly motile tubules that interlink them — can be viewed as a series of storage or 'holding tanks' connected by a reticulated 'pipe' system, where the formation of pipes may be induced and their reticulation changes with time. Hyphae in the growing zone require a constant supply of carbon compounds to sustain growth and provide energy for uptake. This arrives along the hypha from the plant and must travel from base to tip. In contrast nutrients taken up preferentially at or near the tip must travel in the opposite direction. The existence of more than one tubule in the same stretch of hypha would accommodate this requirement for bi-directional flow. However, the frequent branch points and complex behaviour of the vacuole system could also accommodate successive transport of different contents along the same tubule at different times.

Vacuolar tubules and their connections are destroyed by chemical fixation which converts them into a string of vesicles (Orlovich and Ashford, 1993; see also Ashford, 1998). This partly explains why the system had not previously been discovered. Tubules are preserved by freeze substitution, and profiles of tubular vacuoles are found in virtually all cases where hyphae have been freeze-substituted (see Ashford, 1998, for references). The fine structure of hyphae prepared by this technique shows that the tubular vacuole system is even more interconnected than appears from light microscopy of living cells (Shepherd et al., 1993a).

Cytoskeleton affects tubular vacuoles

As might be expected tubular vacuole systems are dependent on an intact cytoskeleton for their integrity and movement. There is good evidence that cytoplasmic microtubules are important for the maintenance of vacuolar tubules, while actin microfilaments are not (Hyde et al., 1999). Microtubule depolymerising drugs suppress the tubulation of vacuoles in the tip region of growing *Pisolithus* colonies. Only about 8% of tip hyphae had vacuolar tubules after treatment with 40 μM oryzalin compared with 85% in non-treated controls. A similar loss of tubular vacuoles was seen with another microtubule-depolymerising drug nocodazole. Hyphae recovered almost completely within an hour of drug removal. Staining of hyphae with α-tubulin antibodies shows longitudinal arrays of microtubules in controls and these disappeared in 40 μm oryzalin, confirming that the drug is causing microtubule depolymerisation as it should (Hyde et al., 1999). Tubular vacuoles lie parallel to bundles of longitudinal microtubules and linkages between them involving fibrillar material are frequently seen (Ashford et al., 2001; Hyde et al., 1999).

Drugs which depolymerise filamentous actin such as cytochalasins B, D (at 100 μM) and latrunculin (at 25 μM) do not have the this effect on the vacuole system. The percentage of tip hyphae containing vacuolar tubules after treatment with these drugs does not differ from that in the controls. This is not due to their failure to disrupt the actin; latrunculin for example is shown to cause disappearance of the actin cap in all treated hyphae (Hyde et al., 1999).

In contrast to the lack of effect on tubule frequency, there is a rather obvious effect of anti-actin drugs on the positioning of the vacuole system. Tubular vacuoles tend to cluster or accumulate in the tip region so that the usual vacuole-free zone about 5 μm long seen in the tips of non-treated hyphae is reduced or eliminated (see Figure 9 of Hyde et al., 1999). About 97% of untreated hyphae possess this vacuole-free zone, whereas in 25 μM latrunculin about 65% of the hyphae do not. Hyphae recover when the drug is removed. When actin caps are disrupted growth ceases. Other treatments which cause cessation or reduction in growth also cause a vacuolar-tubule clustering phenomenon, for example treatment with high salt solutions (Figure 6). Taken together, these observations suggest that the process of vacuole progression is uncoupled from the growth process. Essentially, the vacuoles continue to move forward but in the absence of tip growth, they have nowhere to go and bunch up in the tip region. So, although loss of F-actin does have an effect on vacuole systems, it is clear that it is not primarily responsible for vacuolar tubulation and motility.

Tubular vacuoles contain nutrient elements

It will be obvious from the types of motility exhibited that tubular vacuoles have the capacity to move substances along hyphae. Videos of vacuole motility in individual cells especially at the hyphal tips show that the vacuole system can move fluorescent probes around the same cell. However, if tubular vacuoles are to transfer nutrients for any distance, as required for transfer from hyphal tips to the fungus/root interface they must be able to cross the dolipore septa. There is good evidence that tubular vacuoles from the tip cell cross the first dolipore septum and make incursions into the penultimate cell in *Pisolithus* (Shepherd et al., 1993b) but this is not known for septa between cells more distant from the tips, since these have not been examined.

Vacuoles have long been known to contain high levels of phosphorus, and much of this is stored as polyphosphate as well as orthophosphate. Polyphosphate can be demonstrated in a number of ways. ^{31}P NMR spectroscopy allows its identification in whole mycelium and an assessment of chain length; long-chain polyphosphate which does not spin will not be detected (Martin, 1991; Pfeffer et al., 2001). Shifts in phosphate resonances allow differentiation between pools in the cytoplasm and other compartments, presumed to be vacuoles. Various biochemical tests including polyacrylamide gel electrophoresis against known standards (Ogawa et al., 2000) will allow further assessment of chain length. Toluidine-blue staining at controlled pH, in conjunction with other tests, allows demonstration of polyphosphate *in situ* in vacuoles (Ashford et al., 1975, 1986) but needs confirmation with other methods. Energy dispersive X-ray microanalysis may indicate whether there are high levels of P at a particular location, but does not identify the form. These approaches taken together show that polyphosphate is stored in the vacuoles of growing mycelium of *Pisolithus* sp. The polyphosphate detected has an average chain length of 15 Pi subunits (Ashford et al., 1994); similar to that found in *Hebeloma crustuliniforme* (Martin et al., 1985). (Yeast polyphosphate has a median chain length of about 60 residues and 90% is in the vacuole — see Ogawa et al., 2000.) Some ectomycorrhizal fungi, including *Pisolithus*, are reported to store orthophosphate rather than polyphosphate when growing in axenic culture (see Cairney and Smith, 1992, and references therein), but toluidine blue staining shows that hyphae in the growing zone of *Pisolithus* colonies contain substantial polyphosphate, although with lesser amounts in hyphal tip cells than in more mature cells (Figure 3).

In chemically fixed cells polyphosphate occurs in vacuoles as electron-opaque deposits (variously called granules or globules) surrounded by electron-lucent space. When analysed these deposits are invariably rich in P. If, however, hyphae from an equivalent region are freeze substituted the vacuoles look quite different (Orlovich and Ashford, 1993). They are now filled with a flocculent dispersed material that is more or less homogeneous. Granules or globules are also not usually seen in living hyphae of *Pisolithus* (Cole et al., 1998; Hyde and Ashford, 1997). They are not apparent either in DIC images of non-loaded vacuoles or in fluorescent images of loaded vacuoles (Figure 5). Granules are precipitated in early stages of ethanol dehydration: they first appear at 10% ethanol and further intensify when toluidine blue is added. In bright field they show the characteristic γ-metachromasy, confirming that it is polyphosphate that was precipitated (Orlovich and Ashford, 1993). This direct demonstration of precipitation of polyphosphate into granules by ethanol indicates that the classical EM images of polyphosphate granules surrounded by electron-lucent space are artifacts. Bücking and Heyser (1999) reported globular inclusions in vacuoles in living hyphae and referred to them as polyphosphate granules (Bücking and Heyser, 1999, 2000a); however, these actual vacuolar inclusions have not been demonstrated to contain polyphosphate. Vacuoles may indeed contain inclusions, but many of these are membranous, including multivesicular bodies (Cole et al., 1998). However, in most freeze substituted hyphae of *Pisolithus* and other fungi the vacuole profiles show flocculent dispersed material with either no or very small inclusions (Cole et al., 1998; and see Ashford, 1998, for other references). Vacuoles in the Hartig net of high pressure-frozen freeze-substituted ectomycorrhizas also mostly have a homogeneous content (Frey et al., 1997).

X-ray microanalysis of precipitated granules in conventionally processed *Pisolithus* hyphae shows high levels of P and Ca. In contrast vacuoles of hyphae freeze substituted by anhydrous techniques contain P co-localised with K rather than Ca (Orlovich and Ashford, 1995; Ashford et al., 1999) and so do those of freeze-dried hyphae (Frey et al., 1997; Bücking and Heyser, 1999, 2000b). Freeze substitution by anhydrous techniques is known to retain ions in situ from tests with model systems, if great care is taken to avoid contamination with water (Orlovich and Ashford, 1993; 1995; Pålsgård et al., 1994). It is important to realise that any procedure that makes the tonoplast leaky will change the vacuolar environment and may cause precipitation or alter the form of precipitates. The earlier finding that Ca accompanies vacuolar P in chemically fixed, conventionally-processed hyphae is to be viewed in this light as an artifact (Bücking and Heyser, 1999; Orlovich and Ashford, 1993; 1995).

X-ray maps of hyphal tip cells of *Pisolithus* show that both large spherical and narrower tubular components of the vacuole system contain high levels of P co-localised with K (Cole et al., 1998; Orlovich and Ashford, 1993). In both cases, P and K are precisely localised to the vacuoles, with a homogeneous distribution. This indicates that potentially motile tubular elements as well as the storage vacuoles contain high levels of P. At least some of this P is polyphosphate. This is indicated by the direct demonstration of polyphosphate by toluidine blue staining in large vacuoles as well as in the vesicles arising from interconnecting tubules after their fragmentation by glutaraldehyde (Orlovich and Ashford, 1993).

A similar homogeneous distribution of P and K is seen in the vacuoles of the sheath and Hartig net of mycorrhizas synthesised between *Eucalyptus pilularis* and *Pisolithus* in an agar plate system (Ashford et al., 1999). These vacuoles also stain for polyphosphate, consistent with findings that polyphosphate is detected in high amounts in ectomycorrhizal roots by ^{31}P NMR (reviewed by Pfeffer et al., 2001). In contrast to those at the hyphal tips, these vacuoles in the mycorrhizas also store large amounts of S, but this is not a surprise since the mycorrhizas are known to synthesise hydrophobins (Tagu and Martin 1995, 1996; Tagu et al., 1996). Vacuoles in hyphae of both the growth zones of extramatrical mycelium and ectomycorrhizal roots probably also contain basic amino acids. These are known to be co-localised with polyphosphates in vacuoles of filamentous fungi such as *Neurospora crassa*, as well as in *Saccharomyces cerevisiae* (see Keenan et al., 1998, for references). There is evidence that mycelium of the ectomycorrhizal fungus *Cenococcum graniforme* accumulates arginine and this may be co-localised with polyphosphate (Martin, 1985). So vacuoles contain high levels of at least two and probably three key nutrients, P (at least some of which is polyphosphate), K and N, at both ends of the pathway. The question is what happens between: whether these elements are transported within the vacuole system, or are unloaded from the vacuoles, transported in some other way, and then reloaded into the vacuoles at their destination.

Tubular vacuoles in transport

If tubular motile vacuole systems are to function in long-distance transport of nutrients in mycorrhizas, they would need to be present throughout these systems to link nutrient accumulating hyphal tips to the interface with plant roots. Recent work showing a strong coupling of K, Mg and possibly N with P translocation in *Paxillus involutus* mycorrhizal with *Picea abies* is consistent with the concept of motility generated bulk flow of nutrients within vacuoles (Jentschke et al., 2001).

We have shown that pleiomorphic tubular vacuoles occur throughout all parts of the extramatrical mycelium and in superficial hyphae of ectomycorrhizal roots in syntheses between *Pisolithus* sp. and *Eucalyptus pilularis* (Allaway and Ashford, 2001). Motile tubular vacuoles were present throughout the mycelium connecting the growing front and individual mycorrhizas. Tubular vacuoles were present in hyphae aggregated into cords as well as in single hyphae. However, because of difficulties of imaging on the agar plate system it was not possible to follow a vacuolar continuum between the hyphal tips and individual mycorrhizal roots. Ectomycorrhizas labelled variably with the tracer used, but the surface hyphae and some of the underlying hyphae accumulated the fluorochrome strongly in some cases. Labelling in the surface hyphae of mycorrhizas was in the vacuoles, which were often motile and pleiomorphic. Many other mycorrhizas, especially those which were very hydrophobic, showed labelling in only a few surface hyphae. In undamaged mycorrhizas the inner sheath and Hartig net were unlabelled, the disjunction between labelled and unlabelled hyphae corresponding with the beginning of cementation of the sheath, although the hyphae throughout were alive. We attributed the lack of labelling in cemented parts of the sheath and regions within it (including the Hartig net), and in large parts of some mycorrhizas altogether, to the development of hydrophobicity in these hyphae, so that the lack of fluorochrome labelling reveals a lack of access rather than a lack of vacuoles. This variation in hydrophobicity is undoubtedly significant in modifying the uptake of materials by mycorrhizas (cf. Vesk et al., 2000).

We did not demonstrate long-distance transport of fluorochrome in our experiments, perhaps because a foreign compound such as the fluorescent tracer would be more likely to be accumulated for detoxification than translocated. However, there is certainly the *po-*

tential for the vacuole system to act as a conduit and transport solutes across the mycelium to mycorrhizal roots. Hyphal tips of the growing front are the regions where uptake of nutrients is thought to be greatest and, since the vacuoles of *Pisolithus* hyphal tips as well as those in the ectomycorrhizal sheath are rich in P, it is reasonable to assume that such transport would include this key element required for plant growth.

Mycorrhizas from the field

While tubular vacuole systems can readily be demonstrated in the surface hyphae of synthesised *Eucalyptus pilularis/Pisolithus* mycorrhizas on agar-plate systems, field-collected mycorrhizas have not proved so tractable. Tubular vacuoles could be demonstrated in some fungal coils of ericoid mycorrhizas in *Calluna vulgaris* after treating the hair roots with DFFDA (Figure 7). In beech and oak ectomycorrhizas there was very little uptake of the probe (Figure 8), even though mycorrhizas were obviously young and collected at midsummer in moist conditions, with every indication that the hyphae should be at their most active. In mycorrhizas of both tree species, DFFDA was sometimes taken up into loose hyphae at the edge of the sheath, where the fluorescent product entered the vacuoles as expected, both into narrow tubular or larger elongated vacuoles (Figures 8a, b). Even less frequently, and also with both tree species, the fluorescent probe could be seen in the irregularly shaped large vacuoles of the compact pseudoparenchymatous sheath (Figure 8c). However, the commonest outcome was a more or less complete lack of labelling, even after long exposure to DFFDA. We again interpret this as being primarily due to failure of the probe to gain access to the hyphal plasma membrane in most mycorrhizas, due to the impregnation of hyphal walls and sheath interstices with hydrophobic materials (cf. Vesk et al., 2000).

Conclusion

Motile tubular vacuole systems have the properties and potential to act as conduits for both short and long distance transport. Both tubular and spherical components contain high levels of P and K and probably also N. They occur throughout artificially synthesised ectomycorrhizal systems and can be found in mycorrhizas from the field. It is, however, still to be proved that they do actually transfer P and other elements to the plant interface.

Acknowledgements

We are grateful to Bengt Söderström and the Microbial Ecology Group, Lund University, Sweden for facilities while we were on Study Leave, and to the Swedish Research Council and the Australian Research Council for research support. We thank Sylvia Anton and Bill Hansson, Pheromone Group, Lund University, Sweden, for the use of their microscope, Danielle Davies for assistance, Mayra Gavito for plant material, and Rita Verma for Figures 5 and 6.

References

Allaway W G and Ashford A E 2001 Motile tubular vacuoles in extramatrical mycelium and sheath hyphae of ectomycorrhizal systems. Protoplasma 215, 218–225.

Ashford A E 1998 Dynamic pleiomorphic vacuole systems: are they endosomes and transport compartments in fungal hyphae? Adv. Bot. Res. 28, 119–159.

Ashford A E and Orlovich D A 1994 Vacuole transport, phosphorus and endosomes in the growing tips of fungal hyphae. *In* Pollen-Pistil Interactions and Pollen Tube Growth, Current Topics in Plant Physiology 12. Eds. A G Stephenson and T-h Kao American Soc. Plant Physiologists, Maryland, USA.

Ashford A E Ling Lee M and Chilvers G A 1975 Polyphosphate in eucalypt mycorrhizas; a cytochemical demonstration. New Phytol. 74, 447–453.

Ashford A E, Peterson R L, Dwarte D and Chilvers G A 1986 Polyphosphate granules in eucalypt mycorrhizas: determination by energy dispersive x-ray microanalysis. Can. J. Bot. 64, 677–687.

Ashford A E, Ryde S and Barrow K D 1994 Demonstration of a short chain polyphosphate in *Pisolithus tinctorius* and the implications for phosphorus transport. New Phytol. 126, 239–247.

Ashford A E, Vesk P A, Orlovich D A, Markovina A-L and Allaway W G 1999 Dispersed polyphosphate in fungal vacuoles of *Eucalyptus pilularis/Pisolithus tinctorius* ectomycorrhizas. Fungal Genet. Biol. 28, 21–33.

Ashford A E, Cole L and Hyde G J 2001 Motile tubular vacuole systems. Ch 12. *In* The Mycota, VIII. Biology of the Fungal Cell Eds. R J Howard and N A R Gow. pp 243–265. Springer Ver, Heidelberg.

Bending G D and Read D J 1995a The structure and function of the vegetative mycelium of ectomycorrhizal plants V. Foraging behaviour and translocation of nutrients from exploited litter. New Phytol. 130, 401–409.

Bending G D and Read D J 1995b The structure and function of the vegetative mycelium of ectomycorrhizal plants VI. Activities of nutrient solubilizing enzymes in birch litter colonized by *Paxillus involutus* (Fr.) Fr. New Phytol. 130, 411–417.

Brandes B, Godbold D L, Kuhn A J and Jentschke G 1998 Nitrogen and phosphorus acquisition by the mycelium of the ectomycorrhizal fungus *Paxillus involutus* and its effect on host nutrition. New Phytol. 140, 735–743.

Bücking H and Heyser W 1999 Elemental composition and function of polyphosphates in ectomycorrhizal fungi — an x-ray microanalytical study. Mycol. Res. 103, 31–39.

Bücking H and Heyser W 2000a Subcellular compartmentation of elements in non-mycorrhizal and mycorrhizal roots of *Pinus sylvestris*: an X-ray microanalytical study. I. The distribution of phosphate. New Phytol. 145, 311–320.

Bücking H and Heyser W 2000b Subcellular compartmentation of elements in non-mycorrhizal and mycorrhizal roots of *Pinus sylvestris*: an X-ray microanalytical study. II. The distribution of calcium, potassium and sodium. New Phytol. 145, 321–331.

Cairney J W G and Smith S E 1992 Influence of intracellular phosphorus concentration on phosphate absorption by the ectomycorrhizal basidiomycete *Pisolithus tinctorius*. Mycol. Res. 96, 673–676.

Cole L, Hyde G J and Ashford A E 1997 Uptake and compartmentalisation of fluorescent probes by *Pisolithus tinctorius* hyphae: evidence for an anion transport mechanism at the tonoplast but not for fluid phase endocytosis. Protoplasma 199, 18–29.

Cole L, Orlovich D A and Ashford A E 1998 Structure, function, and motility of vacuoles in filamentous fungi. Fungal Genet. Biol. 24, 86–100.

Cole L, Davies D, Hyde G J and Ashford A E 2000a ER-tracker dye and BODIPY-brefeldin A differentiate the endoplasmic reticulum and Golgi bodies from the tubular-vacuole system in living hyphae of *Pisolithus tinctorius*. J. Microsc. 197, 239–248.

Cole L, Davies D, Hyde G J and Ashford A E 2000b Brefeldin A affects growth, endoplasmic reticulum, Golgi bodies, tubular vacuole system, and secretory pathway in *Pisolithus tinctorius*. Fungal Genet. Biol. 29, 95–106.

Frey B, Brunner I, Walther P, Scheidegger C and Zierold K 1997 Element localization in ultrathin cryosections of high-pressure frozen ectomycorrhizal spruce roots. Plant Cell Environ. 20, 929–937.

Haugland R P 2001 Handbook of Fluorescent Probes and Research Products, 8th ed. Molecular Probes Inc., Eugene, OR. http://www.probes.com

Hyde G and Ashford A E 1997 Vacuole motility and tubule-forming activity in *Pisolithus tinctorius* hyphae are modified by environmental conditions. Protoplasma 198: 85–92.

Hyde G, Davies D, Perasso L, Cole L and Ashford A E 1999 Microtubules but not actin filaments, regulate vacuole motility and morphology in hyphae of *Pisolithus tinctorius*. Cell Motil. Cytoskel. 42, 114–124.

Jentschke G, Brandes B, Kuhn AJ, Schröder W H, Godbold D L 2001 Interdependence of phosphorus, nitrogen, potassium and magnesium translocation by the ectomycorrhizal fungus *Paxillus involutus*. New Phytol. 149, 327–337.

Keenan K A, Utzat C D, Zielinski T K 1998 Isolation and characterization of strains defective in vacuolar ornithine permease in *Neurospora crassa*. Fungal Genet. Biol. 23, 237–247.

Klionsky D J, Herman P K and Emr S D 1990 The fungal vacuole: composition, function, and biogenesis. Microbiol. Rev. 54, 266–292.

Martin F 1985 ^{15}N-NMR studies of nitrogen assimilation and amino acid biosynthesis in ectomycorrhizal fungus *Cenococcum graniforme*. FEBS Lett. 182, 350–354.

Martin F 1991 Nuclear magnetic resonance studies in ectomycorrhizal fungi. Methods Microbiol. 23, 121–148.

Martin F, Marchal J-P, Timinska A, Canet D 1985 The metabolism and physical state of polyphosphates in ectomycorrhizal fungi. A ^{31}P nuclear magnetic resonance study. New Phytol. 101, 275–290.

Ogawa N, DeRisi J and Brown P O 2000 New components of a system for phosphate accumulation and polyphosphate metabolism in *Saccharomyces cerevisiae* revealed by genomic expression analysis. Mol. Biol. Cell 11, 4309–4321.

Orlovich D A and Ashford A E 1993 Polyphosphate granules are an artefact of specimen preparation in the ectomycorrhizal fungus *Pisolithus tinctorius*. Protoplasma 173: 91–105.

Orlovich D A and Ashford A E 1995 Maintenance of ion distribution in frozen dextran droplets during freeze-substitution, embedding and X-ray microanalysis in anhydrous conditions. J. Microsc. 180, 117–126.

Pålsgård E Lindh U and Roomans G M 1994 Comparative study of freeze-substitution techniques for X-ray microanalysis of biological tissue. Microsc. Res. Tech. 28, 254–258.

Perez-Moreno J and Read D J 2000 Mobilization and transfer of nutrients from litter to tree seedlings via the vegetative mycelium of ectomycorrhizal plants. New Phytol. 145, 301–309.

Pfeffer P E, Bago B and Shachar-Hill Y 2001 Exploring mycorrhizal function with NMR spectroscopy. New Phytol. 150, 543–553.

Read D J 1984 The structure and function of the vegetative mycelium of mycorrhizal roots. *In*: The Ecology and Physiology of the Fungal Mycelium. Eds. D H Jennings and A D M Rayner. pp. 215–240. Cambridge University Press, Cambridge.

Rees B, Shepherd V A and Ashford A E 1994 Presence of a motile tubular vacuole system in different phyla of fungi. Mycol. Res. 98: 985–992.

Rost FWD, Shepherd V A and Ashford A E 1995 Estimation of vacuolar pH in actively growing hyphae of the fungus *Pisolithus tinctorius*. Mycol. Res. 99, 549–553.

Shepherd V A, Orlovich D A and Ashford A E 1993a A dynamic continuum of pleiomorphic tubules and vacuoles in growing hyphae of a fungus. J. Cell Sci. 104, 495–507.

Shepherd V A, Orlovich D A and Ashford A E 1993b Cell-to-cell transport via motile tubules in growing hyphae of a fungus. J. Cell Sci. 105, 1173–1178.

Steinberg T M, Newman A S, Swanson J A and Silverstein S C 1987 Macrophages possess probenecid-inhibitable organic anion transporters that remove fluorescent dyes from the cytoplasmic matrix. J. Cell Biol. 105, 2695–2702.

Tagu D and Martin F 1995 Expressed sequence tags of randomly selected cDNA clones from *Eucalyptus globulus–Pisolithus tinctorius* ectomycorrhiza. Mol. Plant Microbe Interact. 8, 781–783.

Tagu D and Martin F 1996 Molecular analysis of cell wall proteins expressed during the early steps of ectomycorrhiza development. New Phytol. 133, 73–85.

Tagu D, Nasse B and Martin F 1996 Cloning and characterization of hydrophobins-encoding cDNAs from the ectomycorrhizal Basidiomycete *Pisolithus tinctorius*. Gene 168, 93–97.

Uetake Y, Kajima T, Ezawa T and Saito M 2001 Tubular vacuoles observed in *Gigaspora margarita* hyphae using laser scanning confocal microscopy. Abstracts ICOM3, Adelaide.

Vesk P A, Ashford A E, Markovina A-L and Allaway W G 2000 Apoplasmic barriers and their significance in the exodermis and sheath of *Eucalyptus pilularis–Pisolithus tinctorius* ectomycorrhizas. New Phytol. 145: 333–346.

Plant and Soil **244:** 189–197, 2002.

Tracking metabolism and imaging transport in arbuscular mycorrhizal fungi

Metabolism and transport in AM fungi

Berta Bago[1,5], Philip E. Pfeffer[2], Warren Zipfel[3], Peter Lammers[4] & Yair Shachar-Hill[4]

[1]*Estación Experimental del Zaidín (CSIC), Calle Profesor Albareda 1, 18008-Granada, Spain.*
[2]*Eastern Regional Research Center, ARS-USDA, 600 East Mermaid Lane, Wyndmoor, PA 19038, USA.*
[3]*Cornell University, Ithaca, NY, USA.*
[4]*New Mexico State University, Las Cruces, NM 88003, USA.*
[5]*Corresponding author* (Present address: Centro de Investigaciones sobre Desertificación (CSIC/UV/GV), Camí de la Marjal s/n, 46470 Albal (Valencia), Spain)*

Key words: arbuscular mycorrhizal fungi, carbon metabolism, monoxenic cultures, multiphoton microscopy, nuclear magnetic resonance spectroscopy, nutrient translocation

Abstract

In the last few years the application of modern techniques to the study of arbuscular mycorrhizas has greatly increased our understanding of the mechanisms underlying carbon metabolism in these mutualistic symbioses. Arbuscular mycorrhizal (AM) monoxenic cultures, nuclear magnetic resonance spectroscopy together with isotopic labeling, and analyses of expressed sequence tags (ESTs) have shed light on the metabolic processes taking place in these interactions, particularly in the case of the mycobiont. More recently, *in vivo* multiphoton microscopy has provided us with some new insights in the allocation and translocation processes which play crucial roles in the distribution of host plant-derived C throughout the fungal colony. In this mini-review we highlight recent advances in these fields, with special attention to the visualization of oleosomes (i.e., lipid bodies) as they move along the long, coenocytic AM fungal hyphae. Volumetric measurements of such oleosomes have allowed us to estimate the flux of triacylglycerides from the intraradical to the extraradical phase of the AM fungal colony. We raise questions and postulate regulatory mechanisms for C metabolism and translocation within the arbuscular mycorrhizal fungal colony.

Transport processes in arbuscular mycorrhizal fungi

Nutrition in arbuscular mycorrhizas is based on the acquisition of soil nutrients by the fungus (George et al., 1995; Jakobsen, 1999; Koide and Schreiner, 1992), and fixation of atmospheric C by the plant (Ho and Trappe, 1973), and on the exchange of these nutrients at specially adapted symbiotic interfaces (Gianinazzi-Pearson et al., 1991; Smith and Read, 1997; Smith and Smith, 1990, 1997). Traditionally much attention has been focused on the mechanisms implied in these two transport processes, nutrient uptake and nutrient transfer (Figure 1; see Ferrol et al., 2002 and Burleigh and Bechman, 2002). In comparison, little is known about other two key classes of transport processes that take place within the arbuscular mycorrhizal (AM) fungus, both of them intracellular: the transport of nutrients into and out of the organelles where they are metabolized and/or stored, and the bidirectional translocation of nutrients (either soil- or plant-derived) along the coenocytic AM fungal hyphae (Figure 1).

Subcellular compartmentation of metabolism in the filamentous fungi is poorly understood, and translocation even less so. Most metabolic studies in fungi have been carried out in yeasts, unicellular organisms in which there is no bulk translocation. Moreover, in most fungal textbooks translocation in filamentous fungi (if at all considered) is usually treated in a very general manner and most often references are to higher

* FAX No: +34-96-127-0967. E-mail: berta.bago@uv.es

fungi, that have septa along their hyphae (e.g., Gow and Gadd, 1984; Jennings, 1995; but see Ashford and Allaway, this volume). Arbuscular mycorrhizal fungi by contrast have coenocytic (i.e., aseptate) multinuclear hyphae (Balestrini et al., 1992), which presents them and AM researchers with unusual challenges.

Carbon metabolism in AM fungi: an overview

The major fluxes of carbon change dramatically during the AM fungal life cycle. The intraradical phase, the extraradical phase and the germination phase each present distinctive characteristics (Bago et al., 2000). We will briefly review here C metabolism in intra- and extraradical hyphae, i.e., in the symbiotic fungal colony.

The intraradical mycelium (IRM) acquires hexose from the root and converts it into trehalose and glycogen

In vivo NMR coupled with ^{13}C-labeling demonstrated that hexose is taken up by the IRM and that trehalose and glycogen are the first substantial fungal C pool labeled from hexose which had been taken up by the intraradical fungal structures (Shachar-Hill et al., 1995). This contrasts with observations in the ectomycorrhizal symbiosis (Martin et al., 1985, 1998), where glucose is metabolized into these compounds via mannitol and where other carbohydrates also accumulate (Pfeffer et al., 1996, 2001). Glycogen and trehalose may serve to buffer cytoplasmic hexose levels (Bago et al., 2000). Recent labeling experiments show that glycogen is also translocated from IRM to ERM, particularly before sporulation (Figure 1; Bago et al., unpublished results). NMR spectroscopy studies, enzymatic determinations, radiorespirometry and molecular biology all indicate the functioning of glycolysis, the tricarboxylic acid (TCA) cycle, and the pentose phosphate pathway (PPP) in AM fungal intraradical structures (Harrier et al., 1998; MacDonald and Lewis 1978; Pfeffer et al., 1999; Saito, 1995; Solaiman and Saito, 1997). The results of recent experiments using ^{14}C-labeling in monoxenic cultures are also consistent with the activity of these pathways (Séjalon-Delmas et al., 2001).

Storage lipids are made in the intraradical mycelium

AM fungi are oleogenic fungi and store large amounts of lipid as triacylglycerides (TAGs) (Beilby, 1983; Beilby and Kidby, 1980; Cooper and Lösel, 1978; Gaspar et al., 1994, 1997; Jabaji-Hare, 1988; Nagy et al., 1980). The synthesis TAG is a substantial sink for carbon in the intraradical hyphae (Lösel and Cooper 1979; Pfeffer et al., 1999; Bago et al., 2000). ^{13}C-labeling and NMR studies indicate that the host-derived hexoses are metabolized by AM fungi (via glycolysis) to triose and acetyl CoA. Labeling patterns suggest that fatty acids are synthesized normally via acetyl CoA carboxylase and the FA synthase complex, then elongated and/or desaturated, and esterified with a glyceryl moiety via acyltransferases to constitute TAGs (Pfeffer et al., 1999; Bago et al., 2000).

The major fluxes of carbon in the IRM thus appear to be: efficient uptake of host-derived hexose, conversion to trehalose and glycogen as interim storage forms and the synthesis of large amounts of storage lipids (Figure 1).

The extraradical mycelium (ERM) is unable to acquire exogenous hexose or synthesize storage lipids

^{13}C-labeling and NMR spectroscopy have demonstrated that, unlike the intraradical fungal phase, extraradical fungal structures cannot absorb exogenous hexose (Pfeffer et al., 1999). Recent ^{14}C-labeling experiments (Séjalon-Delmas et al., 2001) support this view. Glycolytic enzyme activities in the ERM have been found to be low or absent (MacDonald and Lewis 1978; Saito, 1995; Séjalon-Delmas et al., 2001), whereas a substantial gluconeogenic flux fueled by the glyoxylate cycle does exist (Figure 1) (Lammers et al., 2001; Pfeffer et al., 1999). cDNA sequences for isocitrate lyase and malate synthase, the two key enzymes of glyoxylate cycle, have been characterized and their expression in the ERM demonstrated (Lammers et al., 2001). PPP activity also appears to be substantial in the extraradical hyphae (Figure 1), with both enzymatic activity measurements (Saito, 1995) and isotopic labeling data (Pfeffer et al., 1999) suggesting higher flux through this pathway than in the intraradical phase. All these data indicate that gluconeogenic C fluxes in the ERM are very substantial, and that this is mainly fuelled by the consumption of storage lipids (Figure 1). Thus the carbohydrate requirements of the ERM are met via TAG catabolism and gluconeogenesis, and some export of glycogen from the IRM.

Using $^{2}H_2O$-labeling Pfeffer and co-workers (1999) demonstrated that little or no storage lipid production occurred in AM extraradical fungal structures.

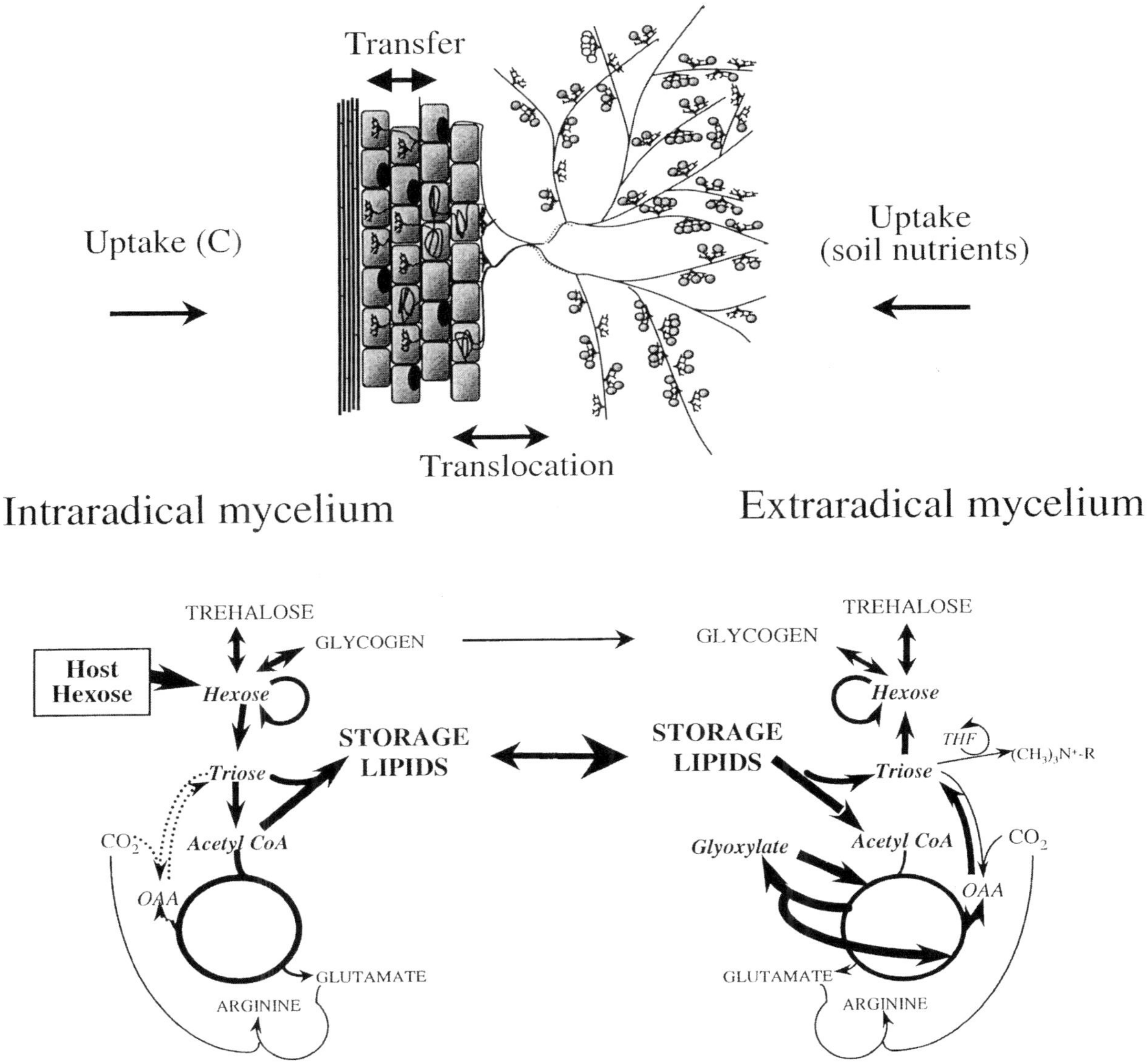

Figure 1. Transport processes occurring along the AM fungal colony (upper pannel) and C metabolic pathways and translocation events (lower pannel) taking place in and between the intraradical and extraradical fungal mycelium.

This has been supported by further experiments (Lammers et al., 2001) in which no label was detected in extraradical fatty acids after ^{13}C-labeled acetate (a direct intermediate in fatty acid synthesis) was supplied to the ERM. When ^{13}C-glycerol (a substrate for both FA and glycerol synthesis) was supplied, only the glyceryl moiety of the TAGs became labeled; again no label in the FA moieties was detected (Bago et al. unpublished results). Supplying any of several precursors to mycorrhizal roots resulted in high levels of labeling in the storage lipids of the ERM (Bago et al., 1999b; Lammers et al., 2001). Thus the lipids needed to sustain metabolism and growth of the ERM are made in the IRM and exported (Figure 1).

Translocation mechanisms must therefore be at work all along the AM fungal colony in order to provide C to the extraradical hyphae. These mechanisms must be well regulated in order to ensure adequate supply of carbon to a mycelium that extends many centimeters, perhaps meters (Smith and Read, 1997) away from the host. To make the regulation of translocation more complex, nutrients acquired from the soil are simultaneously translocated in the opposite direction (i.e., ERM to IRM), by undetermined processes (Bago et al., 2001).

Storage lipid translocation in AM fungi

To get a better insight into the mechanisms involved in the translocation of storage lipids along the AM fungal colony we used fluorescent labeling and *in vivo* multiphoton microscopy (Denk et al., 1990, 1995; Williams et al., 1994; Xu and Webb, 1996; Xu et al., 1996). Taking advantage of the transparent medium, convenient geometry and absence of other microorganisms in AM monoxenic cultures (Bécard and Fortin, 1988; St Arnaud et al., 1996), this approach was previously used to study nuclear behavior in symbiotic and asymbiotic AM fungal hyphae (Bago et al., 1998c, 1999a).

We have therefore followed storage lipid bodies moving in the ERM of AM fungi *Glomus intraradices* and *Gigaspora margarita* using monoxenic cultures of these two fungi with carrot transformed roots and *in vivo* multiphoton microscopy (Bago et al. 2002). Lipid bodies (also called 'oleosomes', Murphy, 1991) were labeled with the neutral lipid-specific fluorochrome Nile Red (Greenspan et al., 1985). Serial optical images ('z-series') were taken through the living tissue, and by using 3D reconstruction software, volumetric measurements of the lipid bodies were carried out (Figure 2a, b). Images at consecutive time points (time series, 't-series') of the oleosome movement along specific sites of runner hyphae were also obtained. Details on fluorochrome staining procedure, as well as on the two-photon microscope setup (i.e., excitation and emission wavelengths) are provided in Bago et al. (2002). Taken together, these data allow an estimate total TAG fluxes along symbiotic AM extraradical hyphae to be made (Bago et al., 2002).

In *G. intraradices* extraradical hyphae the volume of hyphae occupied by lipid globules ranged from 18 to 24% in zones close to the root, i.e., closer to the carbon source. When moving from these zones to the ERM growing front the amount of lipid droplets clearly decreased, to reach its minimum value in zones closer to the hypal apices. In these apical zones the volume of hypha occupied by oleosomes was only 0.5%. This gradient in lipid bodies was specially remarkable at the 'branched absorbing structures' (BAS) level. Since BAS seem to be specially active structures (Bago, 2000; Bago et al., 1998b), as are indeed all apical zones of a fungal colony, we concluded that the observed gradient in lipid bodies along the ERM is due to the progressively higher consumption of C at those apical zones. Observation of oleosome trafficking in *Gi. margarita* extraradical hyphae (Figure 2c–e) showed that in zones closer to the root lipid bodies occupy 35–50% of the total hyphal volume (Figure 2c). Similarly to *G. intraradices*, a gradient in storage lipid bodies was observed for *Gi. margarita* when moving towards the ERM growing front (Figure 2d), almost no oleosomes were seen at hyphal tip level (Figure 2e).

Time series of oleosome movement along different sites of the ERM of both *G. intraradices* and *Gi. margarita* (Bago et al., 2002) revealed that very large quantities of storage lipid bodies move along the so called 'runner hyphae' (hyphae acting as conduits for nutrient translocation and for extending the AM fungal colony radially away from the host; Bago et al., 1998a; Friese and Allen 1991). Most of the movement of such lipid bodies appeared to be carried out *via* cytoplasmic streaming, although on some occasions movement against this massive flow was observed, suggesting the possible implication of cytoskeletal elements in controlling lipid translocation along the ERM. Interestingly, oleosome movement was found to be bidirectional; and this is in agreement with labeling data, which indicate that some storage lipid indeed recirculates along the whole fungal colony and returns to intraradical fungal structures (Shachar-Hill et al., unpublished).

Using both lipid volumetric measurements and the speed with which oleosomes are transported along the runner hyphae (measured in the t-series obtained), we have estimated that mass storage lipid flow in *G. intraradices* and *Gi. margarita* is up to 0.26 and 1.34 μg/h, respectively. These data could suggest at a first glance that *Gi. margarita* is a more C-demanding fungus to the root than *G. intraradices*; however, caution should be used before reaching any further conclusion at the respect, since intraradical colonization features (e.g., number of hyphae per colonized root, total root colonization, *Arum* vs. *Paris* type mycorrhiza) should be considered first.

Global nutrient fluxes in AM fungi

Taking into account the carbon flows described above, it is possible to integrate the different macronutrient fluxes taking place in the symbiotic AM fungal colony. This scheme for C, N and P handling is shown in Fig-

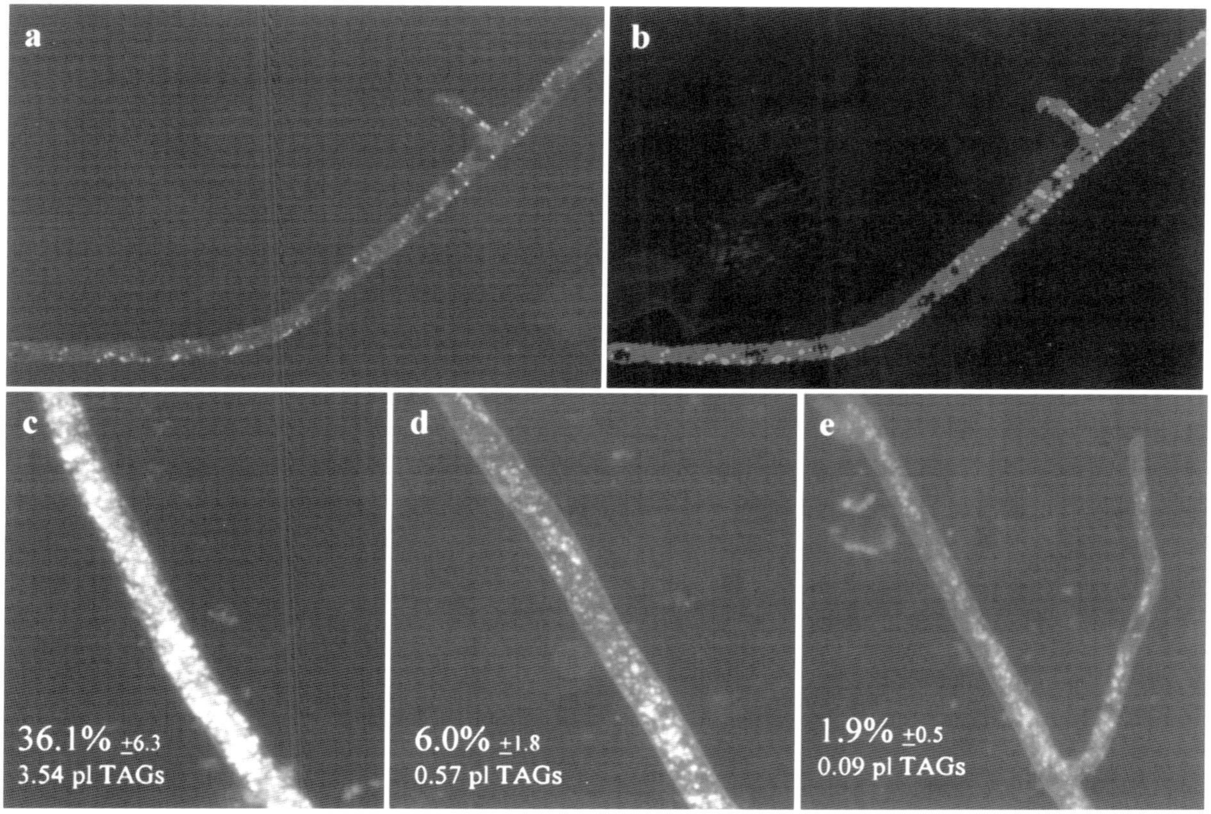

Figure 2. In vivo multiphoton microscopy of storage lipid bodies in extraradical hyphae of the AM fungus *Gigaspora margarita*. (a) 3D reconstruction of a z-series at a given point of a runner hyphae, and (b) the same z-series showing discrimination between lipid bodies (orange color) and cytoplasm (green color) as visualized after applying a software for volumetric measurement of oleosomes. (c–e) z-Series of different sites of a runner hyphae, from zones closer to the root (c), further away from the root (d) and near the hyphal growing front (e). In each micrograph the bottom-right hand side of the image is closer to the root, whereas the top-left hand side is closer to the apex. The percentage of the hyphal volume occupied by lipid droplets in each section is indicated, as well as an estimate of the total triacylglyceride (TAG) content at each site.

ure 3. This model unifies those presented recently for C fluxes (Bago et al., 2000) and postulated for soil nutrient fluxes (Bago et al., 2001), and updates the one recently presented by Pfeffer and co-workers (2001).

According to this scheme (Figure 3) profound metabolic and developmental changes in AM fungi are activated during the establishment of symbiosis with a host root. The signaling and regulation underlying these changes remain unknown, but one of the most crucial changes is probably the differenciation of the fungal partner in two metabolically different but interconnected parts: the intraradical and the extraradical mycelium, with the former being the only location for hexose acquisition. Such hexose fuels metabolism and growth of the IRM, and it is transformed into typical fungal molecules such as trehalose, glycogen and chitin. In order to fuel extraradical fungal development, glycogen is translocated (Figure 3, '1') along newly developed runner hyphae, which would extend the extraradical fungal colony within the soil. Oleosomes containing TAGs are also shipped from the IRM (Figure 3, '2'), some of them recirculating throughout the fungal colony (note the bidirectional arrow of translocation event '2'). These lipid bodies also play key roles in fungal morphogenic and reproductive events such as sporulation.

Extraradical hyphae developing within the substrate take up soil nutrients (mostly P, but also N), and by means of mechanisms that are as yet a matter of inference and postulate (Smith and Read, 1997; Bago et al., 2001) are translocated to the intraradical phase of the fungal colony (Figure 3, '3'), and finally released to the host root. It is possible that amino acids such as Arg are involved in *N*-translocation. In that case,

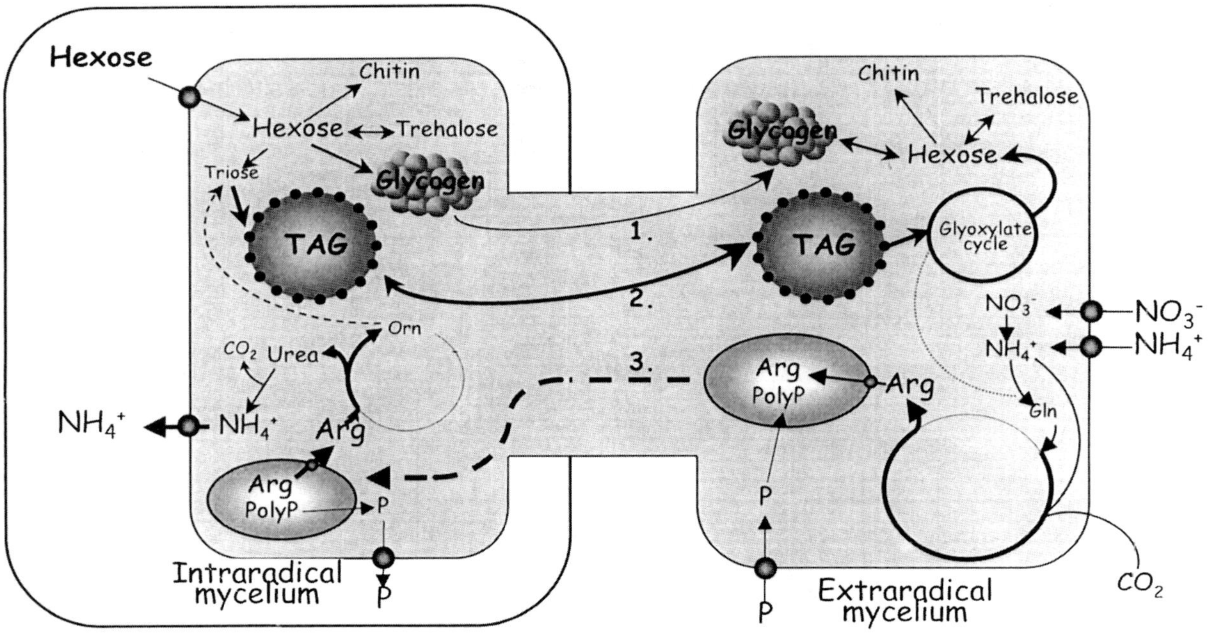

Figure 3. Updated general overview of the translocation events implicating the three major nutrients (C, P and N) which may take place along the AM fungal colony.

carbon and nitrogen translocation along AM fungal hyphae would be linked through the amino acid carbon skeleton. Moreover, since this translocation might be also linked to polyP translocation, as it is the case in ectomycorrhizas (Bütehorn et al., 1999) it could result in that translocation of the three major chemical elements (C, N, P) in AM fungi would be linked. Once released within intraradical fungal structures such carbon skeletons would re-join the C intraradical pool and be used according to fungal needs either for metabolic, translocation or structural purposes.

This working model is consistent with a large number of studies on the AM symbioses, nevertheless it postulates a number of completely untested mechanisms and invites scrutiny and study by a variety of methods.

Possible mechanisms for regulating nutrient translocation in AM fungi

When considering the scheme presented in Figure 3, we might realize that sophisticated translocation processes occur along long hyphal distances that require regulation. How might a coenocytic 'lower' fungus exert control over large bi-directional nutrient flows over long distances?

AM hyphae (both intra- and extraradical) are coenocytic and multinuclear (Balestrini et al., 1992) and these nuclei are distributed very regularly within extraradical hyphae (Figure 4; Bago et al., 1999a). Nuclei are located approximately every 36.0 μm and particular nuclei have been postulated to control particular cellular events (e.g., branching events; Bago et al., 1999a). It is known that transcript levels for beta tubulin vary during the fungal life cycle (Requena et al., 2000) and it may be that at the gene expression level, regulating translocation along the coenocytic AM fungal colony involves an even finer spatial pattern of transcriptional control, with different nuclei in the colony having different transcription profiles.

Nuclear migration is required to distribute nuclei through the hyphal ERM, and this process has been studied using multiphoton microscopy, together with *in vivo* DAPI staining (Figure 4; Bago et al., 1999a). Such migration is bi-directional, apparently independent of the cytoplasmic streaming. These observations put forward once more the probably fundamental role cytoskeletal elements play in the organization and functioning of the AM fungal colony. Microscopy and immunological identification has shed some light on these key components of the AM fungal colony (Aström et al., 1994; Timonen et al., 2001), and the

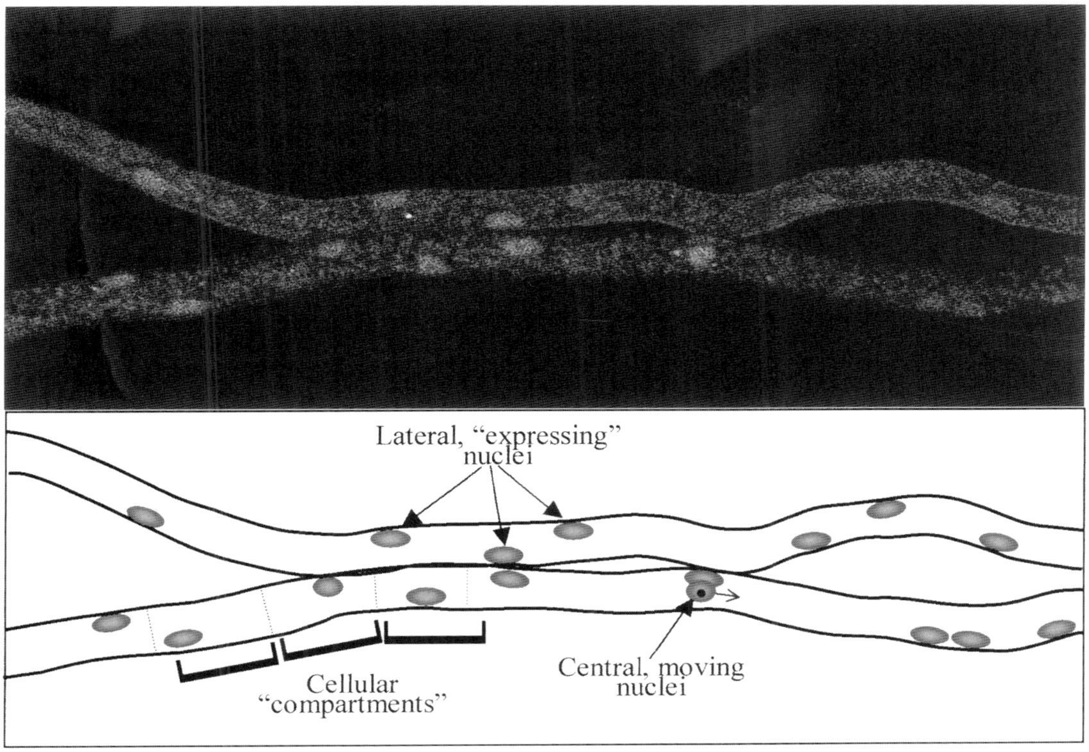

Figure 4. In vivo multiphoton microscopy of fungal nuclei along an extraradical hypha of *Glomus intraradices* (upper pannel) and interpretation of the micrograph (lower pannel), showing the main features of nuclear distribution along ERM in this fungus.

identification of sequences for several key cytoskeletal proteins is described in Jun et al. (2002). Clearly more research is needed to understand translocation processes in AM fungi, and not only as they relate to nutrient distribution along hyphae, but also concerning cellular organelle positioning, and to the functioning of the AM fungal colony.

References

Aström H, Giovannetti M and Raudaskoski M 1994 Cytoskeletal components in the arbuscular mycorrhizal fungus *Glomus mosseae*. MPMI 7, 309–312.

Bago, B 2000 Putative sites for nutrient uptake in arbuscular mycorrhizal fungi. Plant Soil 226, 263–274.

Bago B, Azcón-Aguilar C and Piché Y 1998a Architecture and developmental dynamics of the external mycelium of the arbuscular mycorrhizal fungus *Glomus intraradices* grown under monoxenic conditions. Mycologia 90, 52–62.

Bago B, Azcón-Aguilar C, Goulet A and Piché Y 1998b Branched absorbing structures (BAS): a feature of the extraradical mycelium of symbiotic arbuscular mycorrhizal fungi. New Phytol. 139, 375–388.

Bago B, Zipfel W, Williams RC and Piché Y 1999a Nuclei of symbiotic arbuscular mycorrhizal fungi as revealed by in vivo two-photon microscopy. Protoplasma 209, 77–89.

Bago B, Pfeffer PE, Douds DD Jr, Brouillette J, Bécard G and Shachar-Hill Y 1999b Carbon metabolism in arbuscular mycorrhizal spores as revealed by NMR spectroscopy. Plant Physiol. 121, 263–271.

Bago B, Zipfel W, Williams RC, Chamberland H, Lafontaine J-G, Webb WW and Piché Y 1998c *In vivo* studies on the nuclear behavior of the arbuscular mycorrhizal fungus *Gigaspora rosea* grown under axenic conditions. Protoplasma 203, 1–15.

Bago B, Pfeffer PE and Shachar-Hill Y 2000 Carbon metabolism and transport in arbuscular mycorrhizas. Plant Physiol. 124, 949–957.

Bago B, Pfeffer PE and Shachar-Hill Y 2001 Could the urea cycle be translocating nitrogen in arbuscular mycorrhizal fungi? New Phytol. 149, 4–8.

Bago B, Zipfel W, Williams RC, Jun J, Arreola R, Pfeffer PE, Lammers PJ and Shachar-Hill Y 2002 Translocation and utilization of fungal lipid in the arbuscular mycorrhizal symbiosis. Plant Physiol 128, 108–124.

Balestrini R, Bianciotto V and Bonfante-Fasolo P 1992 Nuclear architecture and DNA location in two VAM fungi. Mycorrhiza 1, 105–112.

Bécard G and Fortin A 1988 Early events of vesicular-arbuscular mycorrhiza formation on Ri T-DNA transformed roots. New Phytol. 108, 211–218.

Beilby JP 1983 Effects of inhibitors on early protein, RNA, and lipid synthesis in germinating vesicular-arbuscular mycorrhizal fungal spores of *Glomus caledonium*. Can. J. Bot. 29, 596–601.

Beilby JP and Kidby DK 1980 Biochemistry of ungerminated and germinated spores of the vesicular-arbuscular mycorrhizal fungus *Glomus caledonium*: changes in neutral and polar lipids. J. Lipid Res. 21, 739–750.

Burleigh S H and Ellegaard Bechman I 2002 Plant nutrient transporter regulation in arbuscular mycorrhiza. Plant Soil 244, 247–251.

Bütehorn B, Gianinazzi-Pearson V and Franken P 1999 Quantification of beta-tubulin RNA expression during asymbiotic and symbiotic development of the arbuscular mycorrhizal fungus *Glomus mosseae*. Mycol. Res. 103, 360–364.

Cooper KM and Lösel DM 1978 Lipid physiology of vesicular arbuscular mycorrhiza. New Phytol. 80, 143–151.

Denk W, Strickler JH and Webb WW 1990 Two-photon laser scanning fluorescence microscopy. Science 248, 73–76.

Denk W, Piston DN and Webb WW 1995 Two-photon molecular excitation in laser scanning fluorescence microscopy. *In* Handbook of Biological Confocal Microscopy. Ed. JB Pawley. pp 445–458. Plenum Pres, New York.

Ferrol N, Barea J M and Azcón-Aguilar 2002 Mechanisms of nutrient transport across interfaces in arbuscular mycorrhizas. Plant Soil 244, 231–237.

Friese CF and MF Allen 1991 The spread of VA mycorrhizal fungal hyphae in the soil: inoculum types and external hyphal architecture. Mycologia 83, 409–418.

Gaspar ML, Pollero RJ and Cabello MN 1994 Triacylglycerol consumption during spore germination of vesicular-arbuscular mycorrhizal fungi. JAOCS 71, 449–452.

Gaspar L, Pollero R and Cabello M 1997 Partial purification and characterization of a lipolytic enzyme from spores of the arbuscular mycorrhizal fungus *Glomus versiforme*. Mycologia 89, 610–6,14.

George E, Marschner H and Jakobsen I 1995 Role of arbuscular-mycorrhizal fungi in uptake of phosphorus and nitrogen from soil. Crit. Rev. Biotechnol. 15, 257–270.

Gianinazzi-Pearson V, Smith SE, Gianinazzi S and Smith FA 1991 Enzymatic studies on the metabolism of vesicular-arbuscular mycorrhizas. V. Is H^+-ATPase a component of ATP-hydrolysing enzyme activities in plant-fungus interfaces? New Phytol. 117, 61–74.

Gow NAR and Gadd GM 1984 The Growing Fungus. Chapman & Hall, London, UK. 472 pp.

Greenspan P, Mayer EP and Fowler SD 1985 Nile Red: a selective fluorescent stain for intracellular lipid droplets. J Cell Biol 100, 965–973.

Harrier LA, Wright F and Hooker JE 1998 Isolation of the 3-phosphoglycerate kinase gene of the arbuscular mycorrhizal fungus *Glomus mosseae* (Nicol. & Gerd.) Gerdemann & Trappe. Curr. Genet. 34, 386–392.

Ho I, Trappe JM 1973 Translocation of ^{14}C from *Festuca* plants to their endomycorrhizal fungi. Nature 244, 30–31.

Jabaji-Hare S 1988 Lipid and fatty acid profiles of some vesicular-arbuscular mycorrhizal fungi: contribution to taxonomy. Mycologia 80, 622–629.

Jakobsen I 1999 Transport of phosphorus and carbon in VA mycorrhizas. *In* Mycorrhiza: Structure, Function, Molecular Biology and Biotechnology, 2nd Edition. Eds. A Varma, B Hock. pp. 305–332. Springer, Berlin.

Jennings DH 1995 The Physiology of Fungal Nutrition. Cambridge University Press, Cambridge.

Jun J, Abubaker J, Rehrer C, Pfeffer P E, Schachar-Hill Y and Lammers P J 2002 Expression in an arbuscular mycorrhizal fungus of genes putatively involved in metabolism, transport, the cyto skeleton and the cell cycle. Plant Soil 244, 141–148.

Koide RT and Schreiner RP 1992 Regulation of the vesicular-arbuscular mycorrhizal symbiosis. Annu. Rev. Plant. Physiol. Mol. Biol. 43, 557–581.

Lammers P, Jun J, Abubaker J, Arreola R, Gopalan A, Bago B, Hernández C, Allen J, Douds DD, Pfeffer PE and Shachar-Hill Y 2001 Gene expression and glyoxylate cycle activity during of an arbuscular mycorrhizal fungus. Plant Physiol. 127, 1287–1298.

Lösel DM and Cooper KM 1979 Incorporation of ^{14}C-labelled substrates by uninfected and VA mycorrhizal roots of onions. New Phytol. 834.

Martin F, Canet D and Marchal P 1985 ^{13}C NMR study of the mannitol cycle and trehalose synthesis during glucose utilization by ectomycorrhizal ascomycete *Cenococcum graniforme*. Plant Physiol. 77, 449–502.

Martin F, Boiffin V and Pfeffer PE 1998 Carbohydrate and amino acid metabolism in the *Eucalyptus globulus–Pisolithus tinctorius* ectomycorrhiza during glucose utilization. Plant Physiol. 118, 627–635.

MacDonald RM and Lewis M 1978 The occurrence of some acid phosphatases and dehydrogenases in the vesicular-arbuscular mycorrhizal fungus *Glomus mosseae*. New Phytol. 80, 135–141.

Murphy DJ 1991 Storage lipid bodies in plants and other organisms. Prog. Lipid Res. 29, 299–324.

Nagy S, Nordby HE and Nemec S 1980 Composition of lipids in roots of six citrus cultivars infected with the vesicular-arbuscular mycorrhizal fungus *Glomus mosseae*. New Phytol. 85, 377–382.

Pfeffer PE and Shachar-Hill Y 1996 Plant/microbe symbioses. *In* Nuclear Magnetic Resonance in Plant Biology. Eds. Y Shachar-Hill and PE Pfeffer. pp. 77–107. APS Press, Rockville, MD.

Pfeffer PE, Douds DD, Bécard G and Shachar-Hill Y 1999 Carbon uptake and the metabolism and transport of lipids in an arbuscular mycorrhiza. Plant Physiol. 120, 587–598.

Pfeffer PE, Bago B and Shachar-Hill Y 2001 Exploring mycorrhizal function with NMR spectroscopy. New Phytol. 150, 543–554.

Requena N, Mann P and Franken P 2000 A homologue of the cell cycle check point TOR2 from *Saccharomyces cerevisiae* exists in the arbuscular mycorrrhizal fungus *Glomus mosseae*. Protoplasma 212, 89–98.

Saito M 1995 Enzyme activities of the internal hyphae and germinated spores of an arbuscular mycorrhizal fungus, *Gigaspora margarita* Becker & Hall. New Phytol. 129, 425–431.

Séjalon-Delmas N, Rogister I, Jauneau A, Roux C and Bécard G 2001 Carbon uptake during symbiotic, asymbiotic and presymbiotic stages of the AM fungus Glomus intraradices. Proceedings of ICOM3, Adelaide, p. P2 137.

Shachar-Hill Y, Pfeffer PE, Douds D, Osman SF, Doner LW and Ratcliffe RG 1995 Partitioning of intermediate carbon metabolism in VAM colonized leek. Plant Physiol. 108, 7–15.

Smith SE, Read DJ 1997 Mycorrhizal Symbiosis. Academic Press, London. 605 pp.

Smith SE and Smith FA 1990 Structure and function of the interfaces in biotrophic symbioses as they relate to nutrient transport. New Phytol. 114, 1–38.

Smith FA and Smith SE 1997 Structural diversity in (vesicular)-arbuscular mycorrhizal symbioses. New Phytol. 137, 373–388.

Solaiman MD and Saito M 1997 Use of sugars by intraradical hyphae of arbuscular mycorrhizal fungi revealed by radiorespirometry. New Phytol. 136, 533–538.

St-Arnaud M, Hamel C, Vimard B, Caron M and Fortin JA 1996 Enhanced hyphal growth and spore production of the arbuscular mycorrhizal fungus *G. intraradices* in an *in vitro* system in the absence of host roots. Mycol. Res. 100, 328–332.

Timonen S, Smith FA and Smith SE 2001 Microtubules of mycorrhizal fungus *Glomus intraradices* in symbiosis with tomato roots. Can. J. Bot. 79, 307–313.

Williams RM, Piston DW and Webb WW 1994 Two-photon molecular excitation provides intrinsic 3-dimensional resolution for laser-based microscopy and microphotochemistry. FASEB J. 8, 804–813.

Xu C and Webb WW 1996 Measurement of two-photon excitation cross sections of molecular fluorophores with data from 690 to 1050 nm. J. Opt. Soc. Am. (B) 13, 481–491.

Xu C, Zipfel W, Shear JB, Williams RM and Webb WW 1996 Multiphoton fluorescence excitation: New spectral windows for biological non-linear microscopy. Proc. Natl. Acad. Sci. USA 93, 10763–10768.

Plant and Soil **244:** 199–210, 2002.

Cytoskeleton in mycorrhizal symbiosis

Sari Timonen[1,3] & R. Larry Peterson[2]
[1]*Department of Biosciences, Division of General Microbiology, P.O. Box 56, (Viikinkaari 9C) 00014 University of Helsinki, Helsinki, Finland,* [2]*Department of Botany, University of Guelph, Guelph, Ontario N1G 2W1, Canada*
[3]*Corresponding author**

Key words: actin filaments, arbuscular mycorrhiza, ectomycorrhiza, microtubules, orchid mycorrhiza

Abstract

An understanding of the role played by the cytoskeleton in formation and function of mycorrhizas has been hampered by the technical difficulty of working with mycorrhizal material. Recently, however, improved labelling techniques suitable for both plant and fungal symbionts in combination with either epifluorescence microscopy or laser scanning confocal microscopy have resulted in new information. As well, molecular methods have made it possible to monitor changes of cytoskeletal elements during mycorrhiza development. Currently we know that the cytoskeletal systems of both plant and fungal partners undergo changes during both ecto- and endomycorrhizal symbiosis. However, little information is available concerning the regulatory factors or the cause and effect relationship of cytoskeletal changes and cellular events. In this article, research involving the cytoskeleton of mycorrhizas is reviewed in detail, whereas basic information of the cytoskeleton of plant and fungal cells is only briefly discussed as background. A brief comparison is also made between the information on mycorrhizas with that of biotrophic pathogenic fungi and the *Rhizobium*–legume symbiosis.

Cytoskeletal components and structures

Currently, cytoskeletal filaments are classified in three categories; actin filaments (microfilaments), intermediate filaments and microtubules. Actin filaments and microtubules are associated with structural and motor proteins, whereas, to date, intermediate filaments have not been shown to be associated with motor proteins. Of the three types of cytoskeletal structures intermediate filaments are the least studied in plants and fungi, although increasing evidence is accumulating concerning their presence in these kingdoms (see, e.g., Menzel, 1993; Yu and Moreno, 1999). Since intermediate filaments have not been reported in mycorrhizal systems they will not be discussed in this review.

Actin filaments and associated proteins

Actin filaments are present in all eucaryotic cells, indicating that from an evolutionary point of view they are very ancient structures. Actin ranks among the five most abundant proteins on earth and it has been suggested that actin filament-based movement is the primary trafficking system in all cells (Kreis and Vale, 1999). Actin filaments consist of 42–45 kDa actin monomers with roughly 375 amino acids (G-actin). The structure of actins is very conserved with most variation in the N-terminus. Actins are usually encoded by gene families and are present in cells as isoforms which may have differential interactions with actin-associated proteins. These filaments are dynamic with complex regulating systems involving a large number of actin binding proteins.

Actin-binding proteins (ABPs) appear to have arisen from the same primordial gene as actin (Vandekerckhove and Weber, 1984). There are, presently, over 60 classes of ABPs and since the mid 1970s the rate of new discoveries has been two or more new classes every year. Surprisingly, most of these classes appear to be universal and it is likely that ABPs unique to certain phyla are an exception rather than the rule (Kreis and Vale, 1999). However, one of the most prominent classes of ABPs, the myosins, appears to have dissimilar subclass sets in different kingdoms.

* FAX No: +358-70859262; Tel.: +358-70859202; E-mail: sari.timonen@helsinki.fi

Myosin types VIII, XI, XIII are only present in plants, whereas all other myosin classes except I, II, V have apparently disappeared from fungi. Presently, representatives of at least six and 11 classes of ABPs have been identified in plants and fungi, respectively. Maximally the speed of myosin mediated transport along actin filaments can reach to 100 μm/s. ABPs take part in regulating the ratio of actin in globular and filamentous forms, modify the actin network, attach actin filaments to membranes, microtubules and other cellular components and create movement along actin filaments.

Microtubules and associated proteins

Microtubules are present in most eucaryotic tissues. Tubulins are considered household proteins and often comprise up to 1% of proteins of the cell. Microtubules have specialised functions particularly in nuclear and cell division and often operate in cells in cooperation with actin filaments. Microtubules are assembled from heterodimers of α- and β-tubulins. Both of these tubulins consist of about 450 amino acids and their molecular weights range from 52 to 55 kDa. Tubulin genes are very conservative with the most variable region in the C-terminus. They are typically present in the genome as isotype families which are particularly large in many plant genomes (Goddard et al., 1994). Certain tubulin isotypes are tissue specific and differential expression of tubulins has been observed during the cell cycle (Ludueña, 1993). Microtubules, similarly to actin filaments, have a more rapidly growing plus end and elaborate regulatory mechanisms.

Microtubule associated proteins (MAPs) are proteins with a high affinity for tubulins. They can be roughly classified into structural MAPs and motor proteins. The most studied of the structural MAPs is γ-tubulin. It is a ca. 58-kDa protein with roughly 35% sequence similarity to α- and β-tubulins. γ-Tubulin is present in microtubule organising centres (MTOCs) where nucleation and assembly of microtubules is initiated (Akashi et al., 1997; Liu et al., 1993). In addition to γ-tubulin several structural MAPs ranging from 40 to 120 kDa have been identified from plants. Even more MAPs have been found from fungi, particularly from yeasts. Microtubule associated motor proteins belong either to dyneins (two types) or kinesins (eight subfamilies). Dyneins move towards the minus end by hydrolysing ATP at speeds of 1.2–2 μm/s. Most kinesins move slowly (0.01–0.07 μm/s) towards the plus end by hydrolysing NTP.

Functions of the cytoskeleton

Plant cells

The cytoskeleton of plant cells has been studied for over 40 years, first by electron microscopy (see Gunning and Hardham, 1982) and later by immunofluorescence and other affinity methods using fluorescent conjugates (Lloyd, 1992). The advent of the latter techniques combined with laser scanning confocal microscopy has extended the knowledge of the roles of the cytoskeleton in plant cells significantly. The literature is voluminous and will not be reviewed in detail here. Briefly, microtubules play important roles in nuclear division and progression of cells through the cell cycle (Baluška and Barlow, 1993; Lloyd et al., 1992; Lloyd, 1992), cytokinesis and cell wall formation (Seagull, 1989), cell elongation and the orientation of cell growth (Yuan et al., 1994), establishment of polarity (Hush et al., 1990), and cytoplasmic streaming along with actin filaments. Frequently, microtubules and actin filaments are associated with the nucleus, endoplasmic reticulum (ER), other cellular organelles, and the plasma membrane (Lichtscheidl et al., 1990; Staehelin, 1997). The outer surface of the nuclear envelope is the main MTOC of plant cells (Lambert, 1993). The maintenance and re-arrangements of the ER system in plant cells is dependent on the actinomyosin system (Staehelin, 1997). Most studies implicate the actin–myosin system as the primary element for vesicle trafficking in plant cells (e.g., Miller et al., 1999; Simon and Pon, 1996), however there are studies where transport of some types of vesicles appear to be microtubule driven (Nick et al., 2000).

Filamentous fungi

The first study of cytoskeletal elements in filamentous fungi appeared in 1970 as observations of actin filaments and microtubules in electron micrographs (Raudaskoski, 1970). In the 1980s more studies were published using the novel method of indirect immunofluorescence (IIF) microscopy (Kilmartin and Adams, 1984; Raudaskoski et al., 1988; Salo et al., 1989) as well as various inhibitor assays. The first report of a fungal cytoskeleton associated protein, myosin, was published by Watts et al., (1985). The major expansion of the research on the cytoskeleton of filamentous fungi took place in the 1990s with the help of more advanced techniques, particularly molecular methods.

In mycelia, actin is typically visualised at septa, in the hyphal tips and around dividing nuclei (Hoch and Staples, 1985). This indicates that actin filaments are involved both in nuclear and cell division as well as generating polarised growth. Indeed, it has been shown that dispersion of actin filaments arrests septum formation (Tucker et al., 1986) and that actin filaments function together with microtubules in nuclear and cell division (Thompson-Coffe and Zickler, 1993). The actin system may be responsible also for at least some of the vesicle trafficking in fungi (Torralba et al., 1998). The actin filament network is indispensable in the morphogenesis and growth of hyphal tips (Drubin et al., 1988; Heath, 1990). Furthermore, actin filaments of fungi are involved in the formation of penetration pegs in pathogenous infection of plant tissues (Bourett and Howard, 1992) and in appressorium formation in a rust fungus (Kwon et al., 1991b; Tucker et al., 1986). Reorganisation of actin is also required in mating of filamentous fungi (Raudaskoski, 1998).

Microtubules have a specialised function together with their associative motor proteins, dyneins, in causing the beating of flagella present in lower fungi (Kreis and Vale, 1999). The importance of microtubules in nuclear division of fungi, where they are the major components of the mitotic spindle and astral microtubules, was the first microtubule-mediated cytoplasmic event recognised (Heath and Heath, 1976). In cell division of dicaryotic fungi, microtubules have also a special role in facilitating nuclear movement from one cell to another in order to maintain dicaryosis (Runeberg et al., 1986). In fungal cells, microtubules move organelles and nuclei in the hyphae and operate in the long-distance transport of vesicles (e.g. Hoch et al., 1987; Inoue et al., 1998; Kwon et al., 1991a; Oakley and Rinehart, 1985; Steinberg et al., 1998). Dispersion of microtubules modifies endoplasmic reticulum as well as disrupts the form and motility of the tubular vacuolar system in the ectomycorrhiza forming fungus *Pisolithus tinctorius* (Ashford et al., 1998). Transport and proper placement of mitochondria require intact microtubules in Basidiomycota (Herr and Heath, 1982; Hoch et al., 1987) but not necessarily in all Ascomycota (Oakley and Rinehart, 1985). Microtubules are indispensable in maintaining the general organisation of fungal cells and their disruption results in an unorganised growth pattern of mycorrhizal fungal hyphae (Niini and Raudaskoski, 1993).

Actin filaments and microtubules frequently appear to interact and cooperate in their various tasks. They influence each other's organisation and stability and share the responsibility of intracellular translocation as well as cell homeostasis (e.g. Kreis and Vale, 1999; Samaj et al., 2000; Torralba et al., 1998).

Cytoskeleton in endomycorrhizal associations

One of the broad categories of mycorrhizal associations is endomycorrhizas, a term used to denote that fungal hyphae penetrate host cell walls and develop fungal structures intracellularly that distinguish each type of endomycorrhiza. The two categories discussed here are arbuscular mycorrhizas (AM) and orchid mycorrhizas, the former characterized by the formation of arbuscules (sometimes vesicles), the latter by hyphal coils called pelotons (see Rasmussen, 2002). In each type, the fungus occupies various proportions of the volume of host cells but is separated from the cell cytoplasm by the elaboration of the host plasma membrane forming a perifungal membrane and an apoplastic compartment between the fungus and host cytoplasm (Bonfante and Perotto, 1995; Peterson et al., 1996, 1998). Considerable alteration in the organization of the cytoplasm occurs during the establishment of both types of mycorrhizas (Bonfante and Perotto, 1995; Peterson et al., 1998) and, therefore, the two have been of interest in terms of studying changes in the cytoskeleton.

Arbuscular mycorrhizas

The first study to indicate that AM fungi affect the cytoskeleton of host cells involved the demonstration that the Tubα3 gene is expressed in cortical cells in which arbuscules develop in maize roots and transformed tobacco roots (Bonfante et al., 1996). The precise role of this gene in this situation is not known but the authors suggest that it may be involved in some way with the formation of the interface between arbuscules and host cell cytoplasm. Recently Bütehorn et al., (1999) have shown that in *Glomus mosseae* a gene encoding β-tubulin was expressed at similar levels in hyphae from germinating sporocarps, extramatrical mycelium, and from colonized root tissue. Interestingly, they found β-tubulin mRNA also in dormant spores and thus suggest β-tubulin is constitutively expressed and could be used as a reference for transcript accumulation of genes.

At the protein level, tomato roots colonised by the AM fungus *Glomus intraradices* have been observed to contain higher amounts of plant α-, β- and

γ-tubulins than uncolonised roots (Timonen, unpublished results). Only in very young tomato radicles does the amount of these cytoskeletal proteins exceed that detected in mycorrhizal roots. In the same study, even in the heavily colonised mycorrhizal roots the plant α-tubulin was always more abundant than fungal α-tubulin. This is in contrast to ectomycorrhizal associations where fungal tubulins dominate in the symbiotic tissues (Timonen et al., 1996). In tomato roots only one ca. 58-kDa γ-tubulin was expressed regardless of the mycorrhizal status. This is similar in size to the constitutively expressed γ-tubulin detected from plant tissues by Petitprez et al. (2001).

Using immunofluorescence methods, Genre and Bonfante (1997) were the first to show that microtubules are rearranged in cortical cells when intracellular hyphae and arbuscules form. Their system, *Nicotiana tabacum* L. roots colonized by the AM fungus *Gigaspora margarita*, provided evidence that bundles of microtubules were located along the length of intracellular hyphae and some linked hyphae to each other or to the cell nucleus. A network of microtubules was seen to envelop the cell nucleus and fine arbuscule branches. A later study (Genre and Bonfante, 1998) provided additional details of the rearrangement of microtubules in colonized cells and added new information on actin filaments in the same symbiotic system. Four types of microtubule patterns were observed in arbuscule-containing cells: long microtubule bundles crossing the cytoplasm among the arbuscule branches and passing through the arbuscule; short microtubules connecting fine arbuscule branches or connecting arbuscule branches either to the cortical region of the cell or the cell nucleus; microtubule bundles in the cortical region of host cells and along trunk hyphae and perinuclear microtubule bundles. Actin filaments, on the other hand, were rarely found adjacent to trunk hyphae, were not found in the cortical region of colonized cortical cells, but lined arbuscule branches and ensheathed collapsed arbuscule branches. Perinuclear actin filaments were always present. Recently Blancaflor et al., (2001) have shown in *Medicago truncatula–Glomus versiforme* symbiosis that the microtubules in cells adjacent to arbuscules or intercellular hyphae are also reorganised. This indicates signal exchange between the symbionts prior to penetration of the fungus through the plant cell wall and active role of the plant cytoskeleton in mycorrhization rather than passive reaction to the physical pressure created by the fungus invaginating the plant plasmalemma.

Based on their previous research related to changes in the cytoskeleton and altered cytology of colonised cells (Bonfante and Perotto, 1995), Genre and Bonfante (1999) explored the localization of MTOCs and clathrin-coated vesicles, both of which may be involved in forming the interface between fungal hyphae and the cytoplasm. Using antibodies to γ-tubulin, a tubulin known to be concentrated in MTOCs, these authors were able to show the presence of MTOCs in the perifungal membrane as well as in the host cell nuclear membrane. Their results support their previous observations of the close association of microtubules with fine arbuscule branches, known to be separated from the cytoplasm by a perifungal membrane. Antibodies to clathrin revealed the presence of this protein along the perifungal membrane as well as along a well-developed tubular endomembrane system. Clathrin-coated vesicles may be involved in the deposition of interfacial matrix material between the fungal cell wall and the perifungal membrane.

In roots with a dimorphic exodermis, AM fungi enter short cells and form a pronounced hyphal coil before growing into the remainder of the cortex. In *Asparagus officinalis* roots, microtubules are closely associated with this coil and as arbuscules develop from hyphae in the inner cortex microtubules are associated with the fine hyphal branches (Matsubara et al., 1999), in a manner similar to that observed previously for tobacco roots (Genre and Bonfante, 1997, 1998).

Most of the observations concerning the interface between AM fungal hyphae and host cell cytoplasm have involved the *Arum*-type mycorrhizal association. Recently, Armstrong and Peterson (2002) have studied the colonization of *Panax quinquefolius* (ginseng) roots by *Glomus intraradices*. A *Paris*-type mycorrhizal association developed with the formation of hyphal coils in the inner cortex from which lateral branches develop, initiating arbuscules (Smith and Smith, 1997). Microtubules (Figure 1a) and actin filaments are closely associated with these hyphal coils and a network of both components of the cytoskeleton forms around the arbuscule branches (Figure 1b) that develop from the coils (Armstrong and Peterson, 2002).

AM are initiated when roots are colonized by aseptate, multinucleate hyphae of fungal species belonging to the order Glomales within the class Zygomycetes (Smith and Read, 1997). To date, there is very little information on the organization of the cytoskeleton in hyphae of this order of fungi. Åström et al., (1994) studied the organization of microtubules and

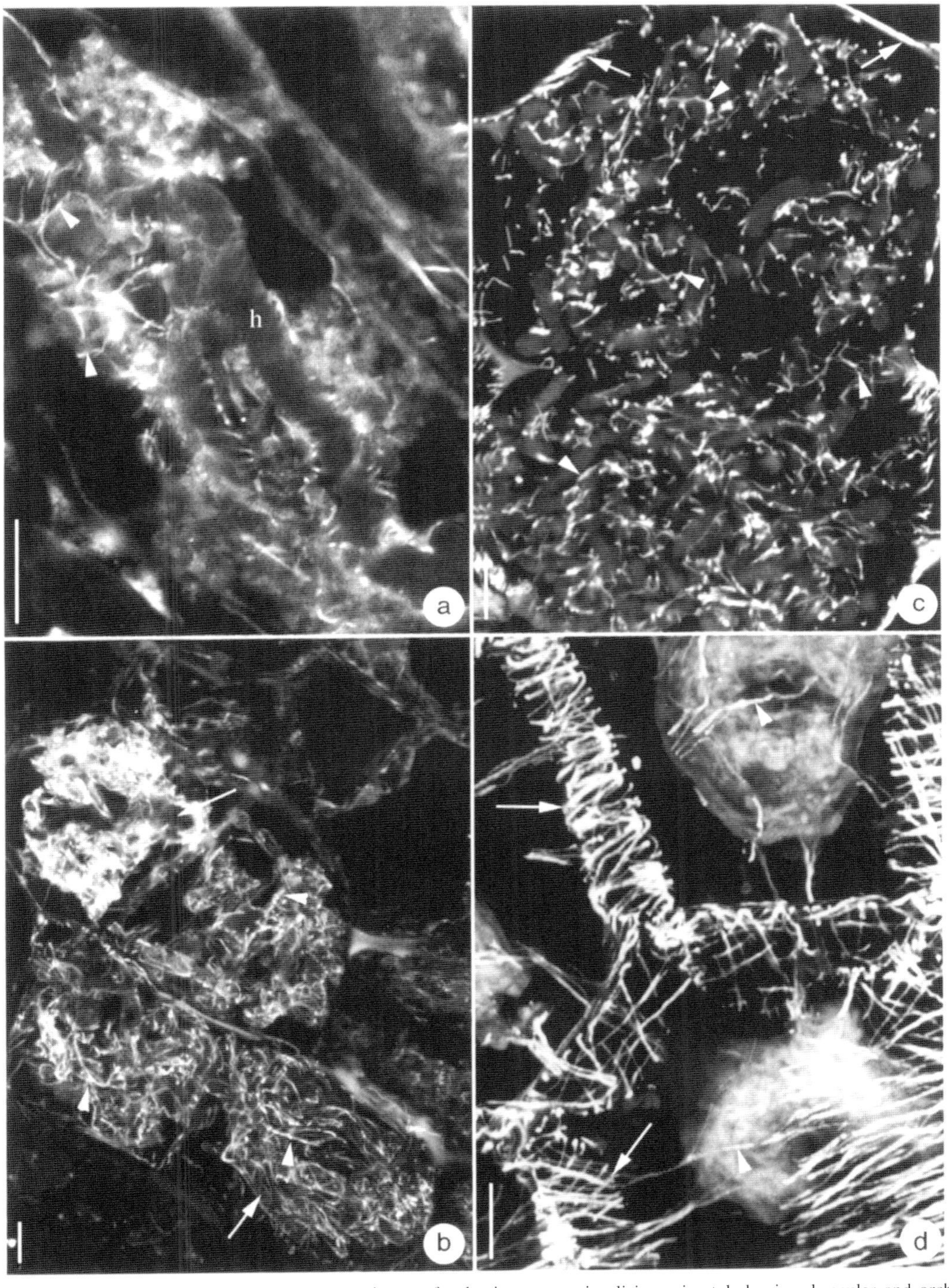

Figure 1. Indirect immunofluorescence (IIF)/laser scanning confocal microscopy visualising microtubules in arbuscular and orchid mycorrhizas, scale bars 10 μm. (a) Hyphal coil in a root cortical cell of ginseng (*Paris*-type AM) showing the close association between the coiled hypha (h) and microtubules (arrowheads). (b) Arbusculate coils in root cells of ginseng showing network of microtubules (arrowheads) surrounding fine arbuscule branches. Microtubules are also associated with hyphal coils (arrows) in this root segment. (c) Coiled peloton hypha of *Ceratobasidium cornigerum* in protocorm cell of the orchid, *Spiranthes sinensis* showing the close association with microtubules (arrowheads). Only a few cortical microtubules (arrow) remain. Photograph: Yukari Uetake. (d) Collapsed pelotons in two protocorm cells of *S. sinensis* with associated microtubules (arrowheads). Cortical microtubules (arrows) have reassembled in these cells. Photograph: Yukari Uetake.

actin filaments by IIF-microscopy in hyphae emanating from germinating spores of *Glomus mosseae*. Microtubules were located within the cortical and central regions of hyphae and they extended into hyphal tips suggesting to the authors that they may be involved in tip growth. Nuclei often appeared to be associated with tracks of microtubules. Actin filaments were only visualized in thin hyphae that proliferated in response to the presence of host roots and their location was confined to the cortical region of hyphal cytoplasm. Plaques and dots of material reacting with antibodies to actin were also present.

There is only one report of the occurrence and orientation of microtubules in an AM fungus when in mycorrhizal symbiosis with host roots (Timonen et al., 2001). In this study, using tomato colonised by *Glomus intraradices,* longitudinal profiles of microtubule bundles were present both in extra- and intraradical hyphae. Microtubules in arbuscules were arranged in either fine cables or bundles of varying thickness. Reticulate organization of microtubules was not seen. Adjacent arbuscules were occasionally connected by hyphae which contained microtubules. Many nuclei were present in arbuscules and these often appeared to be oriented along microtubules. Microtubules were not demonstrated in vesicles although they were present in the subtending hyphae. The organisation of microtubules in intercellular and arbuscular hyphae was surprisingly similar to that of external mycelium at least at the resolution level of this study.

Orchid mycorrhizas

Seeds of orchid species are minute, lack an endosperm, and contain embryos without the organization of a root and shoot apical meristem. In nature and when germinated on agar containing complex carbohydrates, imbibed seeds must be invaded by an appropriate fungal species for embryo development into a protocorm. Seeds of *Spiranthes sinensis* placed on agar medium with cellulose as a carbon source and inoculated with the fungus *Ceratobasidium cornigerum* has been a model system to study the effect of fungal hyphae on the cytoskeleton (Peterson et al., 1998, 2000). In this system it has been shown that uncolonised embryo and protocorm cells have extensive arrays of cortical microtubules as well as perinuclear microtubules (Uetake et al., 1997). As fungal hyphae enter the embryo and protocorm cells, most of the cortical microtubules disappear, perinuclear microtubules are retained, and a population of microtubules develops immediately adjacent to pelotons (Figure 1c). Occasionally microtubules are seen to bridge between segments of the peloton (Uetake et al., 1997; Uetake and Peterson, 1998). As pelotons collapse, microtubules coat the residual hyphal mass and cortical microtubules reappear in host cells (Figure 1d). Protocorm cells can become re-invaded by fungal hyphae and the changes in microtubules are repeated. These results differ substantially from those of Dearnaley and McGee (1996) with protocorms of *Microtis parviflora* colonized by the fungal symbiont, *Epulorhiza repens* in which they reported the disappearance of all microtubules once host cells were invaded by the fungus.

The organization of actin filaments and cables was altered also when embryo and protocorm cells of *S. sinensis* were colonized by *C. cornigerum* (Uetake and Peterson, 1997). Uncolonised cells differed somewhat in the organization and the proportions of actin filaments and cables but in general cortical and perinuclear arrays were present and abundant. With the formation of pelotons, most of the cortical actin filaments disappeared and actin filaments and cables were associated with the coiled hyphae that formed pelotons and the residual hyphal masses after peloton collapse. Often, profiles of actin filaments appeared to bridge a developing peloton and collapsed pelotons to the periphery of the host cell. More recently, Uetake and Peterson (2000) using a combination of scanning confocal laser microscopy, conventional transmission electron microscopy and immuno-gold labelling have shown a close association of endoplasmic reticulum, mitochondria and actin filaments with the perifungal membrane of *C. cornigerum* fungal hyphae in *S. sinensis* protocorm cells.

A few microtubules have been observed by TEM in peloton hyphae in this same orchid system (Uetake and Peterson, 1998) but changes in either fungal microtubules or actin filaments during mycorrhiza formation were not documented.

Cytoskeleton in ectomycorrhizal interactions

Our understanding of cytoskeletal elements and interactions in ectomycorrhizal symbiosis is currently based on remarkably few studies. The first report of microtubules in ectomycorrhizal short roots was published by Edwards and Gessner in 1984. In the electron micrographs, microtubules were shown to line the newly divided apical meristem cells of an

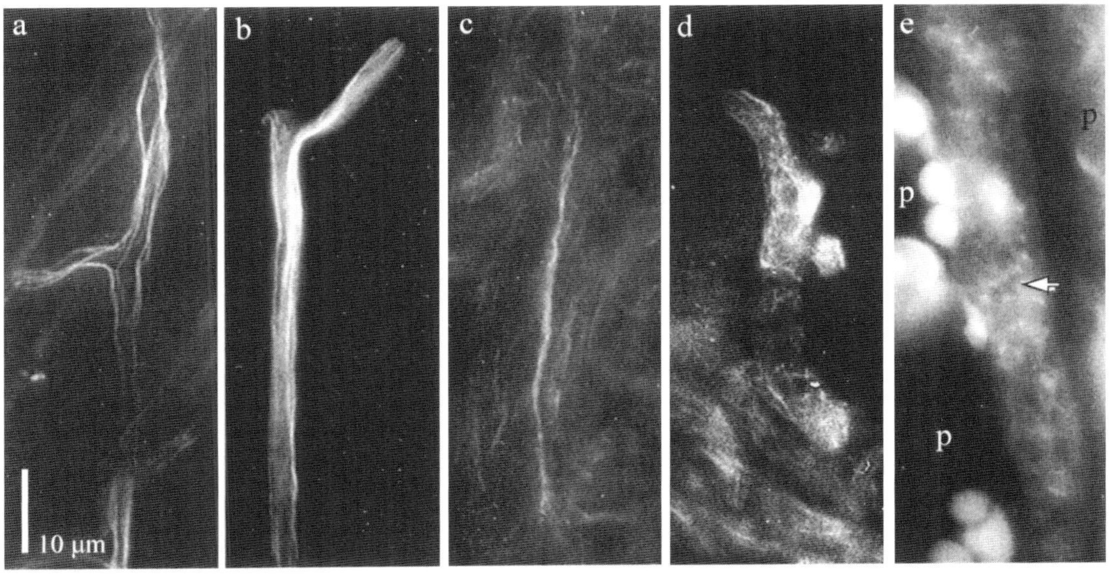

Figure 2. IIF-micrographs of microtubules of ectomycorrhizal fungus *Suillus bovinus*. (a) Tracks of microtubules in external mycelium. (b) Microtubules reaching to the tip of the hypha. (c) Fungal microtubules of the inner mantle of pine–*S. bovinus* mycorrhiza. (d) Reticulate microtubular network in fungal cells of the Hartig net, squash preparation (Timonen et al., 1993). (e) Fungal microtubular network (arrow) in the Hartig net, cryosection. Cortical root cells of the plant (p).

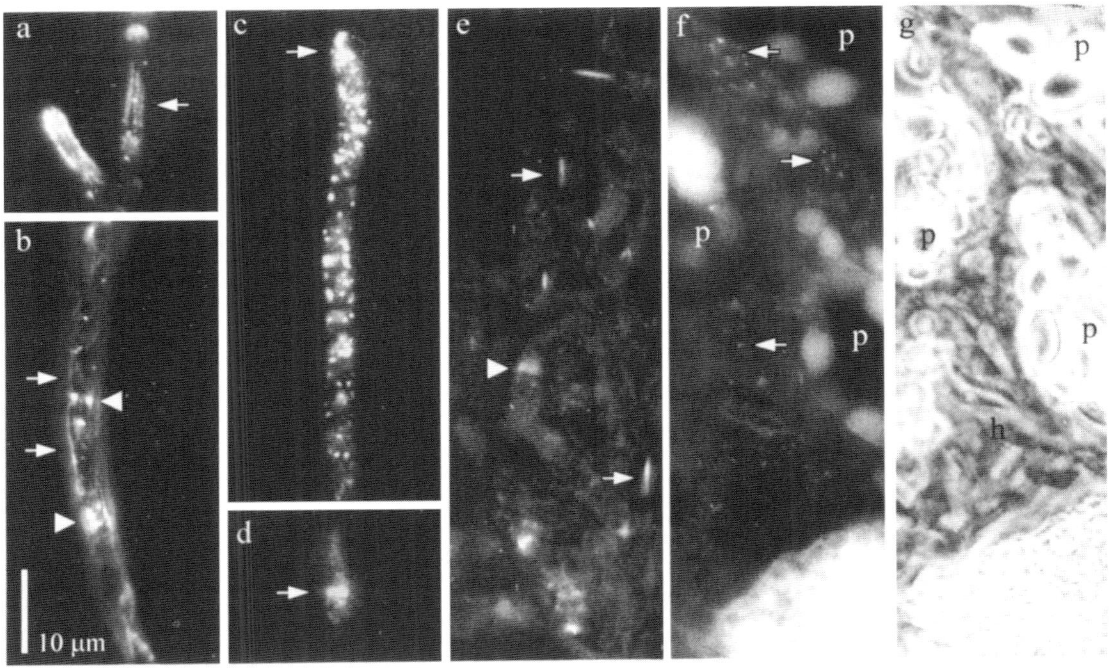

Figure 3. Actin-labelled IIF- micrographs (a, c–f) of ectomycorrhizal fungus *Suillus bovinus*. (a) Short cables (arrow) in free living hyphae. (b) Filaments (arrows) and plaques (arrowheads) in rhodamine phalloidine stained hypha. (c) Actin dots in the cytoplasm of fungal hypha. The tip of the hypha displays most actin signal (arrow). (d) Actin label at a septum (arrow). (e) Actin plaques (arrows) and a labelled septum (arrowhead) in the mantle tissue of pine–*S. bovinus* mycorrhiza (Timonen et al., 1993). (f) Small actin dots in the mycelial tissue of the Hartig net (arrows). (g) Phase contrast micrograph of area displayed in (f). Plant cells full of autofluorescing granules (p). Fungal tissue of the Hartig net (h).

oak root. Another ultrastructural study showing actin filaments in precursory phloem cells of Scots pine–*Suillus bovinus* ectomycorrhizas was published a year later (Warmbrodt and Eschrich, 1985). It was not until 1993 when the cytoskeleton in actual ectomycorrhizal tissues was studied in more detail using immunological methods (Timonen et al., 1993). Cytoskeletal structures were visualised by IIF-microscopy in both plant and fungal tissues of Scots pine–*S. bovinus* ectomycorrhizas. Organisation of actin filaments and microtubules in root cells was similar to that in uncolonised pine root cells except for their absence from cortical cells in the most vigorously colonised areas. In both external mycelium and outer areas of the mantle, microtubules formed long robust tracks attached to each other by thinner cables reaching to the very tip of the hyphae (Figure 2a,b). These cables had been seen previously when microtubules of ectomycorrhizal fungi were studied in pure culture (Salo et al., 1989). In the inner mantle, the microtubular tracks were thinner (Figure 2c) and in the Hartig net fine microtubular cables formed a reticulate structure (Figure 2d,e). These changes may indicate altered function of the cytoskeletal network in the different fungal tissues. In external mycelium of pine mycorrhizal fungi, actin was seen as short cables and frequent plaques (Figure 3a,b). A large portion of actin was present as quite evenly dispersed small dots in the fungal cytoplasm (Figure 3c). Most actin was seen in the hyphal tips and septa (Figure 3c,d) as already shown by Salo et al., (1989). In the mantle some septa and actin plaques could still be seen (Figure 3e) but in the Hartig net all actin seemed to be evenly dispersed as small dots (Figure 3f,g). The dotty appearance of actin was probably due to difficulty in preserving the actin filaments. However, there was a clear difference in the cells of the external mycelium compared with the Hartig net tissue where no septa, tips or plaques could be visualised.

In the study by Timonen et al., (1993), the presence of tubulin and actin was also studied by Western blotting. Both tubulins and actin were present in the mycorrhizal tissues in large quantities but no novel actins or tubulins could be detected. Later, however, pine–*S. bovinus* ectomycorrhizal roots were studied by two-dimensional electrophoresis (Niini et al., 1996). Then, expression of two extra plant α-tubulins was detected in the ectomycorrhizal tissue but no change in plant β-tubulin expression could be seen. This implies that the β-tubulin might be more of a household gene, whereas α-tubulin could be more involved in the morphogenesis of roots. In the same study, four actins were consistently expressed in Scots pine short roots, their abundance increasing at the early stage of mycorrhization. No new actins or tubulins were expressed in the fungus during mycorrhization. In ectomycorrhizal tissues, plant actin was predominant compared with fungal actin in contrast to tubulins where fungal tubulins dominated. Fungal α-tubulin was predominant also in contorta pine–*Suillus variegatus* ectomycorrhizas from 7 d until at least 60 days since colonisation (Timonen et al., 1996). The general amount of actin was increased during mycorrhizal colonisation at the corresponding time and remained high at least for 2 months.

α-Tubulin expression has also been studied on the mRNA level in eucalypt–*Pisolithus tinctorius* ectomycorrhizas (Carnero-Diaz et al., 1994). In this transcription study, the levels of gene expression of one α-tubulin gene (EgTubA1) remained unaffected by mycorrhization by a poorly aggressive fungal strain. However, dramatic up regulation was observed few days after contact with a more aggressive *P. tinctorius* strain. The variation of cytoskeletal effect during ectomycorrhiza colonisation even within fungal species indicates that caution should be taken in reaching conclusions from only a few studies. Nevertheless, all these studies are in agreement concerning up regulation of plant α-tubulin during ectomycorrhizal colonisation.

When actin-encoding genes were studied in *S. bovinus,* two actin genes (*Sbact1* than *Sbact2)* were found (Tarkka et al., 2000). Ten times higher transcription of *Sbact1* than *Sbact2* was observed. Both transcripts were translated into proteins but it was hypothesised that the actin encoded by *Sbact2* could be redundant since translation of it was not detected in another fungus, *Shizophyllum commune*. There was no change in the transcription pattern of these genes during mycorrhization.

The cytoskeleton associated proteins are the dominant factors in the organisation of the pool of cytoskeletal filaments in cells. Consequently, they determine the final structure and function of the cytoskeleton in each cell and are particularly exciting targets for research. Only very recently have these elements been studied in ectomycorrhizal associations. In *Laccaria bicolor*, a gene coding a protein similar to one attaching autophagosomes to microtubules in *Saccharomyces cerevisiae* has been detected (Kim et al., 1999). The expression of this gene was upregulated during ectomycorrhizal colonisation indicating an increase of autophagocytosis-like vesicular transport in ectomycorrhizal symbiosis. Martin et al., (1999) have shown differential expression of mRNAs coding cell wall proteins in *P. tinctorius* mycorrhization. These proteins are suggested to play a role in cytoskeletal regulation. Additionally, the molecular characterisation of two cDNAs coding small GTPases of the Rho subfamily with actin regulating activity (*SbRAC1* and *SbCDC42*) has provided further evidence of actin involvement in morphogenesis of ectomycorrhizal hyphae (Gorfer et al., 2001). CDC42p has been earlier shown to be involved in organising actin-mediated polarised cell growth (Johnson, 1999). In *S. bovinus* CDC42p was localised in association with the plasma membrane in the swollen Hartig net cells as well as at the tips and sites of forming septa of the vegetative hyphae. The location of CDC42p at the sites where actin is typically abundant and active growth and differentiation take place argues that the protein indeed operates in actin-mediated polarisation of fungal hyphae and may have a role in the differentiation of the Hartig net.

Importance and functions of the cytoskeleton in plant microbial interactions

It is evident that mycorrhizal fungi have a profound effect on the cytoskeleton of plant cells and there are several parallels, particularly between endomycorrhizas, and biotrophic pathogenic fungi (Peterson et al., 2000). In AM and orchid associations as well as biotrophic pathogenic fungal systems, the plant defence system has to be overcome, plant cell walls must be breached, and hyphae are modified to establish a nutrient exchange interface. This interface is established by the elaboration of plant-derived plasma membrane and interfacial matrix material around these hyphae. In the case of AM associations and biotrophic fungal pathogens, the interface consists of highly branched arbuscules and haustoria, respectively. In orchids, hyphae branch and coil to form pelotons, again surrounded by plant plasma membrane. It is likely that the cytoskeleton plays a major role in the establishment of the interface.

The legume–*Rhizobium* symbiosis shares several features with the AM symbiosis, as shown particularly through the use of plant mutants (Peterson and Guinel, 2000), and with orchid mycorrhizas. In terms of the cytoskeleton, as in arbuscular and orchid mycorrhizal symbioses, changes occur in host microtubules and actin filaments at all stages of nodule formation (Davidson and Newcomb, 2001a,b). The reorganization of microtubules (Timmers et al., 1999; Turgeon and Bauer, 1985) and actin filaments (Cárdenas et al., 1998) occurs in root hairs as *Rhizobium* attaches to and enters the cell forming infection threads. The early changes in both cytoskeletal components are a response to Nod factors. These triggers for the early changes in the cytoskeleton in mycorrhizal symbioses, however, have not yet been clarified. Later changes in microtubules and actin filaments occur as bacteria are released from infection threads into cortical cells. As in endomycorrhizas, the host elaborates a membrane (the peribacteroid membrane, PBM) around each bacterium, forming PBM–bacteroid complexes called symbiosomes. Actin filaments in mature *Rhizobium*-infected cells form a network that may position at least a portion of the symbiosomes in these cells (Whitehead et al., 1998). In AMs and orchid mycorrhizas, the positioning of arbuscules and pelotons, respectively, may also be controlled by the cytoskeleton.

Conclusions and perspectives

It has been clearly established that the cytoskeleton of both symbionts is modified during the establishment of several types of mycorrhizas. In endomycorrhizas, the organisation of cytoskeletal filaments in plant cells undergoes more extensive changes than that in the fungal structures. In ectomycorrhizas, although there are some modifications to the expression of plant α-tubulin genes, there is little change in the organization of the plant cytoskeleton with the exception of heavily colonized portions of roots where cytoskeletal filaments disappear. However, the fungal cells in the symbiotic tissue of the Hartig net display a novel reticulate organisation not seen in external mycelium. There appears thus to be a profound difference in the two mycorrhizal types. It may reflect difference both in the balance of the regulatory dominance in the two symbionts as well as in the nutrient exchange and translocation mechanisms between the partners. Additional endo- and ectomycorrhizas need to be studied to verify these differences.

Recently, the expression of actin filament and microtubule-associated proteins has been studied in mycorrhizal systems. These studies begin the unravelling of the mechanisms behind the observed organisational changes in the cytoskeletal system. The study of the associated proteins is likely to be the best approach for a better understanding of the cause and effect in cellular regulation of mycorrhizal morphogenesis and function. The new molecular tools such as EST and other genomic libraries as well as transformation of plants and mycorrhizal fungi will be of great help in this task (Bonfante et al., 1996; Pardo et al., 2000; Tagu and Martin, 1995). Additionally, the new method to look specifically for novel plant (Davies et al., 2001) and maybe also fungal cytoskeletal elements may reveal a whole new view to the world of the mycorrhizal cytoskeleton.

Acknowledgements

We thank the Academy of Finland and the Natural Sciences and Engineering Research Council of Canada for financial support. We also thank Marjatta Raudaskoski for her contribution to the work on *Suillus bovinus* ectomycorrhizas and Lewis Melville for assistance with producing the final plates.

References

Akashi T, Yoon Y and Oakley B R 1997 Characterization of gamma-tubulin complexes in *Aspergillus nidulans* and detection of putative gamma-tubulin interacting proteins. Cell Motil. Cytoskel. 37, 149–158.

Armstrong L and Peterson R L 2002 The interface between the arbuscular mycorrhizal fungus *Glomus intraradices* and root cells of *Panax quinquefolius*: a Paris-type mycorrhizal association. Mycologia (in press).

Ashford A, Cole L, Hyde G and Orlovich D 1998 New observations on structure, function and motility of vacuoles in *Pisolitus tinctorius*. *In* Second International Conference on Mycorrhiza. Eds. U Ahonen-Jonnarth, E Danell, P Fransson, O Kårén, B Lindahl, I Rangel and R Finlay. 20. SLU Service/Repro, Uppsala, Sweden.

Åström H, Giovannetti M and Raudaskoski M 1994 Cytoskeletal components in the arbuscular mycorrhizal fungus *Glomus mosseae*. Mol. Plant-Microbe Interact. 7, 309–312.

Baluška F and Barlow P W 1993 The role of the microtubular cytoskeleton in determining nuclear chromatin structure and passage of maize root cells through the cell cycle. Eur. J Cell Biol. 61, 160–167.

Blancaflor, E B, Zhao L and Harrison M 2001 Microtubule organization in root cells of *Medicago truncatula* during development of an arbuscular mycorrhizal symbiosis with *Glomus versiforme*. Protoplasma 217, 154–165.

Bonfante P and Perotto S 1995 Strategies of arbuscular mycorrhizal fungi when infecting host plants. New Phytol. 130, 3–21.

Bonfante P, Bergero R, Uribe X, Romera C, Rigau J and Puigdomenech P 1996 Transcriptional activation of a maize α-tubulin gene in mycorrhizal maize and transgenic tobacco plants. Plant J. 9, 737–743.

Bourett T M and Howard R J 1992 Actin in penetration pegs of the fungal rice blast pathogen *Magnaporthe grisea*. Protoplasma 168, 20–26.

Bütehorn B, Gianinazzi-Pearson V, Franken P 1999 Quantification of β-tubulin RNA expression during asymbiotic and symbiotic development of the arbuscular mycorrhizal fungus *Glomus mosseae*. Mycol. Res. 103, 360–364.

Cárdenas L, Vidali L, Domníquez j, Pérez H, Sánchez F, Hepler P K and Quinto C 1998 Rearrangement of actin microfilaments in plant root hairs responding to *Rhizobium etli* nodulation signals. Plant Physiol. 116, 871–877.

Carnero-Diaz M, Martin F and Tagu D 1994 Cloning and characterization of an alpha-tubulin cDNA from eucalyptus ectomycorrhizas. *In* 4th European Symposium on Mycorrhizas, 11–14 July 1994. Eds. C Axcon-Aguilar and J M Barea. pp. 108. ECSC-EC-EAEC, Granada, Spain.

Davidson A L and Newcomb W 2001a Changes in actin microfilament arrays in developing pea root nodule cells. Can. J. Bot. 79: 767–776.

Davidson A L and Newcomb W 2001b Organization of microtubules in developing pea root nodule cells. Can. J. Bot. 79: 777–786.

Davies E, Stancovic B, Azama K, Shibata K and Abe S 2001 Novel components of the plant cytoskeleton: a beginning to plant 'cytomics'. Plant Sci. 160, 185–196.

Dearnaley J D W and McGee P A 1996 An intact microtubule cytoskeleton is not necessary for interfacial matrix formation in orchid protocorm mycorrhizas. Mycorrhiza 6, 175–180.

Drubin D G, Miller K G and Botstein D 1988 Yeast actin-binding proteins: evidence for a role in morphogenesis. J. Cell Biol. 107, 2551–2561.

Edwards H H and Gessner R V 1984 Light microscopy and transmission electron microscopy of English oak (*Quercus robur*) ectomycorrhizal short roots. Can. J. Bot. 62, 1327–1335.

Genre A and Bonfante P 1997. A mycorrhizal fungus changes microtubule orientation in tobacco root cells. Protoplasma 199, 30–38.

Genre A and Bonfante P 1998. Actin versus tubulin configuration in arbuscule-containing cells from mycorrhizal tobacco roots. New Phytol. 140, 745–752.

Genre A and Bonfante P 1999 Cytoskeleton-related proteins in tobacco mycorrhizal cells: γ-tubulin and clathrin localization. Eur. J. Histochem. 43, 105–111.

Goddard R H, Wick S M, Silflow C D and Snustad P 1994 Microtubule components of the plant cell cytoskeleton. Plant Physiol. 104, 1–6.

Gorfer M, Tarkka M T, Hanif M, Pardo A G, Laitiainen E and Raudaskoski M 2001 Characterization of small GTPases Cdc42 and Rac and the relationship between Cdc42 and actin cytoskeleton in vegetative and ectomycorrhizal hyphae of *Suillus bovinus*. Mol. Plant Microbe Interact. 14, 135–144.

Gunning E S and Hardham A R 1982 Microtubules. Annu. Rev. Plant Physiol. 33, 651–698.

Heath I B 1990 The roles of actin in tip growth of fungi. Int. Rev. Cytol. 123, 95–127.

Heath I B and Heath M C 1976 Ultrastructure of mitosis in the cowpea rust fungus *Uromyces phaseoli* var. *vignae*. J. Cell Biol. 70, 592–607.

Herr F B and Heath M C 1982 The effects of antimicrotubule agents on organelle positioning in the cowpea rust fungus, *Uromyces phaseoli* var. *vignae*. Exp. Mycol. 6, 15–24.

Hoch H C and Staples R C 1985 The microtubule cytoskeleton in hyphae of *Uromyces phaseoli* germlings: Its relationship to the region of nucleation and to the F-actin cytoskeleton. Protoplasma 124, 112–122.

Hoch H C, Tucker B E and Staples R C 1987 An intact microtubule cytoskeleton is necessary for mediation of the signal for cell differentiation in *Uromyces*. Eur. J. Cell Biol. 45, 209–218.

Hush J M, Hawes C R and Overall R L 1990 Interphase microtubule re-orientation predicts a new cell polarity in wounded pea roots. J. Cell Sci.96, 47–61.

Inoue S, Turgeon B G, Yoder O C and Aist J R 1998 Role of fungal dynein in hyphal growth, microtubule organization, spindle pole body motility and nuclear migration. J. Cell Sci. 111, 1555–1566.

Johnson D I 1999 Cdc42: An essential Rho-type GTPase controlling eucaryotic cell polarity. Microbiol. Mol. Biol. Rev. 63, 54–105.

Kilmartin J V and Adams A E M 1984 Structural rearrangements of tubulin and actin during the cell cycle of the yeast *Saccharomyces*. J. Cell Biol. 98, 922–933.

Kim S J, Bernreuther D, Thumm M and Podila G K 1999 LB-AUT7, a novel symbiosis-regulated gene from an ectomycorrhizal fungus, *Laccaria bicolor*, is functionally related to vesicular transport and autophagocytosis. J. Bacteriol. 181, 1963–1967.

Kreis T and Vale R 1999 Cytoskeletal and Motor Proteins. 551 pp. Oxford University Press, New York.

Kwon Y H, Hoch H C and Aist J R 1991a Initiation of appressorium formation in *Uromyces appendiculatus*: Organization of the apex, and the responses involving microtubules and apical vesicles. Can. J. Bot. 69, 2560–2573.

Kwon Y H, Hoch H C and Staples R C 1991b Cytoskeletal organization in *Uromyces urediospore* germling apices during appressorium formation. Protoplasma 165, 37–50.

Lambert A-C 1993 Microtubule-organizing centers in higher plants. Curr. Opin. Cell Biol. 5, 116–122.

Lichtscheidl I K, Lancelle S A and Hepler P K 1990 Actin-endoplasmic reticulum complexes in *Drosera*. Their structural relationship with the plasmalemma, nucleus, and organelles in cells prepared by high pressure freezing. Protoplasma 155, 116–126.

Liu B, Marc J, Joshi H and Palevitz B A 1993 A gamma-tubulin related protein associated with the microtubule arrays of higher plants in a cell cycle-dependent manner. J. Cell Sci. 104, 1217–1228.

Lloyd C W (Ed.) 1992 The Cytoskeletal Basis of Plant Growth and Form. Academic Press, London.

Lloyd C W, Venverloo C J Goodbody K C and Shaw P J 1992 Confocal laser microscopy and three-dimensional reconstruction of nucleus-associated microtubules in the division plane of vacuolated plant cells. J. Microsc. 166, 99–109.

Ludueña R F 1993 Are tubulin isotypes functionally significant. Mol. Biol. Cell 4, 445–457.

Martin F, Laurent P, de C D, Voiblet C, Balestrini R, Bonfante P and Tagu D 1999 Cell wall proteins of the ectomycorrhizal basidiomycete *Pisolithus tinctorius*: Identification, function, and expression in symbiosis. Fungal Genet. Biol. 27, 161–174.

Matsubara Y, Uetake Y and Peterson R L 1999 Entry and colonization of *Asparagus officinalis* roots by arbuscular mycorrhizal fungi with emphasis on changes in host microtubules. Can. J. Bot. 77, 1159–1167.

Menzel D 1993 Chasing coiled coils: Intermediate filaments in plants. Bot. Acta 106, 294–300.

Miller D D, De R N C A, Bisseling T and Emons A M C 1999 The role of actin in root hair morphogenesis: Studies with lipochito-oligosaccharide as a growth stimulator and cytochalasin as an actin perturbing drug. Plant J. 17, 141–154.

Nick P, Heuing A and Ehmann B 2000 Plant chaperonins: A role in microtubule-dependent wall formation? Protoplasma 211, 234–244.

Niini S S and Raudaskoski M 1993 Response of ectomycorrhizal fungi to benomyl and nocodazole: growth inhibition and microtubule depolymerization. Mycorrhiza 3, 83–91.

Niini S S, Tarkka M T and Raudaskoski M 1996 Tubulin and actin protein patterns in scots pine (*Pinus sylvestris*) roots and developing ectomycorrhiza with *Suillus bovinus*. Physiol. Plant. 96, 186–192.

Oakley B R and Rinehart J E 1985 Mitochondria and nuclei move by different mechanisms in *Aspergillus nidulans*. J. Cell Biol. 101, 2392–2397.

Pardo A G, Gorfer M, Hanif M, Tarkka M and Raudaskoski M 2000 Agrobacterium-mediated transformation of mycorrhizal fungi. Plant Physiol. Biochem. 38.

Peterson R L and F C Guinel 2000 The use of plant mutants to study regulation of colonization by AM fungi. *In* Arbuscular Mycorrhizas: Physiology and Function. Eds. Y Kapulnik and D D Douds Jr. pp. 147-171. Kluwer Academic Publishers, Dordrecht.

Peterson R L, Bonfante P, Faccio A and Uetake Y 1996 The interface between fungal hyphae and orchid protocorm cells. Can. J. Bot. 74, 1861–1870.

Peterson R L, Uetake Y and Zelmer C 1998. Fungal symbioses with orchid protocorms. Symbiosis 25, 29–55.

Peterson R L, Uetake Y and Armstrong L N 2000 Interactions between fungi and plant cell cytoskeleton. *In* Current Advances in Mycorrhizae Research. Eds. G K Podila and D. D. Douds. pp. 157–178. APS Press, St. Paul.

Petitprez M, Caumont C, Barthou H, Wright M and Alibert G 2001 Two gamma-tubulin isoforms are differentially expressed during development in *Helianthus annuus*. Physiol. Plant. 111, 102–107.

Rasmussen H N 2002 Recent developments in the study of orchid mycorrhiza. Plant Soil 244, 151–165.

Raudaskoski M 1998 The relationship between B-mating-type genes and nuclear migration in *Schizophyllum commune*. Fungal Genet. Biol. 24, 207–227.

Raudaskoski M 1970 Occurrence of microtubules and microfilaments, and origin of septa in dicaryotic hyphae of *Schizophyllum commune*. Protoplasma 70, 415–422.

Raudaskoski M, Salo V and Niini S S 1988 Structure and function of the cytoskeleton in filamentous fungi. Karstenia 28, 49–60.

Runeberg P, Raudaskoski M and Virtanen I 1986 Cytoskeletal elements in the hyphae of the homobasidiomycete *Schizophyllum commune* visualized with indirect immunofluorescence and NBD-phallacidin. Eur. J. Cell Biol. 41, 25–32.

Salo V, Niini S S, Virtanen I and Raudaskoski M 1989 Comparative immunocytochemistry of the cytoskeleton in filamentous fungi with dikaryotic and multinucleate hyphae. J. Cell Sci. 94, 11–24.

Samaj J, Peters M, Volkmann D and Baluska F 2000 Effects of myosin ATPase inhibitor 2,3-butanedione 2-monoxime on distributions of myosins, F-actin, microtubules, and cortical endoplasmic reticulum in maize root apices. Plant Cell Physiol. 41, 571–582.

Seagull R W 1989 The plant cytoskeleton. Crit. Rev. Plant Sci. 8, 131–167.

Simon V R and Pon L A 1996 Actin-based organelle movement. Experientia 52, 1117–1122.

Smith F A and Smith S E 1997 Structural diversity in (vesicular)-arbuscular mycorrhizal symbioses. New Phytol. 137, 373–388.

Smith S E and Read D J 1997. Mycorrhizal Symbiosis. 2nd Edition. Academic Press, London.

Staehelin L A 1997 The plant ER: a dynamic organelle composed of a large number of discrete functional domains. Plant J. 11, 1151–1165.

Steinberg G, Schliwa M, Lehmler C, Boelker M, Kahmann R and McIntosh J R 1998 Kinesin from the plant pathogenic fungus *Ustilago maydis* is involved in vacuole formation and cytoplasmic migration. J. Cell Sci.111, 2235–2246.

Tagu D and Martin F 1995 Expressed sequence tags of randomly selected cDNA clones from *Eucalyptus globulus–Pisolithus tinctorius* ectomycorrhiza. Mol. Plant Microbe Interact. 8, 781–783.

Tarkka M T, Vasara R, Gorfer M and Raudaskoski M 2000 Molecular characterization of actin genes from homobasidiomycetes: Two different actin genes from *Schizophyllum commune* and *Suillus bovinus*. Gene Amsterdam 251, 27–35.

Thompson-Coffe C and Zickler D 1993 Cytoskeleton interactions in the ascus development and sporulation of *Sordaria macrospora*. J. Cell Sci. 104, 883–898.

Timmers A C J, Auriac M-C, de Billy F and Truchet G 1999 Refined analysis of early symbiotic steps of the *Rhizobium–Medicago* interaction in relationship with microtubular cytoskeleton rearrangements. Development 126, 3617–3628.

Timonen S, Finlay R D, Söderström B and Raudaskoski M 1993 Identification of cytoskeletal components in pine ectomycorrhizas. New Phytol. 124, 83–92.

Timonen S, Söderström B and Raudaskoski M 1996 Dynamics of cytoskeletal proteins in developing pine ectomycorrhiza. Mycorrhiza 6, 423–429.

Timonen S, Smith F A and Smith S E 2001 Microtubules of mycorrhizal fungus *Glomus intraradices* in symbiosis with tomato roots. Can. J. Bot. 79, 307–313.

Torralba S, Raudaskoski M, Pedregosa A M and Laborda F 1998 Effect of cytochalasin A on apical growth, actin cytoskeleton organization and enzyme secretion in *Aspergillus nidulans*. Microbiology Reading 144, 45–53.

Tucker B E, Hoch H C and Staples R C 1986 The involvement of F-actin in *Uromyces* cell differentiation: The effects of cytochalasin E and phalloidin. Protoplasma 135, 88–101.

Turgeon B G and Bauer W D 1985 Ultrastructure of infection-thread development during the infection of soybean by *Rhizobium japonicum*. Planta 163, 328–349.

Uetake Y, Farquhar M L and Peterson R L 1997 Changes in microtubule arrays in symbiotic orchid protocorms during fungal colonization and senescence. New Phytol. 135, 701–709.

Uetake Y and Peterson R L 1997 Changes in actin filament arrays in protocorm cells of the orchid species *Spiranthes sinensis*, induced by the symbiotic fungus *Ceratobasidium cornigerum*. Can. J. Bot. 75, 1661–1669.

Uetake Y and Peterson R L 1998 Association between microtubules and symbiotic fungal hyphae in protocorm cells of the orchid species, *Spiranthes sinensis*. New Phytol. 140, 715–722.

Uetake Y and Peterson R L 2000 Spatial associations between actin filaments, endoplasmic reticula, mitochondria and fungal hyphae in symbiotic cells of orchid protocorms. Mycoscience 41, 481–489.

Vandekerckhove J and Weber K 1984 Chordate muscle actins differ distinctly from invertebrate muscle actins: The evolution of the different vertebrate muscle actins. J. Mol. Biol. 179, 391–414.

Warmbrodt R D and Eschrich W 1985 Studies on the mycorrhiza of *Pinus sylvestris* produced *in vitro* with the basidiomycete *Suillus variegatus*: 2. Ultrastructural aspects of the endodermis and vascular cylinder of the mycorrhizal rootlets. New Phytol. 100, 403–418.

Watts F Z, Miller D M and Orr E 1985 Identification of myosin heavy chain in *Saccharomyces cerevisiae*. Nature 316, 83–85.

Whitehead L F, Day D A and Hardham A R 1998 Cytoskeletal arrays in the cells of soybean root nodules: the role of actin microfilaments in the organization of symbiosomes. Protoplasma 203, 194–205.

Yu W and Moreno D D L E S 1999 The plant nucleoskeleton: Ultrastructural organization and identification of NuMA homologues in the nuclear matrix and mitotic spindle of plant cells. Exp. Cell Res. 246, 516–526.

Yuan M, Shaw P J, Warn R M, Lloyd C W 1994 Dynamic reorientation of cortical microtubules, from transverse to longitudinal, in living plant cells. Proc. Nat. Acad. Sci. USA Cell Biol. 91, 6050–6053.

Plant and Soil **244:** 211–219, 2002.

Endosymbiotic bacteria in mycorrhizal fungi: from their morphology to genomic sequences

D. Minerdi, V. Bianciotto & P. Bonfante[1]
Dipartimento di Biologia Vegetale dell' Università di Torino and Centro di Studio sulla Micologia del Terreno del CNR, Viale Mattioli 25, 10125 Torino, Italy. [1]*Corresponding author**

Key words: arbuscular mycorrhizal fungi, bacterial sequences, endosymbiotic *Burkholderia*, Gigasporaceae, *nif* genes, Tuberaceae

Abstract

Arbuscular mycorrhizal (AM) fungi have been successful in time and space thanks to a long co-evolution with their host plants. In addition to this well known interaction, they also associate with bacteria that reside in the fungal cytoplasm. The chapter mostly focusses on endosymbionts belonging to the genus *Burkholderia* and found in many species of Gigasporaceae. We have used morphological and genetic approaches to investigate these intracellular microrganisms. Some genes related to metabolism, cell colonization events and nitrogen fixation have been characterized and suggest a potential role in the nutritional exchanges between endobacteria, fungi and plants.

Introduction

The genomic projects recently accomplished for *Homo sapiens*, *Caenorhabditis elegans*, *Arabidopsis thaliana* and *Saccharomyces cerevisiae* have contributed a great deal of information, opening exciting perspectives not only on their functional genomics, but also on their evolution. The presence of a remarkable number of prokaryotic genes in Eukaryotes has largely confirmed the endosymbiosis theory (Margulis, 1993) providing sound evidence of the prokaryotic origin of mitochondria and chloroplasts. By contrast, the presence of true bacteria living inside Eukaryotes has been more rarely detected in nature (Douglas, 1994). Bacteria that complete their life cycles within or associated with eukaryotic cells offer examples of symbiosis with a different level of integration between partners. They are increasingly appreciated by ecologists and evolutionary biologists for their huge diversity (Moran and Wernegreen, 2000). These bacteria include chronic pathogens and mutualistic associates which are phylogenetically diverse but often possess distinctive genomic traits including reduced genome size and rapid polypeptide evolution (Moran and Wernegreen, 2000). These resident bacteria may be found inside specialized cells (bacteriocytes) of insects and worms in the Animal Kingdom, and inside specialized organs (nodules) of a restricted number of plant species. The Fungal Kingdom has a limited number of examples: *Geosiphon pyriforme*, a Zygomycete closely related to Glomales, can host cyanobacteria inside characteristic bladders in the apical hyphal region and also harbours BLOs of the glomalean type (Schüßler et al., 1996). In addition to this species, mycorrhizal fungi offer the best known examples of fungal/bacterial association.

Mycorrhizas: a place for multiple interactions

A mycorrhiza is a complex phenomenon comprising fungi belonging to different taxonomic groups and about 90% of land plants (Smith and Read, 1997). Mycorrhizas are successful in colonizing diverse environments and their ecological success reflects a high degree of diversity in the genetic and physiological abilities of the fungal endophytes. About 6000 species in the Zygo-, Asco- and Basidiomycotina have been recorded as mycorrhizal, and the advent of molecular techniques has increased this number, as well as revealing a high degree of inter- and intraspecific genetic diversity (Kowalchuk, 1999). The taxonomic position

* E-mail: p.bonfante@csmt.to.cnr.it

of plant and fungal partners defines the types of mycorrhiza, the main distinction of which is between endomycorrhizas and ectomycorrhizas. In endomycorrhizas the fungal hyphae penetrate the root cells to establish an intracellular symbiosis, whereas in ectomycorrhizas hyphae remain extracellular. In addition to the well known plant/fungal interactions, mycorrhizal roots offer good ecological niches for other microbes: some rhizosphere bacteria adhere tightly to fungal hyphae and are carried along the hyphae towards the roots, while others are directly associated with root surfaces (Bianciotto et al., 2001; Mogge et al., 2000). Since other chapters describe mycorrhizal associations in detail, we focus on bacteria living inside mycorrhizal fungi providing information on their morphology, phylogenetic relationships and genomic sequences.

Symbiotic bacteria live inside most arbuscular mycorrhizal fungi

A peculiar feature of AM fungi is that they host intracellular structures very similar to bacteria, called Bacteria-Like Organisms (BLOs) and first described in the 1970s (Mosse, 1970; Scannerini and Bonfante, 1991). Ultrastructural observations clearly revealed their presence in many field-collected fungal isolates (i.e., *Glomus versiforme, Acaulospora laevis, Gigaspora margarita).* Further investigation of these BLOs was hampered because of their unculturability. Only a combination of morphological observations (electron and confocal microscopy) and molecular analyses have allowed us to identify BLOs as true bacteria and to start unravelling their symbiotic relationship with AM fungi (Bianciotto et al., 1996). Isolate BEG 34 of *Gigaspora margarita* contains a large number of BLOs: about 250 000 live bacteria were counted in a single spore (Figure 1). Ultrastructural observations revealed rod-shaped BLOs in the vacuoles of germinating spores, often associated with large protein bodies (Bonfante et al., 1994). The bacteria have a laminated wall and their cytoplasm is rich in ribosomes (Figure 2). Similar bacteria, some of them undergoing divisions, were found in both germinating and intraradical hyphae, as well as in the arbuscule trunks. Amplification of bacterial 16S RNA gene from total spore DNA followed by direct sequencing indicated that these bacteria are closely related to the genus *Burkholderia*, a group belonging to the β subdivision of the Proteobacteria, and that they are present throughout the fungal life cycle (Bianciotto et al., 1996).

Detection and identification of bacteria in the Gigasporacee

The detection of *Burkholderia* in the spores of the BEG 34 isolate, originally from a New Zealand soil, raised the question as to whether the same bacteria are also harbored by other Glomales, and whether they can be found in isolates from different geographic areas. We focussed our attention on Gigasporaceae, which comprise the genera *Gigaspora* and *Scutellospora*, because they provide a more amenable experimental system. In fact, bacterial DNA can also be amplified from Glomaceae (Bianciotto et al., 1996; Hosny et al., 1999), but more detailed investigations are hampered by the low number of intracellular bacteria and by the extensive microbial contamination of the spore surface (Bianciotto, unpublished observations). Eleven fungal isolates collected from different geographic areas and belonging to six different species of Gigasporaceae were investigated with a combined morphological and molecular approach (Bianciotto et al., 2000). A fluorescent dye was first used to detect bacteria inside the spore cytoplasm, while 16S ribosomal genes were amplified by PCR using universal bacterial primers and primers specific for the endobacteria already identified in *Gi. margarita* BEG 34 (BLO primers). With the exception of the four isolates of *Gi. rosea*, bacteria were detected in the cytoplasm of all fungal isolates and an amplified DNA fragment of the expected size was obtained with universal bacterial primers. The specific BLO primers could further amplify DNA from seven out of eight isolates, belonging to five different species (Bianciotto et al., 2000).

Our observations indicate some interesting features of the endobacteria/AM fungal association. In *Gigaspora*, where four out of the five species included in this genus were analyzed, only *Gi. rosea* lacked endobacteria. This finding demonstrates that endosymbionts are a common feature at least in *Gigaspora* and might be used as molecular markers (Lanfranco et al., 2001). Another interesting finding is that fungal isolates belonging to the same species but derived from distant geographic areas shared a similar situation in terms of bacterial number, shape and 16S rDNA sequence. In *Scutellospora*, both species investigated contained rod-shaped endobacteria that amplified with the BLO primers (Hosny et al., 1999; Bianciotto et

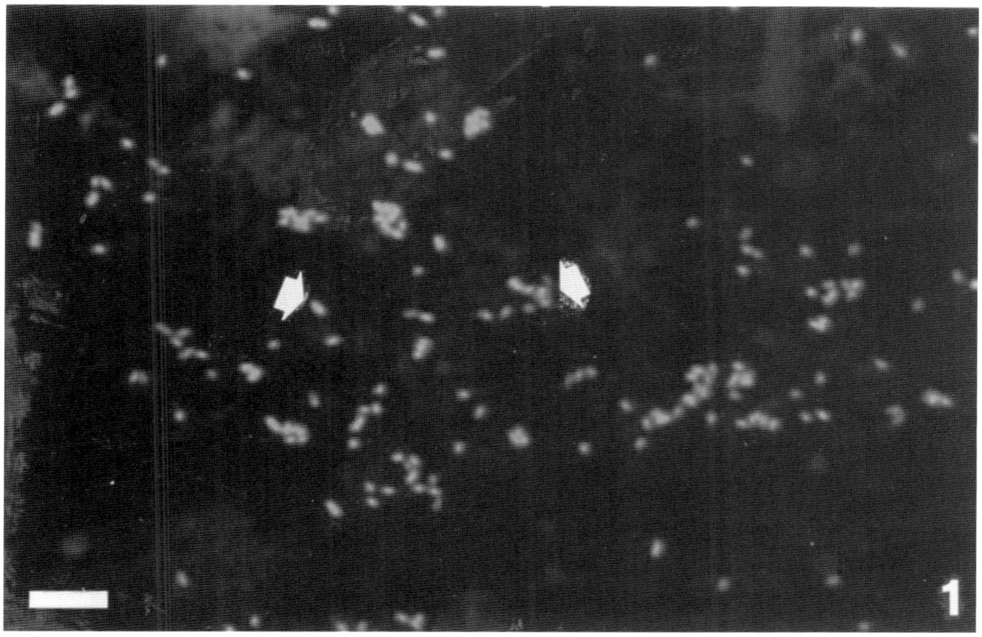

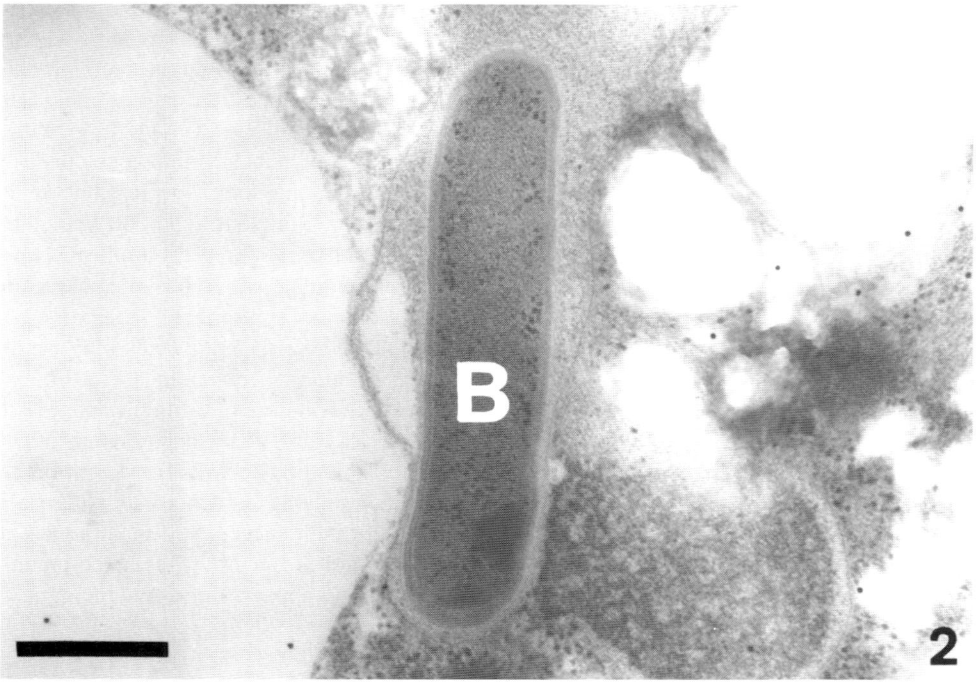

Figure 1. Numerous bacterial endosymbionts in the cytoplasm of a manually crushed spore of the *Gi. margarita* BEG 34 stained with the Bacteria Counting Kit (Molecular Probes) and observed under an Olympus FluoView confocal microscope. Bacteria fluoresce bright green under blue light (arrows) Bar: 7 μm. *Figure 2.* Transmission electron micrograph of a rod-shaped bacterium (B) contained with a layered cell wall in the cytoplasm of *Gi. margarita* BEG 34. Bar: 0.25 μm.

al., 2000), but the analysis of a wider range of species is needed to elucidate how common endobacteria are in this genus, since it comprises almost thirty species. To confirm the topological position of the fungal endobacteria, protocols for *in situ* hybridization, coupled with the use of specific probes, were set up (Bianciotto et al., 2000). They successfully identified bacterial cells inside the fungal cytoplasm of two isolates of *Gi. margarita,* mostly when they were grouped and close to the lipid bodies present in the spores. These experiments provide a confirmation of the taxonomic identity of the endobacteria in the fungal cytoplasm. Margulis and Chapman (1998) recently stated that "Symbiosis is a general term for the cohabitation of organisms of different species, whereas endosymbiosis is a topological condition: one organism lives inside another". Our findings clearly demonstrate that some AM fungi, which are obligate endosymbionts of many plants, are a good example of a further endosymbiosis, since they too host a true bacterial population.

The taxonomic position of the endobacteria of Gigasporaceae

Endobacteria living inside *Gi. margarita* WV205A-5 and in two isolates of *S. persica* clustered together with the *Gi. margarita* BEG 34 endosymbiont, that was originally classified as a sister group of *Burkholderia cepacia* (Bianciotto et al., 2000). The taxonomic rearrangement of *Burkholderia* into two branches, its relationships with the related genus *Ralstonia* and the identification of novel species of *Burkholderia*, have raised basic questions concerning the taxonomic position of the endosymbiotic bacteria of Gigasporaceae. The tree obtained from the alignment of the 16S rDNA suggests that the closest relatives of these endobacteria are still inside *Burkholderia*, as they form a well supported branch nested in this genus and are well separated from *Ralstonia* (Figure 3).

Hypotheses on the establishment of symbiosis between AM fungi and their endobacteria

Intracellular symbioses raise fascinating questions concerning the acquisition of the endosymbionts, their transmission, and the evolution of partner adaptations (Margulis and Chapman, 1998). The presence of endobacteria, or BLOs, in all glomalean families suggests that the ability of AM fungi to establish this type of association appeared very early in evolution. Margulis and Chapman (1998) discuss the importance of endosymbiosis as an evolutionary mechanism and distinguish between permanent and cyclical endosymbioses, the former remaining stable over time whilst the latter involving regular re-association events. In this respect, the type of relationship between AM fungi and their endobacteria remains an open question that will be more properly addressed with the analysis of a wider range of species and isolates. Closely related endobacteria were found in isolates of *Gi. margarita* from distant geographic areas and formed a homogeneous bacterial population in the fungal spores. This situation, which reflects the establishment of a stable bacterial population in the fungal cytoplasm, may be the result of very rare events of bacterial acquisition during evolution, followed by a strictly vertical transmission of the endosymbionts (permanent symbiosis). The asexual reproduction typical of AM fungi, and the coenocytic nature of the mycelium in Zygomycetes are factors that could facilitate this type of transmission. The situation in *Gi. rosea* would suggest that bacteria cannot only be acquired during evolution, but also be lost by the AM fungus.

An alternative and opposite scenario can be also envisaged. The complex situation observed in Gigasporaceae could derive from more temporary but frequent associations of AM fungi with free-living soil bacteria (cyclical symbiosis), with different fungal species selecting their specific bacterial symbionts from the environment. The cell wall that surrounds the fungal hypha represents a physical constraint that prevents endocytosis. Thus the acquisition of bacteria from the environment is a very rare event in fungi. However, Zygomycetes may represent a special case among fungi since *Geosiphon pyriforme* offers a known case of a fungus able to establish cyclical endosymbiotic associations with cyanobacteria (Schüßler et al., 1996). However, on the basis of our accumulating evidence, showing bacteria inside spores through successive fungal generations (V. Bianciotto and G. Becard, personal communication), the first scenario appears more convincing.

What is known about the genome of *Burkholderia*?

Recent developments in microbial genomics have created the exciting possibility of fully cataloging the functional capabilities of symbionts and recognizing the characteristic features of 'resident' genomes: re-

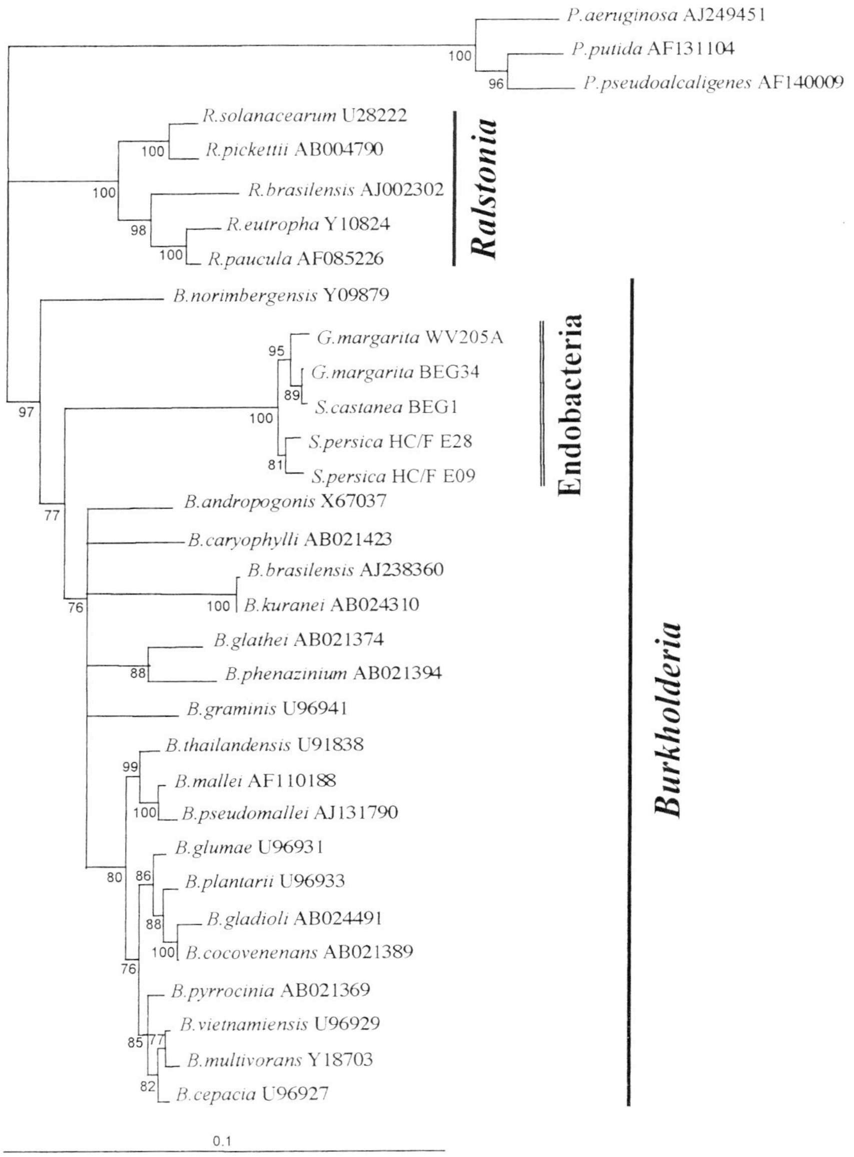

Figure 3. Neighbour-Joining tree obtained from the sequences of the 16S rDNA of the endosymbionts of *Gi. margarita*, *S. castanea*, and *S. persica* isolates and the closest bacterial sequences retrieved by BLAST search. Sequences were aligned by using ClustalX and the alignment was edited with GeneDoc. Neighbor-Joining analysis was performed with ClustalX using Kimura's distance method. The branch comprising species in the genus *Pseudomonas* was used as an outgroup. Branches are shown only when the percentage of bootstrap support (out of 1000 trials) exceeds 70% (from Bianciotto et al., 2000 by permission of American Society for Microbiology, Journal Department).

duced size, adenine-thymine (AT) bias, fast sequence evolution and frequent gene rearrangements (Moran and Wernegreen, 2000). The full genome sequence of *Buchnera*, the bacterial symbionts of pea aphids (Shigenobu et al., 2000), confirms these hypotheses demonstrating that *Buchnera* is an obligate symbiont viable only in its limited niche, the bacteriocyte. As far as *Burkholderia* are concerned, attempts to isolate and grow them from *Gi. margarita* spores have proved unsuccessful, suggesting they are obligate symbionts (V. Bianciotto and E. Lumini, personal communication).

To overcome the problems of a biochemical approach to endobacterial functions, a more feasible genetic approach has been used to identify prokaryotic genes that may be related to specific biological processes or metabolic pathways. A genomic library constructed with the total DNA from spores of *Gi. margarita* (BEG 34), which was demonstrated to be representative of the bacterial genome (van Buuren et al., 1999) was screened with prokaryotic heterologous probes. Genes involved in the colonization process of eukaryotic cells and in the acquisition of nutrients were investigated by using a targeted approach. A gene identified as a *vacB* gene was isolated from the genomic library and characterized (Ruiz-Lozano and Bonfante, 2000). Although VacB was first defined as a virulence factor, it is currently known to be an exoribonuclease RNase R involved in the posttranscriptional processing of mRNAs. It enables the production of the Vir proteins encoded by *vir*G, *ipa*, region-3, region-4, and region-5 operons, and thus modulates the ability of bacteria to adhere to and to penetrate cells, and later to spread with full virulence. Specific primers designed on the nucleotide sequence successfully amplified a PCR fragment of the expected size from DNA extracted from *Gi. margarita* and *S. persica* spores. No amplification was obtained from two isolates of *Gi. rosea*, which are devoid of intracellular bacteria. Endosymbiotic *Burkholderia* therefore seems to possess the molecular determinants required for the colonization of an eukaryotic cell. Interestingly, a corresponding DNA fragment was not amplified from any rhizospheric *Burkholderia* isolates (Ruiz-Lozano and Bonfante, 2000).

AM fungi mostly assist their host plant by providing phosphate (Harrison, 1999). To investigate whether *Burkholderia* can in some way influence P metabolism in AMs, degenerated oligonucleotide primers were designed on conserved regions of a gene coding for a subunit of a bacterial phosphate transporter. A DNA fragment corresponding to a P transporter subunit was found; the insert was used as a probe to screen the genomic library from *Gi. margarita*, leading to the isolation and characterization of the complete operon coding for a high-affinity P transporter (Ruiz-Lozano and Bonfante, 1999). The organization of the *Burkholderia* operon (e.g., gene order, direction of transcription) is the same as in *E. coli* and similar to many other bacteria, thus confirming the notion that this type of transporter is highly conserved. In conclusion, this operon provides the bacteria with the potential for taking up P from the fungal cytoplasm for their basic metabolism. Although this could have a negative impact on the symbiotic efficiency of the *Gi. margarita* isolate, plants colonized by *Gi margarita* have a greater P content when compared to plants colonized by *Gi. rosea* (Ruiz-Lozano and Bonfante, 2001). These data suggest that the bacterial ability for P uptake from the fungal cytoplasm has no significant or negative effects on P uptake and transport by the fungus. Alternatively, we suggested that the potential consumed P may be outweighed by other compensatory mechanisms which lead to an overall positive effect on the fungal P uptake and transport towards the plant.

A very interesting operon: the *nif* operon

The genus *Burkholderia* is characterised by different nutritional strategies and includes members that may be pathogens on plants and humans, saprotrophs, and free-living bacteria. *Burkholderia* associated with the rhizosphere of rice (Gillis et al., 1995) and isolated from banana and pineapple (Margalhaes-Cruz et al., 2001) have been shown to fix nitrogen. Isolates from tomato roots display biocontrol potential on pathogenic fungi and bacteria (Bevivino et al., 1994).

An investigation was conducted to determine if the symbiotic *Burkholderia* spp. possess the nitrogen fixation genes and whether they are expressed. Screening of the genomic library of *Gi. margarita* with *Azospirillum brasilense nif*HDK genes as the probes led to the identification of a 6413-bp region. *Nif*D and *nif*K genes encode for component I (FeMo protein or dinitrogenase), *nif*H encode for component II (Fe protein or dinitrogenase reductase) of the nitrogenase, the enzyme responsible for nitrogen fixation. The analysis of the entire nucleotide sequence revealed the presence of three open reading frames (ORFs) of 900, 1437 and 1557 bp. After tBLASTn and BLASTP searches, the putative amino acid sequences showed very low E-values with the NifH and NifD-NifK Fe-Mo proteins from bacterial and archaeal diazotrophs. The three ORFs are arranged in an operon and their transcription starts at the – 24–12 promoter upstream from the *nif*H gene (Minerdi et al., 2001). RT-PCR experiments — performed on the RNA extracted from non-germinated and germinated spores of *Gi. margarita* — led to an amplified product of the expected size exclusively on RNA extracted from germinating spores (Minerdi et al., 2001).

Phylogenetic trees of the NifK amino acid sequences from the endosymbiont *Burkholderia* and *B. vietnamiensis* TVV75 demonstrated that they belong to a clearly separated group, which also includes *Azospirillum brasilense, Parasponia rhizobium, Bradyrhizobium japonicum*, and *Thiobacillus ferrooxidans*. However, the endosymbiont, *Burkholderia*, was placed closer to *A. brasilense* than to *B. vietnamsiensis* strain TVV75 which, in turn, is closer to *B. japonicum* (Minerdi et al., 2001). This apparent paradox (the endosymbiont *Burkholderia* is more related to *A. brasilense* than to *B. vietnamiensis* as far as *nif* gene sequence similarity is concerned) may be explained by horizontal gene transfer. Hurek et al. (1997) found a high divergence of *nif*H gene in the genus *Azoarcus*, which is responsible for the striking deviation from the ribosomal gene-based phylogeny. The divergence between the two *Burkholderia* NifK sequences might also be the consequence of a horizontal gene transfer, which occurred between micro-organisms belonging to distant taxa (Arber, 1995).

The finding of *nif*HDK genes in the endosymbiont *Burkholderia* and their RNA messengers at least during the germination phase of the spores demonstrates that an important part of the molecular mechanisms required for N_2 fixation is accomplished. There is, therefore, strong evidence in favour of the hypothesis that a fungus, which improves P-uptake, might also fix nitrogen through a specialized bacterial population. This also raises questions about the presence and functionality of putative Nif proteins in the different phases of the fungal life cycle. Biochemical (i.e., Western and immunoblotting) and physiological studies (i.e., acetylene reduction assay) will be required to clarify these points.

Genes identified in the endosymbiotic *Burkholderia* genome using a non-targeted approach

The random sequencing method is one of the first approaches to obtain information on the whole microbe genome. This approach was followed in preliminary experiments to sequence a *Burkholderia* DNA region of about 15K bp: it revealed the presence of ORFs coding 'orphan' proteins (no sequence similarity with proteins available in databases) and putative proteins showing similarity with other bacterial proteins. As shown in Table 1, we found proteins involved in chemotaxis (CheY), signal transduction pathway (McpA), kinases (PrkA), host-specificity proteins (bacteriophage lambda protein J), sporulation proteins (SpoVR) and proteins of unknown functions (Table 1). While the first two hypothetical proteins raise some interesting questions on the capacity of *Burkholderia* to sense the nutrient source (Minerdi et al., 2002), the others suggest the presence of proteins involved in basal cell metabolism and taxis. Interestingly the proteins with no sequence similarity to databases proteins do not show any homology with the ORFs identified in the *Buchnera* genome, which together with *Ricketsia*, represents the only fully sequenced genomes of uncultivable microbes (Shigenobu et al., 2000).

Furthermore, the screening of the library with *A. brasilense his* (histidine biosynthesis) genes did not lead to any positive results in the endosymbiotic *Burkholderia*. This is in contrast with what was found in *Buchnera* where the bacteria provides the host with this essential amino acid (Moran and Wernegreen, 2000).

True bacteria as endosymbionts in ectomycorrhizal fungi?

Associations between mycorrhizal fungi and soil bacteria, particularly plant growth-promoting rhizobacteria (PGPR) such as rhizobia, pseudomonads and *Azospirillum*, have profound effects on plant health through beneficial synergisms. Garbaye (1994) has elegantly shown how bacterial strains increase a root's ability to establish an ectomycorrhizal symbiosis. He has also proposed a new bacterial category to describe this effect: the mycorrhization helper bacteria (MHB). Various bacterial populations have been isolated from truffles (Bedini et al., 1999) and described as microbes living at the ascoma surface as epibionts or among hyphae, as ultrastructurally seen. Bacterial sequences related to the *Cytophaga–Flexibacter–Bacteroides* phylogroup were recently found in the mycelium of *Tuber borchii* (Barbieri et al., 2000) and suggested that they belong to a bacterium associated with this ectomycorrhizal fungus. A. Mello (personal communication) found similar sequences during an investigation on another truffle species, *T. uncinatum*. However, true bacteria or bacteria-like organisms were never found inside the truffle hyphae when these were examined ultrastructurally. The presence of bacteria living inside- or associated to ectomycorrhizal fungi therefore requires further investigation.

Table 1. Genes identified in the endosymbiotic *Burkholderia* genome using a non-targeted approach

ORF	Function	Length (aa)	E-value
1347 (complete)	Methyl-accepting chemotaxis protein McpA	449	2e-61
477 (complete)	Chemotaxis protein CheY	159	1e-14
1947 (complete)	Hypothetical 74.5 kDa protein with strong similarity to PrkA protein	649	0.0
924 (complete)	Hypothetical 49.4 kDa protein in GAPA-RND intergenic region of *E.coli* with similarity to *B. subtilis* YhbH protein	308	7e-52
654 (partial sequence)	Host-specificity protein J of bacteriophage lambda	218	e-131
1296 (partial sequence)	Hypothetical 60.8 kDa protein with similarity to SpoVR protein	432	2e-84
750 (complete)	Hypothetical 22.7 kDa protein with similarity to integral membrane protein	250	1e-11
495 (complete)	No homology with known proteins	165	
423 (complete)	No homology with known proteins	141	
543 (complete)	No homology with known proteins	181	
585 (complete)	No homology with known proteins	195	
444 (complete)	No homology with known proteins	148	
597 (complete)	No homology with known proteins	199	
315 (complete)	No homology with known proteins	105	
441 (complete)	No homology with known proteins	147	
1665 (complete)	No homology with known proteins	555	
543 (complete)	No homology with known proteins	181	

Conclusions

Molecular and morphological analyses demonstrate that some microbes select mycorrhizal fungi as a special niche for accomplishing their life cycle. Microbes, such as the endosymbiont *Burkholderia*, possess genes which may have an impact on bacterial, fungal and plant metabolisms. Recent developments in microbial genomics will help us to catalogue the bacterial functional capabilities and to better understand interactions with the main host, which is another obligate biotroph. The finding of a *nif* operon suggests new applicative scenarios, since the potential capacity of a mycorrhizal fungus to fix nitrogen through a specific bacterial population provides great interest for sustainable agriculture. The natural system consisting of mycorrhizal fungi and nitrogen-fixing bacteria might be an excellent 'biofertilizer' with which to expand crop production and minimize the negative impact of chemical fertilizers.

Acknowledgement

We wish to thank Maria Teresa Della Beffa for the reference list. The research illustrated in this review has been funded by the EU GENOMYCA project, QLK5-CT-2000-01319, by the Italian National Council of Research, and by the National Project Produzione Agricola nella Difesa dell'Ambiente (PANDA).

References

Arber W 1995 The generation of variation in bacterial genomes. J. Mol. Evol. 40, 7–12.

Barbieri E, Potenza L, Rossi I, Sisti D, Giomaro G, Rossetti S, Beimfohr C and Stocchi V 2000 Phylogenetic characterization and *in situ* detection of a *Cytophaga–Flexibacter–Bacteriodes* phylogroup bacterium in *Tuber borchii* Vittad. ectomycorrhizal mycelium. Appl. Environ. Microbiol. 66, 5035–5042.

Bedini S, Bagnoli g, Sbrana C, Leporini C, Tola E, Dunne C, Filippi C, D'Andrea F, O'Gara F and Nuti M P 1999 Pseudomonads isolated from within fruitbodies of *T. borchii* are capable of producing biological control or phytostimulatory compounds in pure culture. Symbiosis 26, 223–236.

Bevivino A, Tabacchioni S, Chiarini L, Carusi M V, Del Gallo M and Visca P 1994 Phenotypic comparison between rhizosphere and clinical isolates of *Burkholderia cepacia*. Microbiology 140, 1069–1077.

Bianciotto V, Bandi C, Minerdi D, Sironi M, Tichy H V and Bonfante P 1996 An obligately endosymbiotic fungus itself harbors obligately intracellular bacteria. Appl. Environ. Microbiol. 62, 3005–3010.

Bianciotto V, Lumini E, Lanfranco L, Minerdi D, Bonfante P and Perotto S 2000 Detection and identification of bacterial endosymbionts in arbuscular mycorrhizal fungi belonging to Gigasporaceae. Appl. Environ. Microbiol. 66, 4503–4509.

Bianciotto V, Andreotti S, Balestrini R, Bonfante P and Perotto S 2001 Mucoid mutants of the biocontrol strain *Pseudomonas fluorescens* CHA0 show increased ability in biofilm formation on mycorrhizal and nonmycorrhizal carrot roots. Mol. Plant Microbe Interact. 14, 255–260.

Bonfante P, Balestrini R and Mendgen K 1994 Storage and secretion processes in the spore of *Gigaspora margarita* Becker & Hall as revealed by high-pressure freezing and freeze sustitution. New Phytol. 128, 93–101.

Douglas A E 1994 Symbiotic Interactions. Oxford University Press, New York, USA. 148 p.

Garbaye J 1994 Helper bacteria: a new dimension to the mycorrhizal symbiosis. New Phytol. 128, 197–209.

Gillis M, Van Van T, Bardin R, Goor M, Hebbar P, Segers P, Kerster K, Heulin T and Fernandez M P 1995 Polyphasic taxonomy in the genus *Burkholderia* leading to an emended description of the genus and proposition of *Burkholderia vietnamensis* sp. nov. for N_2-fixing isolates from rice in Vietnam. Int. J. Syst. Bacteriol. 45, 274–289.

Harrison M J 1999 Molecular and cellular aspects of the arbuscular mycorrhizal symbiosis. Annu. Rev. Plant Physiol. Plant Mol. Biol. 50, 361–389.

Hosny M, van Tuinen D, Jacquin F, Fuller P, Zhao B, Gianinazzi-Pearson V and Franken P 1999 Arbuscular mycorrhizal fungi and bacteria: how to construct prokaryotic DNA-free genomic libraries from the Glomales. FEMS Microbiol. Lett. 170, 425–430.

Hurek T, Egener T and Reinhold-Hurek B 1997 Divergence in nitrogenases of *Azoarcus* spp; proteobacteria of the β subclass. J. Bacteriol. 179, 4172–4178.

Kowalchuk G A 1999 New perspectives towards analysing fungal communities. Curr. Opin. Biotechnol. 10, 247–251.

Lanfranco L, Bianciotto V, Lumini E, Souza M, Morton J B and Bonfante P 2001 A combined morphological and molecular approach to characterize isolates of arbuscular mycorrhizal fungi in *Gigaspora* (Glomales). New Phytol. 152, 169–179.

Margulis L 1993 Symbiosis in Cell Evolution (2nd edition). Freeman, San Francisco, USA.

Margulis L and Chapman M J 1998 Endosymbioses: cyclical and permanent in evolution. Trends Microbiol. 6, 342–345.

Magalhaes Cruz L, Maltempi De Souza E, Weber O B, Baldani J I, Dobereiner J and de Oliveira Pedrosa F 2001 16S ribosomal DNA characterization of nitrogen-fixing bacteria isolated from banana (*Musa* spp.) and pineapple (*Ananas comosus* (L.) Merril). Appl. Environ. Microbiol. 67, 2375-2379.

Minerdi D, Fani R, Gallo R, Boarino A and Bonfante P 2001 Nitrogen fixation genes in an endosymbiotic *Burkholderia* strain. Appl. Environ. Microbiol. 67, 725–732.

Minerdi D, Fani R and Bonfante P 2002 Identification and evolutionary analysis of putative cytoplasmic McpA-like protein in a bacterial strain living in symbiosis with amycorrhizal fungus. Journal Molecular Evolution, in press.

Mogge B, Loferer C, Agerer R, Hutzler P and Hartmann A 2000 Bacterial community structure and colonization patterns of *Fagus sylvatica*. Ectomycorrhizospheres as determined by fluorescence *in situ* hybridization and confocal laser scanning microscopy. Mycorrhiza 5, 271–278.

Moran N A and Wernegreen J J 2000 Lifestyle evolution in symbiotic bacteria: insights from genomics. Tree 15, 321–326.

Mosse B 1970 Honey coloured sessile *Endogone* spores. Changes in fine structure during spore development. Arch. Microbiol. 74, 146–159.

Ruiz-Lozano J M and Bonfante P 1999 Identification of a putative P-transporter operon in the genome of a *Burkholderia* strain living inside the arbuscular mycorrhizal fungus *Gigaspora margarita*. J. Bacteriol. 181, 4106–4109.

Ruiz-Lozano J M and Bonfante P 2000 Intracellular *Burkholderia* of the arbuscular mycorrhizal fungus *Gigaspora margarita* possesses the *vacB* gene, which is involved in host cell colonization by bacteria. Microbial Ecol. 39, 137–144.

Ruiz-Lozano J M and Bonfante P 2001 Intracellular *Burkholderia* strain has no negative effect on the symbiotic efficiency of the arbuscular mycorrhizal fungus *Gigaspora margarita*. Plant Growth Regulation, 34, 347–352.

Scannerini S and Bonfante P 1991 Bacteria and bacteria like objects in endomycorrhizal fungi (Glomaceae). *In* Symbiosis as Source of Evolutionary Innovation: Speciation and Morphogenesis. Eds. L Margulis and R Fester. pp 273–287. MIT Press, Cambridge, MA, USA.

Schüßler A, Bonfante P, Schnepf E, Mollenhaurer D and Kluge m 1996 Characterization of the *Geosiphon pyriforme* symbiosome by affinity techniques: confocal laser scanning microscopy (CLSM) and electron microscopy. Protoplasma 190, 53–67.

Shigenobu S, Watanabe H, Hattori M, Sakaki Y and Ishikawa H 2000 Genome sequence of the endocellular bacterial symbiont of aphids. Nature 407, 81–86.

Smith S E and Read D J 1997 Mycorrhizal Symbiosis. Academic Press, New York, p. 605.

van Buuren M, Lanfranco L, Longato S, Minerdi D, Harrison m J and Bonfante P 1999 Construction and characterization of genomic libraries of two endomycorrhizal fungi: *Glomus versiforme* and *Gigaspora margarita*. Mycol. Res. 103, 955–960.

Plant and Soil **244:** 221–230, 2002.

P metabolism and transport in AM fungi

Tatsuhiro Ezawa[1,3], Sally E. Smith[2] & F. Andrew Smith[2]
[1]*University Farm, Nagoya University, Togo-cho, Aichi 470-0151 Japan.* [2]*Department of Soil and Water and Centre for Plant Root Symbioses, Waite, Campus, Adelaide University, Glen Osmond, South Australia, 5064, Australia*
[3]*Corresponding author**

Received 21 August 2001. Accepted in revised form 15 January 2002

Key words: arbuscular mycorrhizal fungi, P metabolism, polyphosphate, P transport, V-ATPase, vacuoles

Abstract

The arbuscular mycorrhizal symbiosis is mutualistic, based on reciprocal transfer of P from the fungus to the plant and carbon from the plant to the fungus. Thus P is a most important 'currency' in the symbiosis. After absorbing P from the soil solution, the fungi first incorporate it into the cytosolic pool, and the excess P is transferred to the vacuoles. The vacuolar P pool probably plays a central role in P supply to the plant. The main forms of inorganic P in fungal vacuoles are orthophosphate and polyphosphate, but organic P molecules may also be present. Long distance translocation of P from the site of uptake in the external mycelium to the site of transfer to the plant is probably achieved via transfer of vacuolar components. This transport would be mediated either by protoplasmic streaming or the motile tubular vacuole-like system. The site of release of P into the interfacial apoplast and thence to the plant is most probably the fungal arbuscules. The biochemical and biophysical processes involved in P metabolism and transfer between cellular compartments in the symbiosis are currently not well understood. Some recent investigations of substrate specificities of phosphatase-type enzymes in AM fungi and other eukaryotic microorganisms, however, have shed new light on earlier results and permit the construction of a hypothetical scheme of P-flow, including possible regulatory factors. Steps in this scheme are experimentally testable and should stimulate future research.

Introduction

Arbuscular mycorrhizal (AM) fungi colonize the root cortex of the host plant in which the fungi are able to acquire organic C to build 'the infrastructure' for P uptake and transport. The mycorrhizal system is able to take up P more efficiently and transport P over longer distances than the plant root system, overcoming P depletion in the rhizosphere (Zhu et al., 2001). Accordingly, mechanisms of P metabolism and transport are of great interest, not only from the view-point of basic biology, but also from that of applied science. However, biochemical processes in the fungi are difficult to investigate because the fungi are obligate biotrophs. Targeted molecular biological approaches are particularly important to unravel the nature of AM fungi, including information on the expression of genes encoding proteins involved in P uptake, turnover, transport and transfer to the plant. However, for this strategy we need to be in possession of the necessary 'targets'; i.e. we need a testable model of possible pathways of P metabolism. In this review, we attempt to integrate the latest information on P metabolism and transport in AM fungi with that for other eukaryotes, to give a whole picture of the system in order to stimulate future targeted approaches.

* FAX No: +81-5613-8-4473. E-mail: tatsu@agr.nagoya-u.ac.jp

Fungal P uptake, compartmentation and P pools

There is a considerable concentration gradient of inorganic orthophosphate (Pi) between the soil solution and plant/fungal cytoplasm. In the former concentrations rarely exceed 10 μM, whereas in plants the cytoplasmic concentration is probably around 10 mM (Mimura, 1999). To absorb Pi under these conditions, plants and soil microorganisms have developed specialized, energy dependent systems for Pi uptake. Both high-affinity (low K_m and V_{max}) and low-affinity (high K_m and V_{max}) transport systems have been identified and include H^+/Pi symporters, with the necessary proton gradient produced by plasma membrane H^+-ATPases (P-ATPases). Typical K_m values of high–affinity systems in plants (Mimura, 1999) and fungi (Beever and Burns, 1980) are in the range from 1 to 10 μM, and it might be predicted that those of AM fungi would be within this range. However, values of the K_ms in AM fungi vary considerably, depending on the methods used. Thomson et al. (1990) showed that for germ tubes of *Gigaspora margarita*, the K_m of the high-affinity and low-affinity systems were 1.8–3.1 μM and 10.2–11.3 mM, respectively. However, a K_m value of the high-affinity system of extraradical hyphae of AM fungi estimated from physiological–P uptake data in a soil culture system was about one order of magnitude lower, 0.17 μM (Schweiger and Jakobsen, 1999) and that of *GvPT*, a high-affinity Pi transporter expressed in the external mycelium of the AM fungus *Glomus versiforme* was much higher, 18 μM (Harrison and van Buuren, 1995). The high value may be an artefact of the yeast complementation system used to test Pi transporter function for *GvPT* (see Versaw et al., 2002). Additional information on the Pi uptake system can be derived from measurements of the transmembrane potential difference (PD) of germ tubes and extraradical hyphae. Values are rather low [around −47 mV, (Ayling et al., 2000)], compared with other organisms, but using this value and an external Pi concentration of 35 μM gives an $H^+/H_2PO_4^-$ stoichiometry of 2.33. This is higher than the most generally accepted ratio of 2.0, but within the range of published values. However, the much lower K_m for hyphal uptake from soil (c. 0.2 μM; see above) poses energetic problems for this stoichiometry because it shows that there is Pi uptake against very large electrochemical potential gradients.

Another high-affinity system, a Na^+-coupled Pi transporter, is known from animals, bacteria (Reizer et al., 1994) and plants (Reid et al., 2000) and has also been found in eukaryotic microorganisms including yeast, *Saccharomyces cerevisiae,* and the filamentous fungus, *Neurospora crassa*. These two fungi possess both H^+/Pi and Na^+/Pi co-transport systems, and the genes encoding these transporters have been identified (Martinez and Persson, 1998; Versaw and Metzenberg, 1995). In the case of yeast, both the H^+/Pi and Na^+/Pi transporter genes are controlled in the PHO pathway (Martinez and Persson, 1998) which promotes gene transcription in response to Pi deficiency (Lenburg and O'Shea, 1996). The physiological role of these transporters can be interpreted from their different pH optima: the H^+/Pi system operates optimally at neutral pH and the Na^+/Pi system at alkaline pH, suggesting that the organisms can take up Pi efficiently with the two systems over a broad pH range (Martinez and Persson, 1998). It is possible that AM fungi possess a Na^+/Pi transport system as well as the H^+/Pi system, which would contribute to Pi acquisition from alkaline soil.

Following uptake, Pi is first incorporated into the cytosolic Pi pool with the concentration kept constant to maintain various cell functions, such as energy generation and biosynthesis of phospholipid, nucleic acid and precursors of carbohydrate polymers such as UDP-glucose. If the fungi are able to take up more Pi than the amount required for maintenance of cell functions, the excess is transported into the vacuoles, effectively buffering the cytosolic Pi concentration (Klionsky et al., 1990; Shirahama et al., 1996). It has been demonstrated that Pi transport into plant vacuoles depends on an electrical potential gradient (not a chemical gradient for H^+) created by a vacuolar H^+-ATPase (V-ATPase) (Massonneau et al., 2000). In the case of yeast, Booth and Guidotti (1997) demonstrated that isolated vacuoles took up a large amount of Pi by exchange with Pi present in the vacuoles i.e. the Pi uptake was driven by the counterflow of Pi. Indeed, the Pi uptake was considerably reduced in the absence of Pi in the vacuoles. This could not be an acceptable mechanism for net influx unless the stoichiometry was other than 1:1. On the other hand, Kulakovskaya and Kulaev (1997) observed unidirectional transport of Pi into isolated yeast vacuoles, with uptake showing linear kinetics with respect to external Pi concentration up to 10 mM. Both the systems found in yeast were insensitive to protonophoric uncouplers, indicating that the systems were independent of the H^+ gradient. However, it is uncertain if membrane potential, the electric component of the electrochemical gradient created by V-ATPase, is involved in the fungal vacuolar transport, and the transport mechanism is unresolved. In fungi, it may be important that polyphosphate synthesis (described in the next section) coincides with Pi uptake during P accumulation in the vacuole to prevent large-scale efflux. In the case of AM fungi, this vacuolar pool may play a central role in P supply to the host plant.

P metabolism in vacuoles

The vacuole is a most fascinating organelle in the context of symbiotic P metabolism and transport in AM fungi, whereas most biochemical reactions occurring in the cytosol may be essential for survival/growth of AM fungi and occur in common with other fungi, the vacuoles could, in addition, be organized as unique regulatory systems, important for the establishment of the mutualistic relationship with plants.

Inorganic P

Inorganic P is a major component of the vacuolar P pool of AM fungi, with contributions from both Pi and polyphosphate (polyP). PolyP was first visualized in the vacuole as electron-dense and P-rich 'granules' and suggested to be the main form of P transport over long distances in these fungi (Callow et al., 1978; Cox et al., 1980). PolyP is a linear polymer of Pi residues linked by high-energy phosphoanhydride bonds. Although various cellular functions have been suggested (Kornberg et al., 1999), the most important role in fungi is probably as a temporary storage of Pi. The chain-length of polyP in AM fungi is variable (Ezawa et al., 1999), being longer in extraradical than in intraradical hyphae (Solaiman et al., 1999). This may reflect differences in polyP metabolism between the two phases (Ezawa et al., 2001a). Both soluble and insoluble (granular) forms of polyP have been found in nature (Clark et al., 1986). The status of granular polyP in an ectomycorrhizal fungus, *Pisolithus tinctorius* has been questioned. It has been suggested that the granules seen by electron microscopy in the fungus are an artefact produced during the fixation of tissue and that polyP is always dispersed throughout the vacuoles in soluble form (Orlovich and Ashford, 1993). However, Bücking and Heyser (1999) have observed granular polyP in *P. tinctorius*. The matter remains unresolved. Clark et al. (1986) have established a successive extraction method in which short-chain soluble, long-chain soluble and long-chain granular polyP can be fractionated. Using this method, polyP was shown to occur in both soluble and long-chain granular forms in the intra- and extraradical hyphae of the AM fungus *Gigaspora margarita* (Solaiman et al., 1999). Many factors may affect solubility of polyP *in vivo,* including pH, chain-length, concentration of polyP and counter ions such as metal cations and polyamines (Cramer and Davis, 1984). Accordingly, the form of polyP *in vivo* may be quite varied among species and also in the same species grown under different environmental conditions.

PolyP metabolism is best understood from work on prokaryotes, in which many enzymes have been found and characterized (Kornberg et al., 1999). In prokaryotes polyphosphate kinase (PPK) synthesizes polyP using ATP as a phosphoryl donor and also catalyzes a reverse reaction (ATP regeneration from polyP). Polyphosphate glucokinase (PPGK) transfers a terminal Pi residue to glucose and synthesizes glucose 6-phosphate. Exo- and endopolyphosphatases (PPX and PPN) hydrolyze terminal and internal linkages, respectively. The following diagram summarises the catalytic reactions of these enzymes.

Phosphotransferase type

$$\text{PPK: polyP}_{n-1} + \text{ATP} \rightarrow \text{polyP}_n + \text{ADP}$$

$$\text{PPGK: polyP}_n + \text{glucose} \rightarrow \text{polyP}_{n-1} + \text{glucose-6-phosphate}$$

Phosphatase type

$$\text{PPX: polyP}_n + H_2O \rightarrow \text{polyP}_{n-1} + \text{Pi}$$

$$\text{PPN: polyP}_n + H_2O \rightarrow \text{polyP}_{n-m} + \text{polyP}_m \ (m \leq 3)$$

The presence of similar enzymes in eukaryotes and their roles in vacuolar P metabolism is not well understood. Capaccio and Callow (1982), pioneer researchers on the polyP metabolism of AM fungi, showed the activities of all of the above enzymes in *G. mosseae*. However, PPK and PPGK have not been demonstrated in any other eukaryotic microorganisms (Booth and Guidotti, 1995; Kornberg et al., 1999). Recently, we attempted to repeat the work of Capaccio and Callow (1982), but failed to detect either PPK (Ezawa unpublished results) or PPGK (Ezawa et al., 2001b) activity in the AM fungi *Glomus coronatum* and *G. etunicatum*. It is mysterious that PPK, the only enzyme which is involved in polyP synthesis, is apparently missing in the eukaryotes investigated, even in yeast which accumulates polyP up to 10–20% of the dry weight (Kornberg et al., 1999). However, a homologue of the bacterial PPK gene has been found in the slime mould, *Dictyostelium*. The gene was responsible for antifungal drug-resistance and the actual function of the gene product in poly*P* metabolism is yet to be characterized (Kornberg et al., 1999). Ogawa et al. (2000) found four genes involved in polyP accumulation in yeast and hypothesized that polyP might be synthesized through a system analogous to the mitochondrial (F-type) H^+-ATPase, which generates ATP by mechanism energized by the proton gradient. According to their observations, a protein complex translated from the genes, which was apparently different from V–ATPase, might be localized on the yeast tonoplast and be responsible for polyP synthesis, energized in this case by a proton gradient built by V-ATPase. PolyP in AM fungi could be synthesized by a similar system.

The presence of several PPXs has been found for different cell compartments of yeast: the cell envelope (Andreeva and Okorokov, 1993), cytosol (Andreeva et al., 1998a; Wurst and Kornberg, 1994), vacuoles (Andreeva et al., 1998b), nuclei (Lichko et al., 1996) and mitochondria (Lichko et al., 1998). These enzymes showed different pH optima, substrate (chain-length) specificity and sensitivity to various effectors (Kulaev et al., 1997). The cytosolic PPX has been purified and well characterized. It showed a pH optimum at 7.5, specificity for long-chain polyP and was inhibited by polyamine. The only eukaryotic gene involved in polyP metabolism cloned to date is that encoding this enzyme (Wurst et al., 1995). Two types of PPX activity have been found in the AM fungi, *Glomus coronatum* and *G. etunicatum* (Ezawa et al., 2001a). One had a pH optimum near neutrality and exhibited a low K_m with long-chain polyP, similar to the yeast cytosolic PPX (Andreeva et al., 1998a; Wurst and Kornberg, 1994). The other activity showed an acidic pH optimum and a low K_m with short-chain polyP, similar to that found in *G. mosseae* by Capaccio and Callow (1982). This activity was characterized as non-specific acid phosphatase (ACPase) and may be localized in the vacuoles (see below). PPN has been found in various fungi (Kumble and Kornberg, 1996) including AM fungi (Capaccio and Callow, 1982), but only yeast PPN has been purified from the vacuole and characterized (Kumble and Kornberg, 1996). In this case, the main product of polyP hydrolysis was tripolyphosphate, with an optimum pH of 7.5. In AM fungal symbiosis it seems likely that PPX, ACPase and PPN would work together to hydrolyze polyP into Pi completely.

One interesting but very different hypothesis with respect to PPX is its possible function as a Pi pump (transporter) in vacuoles and lysosomes. Kulaev and Kulakovskaya (2000) proposed that cytosolic Pi is taken up through PPX bound to the tonoplast and incorporated directly into polyP by the enzyme. The hypothesis was based upon the observation that ^{32}P-labled Pi was rapidly incorporated into a long-chain polyP in isolated human fibroplast lysosomes (Pisoni and Lindley, 1992). However, the way in which energy is supplied for the polymerization remains obscure.

Organic P

Regulation of cell function and activity requires tight control of the concentration of metabolites, including various P-esters, and surpluses are transported into vacuoles and metabolized there (Klionsky et al., 1990). Among hydrolase-type enzymes found in fungal vacuoles, phosphatases are responsible for the conversion of P-esters into Pi. Non-specific alkaline and acid phosphatases (ALPase and ACPase, respectively) have been demonstrated in AM fungi by electrophoresis (Fabig et al., 1989; Gianinazzi–Pearson and Gianinazzi, 1978), biochemical (Ezawa et al., 1999), histochemical (Ezawa et al., 1995; Tisserant et al., 1993) and cytochemical (Gianinazzi et al., 1979) studies. Both the ALPase and ACPase have been observed in vacuoles of the fungi (Gianinazzi et al., 1979), and most of the activity of these enzymes was associated with the insoluble fraction (Ezawa et al., 1999), possibly the tonoplast, as also revealed in yeast (Klionsky et al., 1990). ALPase activity has been used as a marker for metabolic activity of the fungi (Guillemin et al., 1995; Tisserant et al., 1993) because the enzyme was highly active in mature arbuscules (e.g. Figure 1A), though the actual role of the enzyme has not been clarified (Larsen et al., 1996). Based on its substrate specificity it is unlikely that the ALPase can hydrolyze pyrophosphate compounds such as polyP and any suggestion of a direct role in polyP mobilization must be questioned. Recently, the involvement of ALPase in sugar metabolism has been suggested (Ezawa et al., 1999). AM fungi take up glucose in the symbiotic phase and metabolize it through glycolysis and/or the pentose phosphate pathway (Shachar-Hill et al., 1995; Solaiman and Saito, 1997), in which glucose 6-phosphate (glucose 6–P) is synthesized at the first step (Bago et al., 2000). Glucose 6-P is a key compound in sugar metabolism and is utilized in biosynthesis of various carbohydrates, including trehalose which is a temporary store of carbon in AM fungi (Bécard et al., 1991; Schubert et al., 1992). Glucose 6-P and trehalose 6-P (an intermediary metabolite in trehalose biosynthesis) were the most favorable substrates for the ALPase of AM fungi (Ezawa et al., 1999). In contrast, the ACPase of AM fungi showed no particular substrate specificity: not only phosphomonoesters

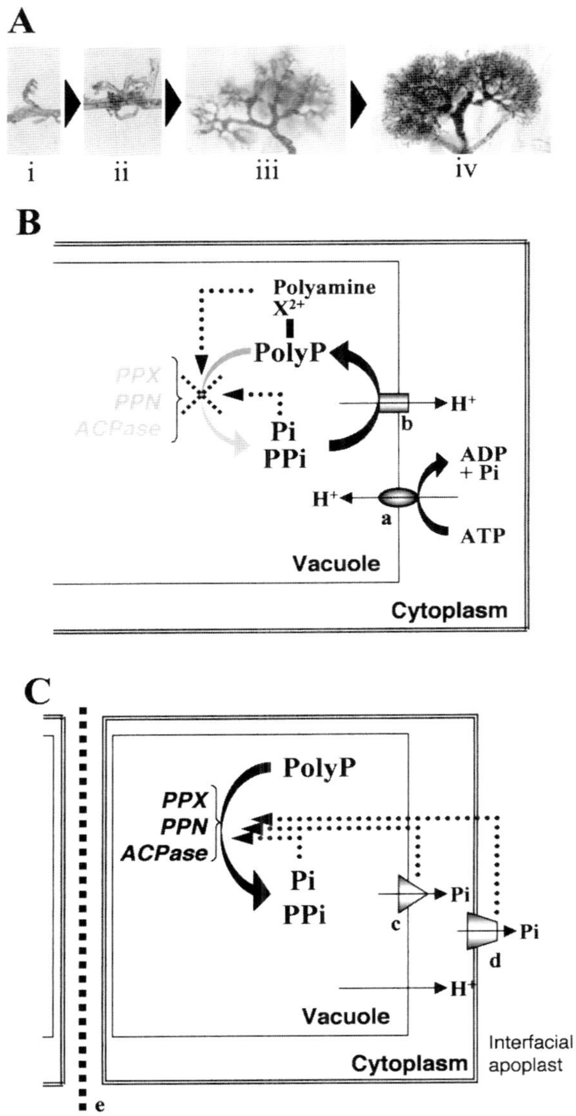

Figure 1. Hypothetical scheme of Pi efflux from arbuscules. Efflux may be regulated at the following three steps, arbuscule formation, polyphosphate hydrolysis and Pi efflux through the membranes (see text). (**A**) Development of an arbuscule from initiation to maturity (i) The first trunk of arbuscule is initiated. (ii) Two major branches are formed. (iii) Fine branches are being formed. (iv) Fully developed (mature) arbuscule showing high activity of alkaline phosphatase. (**B**) Polyphosphate (polyP) metabolism in the initial phase of arbuscule formation. V–ATPase (a) may still be active and able to energize polyP synthesis (b). (**C**) PolyP hydrolysis during maturation of an arbuscule. Inhibition of carbon delivery may inactivate V-ATPase and P-ATPase and net hydrolysis of polyP may occur. The depolarization of membranes due to inactivation of H^+-ATPases may promote Pi efflux through potential difference (PD)-sensitive channels (c, d) on the membranes. In this circumstance, the activity of Pi efflux transporters/channels may be the rate-limiting step. The fungi may minimize the impact by forming a septum (e) in the trunk of each arbuscule. X^{2+}, divalent cations; PPX, exopolyphosphatase; PPN, endopolyphosphatase; ACPase, non-specific acid phosphatase.

but also pyrophosphate compounds can be hydrolyzed. The ALPase and ACPase, which have different pH optima and substrate specificities, may work together to liberate Pi from various P-esters over a wide range of vacuolar pH.

P transport along hyphae

The forms of P involved in long distance transport along the hyphae of AM fungi are probably Pi and polyP, with the possible addition of organic P (ester-P). The size of vacuolar P pools in many organisms appears to respond flexibly to availability of P in the environment, with the vacuoles of plants (Bieleski, 1973) and yeast (Shirahama et al., 1996) acting as a Pi reservoir for the cytoplasm under P deficient conditions. Under relatively P–deficient conditions PolyP contributed between 5 and 17% of the total P in the hyphae of *Gigaspora margarita* (Solaiman et al., 1999). P-deficiency may accelerate Pi turnover in the vacuoles of AM fungi, resulting in low polyP/Pi or organic P/Pi ratios.

Rapid changes in polyP content of AM hyphae in response to P supply have been observed. In *Glomus coronatum*, polyP could not be detected at all in the extraradical hyphae if P was completely withheld one week before harvest, but when mycorrhizal plants were supplied with a solution containing 50 μM Pi about 50% of the hyphae contained polyP (Ezawa et al., 2001a). Furthermore, Boddington and Dodd (1999) reported differences in the amount of polyP occurring in hyphae of different AM fungal species. This interesting observation requires to be confirmed, because the sensitivity and specificity of the stain used to visualize polyP (DAPI) has not yet been evaluated in filamentous fungi.

It has been suggested that the establishment of the symbiotic relationship enables AM fungi to take up and transport Pi in amounts beyond their own requirements, and that the vacuolar system may be responsible for the long distance transport of P along the hyphae. Motile tubular vacuoles have been demonstrated in a range of filamentous fungi (Rees et al., 1994) (see Ashford and Allaway, 2002; Bago et al., 2002). 'The moving vacuoles' can be observed in living cells, most clearly in actively growing young cultures, by loading fluorescent vacuolar tracers. In this system, clusters of vacuoles are interconnected by tubules along which the vacuoles move (Hyde et al., 1997). The tubules are preserved in electron micrographs of freeze-substituted tissue, but not in material fixed by conventional methods. P as well as K is distributed in the tubular vacuoles of the ectomycorrhizal fungus *P. tinctorius,* as revealed by energy-dispersive X-ray analysis (Cole et al., 1998). However, the role of the motile tubular vacuoles have not yet been fully established and we do not know if 'the moving vacuoles' labeled by fluorochrome also have P in the lumen or unique vacuolar enzymes such as V-ATPase. An analogous system to the *Pisolithus* motile tubular vacuoles has been observed in AM fungi (Uetake, personal communication; Dickson, personal communication), but has not yet been fully characterized. However, it is feasible that the system is involved in long distance P transport.

Regulation of arbuscule formation

Arbuscule formation is sensitive to P supply and a low to moderate supply is required. If the P supply is very low, colonization appears to be inhibited (Bolan et al., 1984). Without P fertilization *Glomus intraradices* and *G. coronatum* formed only intercellular hyphae, but not arbuscules in the roots of tomato and mungbean, respectively, and the subsequent application of P triggered arbuscule formation (Rosewarne and Kasiamdari, personal communication). However, nutritionally adequate or high P supply tends to reduce both colonization and arbuscule formation, with the magnitude of the effect varying between plant species and also sensitive to change in environmental parameters such as irradiance. (reviewed in Smith and Read, 1997). These observations indicate that the symbionts, plant and/or fungus, are able to sense the P concentration in the environment and that this may lead to regulation of the expression of genes involved in arbuscule formation. This in turn would have an impact on P transfer to the plant.

Regulation of Pi transfer via arbuscules

The arbuscule has been recognized as the site from which Pi is almost certainly released to the interfacial apoplast and thence absorbed by the plant cortical cell. Supporting evidence includes: (i) ATP-hydrolyzing activity, possibly P-ATPase, was localized on periarbuscular membrane of the host (Gianinazzi–Pearson et al., 1991) and (ii) High-affinity Pi transporters (Rausch et al., 2001; Rosewarne et al., 1999) and P-ATPase (Gianinazzi–Pearson et al., 2000) genes are expressed in arbuscule-containing host cells.These observations suggest that Pi is actively taken up by the plant through the periarbuscular membrane. (iii) ATP-hydrolyzing activity on the fungal plasma membrane of arbuscules is absent (Gianinazzi–Pearson et al., 1991) and (iv) the fungal P transporter gene *GvPT* is not expressed within the roots (Harrison and Van Buuren, 1995). These observations suggest that active nutrient uptake does not occur in the fungal arbuscules.

Although the mechanism of Pi transfer from fungus to plant has not yet been elucidated, it is considered that it may be regulated at the following steps, (i) arbuscule formation; (ii) Pi liberation from the vacuolar P pool; and (iii) efflux through the tonoplast and plasma membrane of AM fungi to the interfacial apoplast. There must be regulatory cross talk between the partners to prevent perturbation in Pi concentration in the fungal mycelium during the Pi transfer phase of arbuscular activity.

Regulation of polyP turnover by V-ATPase

During arbuscule formation, the balance between synthesis and hydrolysis of polyP may be altered via V-ATPase activity. Vacuolar pH, influenced by activity of V-ATPases, may be a regulatory factor in polyP synthesis and hydrolysis. A V-ATPase-defective mutant of yeast is unable to acidify the vacuolar lumen and accumulate polyP (Beauvoit et al., 1991). Bafilomycin A1, a specific inhibitor of V-ATPase, increased vacuolar pH and induced polyP hydrolysis in yeast (Beauvoit et al., 1991). Furthermore, alkalinization of the vacuolar lumen, either by amine or protonophoric uncouplers, induces rapid hydrolysis of polyP in algae (Pick and Weiss, 1991) and yeast (Beauvoit et al., 1991). The rapidity of the response also implies that genes for polyP-hydrolyzing enzymes may be expressed constitutively. PPX/ACP activities in the intraradical hyphae of AM fungi were not affected by the absence or presence of polyP in the hyphae (Ezawa et al., 2001a). These observations suggest that polyP accumulation may be dynamically regulated by the balance between synthesis and hydrolysis. We suggest that the inactivation of V-ATPase may occur during maturation of arbuscules and inhibit synthesis of polyP, resulting in net hydrolysis of polyP.

The V-ATPase is a large and complex enzyme and consists of at least thirteen different types of subunit (Margolles-Clark et al., 1999). Post-transcriptional regulation of V-ATPase activity has been investigated extensively in yeast. The proton pumping activity is regulated by disassembly and reassembly of the cytoplasmic and membrane sectors of the enzyme in as little as 5 min, without any new protein synthesis (Kane and Parra, 2000). The cytoplasmic sector could sense changes in ATP concentration or ATP/ADP ratio in the cytoplasm and regulate disassembly and reassembly of the enzyme. The ATP concentration and ATP/ADP ratios are the reflection of the energy status of a cell. In fact, glucose deprivation decreased the ATP level in cytosol of yeast and disassembled V-ATPase (Kane and Parra, 2000). These observations allow us to hypothesize the following scheme for AM fungal arbuscules: once arbuscule formation is initiated, the system involved in the delivery of the carbon source may be switched off [speculation supported by the absence of P-ATPase activity in arbuscules (Gianinazzi–Pearson et al., 1991, 2000)], and depletion of energy may disassemble the V-ATPase. Energy depletion may also inactivate the P-ATPase and promote Pi efflux via PD-sensitive channels on the fungal plasma membrane and probably also the tonoplast, due to depolarization of the membranes as observed in plants (Mimura, 1999). These events could lead to loss of cell function, but the fungi would be able to minimize the potential impact on the intraradical mycelium by forming a septum in the trunk hypha of each arbuscule (Dickson and Smith, 2001).

Regulation of Pi efflux

Solaiman and Saito (2001) investigated the efflux of Pi from isolated intraradical hyphae of *Gigaspora margarita*. They showed that glucose, which is probably the main carbon source of AM fungi, enhanced polyP hydrolysis and subsequent Pi efflux slightly, but significantly. They hypothesized that carbon uptake and Pi efflux in the fungus might be metabolically linked. On the other hand, Beauvoit et al. (1991) suggested that the polyP hydrolysis induced by glucose in yeast might be for buffering the cytoplasmic Pi concentration, which was decreased during glucose metabolism. The latter interpretation is more likely to fit the results of Solaiman and Saito (2001). However, their experiment provided an important and suggestive result in the context of the regulation of Pi efflux. Net hydrolysis of polyP was much slower than expected from the results of our enzymatic studies (Ezawa et al., 2001a). Maximal activity of polyP-hydrolyzing enzymes in the isolated hyphae of *G. coronatum* can be estimated as about 430 pmol min^{-1} mg^{-1} hyphae from the following information: specific activity, 30 (soluble) and 25 (insoluble) pmol Pi min^{-1} μg^{-1} protein at pH 5.0; protein concentrations per unit hyphae, 6 (soluble) and 10 (insoluble) μg mg^{-1} hyphae; total activity, $30\times6+25\times10=430$. However, net Pi hydrolysis and efflux calculated from the data of Solaiman and Saito (2001) are only 13.4 and 16 pmol min^{-1} mg^{-1} hyphae, respectively. If, as Solaiman and Saito (2001) point out, 'leakage' from the fragmented hyphae is taken into account, the true rate of efflux may be even slower. In their experiment, the majority of polyP (and probably organic P) was retained in the hyphae even after 4 h incubation, so the 'slow hydrolysis' was not due to depletion of substrate. Of course these values are derived from different plant–fungal combinations. In addition, it is uncertain if the enzymes were fully accessible to the substrate and if the polyP synthetic pathway was still sufficiently active in the hyphae to reduce net hydrolysis of polyP. However, the fact that the net hydrolysis and efflux were about 30 times less than the predicted capacity of the hydrolytic enzymes strongly suggests that Pi efflux from the hyphae would be limited by activity of the Pi transporters/channels rather than activity of the enzymes. In general, activity of phosphatase-type enzymes could be physiologically controlled by Pi concentration i.e. a high concentration of end product (Pi) inhibits phosphatases. This could be a mechanism by which Pi transporters/channels regulate hydrolysis of polyP/organic P in the vacuoles.

Information about transporters/channels involved in ion efflux from eukaryotic microorganisms is very limited. Researchers have probably not paid sufficient attention to anion efflux in fungi, because it may seem less important or interesting for understanding the basic system of fungal life. This is in contrast to plants in which anion efflux channels play significant roles in, for example, stomatal closure (Schroeder and Keller, 1992). So far Pi efflux in fungi has only been demonstrated for isolated yeast vacuoles. As described earlier, Booth and Guidotti (1997) demonstrated that the transporters/channels involved in Pi uptake by isolated vacuoles of yeast were actually driven by counterflow exchange with Pi present in the vacuoles. They found that two systems are involved in Pi efflux from the vacuoles: one was the bidirectional transport mediated by the transporters/channels, and the other was unidirectional transport which was much slower than the bidirectional system. Recently, an anion-selective efflux channel on the plasma membrane of a filamentous fungus, *Aspergillus niger*, has been characterized by the patch clamp technique in conjunction with an

UV pulse laser as a tool for ablating the cell wall (Roberts et al., 1997). This technique may have great potential to investigate transporter activity of plant and fungal cells, particularly for those having cell walls that cannot easily be digested by commercial enzymes, such as AM fungi (Ezawa unpublished observation). Furthermore, the laser ablation technique enables selection of the site of patch clamp, so that localized transporter activity along hyphae can be measured. It is uncertain, however, whether the technique can be applied to the tips of finely branched arbuscules from which Pi efflux seems likely to occur.

Conclusions

A hypothetical mechanism of Pi efflux from arbuscules is summarized in Figure 1. Validation of the hypothesis will require: (i) identification of genes that are up- or down-regulated during arbuscule formation, (ii) clarification of the role of V-ATPase activity in polyP synthesis/hydrolysis and its possible inactivation during arbuscule formation and (iii) identification of factors which promote Pi efflux from arbuscules. To achieve these objectives some technical breakthroughs are required. Establishment of an experimental system in which frequency of arbuscules in the roots can be controlled by simple treatments such as P application is necessary to identify genes involved in arbuscule formation by the differential display or screening of cDNA libraries. Even if such genes are identified, we do not yet have a genetic manipulation technique to validate the function of the genes. The particle bombardment method is one possible way of inserting genes of interest into AM fungi, as shown by Forbes et al. (1998) who demonstrated transitory expression of a reporter gene in germinating spores of *Gi. margarita*. Future work should be directed at achieving expression in the symbiotic stages of fungal development, so that the function of genes of interest can be inhibited by introducing antisense genes. A technique for isolation of the intact vacuoles, or at least intact tonoplast, of the fungi must be established to measure V-ATPase activity and investigate the involvement of the enzyme in polyP metabolism. It has been accepted, as described earlier, that arbuscules are the site of Pi release, but it has not been proved directly i.e. P efflux from arbuscules must be visualized or hopefully measured quantitatively. Since the apparent rate of Pi efflux from the intraradical hyphae is slow on the basis of unit hyphal length (see previous section), a more sensitive and specific method to evaluate Pi efflux from arbuscules is required.

Understanding C metabolism (see Bago et al., 2000) may be important to understand P metabolism. In the key steps of primary metabolism, C is metabolized via P-esters. Furthermore, the energy status of cell, which is a reflection of C metabolism, may influence membrane transport via H^+-ATPases and other energy dependent reactions such as polyP synthesis. Thus it is possible that C metabolism may play a key role in the regulation of Pi transfer in the plant-fungus interface. Integration of the pictures of P and C metabolism may help us to understand a whole story of the mechanism of mutualistic relationship.

Acknowledgements

Our collaboration was supported by Japan Society for the Promotion of Science (Postdoctoral Fellowship to TE) and the Australian Research Council (SES and FAS).

References

Andreeva N A, Kulakovskaya T, Sidorov I, Karpov A and Kulaev I S 1998a Purification and properties of polyphosphatase from *Saccharomyces cerevisiae* cytosol. Yeast 14, 383–390.

Andreeva N A, Kulakovskaya T V and Kulaev I S 1998b Purification and properties of exopolyphosphatase isolated from *Saccharomyces cerevisiae* vacuoles. FEBS Letters 429, 194–196.

Andreeva N A and Okorokov L A 1993 Purification and characterization of highly active and stable polyphosphatase from *Saccharomyces cerevisiae* cell envelope. Yeast 9, 127–139.

Ashford A E and Allaway W G 2002 The role of the motile tubular vacuole system in mycorrhizal fungi. Plant Soil 244, 177–187.

Ayling S M, Smith S E and Smith F A 2000 Transmembrane electrical potential difference of germ tubes of arbuscular mycorrhizal fungi responds to external stimuli. New Phytol. 147, 631–639.

Bago B, Pfeffer PE, and Shachar-Hill Y 2000 Carbon metabolism and transport in arbuscular mycorrhizas. Plant Physiol. 124, 949–958.

Bago B, Pfeffer P E, Zipfel W, Lammers P and Schachar-Hill Y 2002 Tracking metabolism and imaging transport in arbuscular mycorrhizal fungi. Plant Soil 244, 189–197.

Beauvoit B, Rigoulet M, Raffard G, Canioni P and Guerin B 1991 Differential sensitivity of the cellular compartments of *Saccharomyces cerevisiae* to protonophoric uncoupler under fermentative and respiratory energy supply. Biochemistry 30, 11212–11220.

Bécard G, Doner L W, Rolin D B, Douds D D and Pfeffer P E 1991 Identification and quantification of trehalose in vesicular–arbuscular mycorrhizal fungi by *in vivo* ^{13}C NMR and HPLC analyses. New Phytol. 118, 547–552.

Beever R E and Burns D J W 1980 Phosphorus uptake, storage and utilisation by fungi. Adv. Bot. Res. 8, 127–219.

Bieleski R L 1973 Phosphate pools, phosphate transport, and phosphate availability. Ann. Rev. Pl. Physiol. 24, 225–252.

Boddington C L and Dodd J C 1999 Evidence that differences in phosphate metabolism in mycorrhizas formed by species of *Glomus* and *Gigaspora* might be related to their life-cycle strategies. New Phytol. 142, 531–538.

Bolan N S, Robson A D and Barrow N J 1984 Increasing phosphorus supply can increase the infection of plant roots by vesicular-arbuscular mycorrhizal fungi. Soil Biol. Biochem. 16, 419–420.

Booth J W and Guidotti G 1995 An alleged yeast polyphosphate kinase is actually diadenosine–$5',5'''$–P–1,P–4-tetraphosphate α, β-phosphorylase. J. Biol. Chem. 270, 19377–19382.

Booth J W and Guidotti G 1997 Phosphate transport in yeast vacuoles. J. Biol. Chem. 272, 20408–20413.

Bücking H and Heyser W 1999 Elemental composition and function of polyphosphates in ectomycorrhizal fungi – an X-ray microanalytical study. Mycol. Res. 103, 31–39.

Callow J A, Capaccio L C M, Parish G and Tinker P B 1978 Detection and estimation of polyphosphate in vesicular-arbuscular mycorrhizas. New Phytol. 80, 125–134.

Capaccio L C M and Callow J A 1982 The enzymes of polyphosphate metabolism in vesicular-arbuscular mycorrhizas. New Phytol. 91, 81–91.

Clark J E, Beegen H and Wood H G 1986 Isolation of intact chain of polyphosphate from *Propionibacterium shermanii* grown on glucose or lactate. J. Bacteriol. 168, 1212–1219.

Cole L, Orlovich D A and Ashford A E 1998 Structure, Function, and, Motility of Vacuoles in Filamentous Fungi. Fungal Genet. Biol. 24, 86–100.

Cox G, Moran K J, Sanders F, Nockolds C and Tinker P B 1980 Translocation and transfer of nutrients in vesicular-arbuscular mycorrhizas. New Phytol. 84, 649–659.

Cramer C and Davis R 1984 Polyphosphate-cation interaction in the amino acid-containing vacuole of *Neurospora crassa*. J. Biol. Chem. 259, 5152–5157.

Dickson S and Smith S E 2001 Cross walls in arbuscular trunk hyphae from after loss of metabolic activity. New Phytol. 151, 735–742.

Ezawa T, Kuwahara S, Sakamoto K, Yoshida T and Saito M 1999 Specific inhibitor and substrate specificity of alkaline phosphatase expressed in the symbiotic phase of the arbuscular mycorrhizal fungus, *Glomus etunicatum*. Mycologia 91, 636–641.

Ezawa T, Saito M and Yoshida T 1995 Comparison of phosphatase localization in the intraradical hyphae of arbuscular mycorrhizal fungi, *Glomus* spp. and *Gigaspora* spp. Pl. Soil 176, 57–63.

Ezawa T, Smith S E and Smith F A 2001a Differentiation of polyphosphate metabolism between the extra- and intraradical hyphae of arbuscular mycorrhizal fungi. New Phytol. 149, 555–563.

Ezawa T, Smith S E and Smith F A 2001b Enzyme activity involved in glucose phosphorylation in two arbuscular mycorrhizal fungi: indication that poly*P* is not the main phosphagen. Soil Biol. Biochem. 33, 1279–1281.

Fabig B, Moawad A M and Achtnich W 1989 Gas-chromatographic separation of organic acids and electrophorectic determination of phosphatases from VA mycorrhizal roots. Z. Pflanzernär. Bodenkd. 152, 261–265.

Forbes P J, Millam S, Hooker J E and Harrier L A 1998 Transformation of the arbuscular mycorrhiza *Gigaspora rosea* by particle bombardment. Mycol. Res. 102, 497–501.

Gianinazzi S, Gianinazzi–Pearson V and Dexheimer J 1979 Enzymatic studies on the metabolism of vesicular–arbuscular mycorrhiza. III. Ultrastructural localization of acid and alkaline phosphatase in onion roots infected by *Glomus mosseae* (Nicol. & Gerd.). New Phytol. 82, 127–132.

Gianinazzi–Pearson V, Arnould C, Oufattole M, Arango M and Gianinazzi S 2000 Differential activation of H^+-ATPase genes by an arbuscular mycorrhizal fungus in root cells of transgenic tobacco. Planta 211, 609–613.

Gianinazzi–Pearson V and Gianinazzi S 1978 Enzymatic studies on the metabolism of vesicular–arbuscular mycorrhiza. II. Soluble alkaline phosphatase specific to mycorrhizal infection in onion roots. Physiol. Plant Pathol. 12, 45–53.

Gianinazzi–Pearson V, Smith S E, Gianinazzi S and Smith F A 1991 Enzymatic studies on the metabolism of vesicular–arbuscular mycorrhizas. V. Is H^+-ATPase a component of ATP-hydrolysing enzyme activities in plant-fungus interfaces? New Phytol. 117, 61–74.

Guillemin J P, Orozco M O, Gianinazzi–Pearson V and Gianinazzi S 1995 Influence of phosphate fertilization on fungal alkaline phosphatase and succinate dehydrogenase activities in arbuscular mycorrhiza of soybean and pineapple. Agri. Ecosystem Environ. 53, 63–69.

Harrison M J and Van Buuren M L 1995 A phosphate transporter from the mycorrhizal fungus *Glomus versiforme*. Nature 378, 626–629.

Hyde G, Cole L and Ashford A E 1997 Mycorrhiza movies. Mycorrhiza 7, 167-169.

Kane P M and Parra K J 2000 Assembly and regulation of the yeast vacuolar H^+-ATPase. J. Exp. Biol. 203, 81–87.

Klionsky D J, Herman P K and Emr S D 1990 The fungal vacuole, composition, function, and biogenesis. Microbiol. Rev. 54, 266–292.

Kornberg A, Rao N N and Ault-Riche D 1999 Inorganic polyphosphate, A molecule of many functions. Ann. Rev. Biochem. 68, 89–125.

Kulaev I S, Andreeva N A, Lichko L P and Kulakovskaya T V 1997 Comparison of exopolyphosphatases of different yeast cell compartments. Microbiol. Res. 152, 221–226.

Kulaev I S and Kulakovskaya T V 2000 Polyphosphate and phosphate pump. Ann. Rev. Microbiol. 54, 709–734.

Kulakovskaya T V and Kulaev I S 1997 Transport of phosphate into vacuoles of *Saccharomyces cerevisiae*. Microbiologia 13, 71–74.

Kumble K D and Kornberg A 1996 Endopolyphosphatases for long chain inorganic polyphosphate in yeast and mammals. J. Biol. Chem. 271, 27146–27151.

Larsen J, Thingstrup I, Jacobsen I and Rosendahl S 1996 Benomyl inhibits phosphorus transport but not fungal alkaline phosphatase activity in a *Glomus*-cucumber symbiosis. New Phytol. 132, 127–133.

Lenburg M E and O'Shea E K 1996 Signaling phosphate starvation. Trends Biochem. Sci. 21, 383–387.

Lichko L P, Kulakovskaya T V and Kulaev I S 1996 Characterization of the polyphosphatase activity of *Saccharomyces cerevisiae* nuclei. Biochemistry (Moscow) 61, 361–366.

Lichko L P, Kulakovskaya T V and Kulaev I S 1998 Membrane–bound and soluble polyphosphatases of mitochondria of *Saccharomyces cerevisiae* – identification and comparative characterization. Biochim. Biophys. Acta Biomembranes 1372, 153–162.

Margolles-Clark E, Tenney K, Bowman E J and Bowman B J 1999 The structure of the vacuolar ATPase in *Neurospora crassa*. J. Bioenergetics Biomembranes 31, 29–37.

Martinez P and Persson B L 1998 Identification, cloning and characterization of a derepressible Na^+-coupled phosphate transporter in *Saccharomyces cerevisiae*. Mol. Gen. Genet. 258, 628–638.

Massonneau A, Martinoia E, Dietz K J and Mimura T 2000 Phosphate uptake across the tonoplast of intact vacuoles isolated from suspension-cultured cells of *Catharanthus roseus* (L.) G. Don. Planta 211, 390–395.

Mimura T 1999 Regulation of phosphate transport and homeostasis in plant cells. Inter. Rev. Cytol. 191, 149–200.

Ogawa N, DeRisi J and Brown P O 2000 New components of a system for phosphate accumulation and polyphosphate metabolism in *Saccharomyces cerevisiae* revealed by genomic expression analysis. Mol. Biol. Cell 11, 4309–4321.

Orlovich D A and Ashford A E 1993 Polyphosphate granules are an artefact of specimen preparation in the ectomycorrhizal fungus *Pisolithus tinctorius*. Protoplasma 173, 91–102.

Pick U and Weiss M 1991 Polyphosphate hydrolysis within acidic vacuoles in response to amine-induced alkaline stress in the halotolerant alga *Dunaliella salina*. Pl. Physiol. 97, 1234–1240.

Pisoni R and Lindley E 1992 Incorporation of [^{32}P]orthophosphate into long chains of inorganic polyphosphate within lysosomes of human fibroblasts. J. Biol. Chem. 267, 3626–3631.

Rausch C, Daram P, Brunner S, Jansa J, Lalol M, Leggewie G, Amrhein N, and Bucher M 2001 A phosphate transporter expressed in arbuscule-containing cells in potato. Nature 414, 462–466.

Rees B, Shepherd V A and Ashford A E 1994 Presence of a Motile Tubular Vacuole System in Different Phyla of Fungi. Mycol. Res. 98, 985–992.

Reid R J, Mimura T, Ohsumi Y, Walker N A and Smith F A 2000 Phosphate uptake in *Chara*, Membrane transport via Na/Pi cotransport. Pl. Cell Environ. 23, 223–228.

Reizer J, Reizer A and Saier Milton H, Jr. 1994 A functional superfamily of sodium/solute symporters. Biochim. Biophys. Acta 1197, 133–166.

Roberts S K, Dixon G K, Dunbar S J and Sanders D 1997 Laser ablation of the cell wall and localized patch clamping of the plasma membrane in the filamentous fungus *Aspergillus* – characterization of an anion-selective efflux channel. New Phytol. 137, 579–585.

Rosewarne G M, Barker S J, Smith S E, Smith F A and Schachtman D P 1999 A *Lycopersicon esculentum* phosphate transporter (LePT1) involved in phosphorus uptake from a vesicular–arbuscular mycorrhizal fungus. New Phytol. 144, 507–516.

Schroeder J I and Keller B U 1992 Two types of anion channel currents in guard cells with distinct voltage regulation. Proc. Natl. Acad. Sci. USA 89, 5025–5029.

Schubert A, Wyss P and Wiemken A 1992 Occurrence of trehalose in vesicular–arbuscular mycorrhizal fungi and in mycorrhizal roots. J. Pl. Physiol. 140, 41–45.

Schweiger P F and Jakobsen I (1999). The role of mycorrhizas in plant P nutrition, fungal uptake kinetics and genotype variation. *In* Plant Nutrition – Molecular Biology and Genetics. Eds. G Gissel-Nielsen and A Jensen. pp 277–289. Kluwer Academic Publishers, Dordrecht.

Shachar-Hill Y, Pfeffer P E, Douds D, Osman S F, Doner L W and Ratcliffe R G 1995 Partitioning of intermediary carbon metabolism in vesicular–arbuscular mycorrhizal leek. Pl. Physiol. 108, 7–15.

Shirahama K, Yazaki Y, Sakano K, Wada Y and Ohsumi Y 1996 Vacuolar function in the phosphate homeostasis of the yeast *Saccharomyces cerevisiae*. Pl. Cell Physiol. 37, 1090–1093.

Smith S E and Read D J 1997 Mycorrhizal symbiosis. Academic Press. San Diego, California.

Solaiman M, Ezawa T, Kojima T and Saito M 1999 Polyphosphates in intraradical and extraradical hyphae of an arbuscular mycorrhizal fungus, *Gigaspora margarita*. Appl. Environ. Microbiol. 65, 5604–5606.

Solaiman M Z and Saito M 1997 Use of sugars by intraradical hyphae of arbuscular mycorrhizal fungi revealed by radiorespirometry. New Phytol. 136, 533–538.

Solaiman M Z and Saito M 2001 Phosphate efflux from the intraradical hyphae of an arbuscular mycorrhizal fungus, *Gigaspora margarita*, *in vitro* and its implication for phosphorus translocation in the hyphae. New Phytol. 151, in press.

Thomson B D, Clarkson D T and Brain P 1990 Kinetics of phosphorus uptake by the germ-tube of the vesicular–arbuscular mycorrhizal fungus *Gigaspora margarita*. New Phytol. 116, 647–654.

Tisserant B, Gianinazzi–Pearson V, Gianinazzi S and Gollote A 1993 *In planta* histochemical staining of fungal alkaline phosphatase activity for analysis of efficient arbuscular mycorrhizal infections. Mycol. Res. 97, 245–250.

Versaw W K and Metzenberg R L 1995 Repressible cation-phosphate symporters in *Neurospora crassa*. Proc. Natl. Acad. Sci. USA 92, 3884–3887.

Versaw W K, Chiou T-J and Harrison M J 2002 Phosphate transporters of *Medicago truncula*ta and arbuscular mycorrhizal fungi. Plant Soil 244, 239–245.

Wurst H and Kornberg A 1994 A soluble exopolyphosphatase of *Saccharomyces cerevisiae* – purification and characterization. J. Biol. Chem. 269, 10996–11001.

Wurst H, Shiba T and Kornberg A 1995 The gene for a major exopolyphosphatase of *Saccharomyces cerevisiae*. J. Bacteriol. 177, 898–906.

Zhu Y-G, Cavagnaro T R, Smith S E and Dickson S 2001 Backseat driving? Accessing phosphate beyond the rhizosphere-depletion zone. Trend Pl. Sci. 6, 194–195.

Plant and Soil **244:** 231–237, 2002.

Mechanisms of nutrient transport across interfaces in arbuscular mycorrhizas

N. Ferrol[1], J.M. Barea & C. Azcón-Aguilar
Departamento de Microbiología del Suelo y Sistemas Simbióticos, Estación Experimental del Zaidín, Consejo Superior de Investigaciones Científicas, Profesor Albareda 1, 18008 Granada, Spain. [1]*Corresponding author*[*]

Key words: arbuscular mycorrhizas, ATPases, gene expression, nutrient transporters, symbiotic interfaces

Abstract

Bidirectional nutrient transfer between the plant and the fungus is a key feature of arbuscular mycorrhizal symbiosis. The major nutrients exchanged between the symbiotic partners are reduced carbon, assimilated through the plant photosynthesis and phosphate, taken up by the fungal hyphae exploring soil microhabitats. This nutrient exchange takes place across the symbiotic interfaces which are bordered by the plant and fungal plasma membranes. This review provides an overview of the current knowledge of the mechanisms underlying nutrient transport processes in the symbiosis, with special emphasis on recent developments in the molecular biology of the plant and fungal primary (H^+-ATPases) and secondary transporters.

Introduction

The mutualistic nature of arbuscular mycorrhizal (AM) symbioses relies on the ability of the fungal mycelium to take up mineral nutrients from the soil solution and to transfer them to the symbiotic roots in exchange for carbohydrates. Plant benefits from AM symbiosis are traditionally recognized as improved access to limiting soil resources, mainly immobile nutrients such as P, Cu, Zn and ammonium. On the other hand, the carbon supplied from the host plant to the fungus is essential to the formation and functioning of arbuscular mycorrhizas and for the completion of the fungal life cycle. Although the mutualism of the AM association is based on this bidirectional nutrient exchange, this does not necessarily mean that nutrient transfer from the fungus to the plant is directly linked to C transfer to the fungus. In fact, different AM fungi in symbiosis with the same host plant, can considerably differ in their C–P exchange ratio and, consequently, in their degree of functional compatibility with the plant (Pearson and Jakobsen, 1993). Additionally, several studies have shown that C transfer can be decoupled from P transfer under certain environmental conditions, as for instance, under conditions of high soil P availability (Graham et al., 1997). In any case, an indirect link between C and P transfer is evident as C supplied to the fungus will support mycelia growth and, indirectly, P uptake from the soil.

This bidirectional transfer of nutrients between the plant and the fungus is believed to take place across the symbiotic interfaces which are bordered by the plant and fungal plasma membranes (Smith and Smith, 1990). Although current knowledge of the molecular mechanisms underlying transport nutrient in arbuscular mycorrhizas is still in its infancy, it is generally accepted that such bidirectional transfer of nutrients across the symbiotic interfaces involves passive efflux of solutes from the donor organism into the interfacial apoplast, followed by active uptake by the receiver organism (Smith and Read, 1997; Woolhouse et al., 1975). A major obstacle to the study of these mechanisms is the difficulty of culturing the fungal partner in the absence of the host plant and of isolating the intact arbuscules that are developed within the cortical root cells. The recent application of molecular tools to the study of the AM symbiosis is enabling the identification of important transport systems, and the results obtained so far support the existence of active mechanisms involved in the transfer of carbon and phosphate, as hypothesized by Woolhouse (1975). This review

* FAX No: +34-958-129600; E-mail: nferrol@eez.csic.es

focuses on current knowledge about the molecular characterization and regulation of transport proteins located in the plant and fungal plasma membranes. The data are discussed in relation to the mechanisms underlying nutrient transport processes in arbuscular mycorrhizas.

Structure and composition of AM symbiotic interfaces

It is widely accepted that in AM symbiosis, as in other biotrophic interactions, the exchange of nutrients between the plant and the fungus takes place across the symbiotic interfaces that are developed during the colonization process of the host plant root (Smith and Smith, 1990). Due to the considerable importance of the symbiotic interfaces for nutrient transport in AM symbiosis, a brief revision of the current knowledge about the structure and composition of the symbiotic interfaces is presented below.

Two different types of symbiotic interfaces can be found, depending on whether the fungus grows inter- or intracellularly in the root system. Intercellular interfaces are created when the fungal hyphae grows within the intercellular spaces of the root cortex, whereas intracellular interfaces are developed when the fungal hyphae penetrates the wall of the root cells. Intracellular structures include coils of fungal hyphae, often formed during initial colonization of the epidermis and (if present) hypodermis in many AM roots, and especially in cortical cells of *Paris*-type mycorrhizas, and arbuscules which are the main intracellular structures in *Arum*-type mycorrhizas (Smith and Smith, 1997). Since very little work has been done on the physiology and molecular biology of *Paris*-types, this chapter focuses on *Arum*-types.

Both inter- and intracellular interfaces have the same basic structure, they are always composed of the membranes of both symbionts separated by a narrow interfacial space. However, there are significant differences among them, concerning mainly to the composition and structure of both plant and fungal walls and to the presence of other materials found in the interfacial apoplast (for extensive revision, see Bonfante and Perotto, 1995; Bonfante, 2001).

When root cells are colonized by AM fungi to form arbuscules the host plasma membrane invaginates and proliferates around the developing fungus. Alexander et al. (1989) have reported a 3.7-fold increase of host plasmalemma in arbuscule-containing cells. The host plasma membrane surrounding the arbuscules is termed periarbuscular membrane and its formation must involve massive synthesis of lipids and proteins, some of the latter being unique to this membrane. The difficulty of isolating the periarbuscular membrane has hampered its biochemical characterization; however, Benabdellah et al. (2000) have shown by 2D-PAGE analysis of plasma membrane proteins isolated from mycorrhizal and non-mycorrhizal roots that development of the symbiosis induces the synthesis of at least 10 new plasma membrane polypeptides. In addition, the amount of phospholipids is higher in membranes isolated from mycorrhizal roots (Benabdellah, 1999). The application of efficient cytological and molecular tools has circumvented the difficulties of biochemical approaches and is enabling the characterization of the periarbuscular membrane. Cytochemical and immunocytochemical analysis have shown that the periarbuscular membrane displays ATPase activity, which is usually absent in the peripheral membrane of the colonized cortical root cell, suggesting that this membrane acquire transport functions that are usually absent in the cortical root cells (Gianinazzi-Pearson et al., 1991, 2000). The nature of other proteins in this membrane remains unknown, although immunocytological studies have shown that the periarbuscular membrane shares components with the peribacteroid membrane in the *Rhizobium*–legume symbiosis (Perotto et al., 1994).

In situ use of affinity probes, such as enzymes, lectins or antibodies has shown that the interfacial apoplast between the periarbuscular membrane and the fungal surface presents many molecules typical of the primary cell wall, such as cellulose, β-1,4-glucans, polygalacturans and hemicelluloses (Bonfante and Perotto, 1995; Gollotte et al., 1997). However, morphological analysis of the interface material indicates that these components are not assembled into a fully structured cell wall (Bonfante and Perotto, 1995). Recent *in situ* hybridisation experiments have shown that activation of some genes in arbuscule-containing cells such as those encoding proteins rich in hydroxyproline (Balestrini et al., 1997), arabinogalactan proteins (van Buuren et al., 1999) and xyloglucan endotransglycolases (Maldonado-Mendoza and Harrison, 1998) might account for the composition and structure of the interfacial matrix.

In relation to the fungal cell wall at the arbuscular level, ultrastructural studies have shown that it is very thin around arbuscule branches and that the chitin, the main component of AM fungal cell wall, has at this level an amorphous nature, in contrast with the rigid and fibrillar structure that presents in the intercellular hyphae (Bonfante-Fasolo et al., 1990). Moreover, these studies have shown that the plant cell wall at the intercellular interfaces suffers little modifications. The observations derived from all these studies, and the fact that arbuscules increase the contact area between symbionts, support the general assumption that the transfer of nutrients takes place preferentially across the arbuscular interfaces (Smith and Smith, 1990). However, some authors have proposed that the presence of other AM interfaces might offer the possibility that different compounds are transferred across the various interfaces (Smith and Read, 1997). Although hyphae and hyphal coils have been usually ignored as potential transfer sites, there is no reason to assume that they do not have transfer functions. The recent observation by confocal microscopy that the intracellular coils in *Paris*-type AM present as great surface area per cortical cell as an arbuscule in *Arum*-types opens the possibility that intracellular coils might have some relevance in nutrient transfer (Dickson and Kolesik, 1999). The possible role of intercellular hyphae in carbohydrate uptake will be discussed in the next section.

It is not clear how far the wall and wall-like materials present in the interfacial apoplast are involved in, or influence, transfer of nutrients across the interface. However, it is likely that nutrient movement between partners is mainly controlled by the bordering plant and fungal plasma membranes, and more specifically by the plasma membrane proteins that mediate controlled exchange of solutes into and out of the adjacent cells. Therefore, elucidation of the mechanisms underlying the controlled transfer of nutrients between the mycorrhizal symbionts requires identification of the specific transport proteins involved in efflux and uptake of solutes by both partners. These membrane proteins have not so far been identified; however, the application of molecular biology techniques to the study of arbuscular mycorrhizas is providing some insights about the transport processes that occur in the symbiotic interfaces. As in other organisms, major advances in the understanding of these processes across the partner membranes have emerged from recent studies of the dominant H^+-pumping ATPase and from the identification of a small range of secondary transporters.

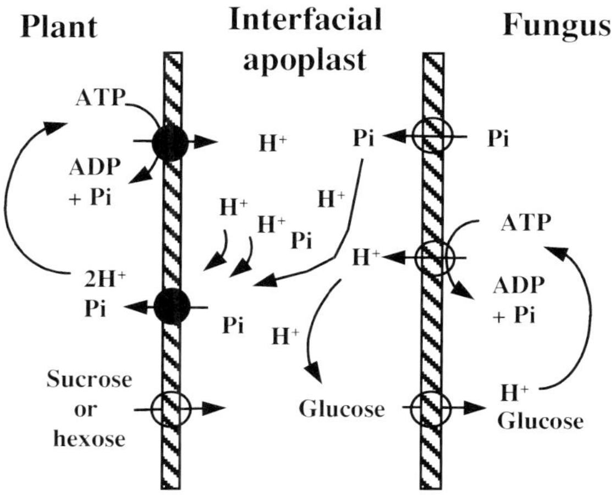

Figure 1. Schematic illustration of the proposed model for the transfer of phosphate and carbon compounds across the arbuscular interface. Plasma membrane H^+-ATPases and secondary transporters that have been experimentally localized in the membranes of the arbuscular interface are indicated by filled circles.

Carbon supply to the fungus

AM fungi, as obligate symbionts, depend for their growth and activity on the supply of carbon compounds by the photosynthetic partner (Jennings, 1995). Consequently, AM symbiosis can cause an important carbohydrate drain: up to 20% of total photoassimilate production can be transferred to the fungal partner (Douds et al., 2000; Graham et al., 2000). There is enough evidence that, as a consequence of mycorrhizal establishment, assimilates are directed towards the root and the rate of net photosynthesis of the host increases (Douds et al., 1988; Kucey and Paul, 1982; Nemec and Vu, 1990; Wright et al., 1998). The fact that cells containing arbuscules are acting as a sink for sucrose is supported by *in situ* hybridisation studies of the expression of genes involved in sucrose hydrolysis such as sucrose synthase and soluble acid invertase (Blee and Anderson, 1998).

Studies using isotopic labelling with nuclear magnetic resonance spectrometry in AM roots (Shachar-Hill et al., 1995) and radiorespirometry measurements on isolated intraradical hyphae (Solaiman and Saito, 1997) have shown that the internal phase of the fungal symbiont can take up and use hexoses, mainly glucose. However, it is still a matter of debate whether the transfer occurs across the intercellular or the intracellular interfaces and how the transfer occurs. Although it has not been experimentally proved, it is generally assumed that the transfer of carbon compounds from the plant to the fungus involves passive efflux across the plant plasma membrane followed by active uptake by the fungal plasma membrane from the interfacial apoplast (Figure 1).

The nature of the membrane transport proteins that mediate these processes remains still unknown. A cDNA clone encoding a hexose transporter whose expression increases in cells containing arbuscules, as well as in adjacent cells, has been cloned in *Medicago truncatula*. Functional characterization of the protein encoded by this gene demonstrated its ability to take up glucose by a H^+ symport process, suggesting that this protein is not involved in the efflux of glucose across the plant membrane at the symbiotic interface (Harrison, 1996). Additionally, a cDNA encoding a novel sugar transporter has been recently cloned in tomato roots following a RT-PCR approach (Ferrol, unpublished results). Preliminary gene expression analysis revealed that this gene was constitutively expressed in tomato roots and that the sugar transporter transcripts increase in leaves of wild-type plants but not in those of mycorrhiza-defective tomato mutants after inoculation with *Glomus mosseae*. Since, as stated before, carbon partitioning increases in mycorrhizal plants, these observations suggest that this transporter could play a role in the symbiosis by increasing the loading into the phloem of the reduced carbon compounds demanded by the microsymbiont for its growth and activity and/or by the increased metabolic activity of arbuscule-containing cells. Ferrol et al. (2002) have also found that development of a successful mycorrhizal interaction increases leaf H^+-ATPase gene expression. This further supports the possibility that phloem loading increases in mycorrhizal plants, since active loading of carbon compounds into the phloem requires the action of a plasma membrane H^+-ATPase (DeWitt et al., 1991; Villalba et al., 1991). Summarising, although sugar transporters that are regulated by the symbiosis have been identified, the plant transporter involved in carbon efflux across the host plasma membrane remains unidentified.

With regard to the mechanisms by which AM fungi take up carbon compounds from the apoplastic interface, it is unclear whether uptake requires an active transport system or whether concentrations of carbon at the interfaces are sufficient to direct uptake by facilitated diffusion. In the ectomycorrhizal fungus *Amanita muscaria*, a hexose transporter that operates by an active hexose-H^+ symport process seems to be responsible for the sugar uptake by the fungus (Nehls et al., 1998; Wiese et al., 2000). Since membranes involved in active solute uptake by H^+ co-transport must be energized by a H^+-ATPase (Palmgren and Harper, 1999; Serrano, 1989), the presence of ATPases in fungal membranes could provide clues about the mechanisms of transport and the sites of glucose transfer. Based on the cytochemical observation that the membrane of the arbuscular branches frequently lacks ATPase activity while the membranes of the intercellular hyphae show high levels of ATPase staining, it was suggested that carbon might be transferred across the intercellular interfaces rather than across the arbuscular interfaces (Gianinazzi-Pearson et al., 1991). However, as pointed out by these authors, this hypothesis must remain tentative until more clear evidence on fungal H^+-ATPase distribution is obtained by other methods. Five PCR fragments encoding plasma membrane H^+-ATPases have been identified in *Glomus mosseae* (Ferrol et al., 2000) and two in *Glomus intraradices* (Ferrol et al., unpublished data). RT-PCR expression analysis revealed that one of the *G. mosseae* genes was expressed in the intraradical structures of the fungus, indicating that active transport processes occur in these structures; however, the specific location of these transcripts remains to be determined (Ferrol et al., unpublished data). Although these studies on H^+-ATPases are providing some insights into the transport mechanisms operating at the fungal membranes, knowledge of the particular mechanism by which AM fungi take up carbon compounds from the apoplastic interface and about the interfaces involved requires the identification and location of the membrane proteins involved.

Transfer of phosphate from the fungus to the plant

Root colonization by the AM fungus is accompanied by the development of a network of fine extraradical hyphae, which increases the absorption rate of slow-diffusing nutrients, mainly phosphate (P), from the soil to the plant (Jackobsen, 1999). Movement of phosphate in the symbiosis begins with its uptake from the soil solution by the phosphate transporters located in the extraradical hyphae, followed by its translocation to the intraradical fungal structures and subsequent transfer to the colonized root cell across the arbuscular interfaces.

Molecular cloning of a phosphate transporter from *Glomus versiforme* has provided evidence about the mechanisms of phosphate uptake by the external hyphae (Harrison and van Buuren, 1995). Functional characterization of this gene in yeast cells indicated that it encodes a high-affinity transporter that operates by an active H^+ symport process. The finding that the transporter transcripts were present in the external mycelium, but not in the intraradical fungal structures, led Harrison and van Buuren (1995) to propose that this transporter may be responsible for the initial uptake of phosphate into the mycorrhiza. Functioning of active secondary transporters requires the action of the plasma membrane H^+-ATPase (Serrano, 1989), therefore, this protein must be present on the plasma membrane of the extraradical hyphae. In this context, Ferrol et al. (unpublished data) have found that H^+-ATPase genes are expressed in the extraradical hyphae of *G. mosseae* and *G. intraradices*. These observations indicate that uptake of P by the extraradical hyphae is mediated by a high-affinity transporter whose ability to function depends on the electrochemical gradient of H^+ generated by the plasma membrane H^+-ATPase.

Uptake of P by extraradical hyphae is followed by the synthesis of large amounts of polyphosphate, part of which is stored in the fungal vacuoles (Cox et al., 1980; Dexheimer et al., 1996). A recent study using *in vivo* ^{31}P nuclear magnetic resonance has shown that the phosphate ions taken up by the fungus are transformed into polyphosphate with a short chain length (Rasmussen et al., 2000). Based on these observations the authors suggest that the process of phosphate translocation along the hyphae of AM fungi may be similar to that of ectomycorrhizal fungi, in which the phosphate appears to be transported mainly as polyphosphate in a motile vacuole system. This is assumed to be followed by hydrolysis of polyphosphate by the action of the phosphatases that have been localized in the vacuoles of the intraradical hyphae (Ezawa et al., 1995; Jabari-Hare et al., 1990; Tisserant et al., 1992) and exopolyphosphatases (Ezawa et al., 2001).

The P transfer across the symbiotic interface begins with efflux from the fungus into the interfacial apoplast, an unusual step, since membrane transport processes for P transport favour absorption over loss (Clarkson, 1985). Several hypothesis have been proposed to explain the high P loss from the arbuscule (Harley and Smith, 1983, Smith and Read, 1997). As individual arbuscules have a lifespan of about 7–11 days, it was accepted in the past that its senescence and partial breakdown would release P to be taken up by the host plant. Nevertheless, measurements of P flux rates across the interface showed that the release of P from senescing arbuscules would be far too slow to account for P transfer to the host (Cox and Tinker, 1976). Nor it is likely that increased leakiness of P by simple diffusion as a result of an increased membrane permeability would support the massive P transfer in AM symbiosis. Consequently, it is believed that transport of phosphate across the arbuscular membrane probably involves specific efflux channels or transporters.

Recent molecular studies of host plasma membrane H^+-ATPases and phosphate transporters have provided evidence that the subsequent uptake of phosphate by the plant may be an active transport process. Up-regulation of host plasma membrane ATPase genes has been demonstrated in barley, tobacco and tomato plants (Ferrol et al., 2002; Gianinazzi-Pearson et al., 2000; Murphy et al., 1997). In tobacco roots, two H^+-ATPase genes were found to be induced in arbuscule-containing cells, by using promoter β-glucuronidase (GUS) fusions (Gianinazzi-Pearson et al., 2000). In addition, immunocytochemical analysis of these roots detected the H^+-ATPase protein in the periarbuscular membrane. Recently, Rosewarne et al. (1999) have identified a tomato phosphate transporter that may be involved in the uptake of the phosphate that is effluxed across the arbuscular membrane, as indicated by the high transcript levels detected by *in situ* hybridisation in arbuscule-containing cells. Based on these observations, it can be concluded that the periarbuscular membrane plays an important role in the absorption by the host plant of the phosphate ions that have been transported by the fungus and released into the symbiotic interface through a H^+-phosphate symporter activated by the plasma membrane H^+-ATPase (Figure 1).

Conclusions

Transport processes are crucial for the functioning of arbuscular mycorrhizal symbioses because they are nutritionally based. The picture of how carbon compounds and mineral nutrients are transferred between the symbionts will be clarified once the major transport proteins had been cloned and located at the interfacial membranes. This will also contribute to a better understanding of the mechanisms regulating AM functioning. During the last years, the application of techniques of molecular biology to mycorrhizal research has brought some information about the molecular mechanisms underlying phosphate uptake by the external hyphae and the periarbuscular membrane. Until now, all the transporters that have been shown or postulated to be involved in the functioning of the symbiosis are proton symporters, which emphasizes the importance of H^+-ATPases for nutrient uptake. From this review it should be evident that our knowledge about transport mechanisms and the molecular structure of the plant and fungal transporters is, at best, scarce. There is in consequence a clear need for further work on the isolation and molecular characterization of these membrane proteins in order to understand the mechanisms underlying nutrient transport processes in arbuscular mycorrhizas.

Acknowledgements

This work was supported by CICyT (PB98-0473 project), Spain.

References

Alexander T, Toth R, Meier R and Weber HC 1989 Dynamics of arbuscule development and degeneration in onion, bean, and tomato with reference to vesicular-arbuscular mycorrhizas in grasses. Can. J. Bot. 67, 2505–2513.

Balestrini R, José-Estanyol M, Puigdomenech P and Bonfante P 1997 Hydroxyproline-rich glycoprotein mRNA accumulation in maize root cells colonized by an arbuscular mycorrhizal fungus as revealed by *in situ* hybridisation. Protoplasma 198, 36–42.

Benabdellah K 1999 Alteraciones bioquímicas inducidas en membranas de células radicales de tomate (*Lycopersicon esculentum* Mill.) por la formación de micorrizas arbusculares. PhD Thesis. University of Granada, Spain.

Benabdellah K, Azcón-Aguilar C and Ferrol N 2000 Alterations in the plasma membrane polypeptide pattern of tomato roots (*Lycopersicon esculentum*) during the development of arbuscular mycorrhiza. J. Exp. Bot. 51, 747–754.

Blee KA and Anderson AJ 1998 Regulation of arbuscule formation by carbon in the plant. Plant J. 16, 523–530.

Bonfante P 2001 At the interface between mycorrhizal fungi and plants: the structural organization of cell wall, plasma membrane and cytoskeleton. *In* The Mycota: Fungal Associations. Ed. B. Hock. pp. 45–61. Springer, Berlin.

Bonfante P and Perotto S 1995 Strategies of arbuscular mycorrhizal fungi when infecting host plants. New Phytol. 130, 3–21.

Bonfante-Fasolo P, Faccio A, Perotto S and Schubert A 1990 Correlation between chitin distribution and cell wall morphology in the mycorrhizal fungus *Glomus versiforme*. Mycol. Res. 94, 157–165.

Clarkson DT 1985 Factors affecting mineral nutrient acquisition by plants. Annu. Rev. Plant Physiol. 36, 77–115.

Cox G and Tinker PB 1976 Translocation and transfer of nutrients in vesicular-arbuscular mycorrhizas. New Phytol. 77, 371–378.

Cox G, Moran KJ, Sanders F, Nockolds C and Tinker PB 1980 Translocation and transfer of nutrients in vesicular-arbuscular mycorrhizas. 3. Polyphosphate granules and phosphorus translocation. New Phytol. 84, 649–659.

De Witt ND, Harper JF and Sussman MR 1991 Evidence for a plasma membrane proton pump in phloem cells of higher plants. Plant J. 1, 121–128.

Dexheimer J, Gérard J, Ayatti H and Ghanbaja J 1996 Etude de l'origine et de la répartition des granules vacuolaires dans les hyphes d'une endomycorhize à vésicules et arbuscules. Acta Bot. Gallica 143, 167–180.

Dickson S and Kolesik P 1999 Visualisation of mycorrhizal fungal structures and quantification of their surface area and volume using laser scanning confocal microscopy. Mycorrhiza 9, 205–213.

Douds DD Jr., Johnson CR and Koch KE 1988 Carbon cost of the fungal symbiont relative to net leaf P accumulation in a split-root VA mycorrhizal symbiosis. Plant Physiol. 86, 491–496.

Douds DD Jr, Pfeffer PE and Shachar-Hill Y 2000 Carbon partitioning, cost, and metabolism of arbuscular mycorrhizas. *In* Arbuscular Mycorrhizas: Physiology and Function. Eds. Y Kapulnik and DD Douds Jr. pp. 107–130. Kluwer Academic Publishers, Dordrecht.

Ezawa T, Saito M and Yoshiada T 1995 Comparison of phosphatase localization in the intraradical hyphae of arbuscular mycorrhizal fungus, *Glomus* spp. and *Gigaspora* ssp. Plant Soil 176, 57–63.

Ezawa T, Smith SE and Smith FA 2001 Differentiation of polyphosphate metabolism between the extra- and intraradical hyphae of arbuscular mycorrhizal fungi. New Phytol. 149, 555–563.

Ferrol N, Barea JM and Azcón-Aguilar C 2000 The plasma membrane H^+-ATPase gene family in the arbuscular mycorrhizal fungus *Glomus mosseae*. Curr. Genet. 37, 112–118.

Ferrol N, Pozo MJ, Antelo M and Azcón-Aguilar C 2002 Arbuscular mycorrhizal simbiosis regulates plasma membrane H^+-ATPase gene expression in tomato plants. J. Exp. Bot. (in press).

Gianinazzi-Pearson V, Smith SE, Gianinazzi S and Smith FA 1991 Enzymatic studies on the metabolism of vesicular-arbuscular mycorrhizas. V: Is H^+-ATPase a component of ATP-hydrolysing enzyme activities in plant-fungus interfaces? New Phytol. 117, 61–79.

Gianinazzi-Pearson V, Arnould C, Oufattole M, Arango M and Gianinazzi S 2000 Differential activation of H^+-ATPase genes by an arbuscular mycorrhizal fungus in root cells of transgenic tobacco. Planta 211, 609–613.

Gollotte A, Cordier C, Lemoine MC and Gianinazzi-Pearson V 1997 Role of fungal wall components in interactions between endomycorrhizal symbionts. *In* Eukaryotism and Symbiosis: Intertaxonomic Combination versus Symbiotic Adaptation. Eds. HEA Schenk, R Herrman, KW Jeon, NE Muller and Schwemmler. pp. 417–427. Springer, Belin, Heidelberg, New York.

Graham JH 2000 Assessing costs of arbuscular mycorrhizal symbiosis in agroecosystems. *In* Current Advances in Mycorrhizae Research. Eds. GK Podila and DD Douds Jr. pp. 127–140. APS Press, St. Paul.

Graham JH, Duncan LW and Eissenstat DM 1997 Carbohydrate allocation patterns in citrus genotypes as affected by phosphorus nutrition, mycorrhizal colonisation and mycorrhizal dependency. New Phytol. 135: 335–343.

Harley JL and Smith SE 1983 Mycorrhizal Symbiosis. Academic Press, London.

Harrison MJ 1996 A sugar transporter from *Medicago truncatula*: altered expression pattern in roots during vesicular-arbuscular (VA) mycorrhizal associations. Plant J. 9, 491–503.

Harrison MJ and van Buuren ML 1995 A phosphate transporter from the mycorrhizal fungus *Glomus versiforme*. Nature 378, 625–629.

Jabari-Hare SH, Therien J and Charest PM 1990 High resolution cytochemical study of the vesicular-arbuscular mycorrhizal association, *Glomus clarum–Allium porrum*. New Phytol. 114, 481–496.

Jackobsen I 1999 Transport of phosphorus and carbon in arbuscular mycorrhizas. *In* Mycorrhiza: Structure, Function, Molecular Biology and Biotechnology. Eds. A Varma and B Hock. pp. 305–332. Springer, Berlin.

Jennings DH 1995 The Physiology of Fungal Nutrition. Cambridge University Press, Cambridge.

Kucey RMN and Paul EA 1982 Carbon flow, photosynthesis, and N_2 fixation in mycorrhizal and nodulated faba beans (*Vicia faba* L.) Soil Biol. Biochem. 14, 407–412.

Maldonado-Mendoza IE and Harrison MJ 1998 A xyloglucan endotransglycosylase (XET) gene from *Medicago truncatula* induced in arbuscular mycorrhizae. In Abstract 2nd Int. Conf. on Mycorrhiza, 5–10 July, 1998, Uppsala, Sweden.

Murphy J, Langridge P and Smith SE 1997 Cloning plant genes differentially expressed during colonization of roots of *Hordeum vulgare* by the vesicular-arbuscular mycorrhizal fungus *Glomus intraradices*. New Phytol. 135, 291–301.

Nehls U, Wiese J, Guttenberger M and Hampp R 1998 Carbon allocation in ectomycorrhizas: Identification and expression analysis of an *Amanita muscaria* monosaccharide transporter. Mol. Plant-Microb. Interact. 14, 334–338.

Nemec S and Vu JCV 1990 Effects of soil phosphorus and *Glomus intraradices* on growth, non-structural carbohydrates, and photosynthetic activity of *Cytrus aurantium*. Plant Soil 128, 257–264.

Palmgren MG and Harper JF 1999 Pumping with plant P-type ATPases. J. Exp. Bot. 50, 883–893.

Pearson JN and Jakobsen I 1993 Symbiotic exchange of carbon and phosphorus between cucumber and three arbuscular mycorrhizal fungi. New Phytol. 124, 481–488.

Perotto S, Brewin N and Bonfante P 1994 Colonization of pea roots by the mycorrhizal fungus *Glomus versiforme* and by *Rhizobium* bacteria: immunological comparison using monoclonal antibodies as probes for plant cell surface components. Mol. Plant-Microbe Interact. 7, 91–98.

Rasmussen N, Lloyd DC, Ratcliffe G, Hansen PE and Jackobsen I 2000 ^{31}P NMR for the study of P metabolism and translocation in arbuscular mycorrhizal fungi. Plant Soil 226, 245–253.

Rosewarne GM, Barker SJ, Smith SE, Smith FA and Schachtman DP 1999 A *Lycopersicon esculentum* phosphate transporter (*LePT1*) involved in phosphorus uptake from a vesicular-arbuscular mycorrhizal fungus. New Phytol. 144, 507–516.

Serrano R 1989 Structure and function of plasma membrane ATPase. Annu. Rev. Plant Physiol. Plant Mol. Biol. 40, 61–94.

Shachar-Hill Y, Pfeffer PE, Douds D, Osman SF, Doner LW and Ratcliffe RG 1995 Partitioning of intermediate carbon metabolism in VAM colonized leek. Plant Physiol. 108, 7–15.

Smith SE and Read DJ 1997 Mycorrhizal Symbiosis. Academic Press, London.

Smith FA and Smith SE 1997 Structural diversity in (vesicular)-arbuscular mycorrhizal symbioses. New Phytol. 137, 373–388.

Smith SE and Smith FA 1990 Structure and function of the interfaces in biotrophic symbioses as they relate to nutrient transport. New Phytol. 114, 1–38.

Solaiman MD and Saito M 1997 Use of sugars by intraradical hyphae of arbuscular mycorrhizal fungi revealed by radiorespirometry. New Phytol. 136, 533–538.

Tisserant B, Gianinazzi-Pearson V, Gianinazzi S and Gollotte A 1992 *In planta* histochemical staining of fungal alkaline phosphatase activity for analysis of efficient arbuscular mycorrhizal infections. Mycol. Res. 97, 245–250.

van Buuren ML, Trieu AT, Blaylock LA and Harrison MJ 1999 Novel genes induced during an arbuscular mycorrhizal (AM) symbiosis formed between *Medicago truncatula* and *Glomus versiforme*. Mol. Plant-Microb. Interact. 12, 171–181.

Wiese J, Kleber R, Hampp R, Nehls U 2000 Functional characterization of the *Amanita muscaria* monosaccharide transporter, AmMst1. Plant Biol. 2, 278–282.

Villalba JM, Lützelschwab M and Serrano R 1991 Immunocytolocalization of plasma membrane H^+-ATPase in maize coleoptiles and enclosed leaves. Planta 185, 485–461.

Woolhouse HW 1975 Membrane structure and transport problems considered in relation to phosphorus and carbohydrate movements and regulation of endotrophic mycorrhizal association. *In* Endomycorrhizas. Eds. FE Sanders, B Mosse and PB Tinker. pp. 209–230. Academic Press, London.

Wright DP, Read DJ and Scholes J 1998 Mycorrhizal sink strength influences whole plant carbon balance of *Trifolium repens* L. Plant Cell Environ. 21, 881–891.

Plant and Soil **244:** 239–245, 2002.

Phosphate transporters of *Medicago truncatula* and arbuscular mycorrhizal fungi

Wayne K. Versaw, Tzyy-Jen Chiou & Maria J. Harrison[1]
The Samuel Roberts Noble Foundation, 2510 Sam Noble Parkway, Ardmore, OK 73401, USA.
[1] *Corresponding author**

Key words: arbuscular mycorrhiza, mineral nutrition, phosphate transporter

Abstract

Most vascular plants acquire phosphate from their environment either directly, via the roots, or indirectly, via a symbiotic interaction with arbuscular mycorrhizal (AM) fungi. The symbiosis develops in the plant roots where the fungi colonize the cortex of the root to obtain carbon from the plant host, while assisting the plant with acquisition of phosphate and other mineral nutrients from the soil solution. As a first step toward understanding the molecular basis of the symbiosis and phosphate utilization, we have cloned and characterized phosphate transporter genes from the AM fungi *Glomus versiforme* and *Glomus intraradices*, and from the roots of a host plant, *Medicago truncatula*. Expression analyses and localization studies indicate that each of these transporters has a role in phosphate uptake from the soil solution.

Abbreviations: AM – arbuscular mycorrhiza

Phosphate availability

Phosphorus is an essential mineral nutrient for the growth and development of plants but its availability is limited by both physical and chemical constraints. Plants acquire phosphorus from soil solution as inorganic phosphate (P_i). However, P_i is not easily accessible to plants due to its low solubility and mobility in soil. A significant fraction (20–80%) of phosphorus in soil is covalently bound to organic molecules and is not directly available to plants (Richardson, 1994). Most of the remaining inorganic fraction exists as insoluble salts or is rapidly adsorbed to clay particles in the soil. As a result, the P_i concentration in soil solution may be as high as 10 μM but more often as low as 1 μM (Marschner, 1995). Some of the mineralized and adsorbed P_i may be released into solution but the presence of P_i depletion zones around roots indicates that the rate of diffusion in soil is much slower than the rate of uptake by plants.

In response to persistent P_i deficiency, plants have evolved several mechanisms to enhance P_i mobilization and its subsequent acquisition. The secretion of phosphatases, organic acids, and protons increase the mobility of soil P_i (Duff et al., 1991; Dunlop and Gardiner, 1983; Lefebvre et al., 1990; Lipton et al., 1987), while enhanced root growth and modification of root morphology increases the root surface area with which to explore a larger volume of soil (Lynch, 1995; Mengel and Kirkby, 1982). In addition, the majority of terrestrial plants are able to form symbiotic associations with arbuscular mycorrhizal (AM) fungi and benefit from the P_i acquired via these fungi (Harrison, 1999; Smith and Read, 1997).

The arbuscular mycorrhiza

Mycorrhiza refers to the symbiotic associations that occur between plant roots and fungi. The most common type of mycorrhizal association is the arbuscular mycorrhiza, which is estimated to occur in 80% of the vascular flowering plant species and involve a limited range of fungi belonging to a single order, the Glomales (Harley and Smith, 1983). The symbiosis is generally of mutual benefit, the plant host provides

* FAX No: +1580-221-7380; Tel.: +1580-223-5810;
E-mail: mjharrison@noble.org

carbon to the obligately biotrophic fungal symbiont, while the fungus assists the plant with the acquisition of P_i and other nutrients from the soil. As a result, AM fungi have significant impact on plant health, growth and yield, particularly for plants growing in nutrient-limiting conditions.

The AM symbiosis initiates when fungal hyphae, arising from spores in the soil or adjacent colonized roots, contact the root surface and form appressoria via which they penetrate the root. Once inside the root, fungal growth is predominantly intercellular proceeding to the inner cortex. Within the inner cortex, hyphae penetrate the cortical cell walls and differentiate to form highly branched structures known as arbuscules. Although the arbuscule develops within a cell, it is essentially apoplastic as the plant plasma membrane remains intact, extending to completely surround it. In these cells, there is an extensive intracellular interface between the two symbionts and it is at this interface that nutrient exchange is thought to occur (Bonfante-Fasolo, 1984; Gianinazzi-Pearson, 1996). Following colonization of the root cortex, the fungus develops an extensive network of hyphae within the soil. This external mycelium has several functions important to the symbiosis including the acquisition of P_i and other nutrients from the soil and their subsequent translocation to the plant, colonization of additional roots, and in many species, the production of spores. The acquisition of P_i from the soil is particularly significant because the fungal hyphae can grow beyond the P_i depletion zone of its host plant to access sources unavailable to the plant. Radiotracer studies have demonstrated that in some cases the fungus is entirely responsible for the uptake of P_i into the mycorrhizal root (Pearson and Jakobsen, 1993a). However, the contribution of the fungus to P_i uptake can also be quite variable depending on the fungal and plant species, indicating that a functional compatibility exists in the symbiosis that is not apparent at the level of colonization (Pearson and Jakobsen, 1993b; Ravnskov and Jakobsen, 1995).

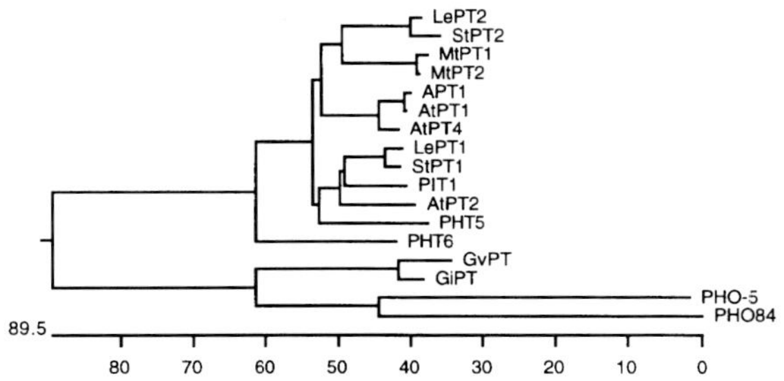

Figure 1. Cladogram of plant and fungal high-affinity P_i transporter sequences. The protein names are followed by their accession numbers. LePT1 (O22548) and LePT2 (O22549) from *Lycopersicon esculentum*; StPT1 (Q43650) and StPT2 (Q41479) from *Solanum tuberosum*; MtPT1 (O22301) and MtPT2 (O22302) from *Medicago truncatula*; APT1 (Q96243), AtPT1 (Q96302), AtPT2 (Q96303), AtPT4 (O04381), PHT5 (O50040) and PHT6 (Q9ZWT3) from *Arabidopsis thaliana*; PIT1 (O22055) from *Catharanthus roseus*; GvPT (Q00908) from *Glomus versiforme*; GiPT (AF359112) from *Glomus intraradices*; PHO-5 (AAA74899.1) from *Neurospora crassa*; PHO84 (P25297) from *Saccharomyces cerevisiae*.

Model systems for the study of phosphate transport in AM symbioses

We selected a model system for molecular and genetic studies of the AM symbiosis comprised of the legume, *Medicago truncatula,* and the AM fungus *Glomus versiforme*. *M. truncatula* is related to alfalfa (*M. sativa*), an important forage crop, but is more amenable to genetic analysis. Some of the key attributes of *M. truncatula* include its small, diploid genome, self-fertility, and rapid generation time (Barker et al., 1990). *G. versiforme* colonizes *Medicago* species efficiently (Bradbury et al., 1991) and produces spores above ground.

The obligate biotrophic nature of AM fungi coupled with the fact that the association occurs underground hampers the study of many aspects of the symbiosis. To facilitate our studies of P_i translocation, we have used an *in vitro* mycorrhizal culture system in which the fungus and roots are cultured in one half of a two-compartment petri plate and the external mycelium ramifies into the other compartment separate from the roots (St-Arnaud et al., 1996). These systems are being used to investigate the molecular processes of P_i transport in AM fungi and in plant roots prior to, and during, the mycorrhizal symbiosis. Our current findings are summarized in this review.

Cloning and characterization of P_i transporters from *G. truncatula* and *G. intraradices*

A cDNA clone, *GvPT*, that shares sequence similarity with the H^+/P_i symporter gene *PHO84* (Bun-ya et al., 1991) of yeast was isolated from a *M. truncatula*/*G. versiforme* cDNA library (Harrison and van Buuren, 1995). *GvPT* encodes a protein of 521 amino acids that shares 47.9% identity with PHO84 and 45% identity with PHO-5, a H^+/P_i symporter from *Neurospora crassa* (Versaw, 1995). Analysis of genomic DNA derived from each of the symbionts demonstrated that *GvPT* represents a gene from *G. versiforme*. As might be expected, the fungal P_i transporters are more closely related to each other than to H^+/P_i symporters of plant origin (Figure 1).

GvPT, like PHO84 and PHO-5, is predicted to be an integral membrane protein containing 12 membrane-spanning domains. However, this general secondary structure is common to many carrier-type transporters with widely varied substrate specificity (Pao et al., 1998). Therefore, to confirm that *GvPT* encodes a functional P_i transporter, a cDNA clone was expressed in a *pho84* yeast mutant. GvPT complemented the *pho84* mutation and mediated high-affinity P_i transport with a K_m of 18 μM. Addition of carbonyl-cyanide-*m*-chlorophenyl-hydrazone (CCCP), which uncouples the proton gradient across the membrane, abolished further P_i uptake, indicating that GvPT phosphate transport activity requires a proton gradient.

The AM fungal-mediated translocation of P_i from soil to plant presumably requires specific transporters that function in at least two fungal structures. External hyphae are expected to mediate uptake from the soil, and arbuscules, located within the root cortical cells are predicted to have an efflux transporter or channel (Smith, 1993). *GvPT* transcripts were detected by RT-PCR in external hyphae, but were barely detectable in internal structures within the root. Previous analysis of P_i uptake by hyphal germ tubes of the AM fungus *Gigaspora margarita* (Thomson et al., 1990) suggested that like other fungi (Beever and Burns, 1980; Tamai et al., 1985), AM fungi possess both high-affinity (K_m 1.8–3.1 μM) and low-affinity (10.2–11.3 mm) P_i uptake activities. The K_m and localization of GvPT are consistent with those of a high-affinity system and suggest a role for GvPT in the acquisition of P_i from soil.

To facilitate use of the root/*G. intraradices in vitro* culture system (St-Arnaud et al., 1996) for analyses of P_i transport, it was necessary first to clone the *G. intraradices* P_i transporter. *GiPT*, a *G. intraradices* homolog of *GvPT*, shares 90% identity at the nucleotide level and 95% identity at the amino acid level with the coding region of *GvPT* (Maldonado-Mendoza et al., 2001). To evaluate the expression of *GiPT* in external hyphae, mycorrhizas were established between *G. intraradices* and either *M. truncatula* or carrot roots in the culture system. The medium in the compartment containing the mycorrhizal root was solidified

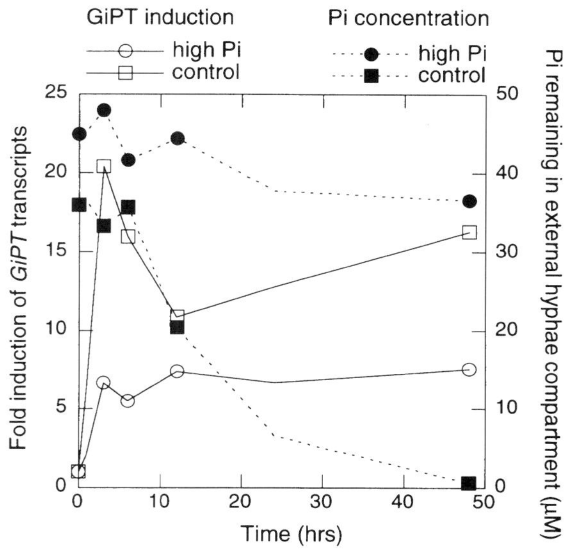

Figure 2. Effect of pretreating carrot/*G. intraradices* mycorrhizal roots with a high concentration of P_i on the induction of *GiPT* transcripts in the external hyphae and the P_i uptake activity of the external hyphae. Established mycorrhizas in one side of a two-compartment plate were treated with high (3.5 mM) or no P_i (control) for 48 h. The external hyphae located in the second compartment were then exposed to fresh medium containing 35 μM P_i and samples of hyphae and medium surrounding the hyphae were removed over the following 48 h (Data taken from Maldonado-Mendoza et al., 2001).

with agar, whereas liquid medium was placed in the second compartment to enable manipulation of the environmental conditions surrounding external hyphae. Both mycorrhizas were functional in terms of P_i transport from the external hyphae to the mycorrhizal roots. Exchange of the medium in the external hyphae compartment with fresh medium containing variable P_i concentrations revealed that the expression of *GiPT* is regulated in response to the abundance of P_i surrounding the external hyphae. Transcript levels are elevated in response to μM levels of P_i, but not to mM levels nor to the absence of P_i. These results indicate that *G. intraradices* has a mechanism for sensing P_i in the environment surrounding the external hyphae and that the turnover of *GiPT* transcripts is rapid. This expression pattern is consistent with a role in the high-affinity uptake of P_i from the soil solution.

In addition, when mycorrhizal roots were pretreated with a high (3.5 mM) level of P_i, the induction of *GiPT* in external hyphae was significantly less than controls, which did not receive a high P_i pretreatment (Figure 2). Furthermore, P_i was not translocated by the external hyphae in the high-P_i treated mycorrhizas

(Figure 2). These results indicate that the P_i status of the mycorrhizal root influences *GiPT* expression and ultimately, P_i uptake by the external hyphae. It is not currently possible to determine the contribution GiPT makes to the P_i uptake activity of external hyphae since additional transporters with a similar function may exist (Maldonado-Mendoza et al., 2001).

Cloning and characterization of P_i transporters from *M. truncatula*

Most vascular plants have two pathways for the uptake of P_i from soil solution, either directly via the roots, or indirectly via association with AM fungi. The kinetics of P_i uptake by roots are often complex, but are generally interpreted as the result of two uptake systems, a high-affinity and a low-affinity system (Mimura, 1999). The high-affinity system is derepressed under conditions of limiting P_i such that it is operating at concentrations in the μM range and therefore requires an active transport mechanism to accumulate P_i against a steep electrochemical gradient. The low-affinity system is expressed constitutively and functions in the mM range using either a passive or active transport mechanism. Uptake of P_i released from AM fungi is predicted to occur at the arbuscule/cortical cell interface via active P_i transporters located in the plant plasma membrane at this interface (Smith and Smith, 1989, 1990). It is possible that the plant P_i transporters operating at the symbiotic interface may be the same as those operating in the nonmycorrhizal root (Rosewarne et al., 1999).

To identify *M. truncatula* P_i transporters that might be active in the AM symbiosis, a *M. truncatula*/*G. versiforme* cDNA library was screened with an Arabidopsis EST that shares similarity with the yeast *PHO84* gene. Two genes, *MtPT1* and *MtPT2*, were identified that share 91.6% identity at the nucleotide level over the coding regions, but contain significantly different 5' and 3' untranslated sequences (Liu et al., 1998b). The encoded proteins share 75–85% identity with P_i transporters identified in other vascular plants (Daram et al., 1998; Kai et al., 1997; Leggewie et al., 1997; Liu et al., 1998a; Mitsukawa et al., 1997; Muchhal et al., 1996; Okumura et al., 1998; Smith et al., 1997), and 29–37% identity with the fungal transporters (Fig. 1). These related plant and fungal P_i transporter sequences comprise the P_i:H^+ symporter (PHS) family, a distinct phylogenetic family within the major facilitator superfamily (Pao et al., 1998).

P_i uptake measurements in yeast expressing *MtPT1* confirmed that this clone encodes a functional P_i transporter with an apparent K_m of 192 μM (Liu et al., 1998b). The uptake activity of MtPT2 was not tested, but based on the high identity (98%) with MtPT1, it is likely that it is also capable of P_i transport. The $K_m$value for MtPT1 is comparable to those of other related plant P_i transporters expressed in yeast (Daram et al., 1998; Leggewie et al., 1997; Liu et al., 1998a; Muchhal et al., 1996), but significantly greater than expected for high-affinity transporters. This may be a consequence of expression in yeast since expression of one transporter gene (*AtPT1*) in tobacco cells showed a K_m of 3.1 μM.

MtPT1 and *MtPT2* are expressed in roots, but not in leaves and the expression of both is greater under conditions of P_i deprivation with *MtPT1* slightly more responsive to P_i deprivation than *MtPT2* (Liu et al., 1998b). Using an antibody specific for MtPT1, we confirmed that MtPT1 protein levels correlate closely with the transcript levels and are elevated in response to P_i deprivation (Chiou et al., 2000). The related P_i transporter genes identified in other plant species show similar expression patterns, although some also have detectable expression in shoot tissues (Daram et al., 1998; Leggewie et al., 1997; Liu et al., 1998a; Muchhal et al., 1996; Smith et al., 1997). These expression patterns are consistent with the derepressible, high-affinity uptake activity measured in roots and cell suspensions (Mimura, 1999). The increased accumulation of these transporters under conditions of P_i deprivation provides an explanation for the observation that under such conditions, P_i uptake activity in roots shows an increase in V_{max} without a change in K_m (Cogliatti and Clarkson, 1983; Dunlop et al., 1997).

The transcript levels of both *MtPT1* and *MtPT2* decrease during the development of a mycorrhizal symbiosis with either *G. versiforme* or *G. intraradices* and the decline is detected early in the colonization process. MtPT1 protein levels also decrease in roots during development of the symbiosis. Therefore it is unlikely that these transporters are involved in the uptake of P_i across the arbuscule–cortical cell interface. The down-regulation of *MtPT1* following colonization of the roots with an AM fungus may result from signals from the fungus, or may be an indirect consequence of increased P_i levels in the plant (Chiou et al., 2000; Liu et al., 1998b). Another P_i starvation-induced gene, *Mt4*, is down-regulated in AM roots

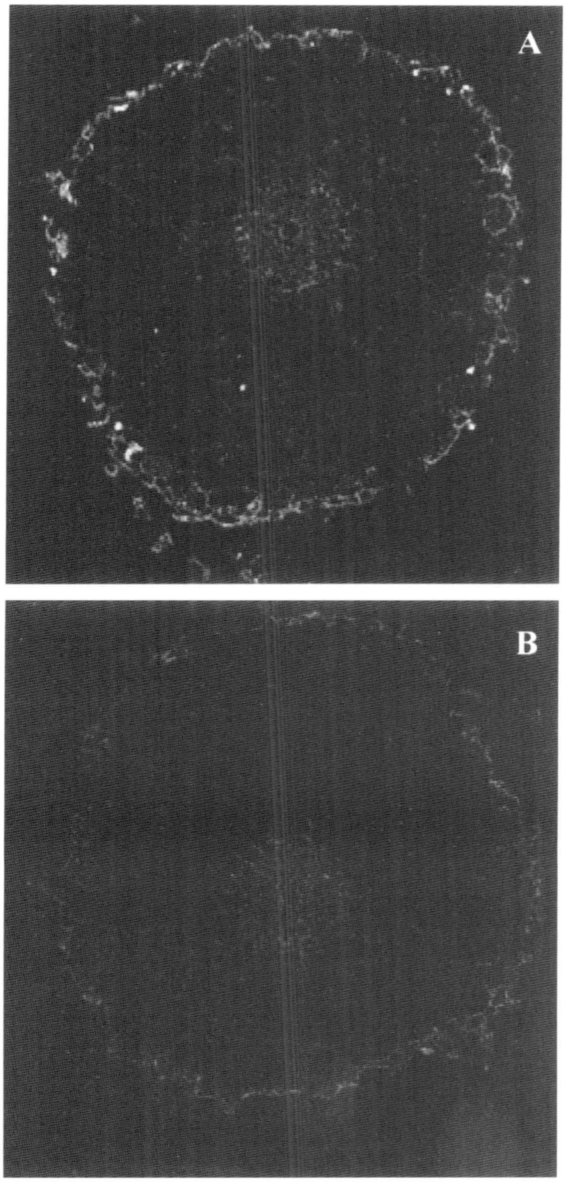

Figure 3. MtPT1 is located in the epidermal cells of *M. truncatula* roots. Transverse sections of *M. truncatula* roots grown in low P_i conditions were hybridized with the MtPT1 antibody (A) or with preimmune serum (B). The primary antibodies were detected with an anti-rabbit antibody conjugated with FITC. The green fluorescent signal in the epidermis of sections indicates the presence of the MtPT1 antibody.

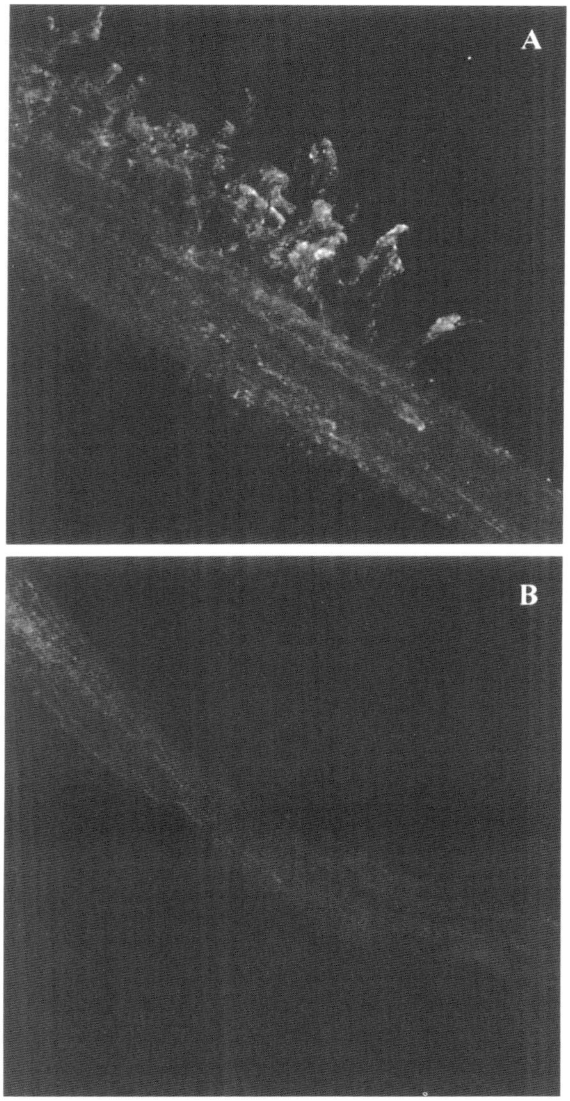

Figure 4. Whole-mount immunolocalization indicates that MtPT1 is present in the root hairs. *M. truncatula* root pieces were hybridized with the MtPT1 antibody (A) or with preimmune serum (B). The primary antibodies were detected with an anti-rabbit antibody conjugated with FITC. The green fluorescent signal in the root hairs indicates the presence of the MtPT1 antibody.

in response to a signal from the fungus (Burleigh and Harrison, 1998).

Examination of the spatial expression patterns of *MtPT1* revealed that transcripts and encoded protein are localized specifically in the epidermal cells (Figure 3) and root hairs (Figure 4) of the root and that the protein is present in the plasma membrane of roots grown under low P_i conditions (Chiou et al., 2000). The expression pattern of MtPT1, coupled with its cellular and subcellular location, suggests that this transporter plays a role in uptake of P_i from the soil solution and not in P_i uptake into the vacuole. The tomato *LePT1* transcripts and encoded protein show similar expression and subcellular localization patterns as *MtPT1* but *LePT1* transcripts are also detected in root cap, stele, and leaf parenchyma cells, as well as cells containing arbuscules (Daram et al., 1998; Liu et al., 1998a; Muchhal and Raghothama, 1999; Rosewarne et al., 1999). These differences suggest overlapping but distinct physiological roles for these transporters.

Conclusions

The cloning of P_i transporters from plants and AM fungi is an essential first step toward understanding the molecular basis of P_i transport in the symbiosis and the mechanisms of P_i utilization in plants. To date, two P_i transporters from AM fungi have been identified and at least 13 P_i transporters from five different plant species have been characterized, including two from *M. truncatula*. These numbers are likely to increase rapidly in the near future, due in part to the advancement of large-scale EST and genomic sequencing projects. These sequence databases will allow rapid screening for novel orthologous transporters, some of which may have escaped detection by hybridization methods. In turn, newly identified transporter genes can be used to probe a wide range of plant and fungal species. The investigation of a diverse group of symbioses may be required to reveal the factors responsible for functional compatibility between symbionts.

References

Barker D G, Bianchi S, Blondon F, Dattée Y, Duc G, Essad S, Flament P, Gallusci P, Génier G, Guy P, Muel X, Tourneur J, Dénarié J and Huguet T 1990 *Medicago truncatula*, a model plant for studying the molecular genetics of the *Rhizobium*-legume symbiosis. Plant Mol. Biol. Rep. 8, 40–49.

Beever R E and Burns D J W 1980 Phosphorus uptake, storage and utilization by fungi. Adv. Bot. Res. 127–129.

Bonfante-Fasolo P 1984 Anatomy and morphology of VA mycorrhizae. *In* VA Mycorrhizae. Eds. C L Powell and D J Bagyaraj. pp 5–33. CRC Press, Boca Raton, FL.

Bradbury S M, Peterson R L and Bowley S R 1991 Interaction between three alfalfa nodulation genotypes and two *Glomus* species. New Phytol. 119, 115–120.

Bun-ya M, Nishimura M, Harashima S and Oshima Y 1991 The*PHO84* gene of *Saccharomyces cerevisiae* encodes an inorganic phosphate transporter. Mol. Cell. Biol. 11, 3229–3238.

Burleigh S H and Harrison M J 1998 Mt4, a phosphorus starvation-inducible cDNA from *Medicago truncatula*, which is down-regulated both by phosphorus fertilization and arbuscular-mycorrhizal colonization. *In* Phosphorous in Plant Biology: Regulatory Roles in Molecular Cellular, Organismic, and Ecosystem Processes. Eds. J P Lynch and J Deikman. pp. 359–360. American Society of Plant Physiologists, Rockville, MD.

Chiou T J, Liu H and Harrison M J 2000 The spatial expression patterns of a phosphate transporter (MtPT1) from *Medicago truncatula* indicate a role in phosphate transport at the root/soil interface. Plant J. 25, 1–15.

Cogliatti H H and Clarkson D T 1983 Physiological changes in, and phosphate uptake by potato plants during development of, and recovery from phosphate deficiency. Physiol. Plant. 58, 287–294.

Daram P, Brunner S, Persson B L, Amrhein N and Bucher M 1998 Functional analysis and cell-specific expression of a phosphate transporter from tomato. Planta 206, 225–233.

Duff S M, Plaxton W c and Lefebvre D D 1991 Phosphate-starvation response in plant cells: *de novo* synthesis and degredation of acid phosphatases. Proc. Natl. Acad. Sci. USA 88, 9538–9542.

Dunlop J and Gardiner S 1983 Phosphate uptake, proton extrusion and membrane electropotentials of phosphorus deficient *Trifolium repens*. J. Exp. Bot. 44, 1801–1808.

Dunlop J, Phung H T, Meeking R and White D W R 1997 The kinetics associated with phosphate absorption by *Arabidopsis* and its regulation by phosphorus status. Aust. J. Plant Physiol. 24, 623–629.

Gianinazzi-Pearson V 1996 Plant cell responses to arbuscular myorrhiza fungi: Getting to the roots of the symbiosis. Plant Cell 8, 1871–1883.

Harley J L and Smith S E 1983 Mycorrhizal Symbiosis. Academic Press, London and New York.

Harrison M J 1999 Molecular and cellular aspects of the arbuscular mycorrhizal symbiosis. Annu. Rev. Plant Physiol. Plant Mol. Biol. 50, 361–389.

Harrison M J and van Buuren M L 1995 A phosphate transporter from the mycorrhizal fungus *Glomus versiforme*. Nature 378, 626–629.

Kai M, Masuda Y, Kikuchi Y, Osaki M and Tadano T 1997 Isolation and characterization of a cDNA from *Catharanthus roseus* which is highly homologous with phosphate transporter. Soil Sci. Plant Nutr. 43, 227–235.

Lefebvre D D, Duff S M, Fife C A, Julien-Inalsingh C and Plaxton W C 1990 Response to phosphate deprivation on *Brassica nigra* suspension cells. Plant Physiol. 85, 315–317.

Leggewie G, Willmitzer L and Riesmeier J W 1997 Two cDNAs from potato are able to complement a phosphate uptake-deficient yeast mutant: Identification of phosphate transporters from higher plants. Plant Cell 9, 381–392.

Lipton D S, Blanchar R W and Blevins D G 1987 Citrate, malate and succinate concentration in exudates from P-sufficient and P-stressed *Medicago sativa* L. seedlings. Plant Physiol. 85, 315–317.

Liu C, Muchhal U S, Uthappa M, Kononowicz A K and Raghothama K G 1998a Tomato phosphate transporter genes are differentially regulated in plant tissues by phosphorus. Plant Physiol. 116, 91–99.

Liu H, Trieu A T, Blaylock L A and Harrison M J 1998b Cloning and characterization of two phosphate transporters from *Medicago truncatula* roots: Regulation in response to phosphate and to colonization by arbuscular mycorrhizal (AM) fungi. Mol. Plant-Microbe Interact. 11, 14–22.

Lynch J 1995 Root architecture and plant productivity. Plant Physiol. 109, 7–13.

Maldonado-Mendoza I E, Dewbre G R and Harrison M J 2001 Expression of a *Glomus intraradices* phosphate transporter gene (*GiPT*) in the extra-radical mycelium of an arbuscular mycorrhiza: regulation in response to phosphate. Mol. Plant-Microbe Interact. 14, 1140–1148.

Marschner H 1995 Nutrient availability in soils. *In* Mineral Nutrition of Higher Plants. Ed. H. Marschner. pp. 483–507. Academic Press, London.

Mengel K and Kirkby E A 1982 Principles of Plant Nutrition. International Potash Institute, Bern.

Mimura T 1999 Regulation of phosphate transport and homeostasis in plant cells. Int. Rev. Cytol. 191, 149–200.

Mitsukawa N, Okumura S, Shirano Y, Sato S, Kato T, Harashima S and Shibata D 1997 Overexpression of an *Arabidopsis thaliana* high-affinity phosphate transporter gene in tobacco cultured cells enhances cell growth under phosphate-limited conditions. Proc. Natl. Acad. Sci. USA 94, 7098–7102.

Muchhal U S and Raghothama K G 1999 Transcriptional regulation of plant phosphate transporters. Proc. Natl. Acad. Sci. USA 96, 5868–5872.

Muchhal U S, Pardo J M and Raghathama K G 1996 Phosphate transporters from the higher plant *Arabidopsis thaliana*. Proc. Natl. Acad. Sci. USA 93, 101519–110523.

Okumura S, Mitsukawa N, Shirano Y and Shibata D 1998 Phosphate transporter gene family of *Arabidopsis thaliana*. DNA Res. 5, 261–269.

Pao S S, Paulsen I T and Saier M H Jr. 1998 Major facilititator superfamily. Microbiol. Mol. Biol. 62, 1–34.

Pearson J N and Jakobsen I 1993a The relative contribution of hyphae and roots to phosphorus uptake by arbuscular mycorrhizal plants, measured by dual labelling with ^{32}P and ^{33}P. New Phytol. 124, 489–494.

Pearson J N and Jakobsen I 1993b Symbiotic exchange of carbon and phosphorus between cucumber and three arbuscular mycorrhizal fungi. New Phytol. 124, 481–488.

Ravnskov S and Jakobsen I 1995 Functional compatibility in arbuscular mycorrhizas measured as hyphal P transport to the plant. New Phytol. 129, 611–618.

Richardson A E 1994 Soil microorganisms and phosphorus availability. Soil Biota 50–62.

Rosewarne G, Barker S, Smith S, Smith F and Schachtman D 1999 A *Lycopersicon esculentum* phosphate transporter (LePT1) involved in phosphorous uptake form a vesicular-arbuscular mycorrhizal fungus. New Phytol. 144, 507–516.

Smith F A and Smith S E 1989 Membrane transport at the biotrophic interface: an overview. Aust. J. Plant Physiol. 16, 33–43.

Smith F W, Ealing P M, Dong B and Delhaize E 1997 The cloning of two *Arabidopsis* genes belonging to a phosphate transporter family. Plant J. 11, 83–92.

Smith S E 1993 Transport at the mycorrhizal interface. Mycorrhiza News 5, 1–3.

Smith S E and Read D J, 1997 Mycorrhizal Symbiosis. Academic Press, San Diego, CA.

Smith S E and Smith F A 1990 Structure and function of the interfaces in biotrophic symbioses as they relate to nutrient transport. New Phytol. 114, 1–38.

St-Arnaud M, Hamel C, Vimard B, Caron M and Fortin J A 1996 Enhanced hyphal growth and spore production of the arbuscular mycorrhizal fungus *Glomus intraradices* in an *in vitro* system in the absence of host roots. Mycol. Res. 100, 328–332.

Tamai Y, Toh-e A and Oshima Y 1985 Regulation of inorganic phosphate transport systems in *Saccharomyces cerevisiae*. J. Bacteriol. 164, 964–968.

Thomson B D, Clarkson D T and Brain P 1990 Kinetics of phosphorus uptake by the germ-tubes of the vesicular-arbuscular mycorrhizal fungus, *Gigaspora margarita*. New Phytol. 116, 647–653.

Versaw W K 1995 A phosphate-repressible, high-affinity phosphate permease is encoded by the *pho-5*$^{+}$ gene of *Neurospora crassa*. Gene 153, 135–139.

Plant and Soil **244:** 247–251, 2002.

Plant nutrient transporter regulation in arbuscular mycorrhizas

Stephen H. Burleigh[1,3] & Iben Ellegaard Bechmann[2]
[1]*Plant Nutrition Programme, Building 301, Risø National Laboratory, Roskilde, DK4000, Denmark;* [2]*Novo Nordisk Engineering A/S, Bagsværd, DK2880, Denmark.* [3]*Corresponding author**

Received 21 August 2001. Accepted in revised form 8 January 2002

Key words: arbuscular mycorrhizas, nutrient transporters, plants, regulation

Abstract

This review discusses the role arbuscular mycorrhizal fungi play in the regulation of plant nutrient transporter genes. Many plant nutrient transporter genes appear to be transcriptionally regulated by a feed-back mechanism that reduces their expression when the plant reaches an optimal level of nutrition. Their down-regulation in mycorrhizal roots, therefore, would be predicted as a result of symbiotic function. A variety of studies on Pi- Zn- and ammonium- or nitrate-transporter genes from two plant species indirectly support this model. For example, one study showed that the expression of the high-affinity Pi-transporter MtPT2 within mycorrhizal roots of *Medicago truncatula* was inversely correlated with the concentration of P within the shoots, which suggested that P supply from the fungus influenced this gene's expression. However, there is some evidence that these plant nutrient transporters are also down-regulated within mycorrhizal roots by a mechanism not involving nutrient supply by the fungus. Carbon demand by the fungus may be one factor that can influence their expression within mycorrhizal roots.

Introduction

Arbuscular mycorrhizal fungi (AMF) are renowned for their ability to improve the nutrition of plants, with phosphate being the most recognized plant nutrient supplied by these fungal symbionts. One estimate suggested that up to 80% of the Pi taken up by a mycorrhizal plant can be supplied by the fungus (Marschner and Dell, 1994). AM fungi can also provide other macro- and micro-nutrients such as N, Zn, Cu and Mg, which can likewise result in improved plant growth (Clark and Zeto, 2000; Smith and Read, 1997). However, AMF do not always improve plant nutrition. While the AM symbiosis has essentially no host specificity as observed in the *Rhizobium*–legume interaction, this symbiosis can be said to have a high level of functional compatibility. One definition of functional compatibility is the degree to which a fungus-plant combination can promote P-uptake in the plant and is exemplified when a particular AM isolate improves the P-uptake of one plant species, but not another when grown under identical environmental conditions (Ravenskov and Jakobsen, 1995). Even plant-fungus combinations that normally improve plant growth can result in plants with reduced nutrition, depending on environmental conditions. For example, some AMF cause growth depressions in host plants when grown on soils containing high levels of P (Graham and Abbott, 2000). Another case where AMF do not improve the nutrition of their host plants involves a 'dilutive effect' observed in some mycorrhizas, whereby the P-induced growth response of the plant apparently increases the demand for other nutrients, which can lead to their deficiency if these nutrients are limiting in the soil (Smith and Read, 1997; Timmer and Leyden, 1980). Figure 1 is an example of this apparent dilutive effect in *M. truncatula* colonized by five species of AMF, whereby shoot P and Cu levels were elevated, but Mg and Mn levels were reduced relative to an uncolonized control. A third and rather surprising response that some plants have to AM colonization is a reduced uptake of certain metals when plants are exposed to relatively toxic levels of these metals (Zhu et al., 2001). The mechanism behind this protective

* FAX No: 4677-4282. E-mail: stephen.burleigh@risoe.dk

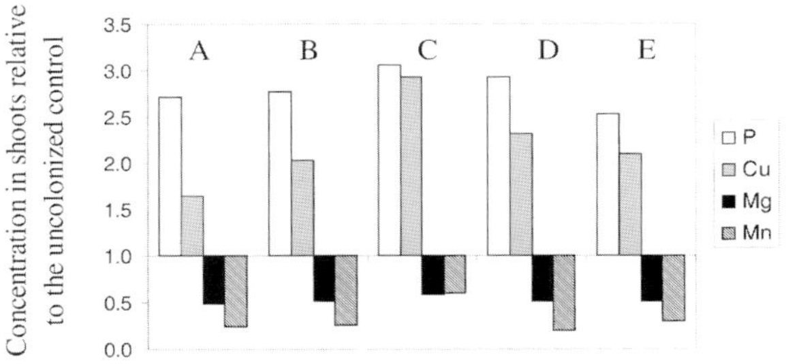

Figure 1. The relative concentration of four inorganic nutrients within the shoots of *Medicago truncatula* colonized by five species of AMF. Concentrations determined using a ICP-AES analyser and presented relative to the uncolonized control (=1.0). Mycorrhizas of *Glomus versiforme*, (A) *G. intraradices*, (B) *G. caledonium*, (C) *G. claroideum*, (D) and *G. mosseae* (E).

effect remains obscure and might be due to the binding of the metal to the fungal mycelium, or possibly due to fungal-induced changes in rhizosphere pH, which could alter the solubility and therefore availability of the metal (Li and Christie, 2001).

Thus, AMF can both increase and decrease the level of specific nutrients within their host plants. With the advent of the cloning of numerous plant genes involved in nutritional processes, these fungal-induced changes in plant nutrition can now be observed at the molecular level. One might predict that the expression of these nutrition-related plant genes would be associated with the nutritional status of the mycorrhizal plant, either up- or down-regulated depending on how well the symbiosis functioned in terms of providing various nutrients to the plant. However, it should be also considered that alterations in their expression could be due to fungal influences not directly related to the symbiotic transfer of these nutrients. For example, studies on the nitrate transporter Nrt2,1 from *Arabidopsis* have found that this gene was regulated not only by the N nutrition of the plant, but also by the plant's carbon (C) nutrition, whereby C stress as a result of reduced photosynthesis down-regulated the expression of this gene (Lejay et al., 1999). Hence, C demand by AMF is one such effect that needs to be considered in models of plant nutrient transporter regulation within mycorrhizal roots. Here, the expression patterns of plant Pi-, Zn- and ammonium- or nitrate-transporter genes are discussed in an attempt to better understand their regulation within mycorrhizal roots.

The regulation of plant phosphate transporter genes within mycorrhizal roots

The study of plant Pi transporter gene expression in mycorrhizas was initiated as soon as the first of these genes were cloned from plants. It was found that high-affinity Pi-transporter genes from both *M. truncatula* and tomato were down-regulated not only in uncolonized roots receiving P-fertilization, but also in mycorrhizal roots not receiving high levels of P fertilizer (Liu et al., 1998; Roswarne et al., 1999). A similar response was observed for the P-starvation-inducible plant gene Mt4 in mycorrhizal roots of *M. truncatula* (Burleigh and Harrison, 1997). These observations lead to the suggestion that their down-regulation by AMF colonization was due do the improved P nutrition of the plant as a result of symbiotic function. In this model, the plant perceives an increase in cellular concentrations of Pi, due to both its own uptake and fungal supply, and correspondingly reduces the expression of its own P transporter genes to maintain an optimal level of nutrition within the plant.

Recently, Burleigh and Jakobsen (unpublished) explored the functional diversity of various plant-fungus combinations to further test this hypothesis. They found that the expression of the high-affinity Pi-transporter MtPT2 and the P-starvation-inducible gene Mt4 within the roots of mycorrhizal *M. truncatula* was negatively, linearly associated with the shoot P concentration of these plants. Hence, their results supported the idea that the down-regulation of P-starvation-inducible plant genes within mycorrhizal roots was primarily an indirect effect as a result of the transfer of Pi from the fungus to the plant (Burleigh and Harrison, 1997; Liu et al., 1998; Rosewarne et al., 1999). Based on the number and diversity of the AMF species used in their study, they suggested that this putative mechanism of regulation could be common among the many AM symbioses.

On a technical note, the expression studies mentioned above used rRNA as a load control in their RNA blot analyses. However, since fungal RNA contributed to the total pool of RNA isolated from the mycorrhizal roots, it was theoretically possible that the reduced expression of these plant genes was, at least in part, due to fungal RNA diluting the samples. However, as discussed by Burleigh (2001) one might expect only about a 10–15% reduction in the signal of plant gene expression as a result of this RNA dilution, since that is an approximate amount of fungal RNA found within a 'typical' colonized root (IE Maldonado–Mendoza and MJ Harrison, pers comm).

The expression of a plant zinc transporter gene within AM roots

A differential screen of an *M. truncatula* mycorrhizal cDNA library identified several plant genes that were down-regulated within roots as a result of AM colonization (Burleigh and Harrison, 1997). One cDNA was identified as a plant zinc transporter and its function was confirmed by the complementation of a yeast Zn-uptake mutant (S Burleigh, unpublished). Expression studies with this gene found that, unlike any of the other zinc transporters characterized thus far from plants, it had the unique feature of being up-regulated in uncolonized roots by Zn fertilization. In roots colonized by a variety of AMF, however, the expression of this gene was down-regulated, which was the opposite response that would be predicted if the AMF had improved the Zn nutrition of their host plants. But when the mineral content was analysed within the shoots of these mycorrhizal plants, it was found that Zn levels were reduced relative to that of the uncolonized control. Hence, these results exemplified the 'dilutive effect' AMF colonization can have on certain plant nutrients, in this case Zn, and were consistent with the idea that the expression of this plant nutrient transporter was influenced by nutritional changes within the plant as a result of symbiotic function.

The expression of a plant nitrate transporter gene within AM roots

In a recent study a putative nitrate transporter gene from *M. truncatula* was identified by a homology-based NCBI database search. Using a quantitative RT-PCR technique it was shown that the expression of this gene was down-regulated in roots colonized by some, but not all AMF (Burleigh, 2001). While the expression of this gene was not compared to the N nutrition of these mycorrhizal plants, the results nonetheless showed that AMF can influence the expression of a diverse selection of plant nutrient transporters, including those that transport both macro- (P, N) and micronutrients (Zn) within the plant, transport elements that are both relatively immobile (P, Zn) and mobile (N) in the soil and that are both down- (P) and up- (Zn) regulated within roots by fertilization with their respective nutrients.

Are plant nutrient transporter genes regulated by other means within mycorrhizas?

Some studies have suggested that the down-regulation of P-starvation-inducible plant genes in mycorrhizal roots might also be due in part to effects other than those involving the P nutrition of the plant. Burleigh and Harrison (1997) found that an alfalfa mycorrhizal mutant that limits colonization to the root surface and thus provides little or no P to the plant nonetheless had reduced expression of a P-starvation-inducible plant gene within its roots. Furthermore, in a study by Rosewarne et al. (1999), the expression of a high-affinity Pi-transporter from tomato was found to be down-regulated in mycorrhizal roots, presumably due to Pi supply by the fungus, but at the same time up-regulated in arbuscule-containing cells, which, as a proposed site of Pi transfer from the fungus to the plant (Smith and Read, 1997), would likely have elevated levels of Pi relative to neighbouring plant cells. Thus, it appeared that AMF could alter the expression of nutrition-related plant genes by two mechanisms, one by nutrient transfer from the fungus to the plant and another by an unknown means not involving nutrient transfer, yet somehow associated with fungal contact.

In a recent study, Burleigh and Jakobsen (unpublished) found that the transcriptional regulation of P-starvation-inducible plant genes within the roots of mycorrhizas was likewise not completely explained by the plant's P nutrition. They found that while the expression of the Mt4 gene within the roots of mycorrhizal *M. truncatula* was negatively correlated with shoot P concentration, its expression was better, negatively correlated with the shoot's fresh weight–dry weight (FW–DW) ratio. The FW–DW ratio is an inverse measure of the plant's dry matter content and elevations in this ratio can be sometimes considered an indicator of C stress in mycorrhizal plants, whereby fungal C demand outpaces photosynthesis and results in a corresponding reduction in photosynthate available for plant growth (Smith and Gianinazzi–Pearson, 1990; Son and Smith, 1988; Tester et al., 1986). It should be remembered that this fungal demand for photosynthate is rather large- up to 20% of the plant's fixed carbon (Johnson et al., 1997).

It can likewise be shown that photosynthesis can influence the expression of a plant Pi-transporter gene. MtPT2 expression was examined in roots of *M. truncatula* grown under light- and P-limitation. In plants exposed to high levels of light, the expression of MtPT2 was up-regulated in the roots of the P-limited

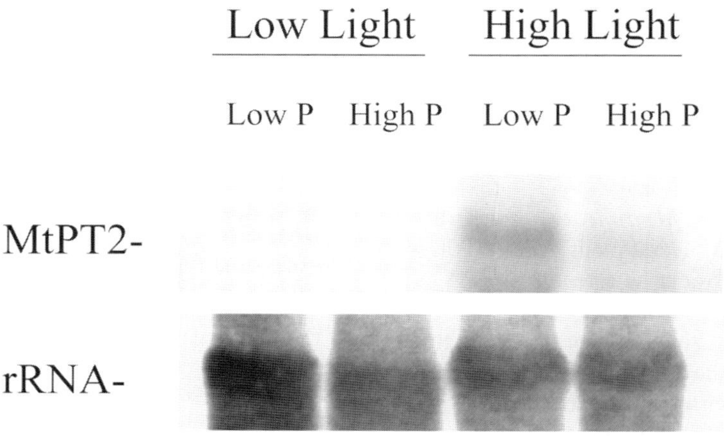

Figure 2. Expression of the Pi-transporter MtPT2 in the roots of *M. truncatula* grown under Pi- and light-limitation. Northern blots of total RNA isolated from roots of plants fertilized with 3.5 (Low P) or 50 (High P) mg kg^{-1} Pi in the form of KH_2PO_4 and exposed to either 10 (Low Light) or 400 (High Light) $\mu E\ m^{-2}\ s^{-1}$ of light 3 days prior to harvest. Blots were probed with ^{32}P-labelled MtPT2 cDNA (upper panel) or ^{32}P-labelled 18S rDNA (lower panel).

plants and down-regulated in the roots of the P-fertilized plants (Figure 2). However, in plants exposed to low levels of light 3 days prior to harvest, expression was down-regulated in the roots of both the P-limited and P-fertilized plants. Physiological symptoms of light-limitation included an elevation in the plant's FW–DW ratio, elongation of the shoot and lighter leaf color. The low-light treatment clearly did not shut down root metabolism, since 18S rRNA was abundantly expressed in these roots. Light effects rather than reduced photosynthate levels might have been responsible for the observed changes in gene expression, although at least in the case of the N-transporter Nrt2,1 from *Arabidopsis,* down-regulation was due to reduced photosynthate levels (Lejay et al., 1999). The role that C-stress plays in the regulation of plant Pi-transporter genes is currently being tested in our laboratory.

Summary

It appears that the transcriptional regulation of plant nutrient transporter genes within mycorrhizal roots is strongly influenced by the plant's nutritional status. This initial conclusion is based on studies involving three plant nutrient transporters, a diverse selection of AMF and two host plant species. In other words, their altered expression in mycorrhizal roots appears to be primarily an indirect effect as a result of nutrient transfer from the fungus to the plant. However, some evidence has also shown, at least in the case of plant Pi-transporter genes, that these fungi can influence their expression by a mechanism not entirely explained by nutrient transfer. In particular, some indirect evidence suggests that their down-regulation within mycorrhizal roots may be somehow linked to the C nutrition of the plant. The biological significance of this proposed dual mechanism of regulation may be that the expression of some plant P-transporter genes are both positively (C) and negatively (Pi) regulated within roots, which might allow the plant greater control to fine-tune their transcription. Considering that mycorrhizal plants often grow under some level of C-stress as a result of fungal C demand (Smith and Gianinazzi–Pearson, 1990; Son and Smith, 1988), one might expect that the expression of plant Pi-transporter genes in roots would be, as observed thus far in two plant species (Liu et al., 1998; Roswarne et al., 1999), down-regulated by AM colonization.

References

Burleigh S H 2001 Relative quantitative RT-PCR to study the expression of plant nutrient transporters in arbuscular mycorrhizas. Plant Sci 160, 899–904.

Burleigh S H and Harrison M J 1997 A novel gene whose expression in *Medicago truncatula* roots is suppressed in response to colonization by vesicular-arbuscular mycorrhizal (VAM) fungi and to phosphate nutrition. Plant Mol. Biol. 34, 199–208.

Clarke R B and Zeto S K 2000 Mineral acquisition by arbuscular mycorrhizal plants. J. Plant Nutr. 23(7), 867–902.

Graham J H and Abbott L K 2000 Wheat responses to aggressive and non-aggressive arbuscular mycorrhizal fungi. Plant Soil 220, 207–218.

Johnson N C, Graham J H and Smith F A 1997 Functioning of mycorrhizal associations along the mutualism-parasitism continuum. New Phytol. 135, 575–585.

Lejay L, Tillard P, Lepetit M, Domingo–Olive F, Filleur S, Daniel–Vedele F and Gojon A 1999 Molecular and functional regulation of two NO_3- uptake systems by N- and C-status of *Arabidopsis* plants. Plant J. 18, 509–519.

Li X and Christie P 2001 Changes in soil solution Zn and pH and uptake of Zn by arbuscular mycorrhizal red clover in Zn-contaminated soil. Chemosphere 42, 201–207.

Liu H, Trieu A T, Blaylock L A and Harrison M J 1998 Cloning and characterization of two phosphate transporters from *Medicago truncatula* roots: regulation in response to phosphate and to colonization by arbuscular mycorrhizal (AM) fungi. Mol. Plant–Microbe Interact. 11, 14–22.

Marschner H and Dell B 1994 Nutrient uptake and the mycorrhizal symbiosis. Plant Soil 159, 89–102.

Ravenskov S and Jakobsen I 1995 Functional compatibility in arbuscular mycorrhizas measured as hyphal P transport to the plant. New Phytol. 129(4), 611–618.

Rosewarne G, Barker S, Smith S, Smith F and Schachtman D 1999 A *Lycopersicon esculentum* phosphate transporter (LePT1) involved in phosphorus uptake from a vesicular-arbuscular mycorrhizal fungus. New Phytol. 144, 507–516.

Smith S E and Gianinazzi–Pearson V 1990 Phosphate uptake and arbuscular activity in mycorrhizal *Allium cepa* L.: effects of photon irradiance and phosphate nutrition. Aust J. Plant Physiol. 17, 177–88.

Smith S E and Read D J 1997 Mycorrhizal Symbiosis. Academic Press, San Diego, USA, 605 p.

Son C L and Smith S E 1988 Mycorrhizal growth responses: interactions between photon irradiance and phosphorus nutrition. New Phytol. 108, 305–314.

Tester M, Smith S E, Smith F A and Walker N A 1986 Effects of photon irradiance on the growth of shoots and roots, on the rate of initiation of mycorrhizal infection and on the growth of infection units in *Trifolium subterraneum* L. New Phytol. 103, 375–390.

Timmer L W and Leyden R F 1980 The relationship of mycorrhizal infection to phosphorus induced copper deficiency in sour orange seedlings. New Phytol. 85, 15–23.

Zhu Y, Christie P and Laidlaw A S 2001 Uptake of Zn by arbuscular mycorrhizal white clover from Zn-contaminated soil. Chemosphere 42, 193–199.

Plant and Soil **244:** 253–261, 2002.

Mycorrhizas and global environmental change: research at different scales

P. L. Staddon[1,2,3], A. Heinemeyer[1] & A. H. Fitter[1]
[1]*Department of Biology, University of York, PO Box 373, York, YO10 5YW, UK.*
[2]*Present address: Risø National Laboratory, Plant Research Department, Postbox 49, Roskilde, DK-4000, Denmark.* [3]*Corresponding author**

Received 21 August 2001. Accepted in revised form 23 April 2002

Key words: climate change, diversity, enhanced [CO_2], mycorrhizas, natural communities, warming

Abstract

Global environmental change (GEC), in particular rising atmospheric CO_2 concentration and temperature, will affect most ecosystems. The varied responses of plants to these aspects of GEC are well documented. As with other key below-ground components of terrestrial ecosystems, the response of the ubiquitous mycorrhizal fungal root symbionts has received limited attention. Most of the research on the effects of GEC on mycorrhizal fungi has been pot-based with a few field (especially monoculture) studies. A major question that arises in all these studies is whether the GEC effects on the mycorrhizal fungi are independent of the effects on their plant hosts. We evaluate the current knowledge on the effects of elevated CO_2 and increased temperature on mycorrhizal fungi and focus on the few available field examples. The value of using long-term and large-scale field experiments is emphasised. We conclude that the laboratory evidence to date shows that the effect of elevated CO_2 on mycorrhizal fungi is dependent on plant growth and that temperature effects seen in the past might have reflected a similar dependence. Therefore, how temperature directly affects mycorrhizal fungi remains unknown. In natural ecosystems, we predict that GEC effects on mycorrhizal fungal communities will be strongly mediated by the effects on plant communities to the extent that community level interactions will prove to be the key mechanism for determining GEC-induced changes in mycorrhizal fungal communities.

Introduction

Global environmental change (GEC) caused by human activities is one of the greatest challenges facing our society (Wyman, 1991). GEC encompasses a wide range of occurrences including rising atmospheric carbon dioxide concentration, leading to global warming, but also increased nitrogen deposition, tropospheric ozone depletion and other impacts. We will focus here on the first two aspects. It is undeniable that atmospheric CO_2 levels are increasing due to fossil fuel burning and land use changes, but a consensus on the role of human activity in global warming has been less forthcoming. However, the Intergovernmental Panel on Climate Change has stated in the last year that human activity is having a measurable effect on the Earth's climate (Kerr, 2001; see also Macilwain, 2000).

Much research into the biological effects of rising atmospheric CO_2 concentrations and temperature has focussed on plant growth and carbon fixation. However, other crucial components of terrestrial ecosystems, especially in the soil, have received less attention. We shall concentrate here on GEC research on mycorrhizal fungi. These fungi form symbiotic associations with plant roots and occur in the vast majority of terrestrial plants (Smith and Read, 1997). In terrestrial ecosystems they can account for a substantial proportion of their hosts photosynthate (e.g. Jakobsen and Rosendahl, 1990) and can mediate competition between plants (Hetrick, 1991). In terms of the global carbon cycle, mycorrhizal fungi could also prove to play a critical role in carbon sequestration in soils.

* FAX No: +45 46774122.
E-mail: philip.louis.staddon@risoe.dk

Little is known about the external phase of arbuscular mycorrhizal fungi but if the extraradical mycorrhizal hyphal density was stimulated under a warmer climate this could act as a negative feedback to rising atmospheric CO_2.

In this review, we will evaluate the research to date on the effects of elevated CO_2 and increased temperature on mycorrhizal fungi, but will pay particular attention to the different hierarchical levels of the observed effects of elevated CO_2 or temperature on mycorrhizal fungi, i.e. whether the observed effect is direct, independent of host-plant response; indirect, but simply a scale-dependent consequence of the host plant's growth response; indirect, resulting from changes in host physiology independently of host growth response; or mediated by more complex mechanisms (such as interspecific interactions) (see Figure 1). The importance of an hierarchy of organisational levels in mycorrhizal systems (e.g. from individual host to plant community) has been recently discussed by Miller and Kling (2000). Without having such a framework in mind, the chances of reaching erroneous conclusions about impacts of key environmental factors on mycorrhizal functioning in ecosystems could be quite high.

Elevated atmospheric carbon dioxide

At the level of an individual plant-mycorrhizal fungus symbiosis

The response of individual mycorrhizal fungal species (especially arbuscular mycorrhizas) to elevated atmospheric CO_2 has been reviewed several times over the past few years (Fitter et al., 2000; Rillig and Allen, 1999; Staddon and Fitter, 1998; Treseder and Allen, 2000). Here, we shall summarise the main points raised in the previous reviews and consider a few more recent papers. Other types of mycorrhizas will also be discussed where appropriate.

Plants grown at elevated CO_2 generally grow faster than when grown at ambient CO_2 (Poorter, 1993). Also, they tend to show an increase in the allocation of carbon to their root system relative to their shoots (Rogers et al., 1996). This could, therefore, result in more carbon being available to symbionts in the roots of plants grown at elevated CO_2 (Díaz et al., 1993). Numerous studies have attempted to demonstrate that this increased below-ground carbon results in the stimulation of mycorrhizal colonisation (see reviews cited above). However, as argued in Staddon and Fitter (1998) and Fitter et al. (2000) much of this research is inconclusive because differential plant growth was not taken into account (a problem inherent to single harvest experiments). Evidence to date shows no plant growth-independent effect of elevated CO_2 on arbuscular mycorrhizal colonisation or extraradical hyphal production. Recent work by Gavito and coworkers (Gavito et al., 2000) also confirms this finding and furthermore shows that mycorrhizal function in terms of phosphorus uptake is unaffected by elevated CO_2, which was previously reported by Staddon et al. (1999). In other words, at elevated CO_2 plants tend to be larger, and their mycorrhizal symbionts are proportionately larger.

Time course studies on ectomycorrhizas (EcM) have shown similar results to those on arbuscular mycorrhizas (AM). For example, Lewis et al. (1994) state that "despite significant effects on root carbohydrate levels, there were generally no significant effects on mycorrhizal colonization" for *Pinus taeda* grown at elevated CO_2. Other research has shown the potential for interspecific variation in the response of both AM (Klironomos et al., 1998) and EcM (Gorissen and Kuyper, 2000) in plants grown at elevated CO_2, although this work was based on a single harvest.

Many contradictions in the reports on the effect of elevated CO_2 on mycorrhizas arise from directly comparing results from pot-based single species plant-mycorrhizal fungus experiments to multispecies systems or by attempting to extrapolate to the field situation. The reason for these contradictions will become clearer in the following sections.

At the multispecies level

The simplest type of mycorrhizal association involves an individual host plant with a single mycorrhizal fungus as discussed above. The next level of complexity is where either several host plant species or more commonly several fungal species are studied. This opens up the possibility for more distal effects of elevated CO_2 on the system; effects which are mediated by interspecific interactions (e.g. competition), which in turn depend on the species-specific responses.

The following scenario can be used to highlight this possibility: a plant in symbiosis with two different mycorrhizal fungal species is grown at elevated CO_2. Assume that it has been previously determined that elevated CO_2 has no effect on the two mycorrhizal fungus species independent of plant growth, when they were grown in individual association with the

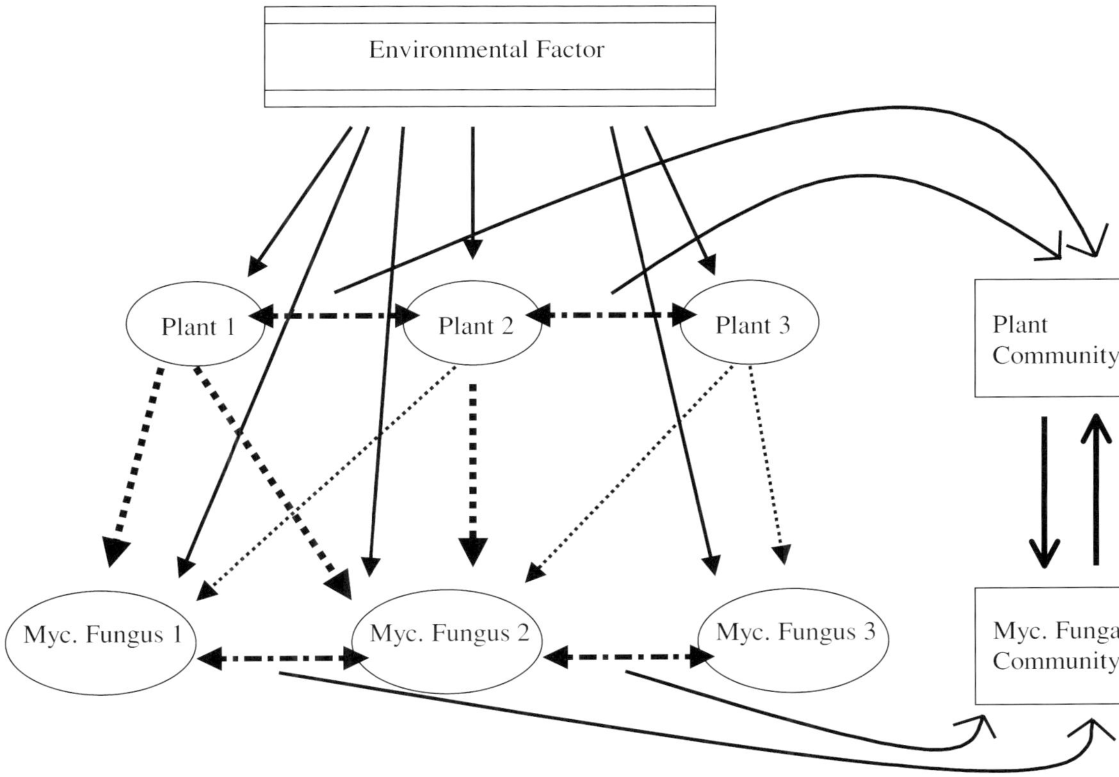

Figure 1. The direct and indirect impacts of an environmental factor on a simplified mycorrhizal fungal community. An environmental factor can have direct effects on both plants and their associated mycorrhizal fungi (→). It can also affect mycorrhizal fungi indirectly via impacts on the host plants; these effects may vary in magnitude depending on the level of association between the host-plant and the mycorrhizal fungus (- - →, the thicker the arrow, the stronger the effect). Interspecific interactions (←·→), at the level of the plants and the mycorrhizal fungi will determine plant and mycorrhizal fungal community structure (⌒). Furthermore, the plant and the mycorrhizal fungal communities both influence each other (⇄).

host plant. The *a priori* prediction would be that there would also be no plant-independent effect of elevated CO_2 on the dual mycorrhizal colonisation. However, it could easily be envisaged that the altered carbohydrate availability in the roots of the host plant might result in a shift in the competitive balance of the two mycorrhizal fungal species if, for example, one has a slightly greater intrinsic growth rate than the other. This could occur even if the same change in carbohydrate availability had no effect on either fungus on its own.

Changes in mycorrhizal fungal species community structure in plants grown at elevated CO_2 have been shown for ectomycorrhizas (EcM) in *Betula papyrifera* (Godbold and Berntson, 1997). Furthermore, changes in the relative proportion of EcM to AM colonisation in *Tsuga canadensis* have also been reported (Godbold et al., 1997). There is also the potential for plant species-specific effects to elevated CO_2 induced changes in mycorrhizal community assemblages (Rillig et al., 1998).

At the community level

There are two crucial points relating to the study of the effects of elevated CO_2 on mycorrhizas at the community level. Firstly, only community level studies (preferably in the field) can reflect what will happen in the 'real world', because we lack the understanding to extrapolate from single-species studies to the community. Secondly, community level studies will reveal an overall effect (e.g. a change in community composition) and can not unambiguously reveal the mechanisms behind any changes in ecosystem structure or function. Based on a few recent papers, we shall concentrate here on what can be achieved from research on the effects of elevated CO_2 at the community level and the inherent difficulties in explaining the mechanisms behind these effects.

There are primarily two types of experimental situation where the effect of elevated CO_2 can be studied at the community level in the field: free-air CO_2 enrichment (FACE) (Prior et al., 1994) or open top chamber (OTC) (Curtis et al., 1989) experiments, where the CO_2 concentration is artificially increased, and natural CO_2 vents or springs (Raschi et al., 1997), which in some cases may have resulted in exposure to elevated CO_2 over several decades or longer. The advantage of the latter is obvious for the length of time involved but the history of the site may not be known. Some FACE experiments (e.g. Schlesinger and Lichter, 2001) have now been running continuously for several years and have generated a great deal of data, especially concerning the response of vegetation (especially agricultural crops and plantation trees) to elevated CO_2. The main advantages of FACE over OTC technology are that under FACE no other environmental parameter is (directly) altered by the set-up and that FACE experiments are much larger in scale. OTCs are, however, much cheaper to set up and operate.

One of the first pieces of research which attempted to look at the effects of elevated CO_2 on mycorrhizas in the field was performed as part of a FACE experiment on cotton (Runion et al., 1994): no difference in mycorrhizal colonisation was detected. On a subsequent FACE experiment at the same site, but this time on sorghum, an increase in extraradical mycorrhizal hyphal length density was reported at elevated CO_2, but there was unfortunately no data for root length density, which could explain this effect (Rillig et al., 2001). Both studies were monocultures, highlighting the lack of funding available for ecological/environmental research when compared with agricultural (and sylvicultural) science. In the next few years, studies are likely to focus on the effect of elevated CO_2 on the below ground component, including mycorrhizas, in FACE experiments in more natural settings such as those located in a desert scrub (Jordan et al., 1999), in a tall grass prairie (Luscher et al., 1998) or in a sheep-grazed pasture (Edwards et al., 2001).

There has also been surprisingly little mycorrhizal work carried out with OTCs apart from that by Rillig and co-workers. In one OTC study, they report an increase in arbuscular mycorrhizal colonisation at elevated CO_2 (Rillig et al., 1999b) and in a second study they report that mycorrhizal colonisation was stimulated in some plant species but not in others (Rillig et al., 1999a). Unfortunately no measure of vegetation response to elevated CO_2 was obtained. It is known that elevated CO_2 may cause changes in ecosystem productivity, plant community structure and species composition (Bazzaz, 1990), and that changes in the plant community are linked to changes in the mycorrhizal fungal community (van der Heijden et al., 1998). Therefore, a change in mycorrhizal colonisation, such as those seen by Rillig and co-workers, could be a consequence of unknown changes in the plant community.

Natural CO_2 springs such as those in New Zealand (Newton et al., 1996) or Italy (van Gardingen et al., 1995) have the potential to reveal how natural terrestrial ecosystems respond to increasing atmospheric CO_2. They also have the advantage of possessing CO_2 gradients. As far as we are aware, only a single study (Rillig et al., 2000) has to date reported on mycorrhizal data from communities near CO_2 vents. They report increasing root colonisation and extraradical hyphal density with increasing CO_2 level. However, as with the OTC data discussed above, there was no quantitative survey of plant species composition, nor measurements of plant productivity or biomass or of root density. Without some attempt to determine the effect of increased atmospheric CO_2 (or any other environmental variable for that matter) on the vegetation, research solely on the mycorrhizal fungal component of an ecosystem is of strictly limited value.

Increased temperature

At the level of an individual plant-mycorrhizal fungus symbiosis

Research investigating temperature effects on the arbuscular mycorrhizal (AM) symbiosis has been reviewed by Daniels Hetrick (1984). However, in most of these studies, effects on the plant could not be separated from those on the fungus as only a single harvest was used (Table 1). Increased fungal growth was possibly due to increased plant growth. Also, treatments were mostly applied at the inoculation stage (Table 1). Fungal growth responses might, therefore, reflect impacts on spore germination and growth of initial infection units (Schenk et al., 1975; Tinker, 1975; Daniels Hetrick, 1984). The fungus itself might respond to temperature directly independently of any plant responses (Fitter et al., 2000). There is an obvi-

Table 1. Effect of temperature on arbuscular mycorrhizal colonisation (percentage root length colonised, RLC) and its external phase in various plant species

Plant species	Fungus species	Location of study	Duration of study	Harvest numbers/ Temperature	Percentage RLC at elevated temperature	Plant responses to elevated temperature	References
Allium cepa	GM?	Gr. ch.[a]	17 wk	9/ 16, 21, 26 °C, d1 treat[g]	increase	increase	Furlan and Fortin (1973)
Sorghum sp.	GF	Gr. hs.[b]	6 wk	1/ (soil temp.) 25, 30, 35 °C, d1 treat.	increase	increase under medium P	Graham and Leonard (1982)
Hordeum vulgare	GH GMa	Gr. hs.	7–9 wk	1/ (soil temp.) ca 11, 26 °C, d1 treat.	increases but low magn[f]	no change	Grey (1991)
Trifolium repens	GC GE	Gr. hs.	18 wk	3/ d1 treat. 18 °C & 150 μmol m^{-2} s^{-1} or 27 °C & 300 μmol m^{-2} s^{-1}	increase	increase	Jensen (1984)
Pascopyrum smithii	(from prairie soil)	Gr. ch.	2 & 4 yr	1 each/ ambient & +4 °C, d1 treat.	decrease but low magn.	not avail[e]	Monz et al. (1994)
Allium cepa	GM?	Gr. ch.	10 wk	1/ ca 14, 23 °C, d1 treat.	increase in RLC_{arb}[c]	increase	Hayman (1974)
Trifolium subterraneum	GC GiC AL	Gr. ch.	15 wk	4-6/ 7 temperatures: 5–37 °C, d1 treat.	not avail. but increase in EMH[d]	not avail.	Tommerup (1983)
T. subterraneum, *Medicago truncatula*	from field soil, >GM?	Gr. ch.	12 days	5/ 4 temperatures: 12–25 °C, d1 treat.	increased number of entry points, depended on root length	increased root length	Smith and Bowen (1979)

Notes:: [a]growth chamber, [b]green house, [c] arbuscular RLC, [d]extraradical mycorrhizal hyphae, elongation only, [e]not available, [f] used very low magnification, [g] treatment started on the day when inoculated with AM fungus, GM (*Glomus mosseae*), GF (*G. fasciculatum*), GH (*G. hoi*), GMa (*G. macrocarpum*), GC (*G. caledonium*), GE (*G. epigaeus*), GiC (*Gigaspora calospora*), AL (*Acaulospora laevis*). ? means possibility of the isolate (inoculum) to be (mainly) *G. mosseae* (i.e. *Endogone* sp.).

ous need of time-course data, allowing allometric analysis (Staddon and Fitter, 1998) to determine whether the response is due to altered plant growth. The only study incorporating plant growth effects on root colonisation of AM fungi was that of Smith and Bowen (1979), but the temperature treatment was imposed at the time of inoculation (Table 1).

Most studies investigating temperature responses of the AM symbiosis have been performed under highly artificial conditions (e.g. high levels of nutrients and in growth chambers) and used only one AM fungus, mostly *Glomus mosseae* or *G. fasciculatum* (Table 1). These conditions do not represent natural conditions, where a root system is often colonized by more than one AM fungus (Merryweather and Fitter, 1998) and where environmental conditions vary both temporally and spatially. Nutrient levels might also influence overall root colonisation responses to temperature as reported for CO_2 (Klironomos et al., 1996). Furthermore, the most commonly used AM fungal species might not reflect the ecologically important species in the field (Helgason et al., 1999).

The need to test temperature effects under field conditions has often been called for (e.g. Miller and Kling, 2000). Yet the only study using a natural AM fungal community was that of Monz et al. (1994). It was also the first published research on the effects of the increase in temperature of the range predicted for global warming on mycorrhizas. Monz et al. (1994) reported a decrease in mycorrhizal colonisation in *Pascopyrum smithii* as a result of a 4 °C increase in temperature but no change in *Bouteloua gracilis*. However, no data was provided on the effect on the host plant, so it could simply be, as for elevated CO_2, a plant-mediated effect.

Nearly all research has focused on intra-radical colonization, yet it is the external phase of the AM fungus which is more likely to respond to temperature (Gavito et al., 2000) as it experiences a wider variation in soil temperatures. Little is known about the nature and dynamics of the EMH (Miller et al., 1995). The only study testing temperature effects on the EMH of a mycorrhizal fungus was carried out by Tommerup (1983). However, only mean hyphal elongation data was given, and there was no data for percent root length colonised (Table 1). If the EMH were to respond directly to temperature their functioning might be changed, possibly leading to altered carbon input into the soil. Furthermore, if EMH respiration acclimated to rising soil temperatures a positive growth response of the EMH might then lead to carbon accumulation in soils under a warmer climate. We do not know whether the EMH might also respond to temperature changes independently of effects on internal hyphal growth. EMH might show less sensitivity to temperature than does internal colonisation, since colonisation can be measured at high levels in natural ecosystems in winter (Merryweather and Fitter, 1995). Compartment studies (Dodd, 1994), commonly used for investigating impacts on ion uptake by the EMH of phosphorus (Jakobsen et al., 1992; Schweiger and Jakobsen, 2000) or nitrogen (Hodge et al., 2001; Mäder et al., 2000) could offer a valuable tool to detect direct temperature responses of the AM fungus.

At the multispecies level

As far as we are aware, only a single paper (Rygiewicz et al., 2000) has been published where the effect of temperature on mycorrhizal community structure has been studied. They report that an increase in temperature by 4 °C above ambient results in the maintenance of a greater number of EcM morphotypes compared to the control. Unfortunately, this paper gives no data on the host plant response. Nonetheless, these results would point to changes in mycorrhizal community structure as a result of global warming.

At the community level

As with elevated CO_2, there is also little work on the effects of temperature on mycorrhizal communities in the field. Similarly, many of the field warming experiments have concentrated on the effects on the vegetation, in particular the above-ground component (e.g. Grime et al., 2000). As with FACE experiments, there is the possibility to use these relatively long-term warming experiments to study below ground aspects including mycorrhizas. Despite the limitations in their design (the soil being warmed directly) they are the best available tools for the study of ecosystem response to climate change.

At the community level, competitive abilities of host plants change with AM fungal colonisation (Fitter, 1977; Hetrick et al., 1992) and this results in plant community changes (van der Heijden et al., 1998). The impact of environmental factors on mycorrhizal fungi at the community level (i.e. temperature and light) is little understood (Abbot and Robson, 1984; Miller 1987; Fitter et al., 2000). To predict the impact of GEC on ecosystem functioning and carbon flow to the soil, we must pay more attention to the ubiquitous mycorrhizal fungi. Future research should determine whether temperature affects internal colonisation, and extraradical mycelium growth and if any temperature responses are independent from effects on the vegetation. Also, the role of AM fungi in carbon flow to the soil under soil warming must be addressed as they are likely to play a key role in determining soil carbon sink capacity under a warmer and CO_2 enriched world.

Conclusions

There is no evidence that elevated atmospheric CO_2 affects mycorrhizal fungi other than by affecting the growth of the host-plants. The potential of indirect effects, mediated for example by increases in soluble carbohydrates in roots has not been clearly demonstrated. Syvertsen and Graham (1999) measured carbohydrate pools in mycorrhizal and non-mycorrhizal citrus plants at ambient and elevated CO_2. There was no effect of elevated CO_2 on mycorrhizal colonisation, but the presence of mycorrhizal fungi decreased root carbohydrate stores; more so at elevated CO_2 than at ambient CO_2, but this was linked to a greater mycorrhizal stimulation of net carbon assimilation at elevated CO_2.

At the ecosystem level, effects of elevated CO_2 on mycorrhizas can be mediated by numerous factors, both biotic (e.g. plant interspecific interactions) and abiotic (e.g. nitrogen availability). As the effects of elevated CO_2 on plants are species-specific, plant community structure is altered, and this in turn will lead to changes in the mycorrhizal fungi community at elevated CO_2. We believe that the overriding factor involved in any effect of elevated CO_2 on mycorrhizas in natural ecosystems will be due to altered plant community structure which inherently leads to changes in all other aspects of the ecosystem.

On the other hand, there is increasing evidence that temperature can have direct effects on mycorrhizal fungi. This is not surprising as all organisms have temperature optima in some sense. Temperature dependence of enzymatic activity is well known and is one obvious reason why temperature can directly affect mycorrhizal fungi. Of course, as temperature also impacts on plants it will also affect mycorrhizal fungi via its effects on their plant hosts. Temperature effects can also be indirect via effects on other environmental factors (e.g. soil moisture). So, as for elevated CO_2, we may find that in natural ecosystems the effects of temperature on mycorrhizas will be, in the main, due to temperature induced changes to plant communities.

Many other aspects of GEC will also impact on mycorrhizas, in some cases direct effects may be evident, but mostly, we predict that the main changes to mycorrhizas in natural ecosystems will result from changes in plant community structure. For example, cloudiness is expected to increase in many areas as a result of climate change (Coughlan and Nyenzi, 1991). The resulting lower photosynthetically available radiation (PAR) will clearly impact on plant productivity and below-ground carbon economy (Fitter et al., 1998) thus leading to altered plant communities, but it would have no foreseeable direct effect on mycorrhizas. Also, it must not be overlooked that the numerous changing environmental factors will also interact with one another and with other factors such as the physicochemical properties of soils. Predicting any detailed impacts of GEC on terrestrial ecosystems, let alone mycorrhizal communities, therefore becomes increasingly problematic. The only certainty is that GEC will change mycorrhizal fungal community structure primarily as a result of changes in plant community structure.

Acknowledgements

Much of the work reported here was funded by the Natural Environment Research Council.

References

Abbott L K and Robson A D 1984 The effect of root density, inoculum placement and infectivity of inoculum on the development of vesicular-arbuscular mycorrhizas. New Phytol. 97, 285–299.

Bazzaz F A 1990 The response of natural ecosystems to rising global CO_2 levels. Annu. Rev. Ecol. Syst. 21, 167–196.

Coughlan M J and Nyenzi B S 1991 Climate trends and variability. *In* Climate Change: Science, Impacts and Policy. Eds. J Jäger and H L Ferguson. pp 71–82. Cambridge University Press, Cambridge, UK.

Curtis P S, Drake B G and Whigham D F 1989 Nitrogen and carbon dynamics in C_3 and C_4 estuarine marsh plants grown under elevated CO_2 *in situ*. Oecologia 78, 297–301.

Daniels Hetrick B A 1984 Ecology of VA mycorrhizal fungi. *In* Vesicular Arbuscular Mycorrhiza. Eds. L I Cornway, D J Powell and J Bagyaraj. pp 35–55. CRC Press, Boca Raton, Florida, USA.

Díaz S, Grime J P, Harris J and McPherson E 1993 Evidence of a feedback mechanism limiting plant response to elevated carbon dioxide. Nature 364, 616–617.

Dodd J C 1994 Approaches to the study of the extraradical mycelium of arbuscular mycorrhizal fungi. *In* Impact of Arbuscular Mycorrhizas on Sustainable Agriculture and Natural Ecosystems. Eds. S Gianinazzi and H Schüepp. pp 147–166. Birkhäuser Verlag, Basel, Switzerland.

Edwards G R, Clark H and Newton P C D 2001 The effects of elevated CO_2 on seed production and seedling recruitment in a sheep-grazed pasture. Oecologia 127, 383–394.

Fitter A H 1977 Influence of mycorrhizal infection on competition for phosphorus and potassium by two grasses. New Phytol. 79, 119–125.

Fitter A H, Graves J D, Self G K, Brown T K, Bogie D S and Taylor K 1998 Root production, turnover and respiration under two grassland types along an altitudinal gradient: influence of temperature and solar radiation. Oecologia 114, 20–30.

Fitter A H, Heinemeyer A and Staddon P L 2000 The impact of elevated CO_2 and global climate change on arbuscular mycorrhizas: a mycocentric approach. New Phytol. 147, 179–187.

Furlan V and Fortin J-A 1973 Formation of endomycorrhizae by *Endogone calospora* on *Allium cepa* under three temperature regimes. Nat. Can. 100, 467–477.

Gavito M E, Curtis P S, Mikkelsen T N and Jakobsen I 2000 Atmospheric CO_2 and mycorrhiza effects on biomass allocation and nutrient uptake of nodulated pea (*Pisum sativum* L.) plants. J. Exp. Bot. 51, 1931–1938.

Godbold D L and Berntson G M 1997 Elevated atmospheric CO_2 concentration changes ectomycorrhizal morphotype assamblages in *Betula papyrifera*. Tree Physiol. 17, 347–350.

Godbold D L, Berntson G M and Bazzaz F A 1997 Growth and mycorrhizal colonization of three North American tree species under elevated atmospheric CO_2. New Phytol. 137, 433–440.

Gorissen A and Kuyper TH W 2000 Fungal species-specific responses of ectomycorrhizal Scots pine (*Pinus sylvestris*) to elevated [CO_2]. New Phytol. 146, 163–168.

Graham J H and Leonard R T 1982 Interaction of light and soil temperature with phosphorus inhibition of vesicular-arbuscular mycorrhiza formation. New Phytol. 91, 683–690.

Grey W E 1991 Influence of temperature on colonization of spring barleys by vesicular arbuscular mycorrhizal fungi. Plant Soil 137, 181–190.

Grime J P, Brown V K, Thompson K, Masters G J, Hillier S H, Clarke I P, Askew A P, Corker D and Kielty J P 2000 The response of two contrasting limestone grasslands to simulated climate change. Science 289, 762–765.

Hayman D S 1974 Plant growth responses to vesicular-arbuscular mycorrhiza VI. Effect of light and temperature. New Phytol. 73, 71–80.

Helgason T, Fitter A H and Young J P W 1999 Molecular diversity of arbuscular mycorrhizal fungi colonising *Hyacinthoides non-scripta* (bluebell) in a seminatural woodland. Mol. Ecol. 8, 659–666.

Hetrick B A D 1991 Mycorrhizas and root architecture. Experientia 47, 355–362.

Hetrick B A D, Wilson G W T and Todd T C 1992 Relationships of mycorrhizal symbiosis, rooting strategy, and phenology among tallgrass prairie forbs. Can. J. Bot. 70, 1521–1528.

Hodge A, Campbell C D and Fitter A H 2001 Arbuscular mycorrhizal fungus accelerates decomposition and acquires nitrogen directly from organic material. Nature 413, 297–298.

Jakobsen I, Abbott L K and Robson A D 1992 External hyphae of vesicular-arbuscular mycorrhizal fungi associated with *Trifolium subterraneum* L. New Phytol. 120, 371–380.

Jakobsen I and Rosendahl L 1990 Carbon flow into soil and external hyphae from roots of mycorrhizal cucumber plants. New Phytol. 115, 77–83.

Jensen A 1984 Influence of inoculum density of two vesicular-arbuscular mycorrhizal fungi and temperature/light intesity on Trifolium repens. Nor. J. Bot. 4, 249–259.

Jordan D N, Zitzer S F, Hendrey G R, Lewin K F, Nagy J, Nowak R S, Smith S D, Coleman J S and Seemann J R 1999 Biotic, abiotic and performance aspects of the Nevada Desert Free-Air CO_2 Enrichment (FACE) Facility. Global Change Biol. 5, 659–668.

Kerr R A 2001 It's official: humans are behind most of global warming. Science 291, 566–566.

Klironomos J N, Rillig M C and Allen M F 1996 Below-ground microbial and microfaunal responses to *Artemisia tridentata* grown under elevated atmospheric CO_2. Funct. Ecol. 10, 527–534.

Klironomos J N, Ursic M, Rillig, M and Allen M F 1998 Interspecific differences in the response of arbuscular mycorrhizal fungi to *Artemisia tridentata* grown under elevated atmospheric CO_2. New Phytol. 138, 599–605.

Lewis J D, Thomas R B and Strain B R 1994 Effect of elevated CO_2 on mycorrhizal colonization of loblolly pine (*Pinus taeda* L.) seedlings. Plant Soil 165, 81–88.

Luscher A, Hendrey G R and Nosberger J 1998 Long-term responsiveness to free air CO_2 enrichment of functional types, species and genotypes of plants from fertile permanent grassland. Oecologia 113, 37–45.

Macilwain C 2000 Global-warming sceptics left out in the cold. Nature 403, 233–233.

Mäder P, Vierheilig H, Streitwolf-Engel R, Boller T, Frey B, Christie P and Wiemken A 2000 Transport of ^{15}N from a soil compartment separated by a polytetrafluoroethylen membran to plant roots via the hyphae of arbuscular mycorrhizal fungi. New Phytol. 146, 155–161.

Merryweather J W and Fitter A H 1995 Phosphorus and carbon budgets: mycorrhizal contribution in *Hyacinthoides non-scripta* (L.) Chouard ex Rothm. under natural conditions. New Phytol. 129, 619–627.

Merryweather J W and Fitter A H 1998 The arbuscular mycorrhizal fungi of *Hyacinthoides non-scripta* I. Diversity of fungal taxa. New Phytol. 138, 117–129.

Miller R M 1987 The ecology of vesicular-arbuscular mycorrhizae in grass- and shrublands. *In* Ecophysiology of VA Mycorrhizal Plants. Ed. G R Safir. pp 135–170. CPR Press, Boca Raton, Florida, USA.

Miller R M, Reinhardt D R and Jastrow J D 1995 External hyphal production of vesicular-arbuscular mycorrhizal fungi in pasture tallgrass prairie communities. Oecologia 103, 17–23.

Miller R M and Kling M 2000 The importance of integration and scale in the arbuscular mycorrhizal symbiosis. Plant Soil 226, 295–309.

Monz C A, Hunt H W, Reeves F B and Elliott E T 1994 The response of mycorrhizal colonization to elevated CO_2 and climate change in *Pascopyron smithii* and *Bouteloua gracilis*. Plant Soil 165, 75–80.

Newton P C D, Bell C C and Clark H 1996 Carbon dioxide emissions from mineral springs in Northland and the potential of these sites for studying the effects of elevated carbon dioxide on pastures. New Zealand J. Agric. Res. 39, 33–40.

Poorter H 1993 Interspecific variation in the growth response of plants to an elevated ambient CO_2 concentration. Vegetatio 104/105, 77–97.

Prior S A, Rogers H H, Runion G B and Hendrey G R 1994 Free-air CO_2 enrichment of cotton: vertical and lateral root distribution patterns. Plant Soil 165, 33–44.

Raschi A, Miglietta F, Tognetti R and van Gardingen P R 1997 Plant Responses to Elevated CO_2: Evidence from Natural Springs. Cambridge University Press, Cambridge, UK. 286 pp.

Rillig M C and Allen M F 1999 What is the role of arbuscular mycorrhizal fungi in plant-to-ecosystem responses to elevated atmospheric CO_2? Mycorrhiza 9, 1–8.

Rillig M C, Allen M F, Klironomos J N, Chiariello N R and Field C B 1998 Plant species-specific changes in root-inhabiting fungi in a California annual grassland: responses to elevated CO_2 and nutrients. Oecologia 113, 252–259.

Rillig M C, Field C B and Allen M F 1999a Fungal root colonization responses in natural grasslands after long-term exposure to elevated atmospheric CO_2. Global Change Biol. 5, 577–585.

Rillig M C, Field C B and Allen M F 1999b Soil biota responses to long-term atmospheric CO_2 enrichment in two California annual grasslands. Oecologia 119, 572–577.

Rillig M C, Hernandez G Y and Newton P C D 2000 Arbuscular mycorrhizae respond to elevated atmospheric CO_2 after long-term exposure: evidence from a CO_2spring in New Zealand supports the resource balance model. Ecol. Letters 3, 475–478.

Rillig M C, Wright S F, Kimball B A, Pinter P J, Wall G W, Ottman M J and Leavitt S W 2001 Elevated carbon dioxide and irrigation effects on water stable aggregates in a *Sorghum* field: a possible role for arbuscular mycorrhizal fungi. Global Change Biol. 7, 333–337.

Rogers H H, Prior S A, Runion G B and Mitchell R J 1996 Root to shoot ratio of crops as influenced by CO_2. Plant Soil 187, 229–248.

Runion G B, Curl E A, Rogers H H, Backman P A, Rodríguez-Kábana R and Helms B E 1994 Effects of free-air CO_2 enrichment on microbial populations in the rhizosphere and phyllosphere of cotton. Agric. For. Meteorol. 70, 117–130.

Rygiewicz P T, Kendall J M and Tuininga A R 2000 Morphotype community structure of ectomycorrhizas on Douglas fir (*Pseudotsuga menziesii* Mirb. Franco) seedlings grown under elevated atmospheric CO_2 and temperature. Oecologia 124, 299–308.

Schenk N C, Graham S O and Green N E 1975 Temperature and light effect on contamination and spore germination of vesicular-arbuscular mycorrhizal fungi. Mycologia 67, 1189–1192.

Schlesinger W H and Lichter J 2001 Limited carbon storage in soil and litter of experimental forest plots under increased atmospheric CO_2. Nature 411, 466–469.

Schweiger P and Jakobsen I 2000 Laboratory and field methods for measurement of hyphal uptake of nutrients in soil. Plant Soil 226, 237–244.

Smith S E and Bowen G D 1979 Soil temperature, mycorrhizal infection and nodulation of *Medicago truncatula* and *Trifolium subterraneum*. Soil Biol. Biochem. 11, 469–473.

Smith S E and Read D J 1997 Mycorrhizal Symbiosis. Academic Press, London, UK. 605 pp.

Staddon P L and Fitter A H 1998 Does elevated atmospheric carbon dioxide affect arbuscular mycorrhizas? TREE 13, 455–458.

Staddon P L, Fitter A H and Graves J D 1999 Effect of elevated atmospheric CO_2 on mycorrhizal colonization, external mycorrhizal hyphal production and phosphorus inflow in *Plantago lanceolata* and *Trifolium repens* in association with the arbuscular mycorrhizal fungus *Glomus mosseae*. Global Change Biol. 5, 347–358.

Syvertsen J P and Graham J H 1999 Phosphorus supply and arbuscular mycorrhizas increase growth and net gas exchange responses of two *Citrus* spp. grown at elevated [CO_2]. Plant Soil 208, 209–219.

Tinker P B H 1975 Effects of vesicular-arbuscular mycorrhizas on higher plants. *In* Symposium of the Society of Experimental Biology. Eds. D H Jennings and D L Lee. Vol 29, 325–349. Cambridge Press, New York, USA.

Tommerup I C 1983 Temperature relations of spore germination and hyphal growth of vesicular-arbuscular mycorrhizal fungi in soil. Trans. Brit. Mycol. Soc. 81, 381–387.

Treseder K K and Allen M F 2000 Mycorrhizal fungi have a potential role in soil carbon storage under elevated CO_2 and nitrogen deposition. New Phytol. 147, 189–200.

Van der Heijden M G A, Klironomos J N, Ursic M, Moutoglis P, Streitwolf-Engel R, Boller T, Wiemken A and Sanders I R 1998 Mycorrhizal fungal diversity determines plant biodiversity, ecosystem variability and productivity. Nature 396, 69–72.

Van Gardingen P R, Grace J, Harkness D D, Miglietta F and Raschi A 1995 Carbon-dioxide emissions at an Italian mineral spring – measurements of average CO_2 concentration and air-temperature. Agric. For. Meteorol. 73, 17–27.

Wyman R L 1991 Global Climate Change and Life on Earth. Chapman and Hall, New York, USA. 282 p.

Plant and Soil **244:** 263–271, 2002.

Is there a role for arbuscular mycorrhizal fungi in production agriculture?

Megan H. Ryan[1,3] & James H. Graham[2]
[1]*CSIRO Plant Industry, GPO Box 1600, Canberra, ACT, 2601, Australia.* [2]*University of Florida, Citrus Research and Education Center, Lake Alfred, FL 33850, U.S.A.* [3]*Corresponding author**

Received 21 August 2001. Accepted in revised form 20 February 2002

Key words: carbon cost, crop production, field studies, glomalin, micronutrients, phosphorus fertilisation

Abstract

This review presents the point of view that arbuscular mycorrhizal fungi (AMF) do not play a vital role in the nutrition and growth of plants in many production-orientated agricultural systems. Highly available soil P often limits AM colonisation and causes the C-costs to the host to outweigh any benefits from colonisation. Even when P availability is low and AM colonisation levels are high, as may occur in organic and biodynamic agricultural systems, AMF may not always contribute to plant growth for reasons not yet understood. AM fungal activity may also be greatly limited by soil fumigation, non-responsive plant varieties, or rotations based primarily on non-mycorrhizal crops or crops of low AM dependency. Thus, profitability may sometimes be enhanced by management practices, such as tillage and P-fertilisation, which limit AM colonisation. Manipulation of agricultural systems to favour AMF must occur only if there is clear evidence that AMF make a positive contribution to yield or are vital for maintenance of ecosystem health and sustainability. A crucial role for AMF in soil structural stability or in enhancing micronutrient concentrations in produce may be sufficient evidence and may eventually compel consideration of AMF responsiveness when breeding new crop varieties.

Introduction

Arbuscular mycorrhizal fungi (AMF) are almost as ubiquitous below-ground as the plant roots they inhabit. AMF are the recipients of wide attention as part of an increasingly popular paradigm that considers an active and diverse soil biological community as essential for increasing the sustainability of agricultural systems (Gianinazzi and Schüepp, 1994). AMF are probably the most abundant fungi in agricultural soils, accounting for somewhere between 5 and 50% of the biomass of soil microbes (Olsson et al., 1999). The ability of AMF to enhance host-plant uptake of relatively immobile nutrients, in particular P and Zn (Thompson, 1987), and their requirement for up to 20% of host photosynthate for establishment and maintenance is well accepted (Graham, 2000; Jakobsen and Rosendahl, 1990). AM colonisation may also protect host roots from certain pathogens and improve water relations, especially under nutrient limitation (Graham, 2001; Sánchez-Díaz and Honrubia, 1994). The extraradical hyphae of AMF may contribute greatly to soil aggregation and structural stability in at least some soil types (Miller and Jastrow, 2000).

While AMF are undoubtedly a vital component of the below-ground ecosystem, quantification of their impact on yield in agricultural systems has been largely unsuccessful, and to this point, the outcomes of field studies have often been contradictory. The contribution of AMF to system sustainability is even further from being quantified. A poor understanding of the complex role of AMF in regulating resource allocation (primarily C and P) between the abiotic and biotic components of the soil system is contributing to this situation (Graham, 2000).

This review will examine the current knowledge on the functioning of AMF in industrialised production agriculture (high input/high output), primarily using selected field-based case studies. This approach will highlight the limitations of our current knowledge and areas that warrant further study.

* FAX No.: +61-2-6246-5399. E-mail: megan.ryan@csiro.au

Table 1. Growth of paired crops, one with reduced AM colonisation following a long fallow, over a number of sites and years in southern Queensland, Australia (Thompson, 1987, 1991)

	Fallow length	Soil available P (Colwell) (mg kg^{-1})	AMF (%)	Early shoot biomass (g $plant^{-1}$)	Yield (t ha^{-1})
Sunflower	Long	10.2	3	0.6	–
	Short	9.0	14	6.9	–
Maize	Long	–	34	–	2.9 (4.9)*
	Short	–	53	–	5.2 (4.7)*
Linseed	Long	–	21	0.11 (0.28)*	–
	Short	–	44	0.17 (0.22)*	–
Wheat	Long	14.0	22	36	–
	Short	16.0	40	37	–
Wheat	Long	–	15	0.10	–
	Short	–	48	0.55	–

*Values in brackets are for crops fertilised with P and Zn.

Effects of management on AMF and crops in the Australian wheatbelt

AM fungal benefit for crop growth is most easily identified on the alkaline heavy clay Vertosol soils (Isbell, 1996) of subtropical southern Queensland, NE Australia (Thompson, 1987, 1991) (Table 1). Many crops in this region, including maize (*Zea mays*), sorghum (*Sorghum* spp.), sunflower (*Helianthus annuus*), chickpea (*Cicer arietinum*) and linseed (flax) (*Linum usitatissimum*), when grown after long periods of bare fallow may exhibit poor growth and P and Zn deficiency. This syndrome, called Long Fallow Disorder (LFD), is associated with low AM colonisation (Thompson 1987, 1991). In a field experiment, the differing growth of linseed following a variety of previous crops was closely related to AM inoculum density, with poorest growth after a long fallow and after non-mycorrhizal canola (*Brassica napus* var. *napus*) (Thompson, 1991).

Farmers in this region have, until recently, generally not applied P fertilisers, and many soils have very low available P. Thus, while glasshouse experiments showed LFD to be alleviated through inoculation of soil with indigenous AMF, under glasshouse and field conditions LFD was also remedied with P and Zn fertilisation (Thompson, 1991, 1996). Thompson's research alerted farmers in southern Queensland to consider AM fungal activity when planning crop rotations and farm management practices.

Wheat (*Triticum aestivum*) and other cereals including barley (*Hordeum vulgare*) and oats (*Avena sativa*) with relatively high specific root lengths (length/dry mass) were mostly reported by Thompson to be unaffected by reduced AM colonisation, with some exceptions (Table 1). Crops that responded negatively to a reduction in AM colonisation in low-P long fallow or fumigated soils generally had a relatively low specific root length and high mycorrhizal dependency.

In Danish fields, Thingstrup et al. (1998) confirmed the relatively high dependency of linseed on AMF at Olsen available P<40 mg kg^{-1} (Rayment and Higginson, 1992) using dazomet fumigant to suppress colonisation. However, in two Finnish soils with around 60 mg kg^{-1} of available P, crop biomass at harvest was generally increased when benomyl fumigant was applied; increases being up to 8% for barley and 42% for linseed (Kahiluoto et al., 2001). Thus, application of P and Zn fertilisers into soils of low fertility following long fallows or applications of fumigants may not only alleviate the negative effects of low AM colonisation and fertility, but crops may respond better than in soils with higher AM fungal inoculum levels (Thingstrup et al., 1998; Table 1).

On the acidic Kandosol earths in temperate New South Wales and Victoria, SE Australia, long fal-

Table 2. Characteristics of wheat and field pea sown in autumn, with and without P-fertiliser, following either long fallow or linola in an experiment in southern New South Wales, Australia (M H Ryan and J F Angus, unpublished data, 2000)

	Pre-treatment	Fertiliser P (kg ha^{-1})	AMF* (%)	Shoot biomass* (t ha^{-1})	Shoot P* (kg ha^{-1})	Yield (t ha^{-1})	Grain Zn (mg kg^{-1})
Wheat	Long fallow	None	46	0.95	2.0	3.3	21
	Linola	None	65	0.94	2.1	2.9	28
	Long fallow	20	12	2.42	6.6	4.8	14
	Linola	20	25	1.98	6.6	4.9	18
		$LSD_{0.05}$	13	0.40	1.4	0.5	2
Field pea	Long fallow	None	18	0.40	1.1	2.1	26
	Linola	None	41	0.45	1.3	2.0	35
	Long fallow	20	5	0.80	3.9	3.0	20
	Linola	20	10	0.80	3.2	3.4	24
		$LSD_{0.05}$	14	0.14	0.6	0.6	2

*109 days after sowing.

lows are generally beneficial to the growth of following crops through water and mineral N conservation (Cooke et al., 1985). In southern New South Wales, the role of AMF in P-uptake and growth of wheat and field pea (*Pisum sativum*) was examined following long fallow, to reduce AM inoculum levels, and linola (low linolenic acid linseed), to augment inoculum levels. The crops were grown both without P-fertiliser and with 20 kg ha^{-1} of P as superphosphate, as is regular commercial practice in this region, in a field with low soil available P (11 mg kg^{-1} Colwell; see Rayment and Higginson, 1992). For both wheat and field pea, long fallow reduced AM colonisation, but not crop growth or P-uptake, and P-fertiliser increased shoot biomass, P-uptake and yield (Table 2). Thus, in contrast to southern Queensland, reduced AM colonisation did not have a negative impact on growth or P-uptake of the crops, even when unfertilised with P. As field pea are the only AMF-hosting non-cereal crop widely grown in southern New South Wales, farmers apparently do not need to consider AMF when planning farm management practices.

Of interest, though, are the enhanced grain Zn concentrations that resulted from higher AM colonisation (Table 2). There is rising concern about the implications for human health of low concentrations of Zn, and other micronutrients, in produce from modern agricultural systems (Welch and Graham, 1999). The ability of AMF to enhance micronutrient uptake has often been noted, although generally under glasshouse conditions (Marschner and Dell, 1994), and may now require serious consideration in agricultural systems.

On the alkaline Vertosol soils in Victoria, SE Australia, examination of wheat following non-mycorrhizal brassicas and AM colonised crops again showed no relationship between wheat yield and AM colonisation level over a number of field experiments and a large range of soil available P (Ryan et al., unpublished data, 1998–2000). Indeed, yields of wheat following non-mycorrhizal canola in southern New South Wales and Victoria are, on average, 19% greater than for wheat following wheat (Angus et al., 1999). This increase is attributed to better N-nutrition due to reduced infection by soil-borne fungal pathogens and healthier root systems (Kirkegaard et al., 2000). However, a positive impact from reduced AM colonisation, and hence a reduced C-drain, on growth of wheat following canola cannot be disregarded.

While the uniquely important role for AMF in NE Australia may largely reflect the traditionally low rates of P-fertiliser applied in this region, explanation for the lack of a similar role in P-limited conditions further south in temperate New South Wales and Victoria is still required. Possibilities include differences in: crop growth rates and P-requirements during the season; rainfall, soil moisture levels and soil P-availability during the season; limitations on crop growth by soil-borne disease or micronutrients where uptake is mediated by AMF; or AMF community composition (Jakobsen et al., 2001). In particular, the low temperatures experienced by crops in SE Australia, which are predominantly autumn/winter sown, may adversely affect AMF functioning (Cooper and Tinker, 1981).

Mediation of the tillage effect on maize in fertile soils in North America by AMF

Twenty years of research in southern Ontario, Canada, on the interactions of tillage intensity with P-nutrition and growth of maize has clearly demonstrated that enhanced early P-absorption by no-tillage maize in the field is correlated with higher AM colonisation (Miller, 2000). In glasshouse experiments, P-uptake by non-mycorrhizal canola and spinach (*Spinacea oleracea*) was not reduced by disturbance of soil, but plant P-uptake was reduced when fungicides were applied to AM colonised maize roots (Evans and Miller, 1988). Greater P-uptake by maize in undisturbed soil at times occurred even though AM colonisation levels were no different from maize growing in disturbed soil. Further experiments showed that disruption to the external hyphal network of the AMF was the crucial factor responsible for reduced P-uptake in disturbed soil (Miller, 2000). Thus, the maintenance of an intact hyphal network in no-tillage systems provides both a source of inoculum of high potential and a means of nutrient acquisition early in the growth of following crops.

Maize yields under no-tillage were, however, higher than under conventional tillage only at sub-optimal P levels. At higher available soil P, more typical of midwestern North America (>50 mg kg^{-1}), early root colonisation and P-uptake were enhanced, but yields were consistently lower than under conventional tillage (Miller, 2000). Likewise, in a high-P organically managed soil in Pennsylvania, U.S.A., high AM colonisation and shoot P concentrations in no-tillage maize were again associated with lower yields than under conventional tillage (Galvez et al., 2001). These studies note that many physical and biological factors in no-tillage crop systems could contribute to reduced root health, and thereby increase shoot P concentrations whilst decreasing yield. When no-tillage is coupled with stubble retention, yield may be reduced by higher weed competition, and poor temperature and light conditions for emerging seedlings (Galvez et al., 2001). In New South Wales, Australia, poor growth of no-tillage wheat has been linked to greater activity by inhibitory rhizosphere *Pseudomonas* spp. than under conventional tillage (Simpfendorfer et al., 2001). Nevertheless, the responses of maize to tillage in fertile high-P soils supports the hypothesis that if symbiotic costs are controlled by a reduction in early season activity of AMF, positive yield effects will develop due to reallocation of C to shoot production during subsequent exponential growth stages (Graham, 2000; Graham and Eissenstat, 1998).

High input horticultural systems

In horticultural systems involving crops such as vegetables and citrus, most growers are unwilling to risk low production through reduced fertiliser and biocide inputs. For intensively managed tomato (*Lycopersicon esculentum*) and pepper (*Capsicum annuum*) crops, the typical management strategy is to apply high rates of P-fertiliser and to fumigate with methyl bromide prior to planting to reduce soil-borne pathogens (Chellemi, 2000). With past reliance on methyl bromide for pathogen control, the question of whether to manage the positive attributes of AMF for biocontrol was a moot point based on the assumption that colonisation would be severely limited. Worldwide phase-out of methyl bromide by 2005 could conceivably increase reliance on AMF to maintain host tolerance to pathogens, as diseases caused by *Phytophthora, Pythium, Fusarium* and *Sclerotium rolfsii* are now resurgent in the absence of a single stand-alone replacement (Chellemi, 2000). However, the interactions of AMF with soil-borne pathogens in the field have been barely investigated and, so far, biological control interactions appear relatively minor and dependent on the host-pathogen complex (Graham, 2001; Larsen and Bodker, 2001).

Moreover, AM colonisation is generally severely limited by the high P-inputs in vegetable systems. In Florida, U.S.A., production of tomato and peppers in sandy soils mimics hydroponics. Even with alternatives to methyl bromide that are less suppressive to AMF (e.g. solarisation, plastic mulches) or the use of strategies to augment AM fungal inoculum (e.g. pre-planted cover crops), colonisation ranges from very low to non-detectable (Table 3). This result was

Table 3. Impact on AM colonisation of management of tomato and pepper crops in SE Florida, U.S.A., with Sunn Hemp (*Crotalaria juncea*) and Iron Clay Pea (*Vigna unguiculata*) cover crops and alternatives to methyl bromide (J H Graham and D O Chellemi, unpublished data, 1998 and 1999) (mean±s.e.)

Pre-treatment	AM colonisation (%)		
	Cover crop	Tomato	Pepper
None	–	1.5±1.3	–
Solarisation	–	0.8±0.9	1.3±1.1
Methyl bromide	–	0.8±1.3	0.7±0.8
Solarisation	–	0.4±0.3	0.0±0.0
White plastic	–	0.4±0.1	0.0±0.0
Sunn Hemp	2.5±2.2	4.2±1.4	3.6±2.5
Iron Clay Pea	4.3±4.3	1.1±0.5	2.4±0.8
Solarisation	–	–	0.0±0.0
White plastic	–	–	0.0±0.0
Methyl bromide	–	–	0.0±0.0
Sunn Hemp	24.3±3.4	–	0.0±0.0
Iron Clay Pea	9.5±3.0	–	0.3±0.3

not unexpected as the prevailing P availability in the system was high (Bray 1 P>100 mg kg^{-1}).

If AMF are to play a positive role in the areas of disease suppression (Larsen and Bodker, 2001) or tolerance of stress after transplant of greenhouse transplants (Biermann and Linderman, 1983), P-inputs must be lowered substantially in horticultural systems. The incentive to do this is weaker than in broadacre agricultural crops as P-fertiliser is a very low cost component of vegetable production. An economic analysis by Olsen et al. (1999b) of pepper production in the Bundaberg district of southern Queensland, Australia, found that reducing P-inputs to favour AM fungal activity resulted in only a 2.1% savings in production costs. Indeed, in a glasshouse experiment, maintenance of the integrity of the extraradical hyphal network in soil above the threshold of P-supply for maximum yield of pepper and tomato resulted in substantial growth depression due to apparent C-cost without benefit (Olsen et al., 1999a), as also demonstrated for citrus (Graham and Eissenstat, 1998).

This negative response of vegetables to an intact soil hyphal network is reminiscent of that in maize in Ontario, Canada, (Miller, 2000) and Pennsylvania, U.S.A., (Galvez et al., 2001) in high P soils and the decrease in harvest dry weight of linseed in fertile Finnish fields when fumigant was not applied (Kahiluoto et al., 2001). Thus, unless P-supply is balanced carefully with crop requirements, management practices favouring AM fungal activity may risk crop growth depression and reduce profits from high-value crops like vegetables or eliminate the slim profit margin from broadacre crops. In such situations where annual crops are involved, reduction of colonisation through tillage may be appropriate, remembering that if the inoculum level is high, more disturbance will be required (Jasper et al., 1991).

Agricultural systems with low P inputs

Organic, biodynamic and other alternative agricultural systems are often considered prototypes for agricultural systems with greater sustainability than current conventional systems. Alternative systems are characterised by the avoidance of chemical biocides and soluble inorganic fertilisers, but still generally maintain the high throughputs of nutrients that characterise modern industrialised agricultural systems. It is often assumed that AMF will play a more pivotal role in alternative systems than in conventional systems and that a larger, more diverse and more beneficial AM fungal community will develop to compensate, to some degree, for the reduced use of soluble P-fertilisers.

Numerous studies have reported alternative farms to have greater levels of AM colonisation than nearby conventional farms, generally associated with lower available soil P due to reduced applications of soluble P-fertiliser (Mäder et al., 2000; Ryan, 1999; Ryan et al., 2000). Scullion et al. (1998) compared the ability of AMF isolated from organic and conventional farms in the U.K. to enhance P-uptake and growth of white clover (*Trifolium repens*) and leek (*Allium ameloprasum*) in three soils. The results suggested that the AMF present on organic farms where soil available P was low might be more effective than those on conventional farms.

Ryan et al. (2000) examined AM colonisation levels in permanent perennial irrigated pasture on ten paired biodynamic and conventional dairy farms in Victoria, Australia. The biodynamic farms had received no P-fertiliser for an average of 17 years, whereas the conventional farms averaged 27 kg P ha^{-1} yr^{-1} as soluble P-fertiliser. Colonisation of roots of pasture species by AMF was higher on the biodynamic farms, but there was a strong inverse relationship between colonisation level and both the concentration of P in shoots and available P in soil. No other soil or pasture characteristics correlated with AM colonisation levels. A glasshouse experiment, using soil from three farm pairs then considered the response of white clover (*T. repens*) and perennial ryegrass (*Lolium perenne*) to four levels of P and N as a bioassay of the ability of the soil biological community to provide these nutrients to pasture (Ryan and Ash, 1999). There was no indication that the biodynamic soils had developed different processes to enhance plant nutrition, or that the indigenous AM fungal communities differed in their tolerance to applications of soluble nutrients.

Similarly, on an organic wheat farm in New South Wales, Australia, 30 years of replacing superphosphate with poorly soluble rock phosphate had lowered soil available P, increased AM colonisation, but reduced yields compared with a neighbouring conventional farm (Dann et al., 1996; Ryan et al., 1994). The response of wheat grown on each farm to application of four rates of superphosphate and rock phosphate indicated crops on the organic farm did not have greater access to P from the rock phosphate. Maximum yields were achieved on both farms with applications of soluble P-fertiliser, with the yield plateau in response to both fertilisers being higher on the conventional farm than on the organic farm.

Therefore, the premise that AMF are a substitute for P-fertiliser is not necessarily valid, even given long periods of time for AMF to adapt to a low-P agricultural system. The low yields that inevitably result from the elimination of soluble P-fertilisers, the necessary prerequisite for the high levels of AMF in these systems, may be undesirable for economic and environmental reasons. Meaningful gains in plant P-efficiency in conventional and alternative systems may be more easily achieved through selection of cultivars to modify root architecture and to increase seed P content, or utilisation of break crops to reduce root disease (Angus et al., 1999; Bolland and Baker, 1988; Snapp et al., 1995).

Crop breeding and responsiveness to AMF

The poor response to AM colonisation in many agricultural systems may be exacerbated by past breeding of crops for resistance to fungal and bacterial pathogens depressing AM colonisation and responsiveness to AMF (Toth et al., 1990). This selection pressure may be high enough to produce an incompatible reaction between the host plant and certain AMF. Recent demonstration of a hypersensitive response of alfalfa (*Medicago sativa*) to *Gigaspora margarita* is the most extreme evidence so far for such selection (Douds et al., 1998). For wheat, modern breeding programs may have yielded cultivars highly dependent on fertiliser supply and non-responsive to AM colonisation (Hetrick et al., 1993). However, a high degree of variability in mycorrhizal dependency, from highly negative to highly positive, occurs within both modern hexaploid varieties and their ancestors (Hetrick et al., 1993). A high degree of variation in AM colonisation levels exists among genotypes of other crops (Smith et al., 1992). Thus, there may be considerable potential for reselection of crops under reduced-P agricultural conditions to increase dependency on AMF, and allow predictable positive responses to AMF to be obtained.

Management of AMF to maintain soil structure

In agricultural systems the formation and maintenance of soil structure will be influenced by soil properties, root architecture, management practices (tillage, fertilisation) and, it now appears, characteristics of AMF and AMF-host compatibility. The extraradical hyphae (ERH) of AMF and roots may be viewed as a 'sticky string bag' wherein hyphae and roots entangle and enmesh soil particles to form macroaggregate structures (Miller and Jastrow, 2000). The stable glue holding the hyphae to soil particles is glomalin, a glycoprotein deposited on the cell wall of the ERH (Wright and Upadhyaya, 1998). Miller and Jastrow (2000) suggest the hydrophobic glycoprotein coating the hyphae and adjacent soil particles enables the ERH to survive in gas water interfaces and may reduce macroaggregate disruption during wetting and drying cycles. By their estimation, ERH and glomalin together contribute up to 15% of soil organic C in a grassland.

Similar estimates of the glomalin contribution have not yet been reported for crop systems. Although, Wright et al. (1999) correlated glomalin with stabilisation of soil aggregates in a maize cropping system after a 3-year transition from conventional tillage to no-tillage and there are also indications that some crop rotations favour glomalin production and aggregate stabilisation more than others (Wright and Anderson, 2000). Thus, management of cropping systems to enhance soil stability and reduce erosion may often benefit from consideration of the factors controlling production and maintenance of ERH and glomalin.

The importance of ERH and glomalin to soil stability will, however, vary with soil type and environment. For instance, in sandy soils ERH length may be inadequate to enmesh fine particles into macro-aggregates (Degens, 1997). However, even in situations where AMF are not important for soil stability, they may still make substantial contributions to below-ground carbon pools (Miller and Jastrow, 2000; Olsson et al., 1999) and hence, greatly influence below-ground ecosystem processes. This role will alter, and maybe increase in significance, in response to rising concentrations of atmospheric CO_2 (Rillig et al., 2000).

Conclusions

Reviewing the results from field-based studies it appears that AMF are generally not crucial for the nutrition, growth or health of plants in many production agricultural systems. In contrast, while a role for AMF in soil aggregation has long been known, until the identification of glomalin, the magnitude of this contribution was not fully appreciated. The ability of AMF to improve food nutritional value through enhanced uptake of micronutrients, especially Zn, currently merits more attention.

Intensive horticultural systems with high inputs of P-fertiliser and fungicides may reduce formation of AMF and negate any potential benefits from AMF for crop nutrition and root health. Under the moderate P-levels that prevail in the majority of field crop systems, early season colonisation by AMF may often be parasitic, creating a C-drain on crops and reducing yields. Even if P is deficient for plant growth, AMF can not be assumed to enhance growth as in some regions the ability of AMF to enhance host growth appears constrained by factors we currently do not understand. Hence crop responses to AMF are often unpredictable and contrary to their reputation as 'biological fertilisers'.

The role of AMF in agricultural systems based around localised food production and distribution, with low nutrient inputs and outputs, a high degree of internal nutrient cycling and traditional crop varieties may be greater than in the production-orientated agricultural systems examined in this paper. Redesign of these systems to improve the cost/benefit of AMF to plant growth and nutrition must involve an integrated approach. For conventional systems, this could involve a reduction in P-fertilisation; a change to cultivars more responsive to AMF; and perhaps even mitigation of AM colonisation, through means such as tillage, when appropriate. Deliberate enhancement of AM colonisation should proceed with caution since crops and current cultivars vary widely in their degree of dependency and a positive response to AMF may only occur under conditions of severe P-limitation to a dependent crop. Moreover, management practices that reduce AM fungal activity can also enhance environmental sustainability, and crop nutrition and growth, through other means: an example being the use of biofumigant *Brassica* crops to reduce disease and improve efficiency of water and N-use in following crops.

In conclusion, more research is required to determine the role(s) that AMF will play in industrialised agricultural systems. AMF are ubiquitous in these systems and the potential for reselecting crops that are more responsive to AMF appears great. This realisation demands attention based on the long-term implications of AM fungal activity for deposition of glomalin in soils and the probable roles of AMF in enhancing micronutrient concentrations in produce, root health, efficient cycling of non-renewable P and maintenance of the below-ground ecosystem.

Acknowledgements

Thanks to John Kirkegaard, Sabine Ravnskov, Jason Olsen and Pete Timmer for commenting on the manuscript and Dan Chellemi for providing unpublished data. MHR's research was supported by the Grains Research and Development Corporation (GRDC) and JHG's research was supported in part by the Florida Agricultural Research Station and is approved for publication as Journal Series No. R-08228.

References

Angus J F, Desmarchelier J M, Gardner P A, Green A, Hocking P J, Howe G N, Kirkegaard J A, Marcroft S, Mead A J, Pitson G D, Potter T D, Ryan M H, Sarwar M, van Herwaarden A H and Wong P T W 1999 Canola and Indian mustard as break crops for wheat. *In* Proceedings of the 10th International Rapeseed Congress, September 26–29, Canberra, Australia, published on CD.

Biermann B J and Linderman R G 1983 Increased geranium growth using preplant inoculation with a mycorrhizal fungus. J. Am. Soc. Hortic. Sci. 108, 972–976.

Bolland M D A and Baker M J 1988 High phosphorus concentrations in seed of wheat and annual medic are related to higher rates of dry matter production of seedlings and plants. Aust. J. Exp. Agric. 28, 765–770.

Chellemi D O 2000 Adaptation of approaches to pest control in low input agriculture. Crop Protection 19, 855–858.

Cooke J W, Ford G W, Dumsday R G and Willatt S T 1985 Effect of fallowing practices on the growth and yield of wheat in south-eastern Australia. Aust. J. Exp. Agric. 25, 614–627.

Cooper K M and Tinker P B 1981 Translocation and transfer of nutrients in vesicular-arbuscular mycorrhizas IV. Effect of environmental variables on movement of phosphorus. New Phytol. 88, 327–339.

Dann P R, Derrick J W, Dumaresq D C and Ryan M H 1996 The response of organic and conventionally grown wheat to superphosphate and reactive rock phosphate. Aust. J. Exp. Agric. 36, 71–78.

Degens B P 1997 Macro-aggregation of soils by biological bonding and binding mechanisms and the factors affecting these: a review. Aust. J. Soil Res. 35, 431–459.

Douds D D, Galvez L, Becard G and Kapulnik Y 1998 Regulation of arbuscular mycorrhizal development by plant host and fungus species in alfalfa. New Phytol. 138, 27–35.

Evans D G and Miller M H 1988 Vesicular-arbuscular mycorrhizas and the soil-disturbance-induced reduction of nutrient absorption in maize I. Causal relations. New Phytol. 110, 67–74.

Galvez L, Douds D D, Drinkwater L E and Wagoner P 2001 Effect of tillage and farming system upon VAM fungus populations and mycorrhizas and nutrient uptake of maize. Plant Soil 228, 299–308.

Gianinazzi S and Schüepp H 1994 Impact of Arbuscular Mycorrhizas on Sustainable Agriculture and Natural Ecosystems. Birkhauser Verlag, Basel. 226 p.

Graham J H 2000 Assessing costs of arbuscular mycorrhizal symbiosis in agroecosystems. *In* Current Advances in Mycorrhizal Research. Eds. G K Podila and D D Douds Jr., p 127–140. APS Press, St. Paul, MN.

Graham J H 2001 What do root pathogens see in mycorrhizas? New Phytol. 148, 357–359.

Graham J H and Eissenstat D M 1998 Field evidence for carbon cost of citrus mycorrhizas. New Phytol. 140, 103–110.

Hetrick B A D, Wilson G W T and Cox T S 1993 Mycorrhizal dependence of modern wheat cultivars and ancestors: a synthesis. Can. J. Bot. 71, 512–518.

Isbell R F 1996 The Australian Soil Classification. CSIRO Publishing, Melbourne. 143 pp.

Jakobsen I and Rosendahl L 1990 Carbon flow into soil and external hyphae from roots of mycorrhizal cucumber plants. New Phytol. 115, 77–83.

Jakobsen I, Gazey C and Abbott L K 2001 Phosphate transport by communities of arbuscular mycorrhizal fungi in intact cores. New Phytol. 149, 95–103.

Jasper D A, Abbott L K and Robson A D 1991 The effect of soil disturbance on vesicular-arbuscular mycorrhizal fungi in soils from different vegetation types. New Phytol. 118, 471–476.

Kahiluoto H, Ketoja E, Vestberg M, and Saarela I 2001 Promotion of AM utilization through reduced P fertilization 2. Field studies. Plant Soil 231, 65–79.

Kirkegaard J A, Sarwar M, Wong P T W, Mead A, Howe G and Newell M 2000 Field studies on the biofumigation of take-all by *Brassica* break crops. Aust. J. Agric. Res. 51, 445–456.

Larsen J and Bodker L 2001 Interactions between pea root-inhabiting fungi examined using signature fatty acids. New Phytol. 149, 487–493.

Mäder P, Edenhofer S, Boller T, Wiemken A and Niggli U 2000 Arbuscular mycorrhizae in a long-term field trial comparing low-input (organic, biological) and high-input (conventional) farming systems in a crop rotation. Biol. Fertil. Soils 31, 150–156.

Marschner H and Dell B 1994 Nutrient uptake in mycorrhizal symbiosis. Plant Soil 159, 89–102.

Miller M H 2000 Arbuscular mycorrhizae and the phosphorus nutrition of maize: a review of Guelph studies. Can. J. Plant Sci. 80, 47–52.

Miller R M and Jastrow J D 2000 Mycorrhizal fungi influence soil structure. *In* Arbuscular Mycorrhizas: Physiology and Function. Eds. Y Kapulnik and D D Douds. pp. 3–18. Kluwer, Dordrecht.

Olsen J K, Schaefer J T, Edwards D G, Hunter M N, Galea V J and Muller L M 1999a Effects of mycorrhizae, established from an existing intact network, on the growth response of capsicum (*Capsicum annuum* L.) and tomato (*Lycopersicon esculentum* Mill.) at five rates of applied phosphorus. Aust. J. Agric. Res. 50, 223–237.

Olsen J K, Schaefer J T, Edwards D G, Hunter M N, Galea V J and Muller L M 1999b Effects of a network of mycorrhizae on capsicum (*Capsicum annuum* L.) grown in the field with five rates of applied phosphorus. Aust. J. Agric. Res. 50, 239–252.

Olsson P A, Thingstrup I, Jakobsen I and Bååth E 1999 Estimation of the biomass of arbuscular mycorrhizal fungi in a linseed field. Soil Biol. Biochem. 31, 1879–1887.

Rayment G E and Higginson F R 1992 Australian Laboratory Handbook of Soil and Water Chemical Methods. Inkata Press, Melbourne. 330 pp.

Rillig M C, Hernández G Y and Newton P C D 2000 Arbuscular mycorrhizae respond to elevated atmospheric CO_2 after long-term exposure: evidence from a CO_2 spring in New Zealand supports the resource balance model. Ecol. Lett. 3, 475–478.

Ryan M H 1999 Is an enhanced soil biological community, relative to conventional neighbours, a consistent feature of alternative (organic and biodynamic) agricultural systems? Biol. Agric. Hort. 17, 131–144.

Ryan M H and Ash J E 1999 Effects of phosphorus and nitrogen on growth of pasture plants and VAM fungi in SE Australian soils with contrasting fertiliser histories (conventional and biodynamic). Agric. Ecosyst. Environ. 73, 51–62.

Ryan M H, Chilvers G A Dumaresq D C 1994 Colonisation of wheat by VA-mycorrhizal fungi was found to be higher on a farm managed in an organic manner than on a conventional neighbour. Plant Soil 160, 33-40.

Ryan M H, Small D R and Ash J E 2000 Phosphorus controls the level of colonisation by arbuscular mycorrhizal fungi in conventional and biodynamic irrigated dairy pastures. Aust. J. Exp. Agric. 40, 663–670.

Sánchez-Díaz M and Honrubia M 1994 Water relations and alleviation of drought stress in mycorrhizal plants. *In* Impact of Arbuscular Mycorrhizas on Sustainable Agriculture and Natural Systems. Eds. S Gianinazzi and H Schüepp. pp. 167–178. Birkhäuser Verlag, Basel.

Scullion J, Eason W R and Scott E P 1998 The effectivity of arbuscular mycorrhizal fungi from high input conventional and organic grassland and grass-arable rotations. Plant Soil 204, 243–254.

Simpfendorfer S, Kirkegaard J A and Heenan D P 2001 Involvement of root inhibitory *Pseudomonas* spp. in the poor early growth of direct drilled wheat: studies in intact cores. Aust. J. Agric. Res. 52, 845–853.

Smith S E, Robson A D and Abbott L K 1992 The involvement of mycorrhizas in assessment of genetically dependent efficiency of nutrient uptake and use. Plant Soil 146, 169–179.

Snapp S, Koide R and Lynch J 1995 Exploitation of localized phosphorus-patches by common bean roots. Plant Soil 177, 211–218.

Thingstrup I, Rubæk G, Sibbesen E and Jakobsen I 1998 Flax (*Linum usitatissimum* L.) depends on arbuscular mycorrhizal fungi for growth and P uptake at intermediate but not high soil P levels in the field. Plant Soil 203, 37–46.

Thompson J P 1987 Decline of vesicular-arbuscular mycorrhizas in long fallow disorder of field crops and its expression in phosphorus deficiency in sunflower. Aust. J. Agric. Res. 38, 847–867.

Thompson J P 1991 Improving the mycorrhizal condition of the soil through cultural practices and effects on growth and phosphorus uptake in plants. *In* Phosphorus Nutrition of Grain Legumes in the Semi-Arid Tropics. Eds. C Johansen, K K Lee and K L Sahrawat. pp. 117–137. ICRISAT, Andhra Pradesh.

Thompson J P 1996 Correction of duel phosphorus and zinc deficiencies of linseed (*Linum usitatissimum* L.) with cultures of vesicular-arbuscular mycorrhizal fungi. Soil Biol. Biochem. 28, 941–951.

Toth J D, Toth D, Starke D and Smith D R 1990 Vesicular-arbuscular mycorrhizal colonization of *Zea mays* affected by breeding for resistance to fungal pathogens. Can. J. Bot. 68, 1039–1044.

Welch R M and Graham R D 1999 A new paradigm for world agriculture: meeting human needs. Productive, sustainable, nutritious. Field Crops Res. 60, 1–10.

Wright S F and Anderson R L 2000 Aggregate stability and glomalin in alternative crop rotations for the central Great Plains. Biol. Fert. Soils 31, 249–253.

Wright S F and Upadhyaya A 1998 A survey of soils for aggregate stability and glomalin, a glycoprotein produced by hyphae of arbuscular mycorrhizal fungi. Plant Soil 198, 97–107.

Wright S F, Starr J L and Paltineanu I C 1999 Changes in aggregate stability and concentration of glomalin during tillage management transition. Soil Sci. Soc. Am. J. 63, 1825–1829.

Plant and Soil **244:** 273–279, 2002.

Inoculation with arbuscular mycorrhizal fungi: the status quo in Japan and the future prospects

Masanori Saito[1,3] & Takuya Marumoto[2]
[1]*Department of Grassland Ecology, National Institute of Livestock and Grassland Science, Nishi-nasuno, Tochigi, 329-2793, Japan;* [2]*Faculty of Agriculture, Yamaguchi University, Yoshida, Yamaguchi, 753-8515, Japan;* [3]*Corresponding author**

Received 21 August 2001. Accepted in revised form 12 December 2001

Key words: charcoal, degraded soil, inoculum, revegetation, VA mycorrhizal fungi

Abstract

Inoculation of arbuscular mycorrhizal (AM) fungi has potential benefits in not only sustainable crop production but also environmental conservation. However, the difficulty of inoculum production due to the obligate biotrophic nature of AM fungi has been the biggest obstacle to putting inoculation into practice. Nevertheless, several companies have sought to produce inoculum of AM fungi. Firstly in this review, the present status of inoculum production and its use in Japan is described. Secondly, although the effectiveness of inoculation is primarily limited by environmental and biological factors, some possible ways to improve inoculation performance are discussed. Approaches include use of chemicals to increase spore germination and colonization and soil application of charcoal to provide a microhabitat for AM fungi to colonize and survive.

Introduction

Inoculation with arbuscular mycorrhizal (AM) fungi has potential value for improved crop production, and numerous trials have been conducted since the 1970s (Gianinazzi et al., 1990; Jarstfer and Sylvia, 1993, 1999; Kurle and Pfleger, 1994; Menge, 1984; Powell, 1984; Safir, 1994; Sieverding, 1991). However, the difficulty in inoculum production due to the obligate biotrophic nature of AM fungi has been the biggest obstacle to putting inoculation into practice. At present, nevertheless, several companies all over the world have commercialized the inoculum of AM fungi (http://dmsylvia.ifas.ufl.edu/commercial.htm).

Sustainability of agricultural systems has been disputed because of environmental pollution and deterioration due to 'modern' and 'intensive' agriculture (Matson et al., 1997). The significance of AM fungi to the sustainability of the environment has been realized in arable and natural lands because AM fungi affect the plant community structure in various ways (Van der Heijden et al., 1998). Therefore, for rehabilitation of deteriorated or degraded land, inoculation with AM fungi can be considered an effective option (Miller and Jastrow, 1992; Pfleger et al., 1994). Such a trend may encourage the inoculum production companies.

The intention of this is not to review numerous papers on inoculum production of AM fungi, but to present the current status of inoculum production and its use in Japan. Secondly, we discuss some possible ways to improve inoculation performance and the future prospects in inoculation technology.

Inoculum production in Japan

In the 1940s in Japan, the growth promoting effect of endomycorrhizal fungi on plant growth was established by a pioneering scientist, T. Asai (Asai, 1943, 1944). However, his work was neglected until research on inoculum production of AM fungi boomed in the 1980s in Japan (Ogawa, 1987). In the early 1990s some companies started commercial production of inoculum of AM fungi.

* FAX No: +81-287-36-6629. E-mail: msaito@affrc.go.jp

The Ministry of Agriculture, Forestry and Fisheries (MAFF) of the Japanese Government has promoted the introduction of various technologies to reduce agrochemical inputs to arable lands for sustainable agriculture. MAFF recognized that AM inocula are useful to reduce phosphate fertilizer application, and in 1996 approved AM inocula as soil amendments by an ordinance of the Soil Productivity Improvement Law. The ordinance regulates the quality of 12 types of amendments such as zeolite, peat, and other organic/inorganic materials that are effective for the improvement of soil productivity. The ordinance specifies that a quality guarantee be labeled on the outside of the product container. In the case of AM fungi, the following items are required; (i) name and address of producer, (ii) raw materials, (iii) symbiotic efficiency, (iv) efficacy, (v) recommended application rate, (vi) storage conditions, (vii) expiration date. In item (ii), carrier material, such as peat or zeolite, is indicated. In item (iii), the symbiotic efficiency is expressed as percentage of colonization by the inoculum of a specific test plant such as Welsh onion under standard conditions. In item (iv), the producer should note that the inoculum is not effective for some crops species belonging to the Brassicaceae and Chenopodiaceae, and that the inoculum may not be effective in soils rich in available phosphate. This quality guarantee is important to expel poor quality microbial inocula from the market place.

At present, three companies, Central Glass Co., Idemitsu Kosan Co., and Osaka Gas Co., produce AM inoculum. The MAFF statistics indicate that 28 – 83 tons of the inocula were supplied per year from 1997 to 1999. Much of inoculum was supplied for non-agricultural applications such as rehabilitation of degraded or devegetated landscapes. Because enduse of the inocula is not within the framework of the ordinance, the above statistics do not include the supply for such non-agricultural objectives.

Although the detailed procedure for inoculum production is proprietary, these inocula are produced under glasshouse conditions based upon the pot culture technique (Jarstfer and Sylvia, 1993). One company uses expanded clay as a potting medium (Dehne and Backhaus, 1986). Others extract and concentrate the propagules of AM fungi from potting media, and carriers such as peat are mixed with the propagules. The inocula are mainly sold to horticulture farmers. Some formulations are specific for each crop species by taking into account factors such as host plant species, AM fungal species, and the handling of inoculum into horticultural practices.

The cost of inoculum production is a serious problem because the inocula are not competitive in price with phosphorus fertilizer. Even if farmers understand the significance of sustainable agricultural systems, the reduction of phosphorus inputs by using AM fungal inocula alone cannot justify the use of the inocula except in the case of high value crops. Another serious problem is control of phytopathogenic micro-organisms. At present, the inoculum produced is not completely free from pathogens, even though the producers attempt to control pathogens with various agrochemicals. Farmers are very aware of the risk of pathogens, so they do not accept inoculum containing host root residues. Although pieces of root colonized with AM fungi, especially *Glomus intraradices* and related species, function well as propagules, the companies remove such residue of host roots from their products.

Rehabilitation of a volcanic deposit-affected area

Inocula of AM fungi are expected to be substantially beneficial in the establishment of vegetation in degraded or bare landscapes (Miller and Jastrow, 1992; Pfleger et al., 1994; Requena et al., 2001). Currently in Japan, AM fungal inocula has been applied most successfully in revegetation of land by devastated by volcanic activities (Marumoto et al., 1996, 1999).

In the Japanese islands, active volcanos sometimes seriously damage large areas and destroy human activities. Newly deposited volcanic materials are very low in nutrients available for plants and very susceptible to erosion. The revegetation process is therefore slow, so the acceleration of revegetation is required not only for environmental conservation but also for the prevention of erosion. In 1990, Mt. Fugendake (32° 45′ N, 130° 19′ E) began volcanic activities after 200 years of inactivity. During 1991 – 1994, several large pyroclastic flows completely destroyed the previous vegetation on the mountain slope. More than 1000 ha of the mountain slope and the base of the mountain were covered with a thick pyroclastic deposit.

Immediately after volcanic activities ceased in 1995, a revegetation project was started. Because the area to be revegetated was on a steep slope of the volcano and was still a hazard, the workers were not able to access the revegetation project site during the first a few years. Therefore, revegetation materials were

Figure 1. Revegetation in a volcano-devastated area of Mt. Fugendake, Japan. (A) Application of bags containing AM fungal inoculum on the pyroclastic flow. (B) Germination of plants from seeds contained in the bag. (C) A landscape of the pyroclastic flow 3 weeks after the application. (D) A landscape in the same site as (C) 3 years after the application.

applied from the air by a helicopter in 1995. A bag of unwoven polyester fabric, weighing about 2 kg, contained plant seeds, AM fungal inocula, slow-release chemical fertilizer, and some carriers such as peat moss. Seeds of various wild grass and shrub species were used: *Miscanthus sinensis*, *Artemisia princeps*, *Lespedeza cuneata* and others. *Gigaspora margarita* and *Glomus* sp. were used as the fungal inocula. About 3000 bags per ha were applied to the target-area (Figure 1 A, B). The grass plants that germinated from the bag were highly colonized with AM fungi. Recently, the inoculated fungal species were still proliferating 6 years after application (unpublished). Thus, the AM fungi and some nutrients in the bag supported the growth of the plants contained in the bag in the nutrient poor pyroclastic flow. The site where the bags were located became a base from which the plants revegetated the site and prevented serious erosion (Figure 1 C, D). This 'bag' method reduced the amount of revegetation materials, including inoculum, and established plant coverage more effectively than broadcast application of revegetation materials to the whole area. Because the local government urgently requested the revegetation project, the cost for inoculum was not taken into account in the program.

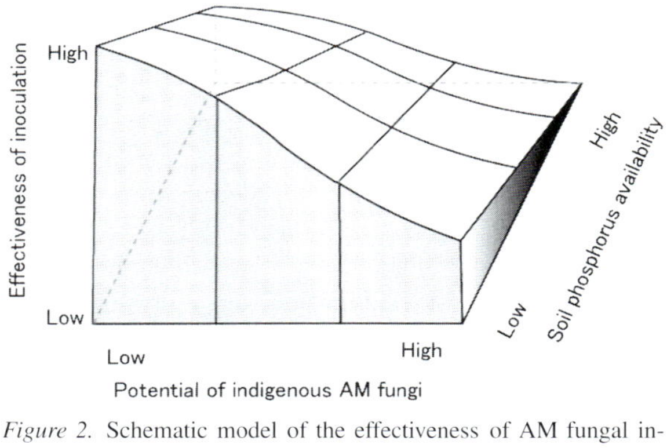

Figure 2. Schematic model of the effectiveness of AM fungal inoculation in relation to indigenous AM fungal potential and soil phosphorus availability.

Improvement in inoculation performance

The effectiveness of AM fungal inoculation is affected by various environmental and biological factors, especially the phosphorus availability in soil and the inoculum potential of indigenous AM fungi (Gianinazzi et al., 1990). In soil low in phosphorus availability and indigenous AM fungi, the effectiveness of inoculation is expected to be the greatest (Figure 2). This was demonstrated in the revegetation project described above. Field trials in Japan with a commercial inoculum indicate that the efficacy is generally highest in Andisols which show high phosphate fixing capacity (Ueda and Kubo, pers. comm.). On the other hand, in soils rich in available phosphorus, the effectiveness is reduced low and the inoculation might even reduce the crop performance (Peng et al., 1993).

Although the effectiveness of inoculation is primarily limited by inoculum potential and P availability, there are possible ways to improve inoculation performance. The inoculation process can be divided into three stages: (i) spore germination, (ii) colonization, and (iii) growth of extraradical hyphae and sporulation. The potential improvement of inoculum performance is discussed for each stage of this process.

Spore germination

The propagules (mainly spores) in the inoculum should be active and should immediately initiate growth after inoculation. Many chemical compounds and unidentified fractions of various extracts have been found to stimulate spore germination and hyphal extension (Azcón–Aguilar et al., 1999; Hepper, 1984; Ishii et al., 1997; Nagahashi and Douds, 2000; Nair et al., 1991). Addition of these compounds to the inoculum may improve the germination rate and increase the colonization potential of the inoculum (Tawaraya et al., 1998).

Spores of AM fungi often show dormancy, although this phenomenon has not yet been critically investigated. Storing the spores for a period from some weeks to some months under cold temperature usually breaks spore dormancy (Hepper, 1984; Safir et al., 1990). *G. margarita* often shows spore dormancy and a very low germination rate unless elevated temperature treatment is used to break the dormancy (Figure 3). Breaking the spore dormancy technique can increase initial colonization and may enhance the effectiveness of the inoculation (Miyamoto et al., 1994).

Root colonization

Arbuscular mycorrhizal colonization is comprised of a series of complicated processes from recognition between the fungi and host plant to arbuscule formation (Nagahashi, 2000). Root exudates contain compounds that promote colonization by AM fungi (Nagahashi, 2000; Tawaraya et al., 1998). Since the colonization process may depend on a balance among phytohormones (Ludwig–Müller, 2000), and some phytohormones affect colonization by AM fungi (Ghachtouli et al., 1996). Application of a flavonoid, formononetin at time of inoculation stimulated colonization by AM fungi (Fries et al., 1998; Koide et al., 1999). Application of such compounds at transplanting may increase inoculation performance.

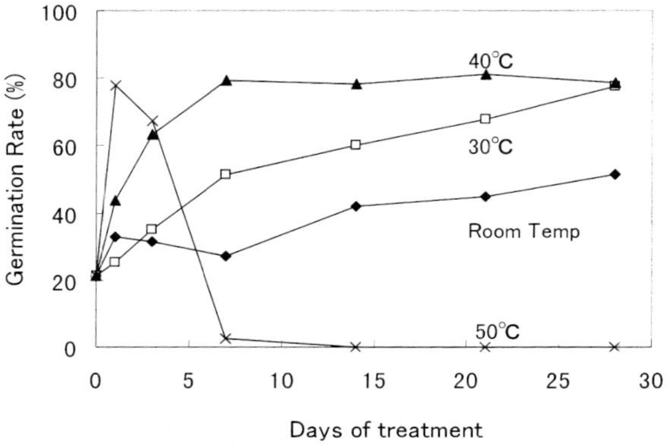

Figure 3. Effect of temperature regime on spore germination of *Gigaspora margarita* (modified from Miyamoto et al., 1994). The spores collected from a pot culture medium were mixed with calcined attapulgite and were kept for 1 – 28 days at different temperatures. After the treatment, 200 spores were extracted by wet-sieving, and the germination rates for 2 weeks at 30 ° C were examined.

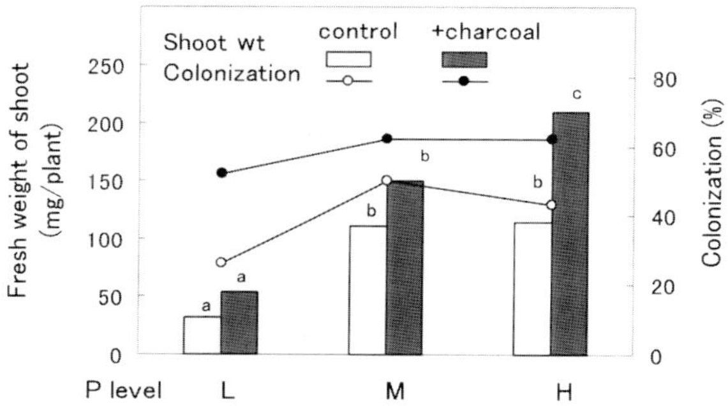

Figure 4. Effect of charcoal application on the growth of alfalfa and indigenous AM fungi in a pot experiment. Alfalfa was grown in a clay loam infertile Andisol with N (2 g m^{-2}) and K (8.3 g m^{-2}) fertilization under different phosphate fertilizer application levels. L, M and H in P level was equivalent to 0, 2.2 and 4.4 g P m^{-2}, respectively. Charcoal made of a mixture of barks from deciduous trees was applied at a rate of 1 kg m^{-2} (modified from Nishio and Okano, 1991).

Growth of extraradical hyphae and sporulation

Nutrient absorption of mycorrhizal plants from soil depends upon the function of extraradical hyphae. Soil management that assures the extraradical hyphae to be fully functioning may enable improvement of inoculation performance. Application of charcoal to soil stimulated the colonization of crops by indigenous AM fungi (Nishio and Okano, 1991; Ogawa et al., 1983; Saito, 1990) (Figure 4). The effect of charcoal was ascribed to its physioco-chemical properties. Charcoal is porous, weakly alkaline, and does not serve as a substrate for saprophytes. AM fungi sensitive to competition from saprophytes can easily extend their extraradical hyphae into charcoal buried in soil and sporulate in the particles (Ogawa, 1987). Charcoal particles act as a micro-habitat for AM fungi to survive and later grow into the soil, which makes charcoal suitable as a carrier of AM fungal inoculum (Ogawa, 1989). Charcoal can be applied in a large scale in infertile soils rich in indigenous AM fungi, but its cost is not competitive with inorganic fertilizers at least in Japan.

Although many inoculation studies have been conducted, there is still little known about how AM fungi proliferate and survive in soil and how they affect the following crop. In arable lands, various crops are cultivated in sequence. Absence of a host plant, cultivation of non-host crops, or long fallow negatively affect the population of AM fungi in soil (Kurle and Pfleger, 1994; Safir, 1994; Thompson, 1987). On the other hand, some mycotrophic crops increase the growth of the succeeding mycotrophic crop (Arihara and Karasawa, 2000; Karasawa et al., 2001). Irrespective of inoculation, therefore, crop sequence should be taken into account when predicting the responses of AM fungi in a sustainable agricultural system.

Future prospects

Commercial inoculum production of AM fungi has been increasing during the past decade, although the future prospect of the business is still uncertain. The obligate biotrophic nature of AM fungi inevitably raises the cost for inoculum production. Soilless culture systems seem to be promising, because this enables the production of clean spores (Jarstfer and Sylvia, 1999). A dual culture of transformed carrot hairy root with AM fungi is an excellent system for providing the sterile spores for research purposes (Glomales *in vitro* collection; http://www.mbla.ucl.ac.be/ginco-bel/anglais/cadre.htm) but is unsuitable for commercial inoculum production. Even if mass production system of the dual culture of lower production cost was developed (Jolicoeur et al., 1999), it is uncertain whether the use of spores produced from roots transformed by a pathogenic bacterium, *Agrobacterium rhizogenes*, would be allowed for inoculation in field.

In the field application of any microbial inoculum, it is essential to verify that the inoculated microorganisms caused the plant response in field. Various molecular techniques have been developed to distinguish the inoculated strain from other indigenous strains. However, these techniques to identify AM fungi are still not routinely used because of genetic heterogeneity in AM fungi. Although ITS (internal transcribed spacer) sequence of rDNA is widely used for discrimination among strains in other fungal taxa, application of the ITS sequence for identification of AM fungi is complicated. For example, ITS sequences of 3 isolates of *G. margarita* including BEG34 and two Japanese isolates showed that the sequences were too diverse to discriminate the isolates; the accession number of the sequence data used were AJ006838–50 for isolates BEG34, AB048619–29 for MAFF520052, and AB048607–18 for MAFF520054, respectively (Saito, 2000). Single stranded conformation polymorphism (SSCP) of the large ribosomal subunit gene may be promising to discriminate isolates of several *Glomus* spp. (Kjøller and Rosendahl, 2000), while the sequence diversity in this region is very high (Clapp et al., 2001). We need more reliable molecular techniques to trace the inoculated fungi. Therefore, from the standpoint of not only basic biological interest but also application, molecular genetics of these multinucleate fungi is of high research priority.

It is now well recognized that inoculation of AM fungi has a potential significance in not only sustainable crop production, but also environmental conservation. However, the status quo of inoculation is far from practical technology that can be widely used in the field. Together, a basic understanding of the biology of AM fungi and an improvement in inoculum production and inoculation technology are required to advance management of these fungi.

Acknowledgements

The preparation of this paper was supported in part by

Promotion of Basic Research Activities for Innovative Biosciences (PROBRAIN), Bio-oriented Technology Research Advancement Institution, Japan. The authors are grateful to Central Glass Co. Ltd., Idemitsu Kosan Co. Ltd., and Osaka Gas Co. Ltd. for valuable information on inoculum production and to Prof. R. Koide for his critical reading of this manuscript. MS is grateful to Dr. Wu Chi–Guang for fruitful discussions on inoculum production.

References

Arihara J and Karasawa T 2000 Effect of previous crops on arbuscular mycorrhizal formation and growth of succeeding maize. Soil Sci.Plant Nutr. 46, 43–51.

Asai T 1943 Die Bedeutung der Mykorrhiza fur das Pflanzenleben. Jpn. J. Bot. 12, 359–436.

Asai T 1944 Uber die Mykorrhizenbildung der Leguminosen-Pflanzen. Jpn. J. Bot. 13, 463–485.

Azcón–Aguilar C, Bago B and Barea J 1999 Saprophytic growth of arbuscular mycorrhizal fungi. *In* Mycorrhiza. Eds. A Varma and B Hock. pp. 391–408. Springer-Verlag, Berlin.

Clapp J P, Rodriguez A and Dodd J C 2001 Inter- and intra-isolate rRNA large subunit variation in *Glomus coronatum* spores. New Phytol. 149, 539–554.

Dehne H W and Backhaus G F 1986 The use of vesicular-arbuscular mycorrhizal fungi and plant production. I. Inoculum production. J. Plant Dis. Protect. 93, 415–424.

Fries L L M, Pacovsky R S and Safir G R 1998 Influence of phosphorus and formononetin on isozyme expression in the *Zea mays- Glomus intraradices* symbiosis. Physiol. Plant. 103, 172–180.

Ghachtouli N E, Martin–Tanguy J, Paynot M and Gianinazzi S 1996 First report of the inhibition of arbuscular mycorrhizal infection of *Pisum sativum* by specific and irreversible inhibition of polyamine biosynthesis by giberellic acid treatment. FEBS Lett. 385, 189–192.

Gianinazzi S, Trouvelot A and Gianinazzi–Pearson V 1990 Conceptual approaches for the rational use of VA endomycorrhizae in agricultrure: Possibilities and limitations. Agric. Ecosyst. Environ. 29, 153–161.

Hepper C M 1984 Isolation and culture of VA mycorrhizal (VAM) fungi. *In* VA Mycorrhiza. Eds. C L Powell and D J Bagyaraji. pp. 95–112. CRC Press, Boca Raton, FL.

Ishii T, Narutaki A, Sawada K, Aikawa J, Matsumoto I and Kadoya K 1997 Growth stimulatory substances for vesicular-arbuscular mycorrhizal fungi in Bahia grass (*Paspalum notatum* Flugge.) roots. Plant Soil 196, 301–304.

Jarstfer A G and Sylvia D M 1993 Inoculum produciton and inoculation strategies for vesicular-arbuscular mycorrhizal fungi. *In* Microbial Ecology. Application in Agricultural and Environmental Management. Ed. F B Metting Jr. pp. 349–377. Marcel Dekker, New York.

Jarstfer A G and Sylvia D M 1999 Aeroponic culture of VAM fungi. *In* Mycorrhiza. Structure, Function, Molecular Biology and Biotechnology. 2nd edition. Eds. A Varma and B Hock. pp. 427–441. Springer-Verlag, Berlin.

Jolicoeur M, Williams R D, Chavarie C, Fortin J A and Archambault J 1999 Production of *Glomus intraradices* propagules, an arbuscular mycorrhizal fungus, in an airlift bioreactor. Biotech. Bioengineer. 63, 224–232.

Karasawa T, Kasahara Y and Takebe M 2001 Variable response of growth and arbuscular mycorrhizal colonization of maize plants to preceding crops in various types of soils. Biol. Fertil. Soils 33, 286–293.

Kjøller R and Rosendahl S 2000 Detection and arbuscular mycorrhizal fungi (Glomales) in roots by nested PCR and SSCP (Single Stranded Conformation Polymorphism). Plant Soil 226, 189–193.

Koide R T, Landherr L L, Besmer Y L, Detweiler J M and Holcomb E J 1999 Strategies for mycorrhizal inoculation of six annual bedding plant species. Hortscience 34, 1217–1220.

Kurle J E and Pfleger F L 1994 The effects of cultural practices and pesticides on VAM fungi. *In* Mycorrhizae and Plant Health. Eds. F L Pfleger and R G Linderman. pp. 101–131. APS Press, St. Paul, MN.

Ludwig–Müller J 2000 Hormonal balance in plants during colonization by mycorrhizal fungi. *In* Arbuscular Mycorrhizas: Physiology and Function. Eds. Y Kapulnik and D D Douds. pp. 263–286 Kluwer Academic Publishers, Dordrecht, The Netherlands.

Marumoto T, Okabe H, Ezaki T, Nishiyama M and Yamamoto K 1996 Application of symbiotic micro-organisms to soil conservation and reforestation. Proceeding of BIO JAPAN, 242–250.

Marumoto T, Kohno N, Ezaki T and Okabe H 1999 Reforestation of volcanic devastated land using the symbiosis with mycorrhizal fungi. Soil Microorganisms 53, 81–90.

Matson PA, Parton W J, Power A G and Swift M J 1997 Agricultural intensification and ecosystem properties. Science 277, 504–509.

Menge JA 1984 Inoculum production. *In* VA Mycorrhiza. Eds. C L Powell and D J Bagyaraji. pp. 187–203. CRC Press, Boca Raton, FL.

Miller R M and Jastrow J D 1992 The application of VA mycorrhizae to ecosystem restoration and reclamation. *In* Mycorrhizal Functioning. Ed. M J Allen. pp. 438–467. Chapman and Hall, New York.

Miyamoto H, Sakai M and Miyamoto J 1994 Production of spores of VA mycorrhizal fungi. Japanese Patent Pub. No. 06-277048.

Nagahashi G 2000 *In vitro* and *in situ* techniques to examine the role of root exudates duringAM fungus-host interactions. *In* Arbuscular Mycorrhizas: Physiology and Function. Eds. Y Kapulnik and D D Douds. pp. 287–300. Kluwer Academic Publishers, Dordrecht, The Netherlands.

Nagahashi G and Douds D D 2000 Partial separation of root exudate components and their effects upon the growth of germinated spores of AM fungi. Mycol.Res. 104, 1453–1464.

Nair M G, Safir G R and Siqueira J O 1991 Isolation and identification of vesicular-arbuscular mycorrhiza-stimulatory compounds from clover (*Trifolium repens*) roots. Appl. Environ. Microbiol. 57, 434–439.

Nishio M and Okano S 1991 Stimulation of the growth of alfalfa and infection of roots with indigenous vesicular-arbuscular mycorrhizal fungi by the application of charcoal. Bull. Natl. Grassl. Res. Inst. 45, 61–71.

Ogawa M 1987 Sakumotsu to tsuchi wo tsunagu kyousei-biseibutsu (Symbiotic Microorganisms liking Crop with Soil). Nobunkyo Pub., Tokyo, Japan. 241 pp.

Ogawa M 1989 Inoculation methods of VAM fungi: Charcoal ball method and rice hulls method. *In* Recent Advances in Microbial Ecology. Eds. T Hattori T et al., pp. 247–252. Japan Scientific Societies Press, Tokyo.

Ogawa M, Yambe Y and Sugiura G 1983 Effect of charcoal on the root nodule and VA mycorrhiza formation of soybean. Abst. 3rd Int. Mycol. Cong. pp. 578. Tokyo.

Peng S, Eissenstat D M, Graham J H, Williams K and Hodge N C 1993 Growth depression in mycorrhizal citrus at high-phosphorus supply. Plant Physiol. 101, 1063–1071.

Pfleger F L, Stewart E L and Noyd R K 1994 Role of VAM fungi in mine land reclamation. *In* Mycorrhizae and Plant Health. Eds. F L Pfleger and R G Linderman. pp. 47–81. APS Press, St. Paul, MN.

Powell C L 1984 Field inoculation with VA mycorrhizal fungi. *In* VA Mycorrhiza. Eds. C L Powell and D J Bagyaraji. pp 205–222. CRC Press, Boca Raton, FL.

Requena N, Perez–Solis E, Azcón–Aguilar C, Jeffries P and Barea J 2001 Management of indigenous plant-microbe symbioses aids restoration of desertified ecosystems. Appl. Envrion. Microbiol. 67, 495–498.

Safir G R 1994 Involvement of cropping systems, plant produced compounds and inoculum production in the functioning of VAM fungi. *In* Mycorrhizae and Plant Health. Eds. F L Pfleger and R G Linderman. pp. 239–259. APS Press, St. Paul, MN.

Safir G R, Coley S C, Siqueira J O and Carlson P S 1990 Improvement and synchronization of VA mycorrhiza fungal spore germination by short-term cold storage. Soil Biol. Biochem. 22, 109–111.

Saito M 1990 Charcoal as a micro-habitat for VA mycorrhizal fungi, and its implication. Agric. Ecosyst. Environ. 29, 341–344.

Saito M 2000 Use of VA mycorrhizal fungi. *In* Biseibutsu no Shizaika: Kenkyu no Saizennsen (Micro-organisms Resources: Its Characteristics and Utilization). Eds. T Suzui, N Okada, H Kunimi, T Makino, M Saito and K Miyashita. pp. 57–70. Soft Science Inc., Tokyo, Japan.

Sieverding E 1991 Vesicular-arbuscular mycorrhiza management in tropical agrosystems. Deutsche Gesellschaft fuer Technische Zusammenarbeit GmbH, Eschborn. 371 pp.

Tawaraya K, Watanabe S, Yoshida E and Wagatsuma T 1996 Effect of onion (*Allium cepa*) root exudates on the hyphal growth of *Gigaspora margarita*. Mycorrhiza 6, 57–59.

Tawaraya K, Hashimoto K and Wagatsuma T 1998 Effect of root exudate fractions from P-deficient and P-sufficient onion plants on root colonisation by the arbuscular mycorrhizal fungus *Gigaspora margarita*. Mycorrhiza 8, 67–70.

Thompson J P 1987 Decline of vesicular-arbuscular mycorrhizae in long fallow disorder of field crops and its expressionn in phosphorus deficiency of sunflower. Aust. J. Agric. Res. 38, 847–867.

Van der Heijden M G A, Klironomos J N, Ursic M, Moutoglis R, Streitwolf–Engel T, Boller T, Wiemken A and Sanders I R 1998 Mycorrhizal fungal diversity determines plant biodiversity, ecosystem variability and productivity. Nature 396, 69–72.

Plant and Soil **244:** 281–290, 2002.

Host-specificity of AM fungal population growth rates can generate feedback on plant growth

James D. Bever
*Department of Biology, Indiana University, Jordan Hall, Bloomington IN 47405-3700, USA**

Key words: arbuscular mycorrhizal fungi, community dynamics, feedback, host-specificity, population growth rates

Abstract

While the mutualistic interaction between plants and AM fungi is of obvious importance to ecosystem processes, the factors influencing the ecological and evolutionary dynamics within this interaction are poorly understood. The mutual interdependence of plant and AM fungal relative growth rates could generate complex dynamics in which the composition of the AM fungal community changes due to association with host and this change in fungal composition then differentially feeds back on plant growth. I first review evidence for feedback dynamics and then present an approach to evaluating such complex dynamics. I specifically present evidence of host-specific differences in the population growth rates of AM fungi. Pure cultures of AM fungi were mixed to produce the initial fungal community. This community was then distributed into replicate pots and grown with one of four co-occurring plant species. Distinct compositions of AM fungal spores were produced on different host species. The AM fungal communities were then inoculated back onto their own host species and grown for a second growing season. The differentiation observed in the first generation was enhanced during this second generation, verifying that the measure of spore composition reflects host-specific differences in AM fungal population growth rates. In further work on this system, I have found evidence of negative feedback through two pairs of plant species. The dynamic within the AM fungal community can thereby contribute to the coexistence of plant species.

Introduction

Arbuscular mycorrhizal (AM) fungi can facilitate plant growth by increasing access to soil resources. The interaction is a well-established mutualism, as the fungus, in turn, is wholly dependent on carbohydrates from the plant. The ecological importance of this interaction is broadly appreciated, as the majority of plant species associate with AM fungi, including most agricultural plants (Smith and Read, 1997) and the presence of these fungi have been shown to influence interspecific competition (Grime et al., 1987; Hartnett et al., 1993; Hartnett and Wilson, 1999), the trajectory of succession (Allen and Allen, 1990; Janos, 1980; Medve, 1984), and the stabilization of soil aggregates (Miller and Jastrow, 2000).

AM fungi have been found to associate with plants with relatively low specificity and therefore the association is thought to be largely non-specific (Hoeksema, 1999; Law, 1985; Smith and Read, 1997). However, there is growing evidence of host-specific differences in plant response to AM fungi and in fungal response to plants. The extent of plant growth promotion by AM fungi depends upon the specific plant and fungal combinations (Adjoud et al., 1996; Streitwolf-Engel et al., 1997; Van der Heijden et al., 1998). Conversely, measures of growth of AM fungal species also depends on the associated host plant species (Bever et al., 1996; Eom et al., 2000; Johnson et al., 1992b; Sanders and Fitter, 1992).

This level of specificity of AM fungi can have important consequences on plant ecology. Specifically, as a result of the specificity of plant response to the AM fungi, the diversity and composition of the AM fungal community has been shown to exert large effects on plant diversity and composition (Van der Heijden et al., 1998). Conversely, the composition of the plant community has similar large effects on the diversity and composition of the AM fungal community (Bever et al., 1996; Eom et al., 2000; Johnson

* FAX No: +1-812-855-6705. E-mail: jbever@indiana.edu

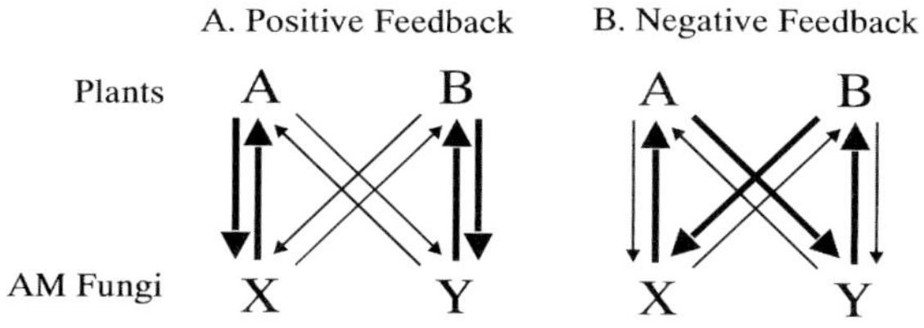

Figure 1. Representation of fitness sets that would produce positive and negative feedback within the interaction of plants and AM fungi. In these figures, the thickness and direction of the arrows represent the relative benefit that two plant types (A and B) and two AM fungal types (X and Y) receive from their association. In the case of a highly symmetric fitness relationship between plants and AM fungi (a), an initial abundance of plant A will result in an increase in representation of AM fungus X, which will increases the growth rate of plant A and thereby generate a positive feedback that can lead to the loss of diversity. Alternatively, in the case of a highly asymmetric fitness relationship (b), an initial abundance of plant A will increase the representation of AM fungus Y and thereby boost the performance of plant B, resulting in a negative feedback on plant A (redrawn from Bever et al. 1997).

et al., 1992b; Sanders and Fitter, 1992). Because of the mutual interdependence of plant and fungal growth rates, it is difficult to project the long-term dynamics and stability of the community.

The mutual interdependence of plant and AM fungal population growth rates could generate complex community dynamics between the plant and fungal guilds. Host-specific changes in the AM fungal community could lead to increases in the relative growth rates of the most abundant plant (i.e., positive feedback) or to decreases in the relative growth rates of the most abundant plant species (i.e., negative feedback) (Bever, 1999; Bever et al., 1997, in press). Positive feedback is generated by symmetric fitness relations between plant and AM fungi while highly asymmetric fitness relations will generate negative feedback (Figure 1). These two dynamics lead to very different predictions for the community. Positive feedback causes a strengthening of the mutualism between plant and fungal guilds, but a decline in species diversity. Negative feedback causes a weakening of the mutualism, but can contribute to the coexistence of competing plant species (Bever, 1999).

Evidence of negative feedback through the AM fungal community

Currently, there is little evidence of such complex community feedback dynamics between plants and AM fungi, or within any other mutualism. Several studies have identified patterns consistent with negative feedback through the AM fungal community (Bever, 1994; Bever et al., in press; Johnson et al., 1991; Kiers et al., 2000). Johnson et al. (1991) identified a shift in AM fungal spore composition due to cropping sequence in the midwestern United States. Specifically, different species of AM fungi increased with repeated cropping of corn compared to repeated cropping of soybeans. Corn and soybeans are regularly rotated in the Midwestern US because of yield decline following repeated monocropping. This yield decline could result from the shift in AM fungal community composition (Johnson et al., 1992a). A similar correlation has been observed between host-specific degradation of the soil community and shifts in the AM fungal composition in old field weeds in North Carolina (Bever, 1994; Bever et al., 1996, 1997). However, in both of these cases, the plant growth effects could result from accumulation of host-specific root pathogens. In the case of the corn–bean rotation host-specific root feeding nematodes are known to be a potential driver of the yield decline (Agrios, 1997). In the North Carolina system, both host-specific accumulation of root pathogens in the genus *Pythium* (Mills and Bever, 1998) and host-specific shifts in the composition of rhizosphere bacteria have been shown to be potential agents of negative feedback (Westover and Bever, 2001).

Two studies, one in the North Carolina old field (Bever, 1994) and the second in a tropical forested system (Kiers et al., 2000), observed negative feedback in experiments that attempted to separate the effects of soil pathogens from other components of the soil community by inoculating with roots and controlling for non-mycorrhizal species with microbial washes. Specifically, they found that plants grew more poorly when grown with their own live roots than when grown with each other's roots. The microbial wash, however, is unlikely to equalize the density of root pathogens and, since root pathogens are known to have strong effects in both of these systems (Augspurger and Kelley, 1984; Mills and Bever, 1998; Westover and Bever, 2001), these studies cannot eliminate the possibility that the differential effects of root inocula on plant growth was due to root pathogens.

A third approach to investigating the feedback dynamics through the AM fungal community is to make inference into the dynamics from measures of the interdependence of plant and fungal growth rates. While measurement of specificity of plant response to AM fungi is common, concurrent measurement of AM fungal growth response is rare. This likely reflects the difficulties that measurement of AM fungal growth presents. AM fungi can reproduce vegetatively through hyphal extension and by the production of resting propagules in the form of asexual spores. The hyphal structures, which are embedded in the root and soil, cannot be identified morphologically and are hard to quantify. AM fungal community composition therefore has been inferred from counts of the spores, an imperfect measure. One such study compared the growth responses of six plants and five AM fungi that co-occur within an old field in North Carolina (Pringle, 2001). In this study, plant growth and fungal sporulation were measured over a 5-month period. In reanalysis of this data in the context of the feedback model (Bever, 1999), we found evidence of highly asymmetric fitness relations between plants and AM fungi that would generate negative feedback (Bever et al., in press). Because this study started with pure cultures of AM fungi, it avoids the problems of separating the effects of pathogens from that of mycorrhizal fungi. However, the prediction of negative feedback depends upon the accuracy of counts of spores as estimates of fungal population growth rates.

Clearly, evaluation of feedback effects through the AM fungal community is made difficult both by the separation of effects of AM fungi from the effects of other soil organisms and by the measurement of the relative growth rates of AM fungal populations. Here, I present an approach to testing AM fungal community feedback that overcomes these difficulties. I do this by first manipulating replicate laboratory AM fungal communities and then testing the consequences for plant growth (Bever et al., in press). Specifically, a diverse community of AM fungi was reconstructed from pure cultures to eliminate plant pathogens. The initial AM fungal community was then trained on individual plant hosts and then the growth responses of these plant species were evaluated in response to inoculation by the trained fungal communities. The inoculum in this test experiment includes both the spores and hyphal structures of AM fungi and therefore accurately represents the entire fungal community. I test the interactions between four plant species and co-occurring AM fungi from an old field in North Carolina (Bever et al., 1996, 2001, in press). Through this work, I have demonstrated negative feedback through two pairs of plant species (Bever, unpublished ms.). In the present paper, I detail the evidence of host-specificity of AM fungal population growth rates and the resulting differentiation of the AM fungal community.

Methods

The overall approach involved the reconstruction of an AM fungal community free of plant pathogens, followed by the training of this fungal community on plant hosts and finally testing the growth response of these plants to inoculation by these trained fungal communities (Bever et al., in press). Specifically, replicate AM fungal communities were grown on individual host plants for one growing season, during which time the composition of the fungal community may differentiate due to host-specific differences in fungal growth rates (Figure 2). I monitored this differentiation by examining the production of AM fungal spores. To confirm these inferences into the AM fungal community differentiation, I reinoculated the AM fungal communities on their own hosts and monitored AM fungal differentiation after a second growing season (Figure 2). If the differentiation observed after the first growing season is enhanced during this second growing season, this confirms that the spore counts do reflect the differentiation of the AM fungal community.

Isolation of plants and AM fungi

These experiments were performed using co-occurring plants and AM fungi from a North Carolina grassland. The site has a diverse community of plants and fungi (Bever et al., 2001). Four plant species and eight AM fungal species were used in this study, all of which were isolated from within a 75-m^{-2} area (Bever et al., 1996). The four plant species included two herbs, *Allium vineale* and *Plantago lanceolata* and two grasses, *Anthoxanthum odoratum* and *Panicum sphaerocarpon*. Six genotypes of each of the plant species used in this study were clonally propagated. *Anthoxanthum* and *Panicum* were cloned and cleaned as described in (Bever et al., 1996). Clones of *Allium* were replicated with their asexually produced bulbils and clones of *Plantago* were grown from leaf cuttings dipped in IAA.

The eight fungal species included *Acaulospora morrowiae*, *Archaeospora trappei*, *Gigaspora gigantea*, *Gi. decipiens*, *Scutellospora calospora*, *S. pellucida*, *S. reticulata*, and an undescribed species of *Glomus* (identified as *Gl. D1* in Bever et al. (1996)). Each of these species has been deposited in the *International Culture Collection of Arbuscular and Vesicular–Arbuscular Mycorrhizal Fungi* (INVAM) as described in (Bever et al., 1996). All fungal cultures

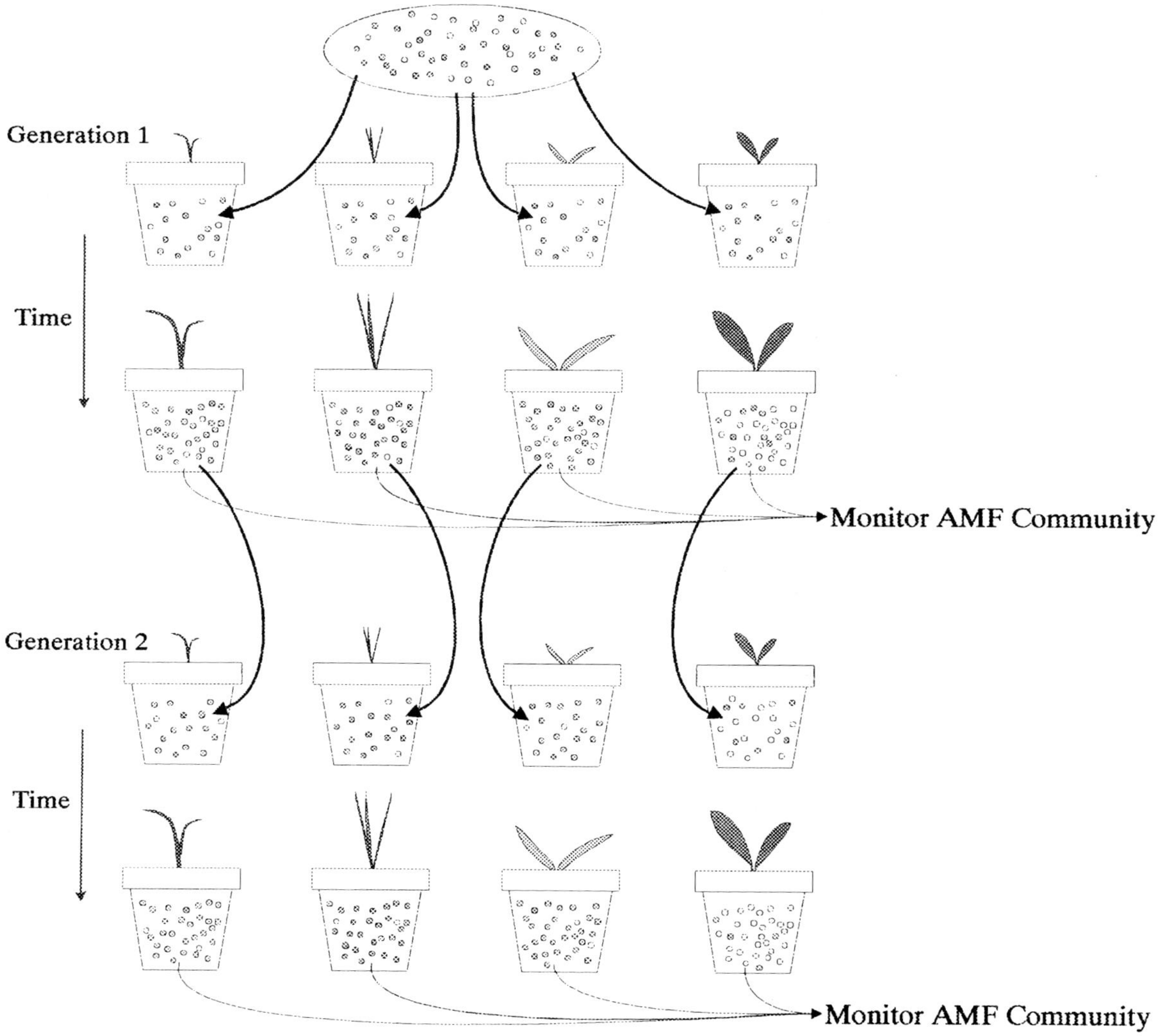

Figure 2. Experimental design. Pure cultures of AM fungi were mixed to produce the initial fungal community. This community was then distributed into replicate pots into which soil-microbe free replicates of four plant species were planted. These were grown for 4.5 months during which the AM fungal species infected their hosts' roots, proliferated, and then sporulated. At the end of this time, the AM fungal spore composition was monitored. To verify that the differentiation observed after the first generation reflects differences in AM fungal growth rates, the AM fungal communities were inoculated back onto their own host genotype and grown for a second growing season, after which the AMF community composition was monitored.

were started from a single spore to ensure that they were free of pathogens. The cultures were grown on Sorghum in autoclaved 1:1 mixture of field soil and sand for 4.5 months in a growth chamber at which time there was abundant sporulation. The cultures were then dried and stored for several months in the refrigerator to break dormancy of spores prior to use.

Laboratory microcosms

The overall design of the microcosm experiment is as depicted in Figure 2. The initial AM fungal community was created by mixing the eight pure cultures in equal proportion. Replicate 600-cc pots were filled with 400 cc 1:1 mixture of sterile soil and sand and inoculated with 200 cc of the initial AM fungal community. Into these pots were planted single, soil-microbe-free plants of one of six genotypes of each of

P. lanceolata, *A. odoratum* and *P. sphaerocarpon*, each replicated three times. For *Allium* there were 5 genotypes three of which were replicated five times and two replicated four times (for a total of 77 AM fungal communities). The plants were grown in a growth chamber with conditions set to mimic Spring in North Carolina (12 h day length, 27 and 20 °C day and night). After 4 months, when the perennial plants begin to dieback, watering was reduced and AM fungi sporulated (Bever et al., 1996). This pattern of sporulation is consistent with the seasonal patterns of spore production observed in the field at this site (Bever et al., 2001; Pringle, 2001; Schultz, 1996). The reduced watering and subsequent harvest was staggered, with the water being reduced in the first replicate at 4 months and the watering of the second replicate reduced after 4.5 months and the watering of the third replicate reduced after 5 months. The three times of harvest allow testing of whether AM fungal sporulation patterns reflect shifts in the phenology of sporulation of individual AM fungal species. These soils were stored dry in the refrigerator for several months prior to use in order to break dormancy of spores. The second generation of the experiment was started as the first, with each AM fungal community being grown with its same plant genotype.

Monitoring of the AM fungal community composition

The composition of the AM fungal community within each pot was monitored through the extraction, identification and enumeration of freshly produced spores, as in (Bever et al., 1996). Spore biovolume was then estimated from spore counts of individual fungal species by weighting these counts by the average volume of individual spores.

Statistical analysis

Differences in production of total spore biovolume were tested using univariate analysis of variance using the general linear models procedure in SAS (SAS, 1990). Host-specific differentiation of AM fungal communities was evaluated using multivariate profile analysis on rank of spore densities as detailed in Bever et al. (1996). The conversion to ranks improves the homogeneity of variance of the skewed distribution of spore counts. Profile analysis directly tests for shifts in the spore composition of the AM fungal community through the significance of the interaction term between plant species and response profile (Bever et al., 1996). The strong interaction terms of profile, plant species and harvest would suggest that sporulation patterns are sensitive to harvest date and may reflect shifts in AM fungal phenology rather than host-specific shifts in population growth rates (Bever et al. 1996). For all multivariate analyses, Wilk's Λ was used for testing hypotheses, though all of the multivariate tests gave similar results. The multivariate analyses were followed by univariate analyses of variances of the ranks of spore counts of individual fungal species. All analyses tested the effects of plant species, harvest dates and the interaction of plant species and harvest date. Comparisons of means after the univariate analyses were corrected for multiple comparisons using the Tukey–Kramer method (Sokal and Rohlf, 1995).

The consistency of the differentiation of spore composition between the two generations was tested using two approaches. First, an increase in the value of the Wilk's Λ F statistic for the interaction of the plant species and response profile indicates that differentiation is larger in the second generation than the first. Secondly, I conducted analyses of covariance including spore counts from the first generation (and the interaction of these spore counts and plant species) as predictors of the spore counts in the second generation. As I knew that these spore counts were correlated with plant species, I used backward elimination of predictors to identify which predictors were strongest. I started with the full model of predictors and eliminated the predictor with the highest F value and reran the analysis until all predictors were significant at $p \leq 0.1$.

Results

AM fungal community differentiation: generation 1

After the first generation of training, I found strong evidence of host-specific differences in the AM fungal communities. The total biovolume of AM fungal spores produced after the first generation was greatest with *Allium* and least with *Anthoxathum*, but was not different between *Panicum* and *Plantago*. For all hosts, spore biovolume was largely made up by four fungal species, *Acaulospora morrowiae*, *Archaeospora trappei*, *Gigaspora gigantea* and *Scutellospora calospora*. Spores of *Gi. decipiens*, *Gl.* D1, *S. pellucida* and *S. reticulata* were encountered infrequently and made up a negligable component of the total biovolume. Because of their low level of sporulation, I

Table 1. Multivariate profile analysis of AM fungal community composition for generation 1 and generation 2. The compositions of the AM fungal communities were highly different between hosts as indicated by the strong effect of plant species. This differentiation of the AM fungal communities increased from generation 1 to generation 2, indicating that host-specific differences in spore abundance reflects host-specific differences in relative rates of AM fungal population growth

Source	Generation 1 Interaction with Profile			Generation 2 Interaction with Profile		
	No. d.f.	Den. d.f.	Wilk's Λ *F*	No. d.f.	Den. d.f.	Wilk's Λ *F*
Plant species	9	150	8.97****	9	150	14.78****
Harvest	6	124	2.98*	6	124	3.52**
Plant species × harvest	18	175	1.57	18	175	1.21

Significance conventions: $^{*}P<0.05$, $^{**}P<0.01$, $^{***}P<0.001$, $^{****}P<0.0001$.

was not able to analyze the abundance of these species statistically and analyses of composition focus on the four dominant species.

The composition of the AM fungal community was highly divergent between plant species as revealed by the high significance of the interaction of AM fungal sporulation profile with plant species (Table 1). The host specific differences in sporulation were similar across the three harvests as indicated by the insignificant and small interaction term of plant species and harvest (Table 1). This suggests that the shifts in sporulation reflect host-specific differences in population growth rates rather than host-specific differences in the phenology of sporulation.

Gi. gigantea sporulated most abundantly with *Allium* and least abundance with *Anthoxanthum* (Table 2 and Figure 3). *S. calospora* also sporulated most abundantly with *Allium*, followed by *Plantago*, and then *Panicum* and *Anthoxanthum* (Table 2 and Figure 3). *Ar. trappei* sporulated most abundantly with *Allium* and *Panicum,* followed by *Plantago* and then *Anthoxanthum* (Table 2 and Figure 3). *Ac. morrowiae* sporulated most abundantly with *Panicum*, followed by *Allium*, *Plantago* and *Anthoxanthum* (Table 2 and Figure 3).

AM fungal community differentiation: generation 2

The sporulation patterns observed in the first generation were reinforced during the second generation. The species of AM fungi that were infrequently observed after the first generation were even more infrequent after the second generation. Spores of *S. reticulata* and *G. decipiens* were not found and only a few spores of *S. pellucida* and *Gl.* D1 were observed. The differences in spore volume remained as in the first generation. The differentiation of the AM fungal communities was enhanced as demonstrated by the increased *F* value and significance of the interaction of AM fungal sporulation profile with plant species (from 8.97 to 14.78, Table 1). Again, the interaction of host species and harvest was not significant. The specific patterns of sporulation were generally similar to that observed in the first generation (Table 2 and Figure 3).

The continual decline of the rare fungal species and the increased differentiation of the common fungal species both indicate that the differentiation observed in the AM fungal community does reflect host-specific differences in AM fungal population growth rates. This interpretation is reinforced by re-analysis of spore abundance in the second generation using spore abundance in the first generation as a covariate. For each of the four common fungal species both spore abundance in generation 1 and host in generation 2, or the interaction of these two factors are significant predictors of spore abundance in the second generation (Table 3). The significance of previous generations' spore counts in predicting spore counts in future generations demonstrates that spore counts in generation 1 do indeed reflect fungal population growth rates.

Discussion

In this report, I demonstrate that AM fungal species, though associating with all hosts, have host-specific differences in their population growth rates. All four of the common AM fungal species showed host-specific differences in their sporulation rates after the first growing season resulting in differentiation of the AM fungal spore composition (Table 1 and Figure 3). Previous studies have also found host-specific differentiation of spore composition (Bever et al., 1996; Eom et al., 2000; Johnson et al., 1992b; Sanders and Fitter, 1992). The present study, however, is the first to demonstrate that such differentiation is increased over successive growing seasons (Table 1 and Figure 3), thereby providing strong evidence that changes in spore composition reflect underlying differences in AM fungal population growth rates. Moreover, both species of plant host and spore abundance in the previous generation are good predictors of AM fungal spore composition in the second generation (Table 3). These patterns cannot be explained under the alternative hypotheses of host-specific differences in phenology of

Table 2. Analysis of variance of spore counts. This table presents the analyses of the rank of the spore counts for the four common fungal species. The *F* value is presented.

Source	d.f.	Generation 1				Generation 2			
		Gi. gig.	*S. cal.*	*Ac. mor.*	*Ar. trap.*	*Gi. gig.*	*S. cal.*	*Ac. mor.*	*Ar. trap.*
Plant species	3	35.4****	31.9****	53.3****	43.8****	113.7****	25.5****	20.5****	19.5****
Harvest	2	0.1	1.8	12.6****	8.7***	0.6	0.4	10.3****	3.0
Plant species × harvest	6	1.8	0.4	0.5	3.1*	1.7	0.9	1.5	1.2

Significance conventions: * $P<0.05$, ** $P<0.01$, *** $P<0.001$, **** $P<0.0001$.

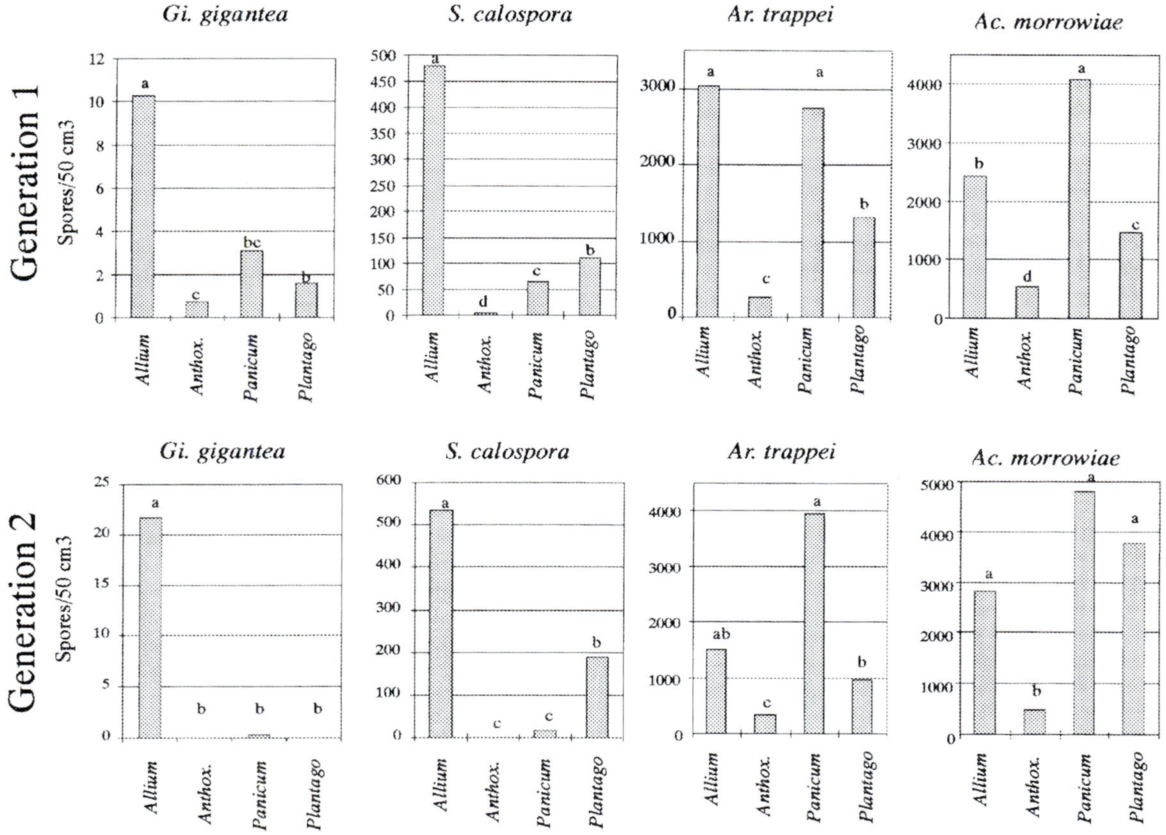

Figure 3. Spore composition after the first and second generation of training. The top row of figures give the average spore abundance of the four common AM fungal species after the first generation of training. The second row of figures gives the average spore abundance for these species after the second generation of training. Within each individual histogram, different letters indicate significant differences between means ($P<0.05$).

sporulation or of sporulation being meaningless to AM fungal population growth.

We note that confidence in the relationship between changes in spore composition and underlying differences in AM fungal population growth rate may be system specific. Species of *Acaulospora*, *Gigaspora* and *Scutellospora* from our field site have shown consistent sporulation pattern over successive years, that is likely associated with the seasonal behavior of their hosts (Bever et al., 2001; Pringle, 2001; Schultz, 1996), suggesting that, for these fungi at this site, sporulation is an important part of their life history. Other AM fungi at this site, namely species of *Glomus*, do not show this consistent pattern (Pringle, 2001). It is possible that for these species or in other

Table 3. Analysis of co-variance of spore counts from the second generation. This table presents the analyses of the rank of the spore counts from the second generation for the four common fungal species. For each fungal species the spore counts in the previous generation (indicated by the species name followed by 1) are used as covariates. The model that resulted from the backward elimination of predictors is presented. A significant effect of prior spore density supports hypothesis that spore counts reflect population growth rates. With both prior spore density and plant species predicting sporulation in generation 2, the results support the idea that differentiation continued between generations, a result that is only consistent with the hypothesis of host-specific differences in population growth rates.

Source	d.f.	*F*
Gi. gigantea (Generation 2)		
Gi. gig. 1	1	2.6†
Plant species	3	77.0****
S. calospora (Generation 2)		
S. cal. 1	1	28.8****
Plant species	3	2.2†
S. cal. 1 × Plant species	3	3.4*
Ac. morrowiae (Generation 2)		
Ac. mor. 1	1	1.3
Plant species	3	6.6***
Harvest	2	8.4***
Ac. mor. 1 × Plant species	3	4.1*
Ar. trappei (Generation 2)		
Ar. trap. 1	1	2.9†
Plant species	3	5.1**

Significance conventions: †$P<0.1$, *$P<0.05$, **$P<0.01$, ***$P<0.001$, ****$P<0.0001$.

systems, sporulation may not correlate as strongly with population growth rates.

The present study does not identify the mechanism of host-specific differences in the AM fungal population growth rates. However, comparison of the patterns of differentiation of the AM fungal community observed in the present study to those patterns previously observed in an independent study of these same plant and fungal species (Bever et al., 1996) provides important clues about the factors involved in these differences. Specifically, some of the patterns are consistent between the two studies. For example, *S. calospora* sporulated in greater abundance with *Plantago* than with *Panicum* in both of these studies. However, other patterns are markedly different. *S. calospora*, for example, sporulated most abundantly with *Allium* in the present study, but sporulated poorly with *Allium* in the previous study. Such differences in specificity likely result from differences between the two studies, namely the removal of the other components of the soil community from the present study.

For example, in the earlier study, sporulation with *Allium* was dominated by *Ac. colossica* (Bever et al., 1996). Field sampling at our study site also indicate strong spatial correlation of the local abundance of *Allium* and *Ac. colossica* (Bever et al., 1996; Schultz, 1996) and the seasonalities of this plant and fungus appear to be closely matched, with both plant and fungus being physiologically active in the cool-season (Bever et al., 2001; Pringle, 2001; Schultz et al., 1999), suggesting a strong positive relationship between this species of plant and fungus. However, as *Ac. colossica* is apparently an obligately cool-season fungus, it was not easily cultured in the lab (Schultz et al., 1999) and I was therefore not able to include it in the present study. Perhaps with the absence of this fungus, or of other components of the soil community, from the present study, fungal species that had previously sporulated relatively poorly with *Allium* now sporulated most abundantly with *Allium*. *S. calospora* and *G. gigantea* sporulated more abundantly with *Plantago* than *Allium* in the presence of *Ac. collosica*, but sporulated in greater abundance with *Allium* than *Plantago* with *Ac. collosica* removed in the present study (Figure 3). This suggests that the relative growth rates of AM fungal species depends not only on the identity of the host with which they are associated, but also on the identity of the other components of the AM fungal community or other components of the soil community. That is, the host-specificity of the AM fungal population growth rates is context dependent. This context-dependence may result from the differences in interspecific competition between individual species of AM fungi.

Regardless of the mechanism, the host-specificity of AM fungal population growth rates has important consequences on our understanding of the ecology of the plant–AM fungal interaction. One simple consequence of such specificity is that the composition and diversity of the AM fungal community would be expected to change with the local composition of the plant community. The extended dominance by a single plant species might result in decreased diversity of the fungal community, due to the loss of individual fungal species that are less competitive in association with that host. Conversely, a more diverse plant community would be expected to maintain greater diversity in the fungal community. There then should be a correlation between plant and fungal diversity, a result observed within our study site (Schultz, 1996).

As plants respond differently to individual species of AM fungi, the change in diversity and composition within the AM fungal community should feed back on plant community dynamics. In further experimentation on this system, I have found evidence of negative feedback between two pairs of plant species (Bever, unpublished ms.). Such negative feedback results from an asymmetry in the delivery of benefit between plant and fungal species (Bever, 1999; Bever et al., in press). The results of this study suggest that dynamics within the AM fungal community can contribute to the maintenance of diversity within plant communities. A positive effect of AM fungal diversity on plant diversity has recently been demonstrated in two grassland systems (van der Heijden et al., 1998; but see Wardle, 1999) and negative feedback represents one of several possible mechanisms generating this effect (Bever et al., 2001). Together with other recent demonstrations of microbial control over plant community diversity and dynamics (Clay and Holah, 1999; Mills and Bever, 1998; Packer and Clay, 2000; van der Putten et al., 1993), these studies call for heightened attention to the potential importance of microbes in plant community processes.

Acknowledgements

I thank J. Morton, J. Antonovics, P. Schultz, A. Bennett, and K. Mills for advice and assistance. This project was supported by USDA grants 92-37101-7461 and 94-37101-0354, and NSF grants DEB-9996221 and DEB-0049080.

References

Adjoud D, Plenchette C, Halli-Hargas R and Lapeyrie F 1996 Response of 11 eucalyptus species to inoculation with three arbuscular mycorrhizal fungi. Mycorrhiza 6, 129–135.

Agrios G N 1997 Plant Pathology. 635 pp. Academic Press, San Diego, CA.

Allen M F and Allen E B 1990 The mediation of competition by mycorrhizae in successional and patchy environments. *In* Perspectives on Plant Competition. Ed. J B Grace and D Tilman. pp. 367–389. Academic Press, New York.

Augspurger C K and Kelley C K 1984 Seedling survival among tropical tree species: interactions of dispersal distance, light-gaps and pathogens. Ecology 65, 1705–1712.

Bever J D 1994 Feedback between plants and their soil communities in an old field community. Ecology 75, 1965–1977.

Bever J D 1999 Dynamics within mutualism and the maintenance of diversity: inference from a model of interguild frequency dependence. Ecol. Lett. 2, 52–61.

Bever J D, Morton J B, Antonovics J and Schultz P A 1996 Host-dependent sporulation and species diversity of arbuscular mycorrhizal fungi in a mown grassland. J. Ecol. 84, 71–82.

Bever J D, Westover K M and Antonovics J 1997 Incorporating the soil community into plant population dynamics: the utility of the feedback approach. J. Ecol. 85, 561–573.

Bever J D, Schultz P A, Pringle A and Morton J B 2001 Arbuscular mycorrhizal fungi: more diverse than meets the eye, and the ecological tale of why. Bioscience. 51, 923–931.

Bever J D, Pringle A and Schultz P A in press Dynamics within the plant-AM fungal mutualism: testing the nature of community feedback. *In* Ecology of Mycorrhizae. Eds. MGA Van der Heijden and IR Sanders. Springer, Berlin.

Clay K and Holah J 1999 Fungal endophyte symbiosis and plant diversity in successional fields. Science 285, 1742–1744.

Eom A-H, Hartnett D C and Wilson G W T 2000 Host plant species effects on arbuscular mycorrhizal fungal communities in tallgrass prairie. Oecologia 122, 435–444.

Grime J P, Macky J M, Hillier S H and Read D J 1987 Mechanisms of floristic diversity: evidence from microcosms. Nature 328, 420–422.

Hartnett D C and Wilson G W T 1999 Mycorrhizae influence plant community structure in tallgrass prairie. Ecology 80, 1187–1195.

Hartnett D C, Hetrick B A D, Wilson G W T and Gibson D J 1993 Mycorrhizal influence on intra- and inter-specific neighbour interactions among co-occurring prairie grasses. J. Ecol. 81, 787–795.

Hoeksema J D 1999 Investigating the disparity in host specificity between AM and EM fungi: lessons from theory and better-studied systems. Oikos 84, 327–332.

Janos D P 1980 Mycorrhizae influence tropical succession. Biotropica 12S, 56–64.

Johnson N C, Pfleger F L, Crookston R K, Simmons S R and Copeland P J 1991 Vesicular arbuscular mycorrihizas respond to corn and soybean cropping history. New Phytol. 117, 657–663.

Johnson N C, Copeland P J, Crookston R K and Pfleger F L 1992a Mycorrhizae — possible explanation for the yield decline with continuous corn and soybean. Agron. J. 84, 387–390.

Johnson N C, Tilman D and Wedin D 1992b Plant and soil controls on mycorrihizal fungal communities. Ecology 73, 2034–2042.

Kiers E T, Lovelock C E, Krueger E L and Herre E A 2000 Differential effects of tropical arbuscular mycorrhizal fungal inocula on root colonization and tree seedling growth: implications for tropical forest diversity. Ecol. Lett. 3, 106–113.

Law R 1985 Evolution in a mutualistic environment. *In* The Biology of Mutualism. Ed. D H Boucher. pp. 145–170. Croom Helm, London.

Medve R J 1984 The mycorrhizae of pioneer species in disturbed ecosystems in western Pennsylvania. Am. J. Bot. 71, 787–794.

Miller M A and Jastrow J D 2000 Mycorrhizal fungi influence soil structure. *In* Arbuscular Mycorrhizas: Molecular Biology and Physiology. Ed. D D Douds. Kluwer Academic Press, Dordrecht.

Mills K E and Bever J D 1998 Maintenance of diversity within plant communities: soil pathogens as agents of negative feedback. Ecology 79, 1595–1601.

Packer A and Clay K 2000 Soil pathogens and spatial patterns of seedling mortality in a temperate tree. Nature 404, 278–281.

Pringle A 2001 Ecology and Genetics of Arbuscular Mycorrhizal Fungi. 330 pp. Duke University, Durham, NC.

Sanders I R and Fitter A H 1992 Evidence for differential responses between host-fungus combinations of vesicular-arbuscular mycorrihizas from a grassland. Mycol. Res. 96, 415–419.

SAS 1990 SAS/STAT User's Guide. SAS Institute Inc., Cary, NC. 1686 pp.

Schultz P A 1996 Arbuscular Mycorrhizal Species Diversity and Distribution in an Old Field Community. Duke University, Durham, NC.

Schultz P A, Bever J D and Morton J B 1999 *Acaulospora colossica* sp. nov. from an old field in North Carolina and morphological comparisons with similar species, *A. laevis* and *A. koskei*. Mycologia 91, 676–683.

Smith S E and Read D J 1997 Mycorrhizal Symbiosis. 605 pp. Academic Press, San Diego, CA.

Sokal R R and Rohlf F J 1995 Biometry. 887 pp. W. H. Freeman, New York.

Streitwolf-Engel R, Boller T, Wiemken A and Sanders I R 1997 Clonal growth traits of two *Prunella* species are determined by co-occurring arbuscular mycorihizal fungi form a calcareous grassland. J. Ecol. 85, 181–191.

Van der Heijden M G A, Klironomos J N, Ursic M, Moutoglis P, Streitwolf-Engel R, Boller T, Wiemken A and Sanders I R 1998 Mycorrhizal fungal diversity determines plant biodiversity, ecosystem variability and productivity. Nature 396, 69–72.

Van der Putten W H, Van Dijk C and Peters B A M 1993 Plant-specific soil-borne diseases contribute to succession in foredune vegetation. Nature 362, 53–56.

Wardle D A 1999 Is 'sampling effect' a problem for experiments investigating biodiversity-ecosystem function relationships. Oikos 87, 403–407.

Westover K M and Bever J D 2001 Mechanisms of plant species coexistence: complementary roles of rhizosphere bacteria and root fungal pathogens. Ecology 82, 3285–3294.

Plant and Soil **244:** 291–305, 2002.

Ecological role of hypogeous ectomycorrhizal fungi in Australian forests and woodlands

Andrew W. Claridge
*New South Wales National Parks and Wildlife Service, Threatened Species Unit, Southern Directorate, PO Box 2115, Queanbeyan, New South Wales 2620, Australia**

Key words: Australia, ecology, forests, fungi, woodlands

Abstract

The Australian continent is characterised by a harsh climate and highly weathered, nutrient-poor soils. Trees and shrubs in these stressful environmental conditions typically form ectomycorrhizae with a variety of fungi, many of which form hypogeous (underground) fruit-bodies. The total number of hypogeous fungi Australia-wide is unknown, although recent systematic studies in the far south-eastern corner of the country suggest that they may number well over a thousand. Similar surveys elswhere are urgently required to clarify the situation. The precise ecological role of many hypogeous fungi remains to be determined, although most presumably facilitate nutrient and water uptake on behalf of their mycorrhizal partners. Others may also protect their plant host from root pathogens. One key function of hypogeous fungi is the role their fruit-bodies play as a food resource for a large range of terrestrial mammals. For a few animals, hypogeous fungi form the single most important dietary item year-round, whereas for others they may only be of seasonal or supplementary value. The extent to which fungi form part of the diet of any mammal species is reflected in the various levels of adaptation toward acquiring, then processing and digesting these cryptic and nutritionally challenging foodstuffs.

Introduction

The Australian environment has several distinctive features that, combined, set it apart from other major regions of the world. Firstly, it is characterised by a mainly arid climate. More than two-thirds of the land mass receives less than 500 mm of precipitation per annum and a third less than 250 mm. Rainfall is typically seasonal, with more than 80% of the continent having at least 3 months in each year which are without any effective precipitation: in the southern half the general pattern is of humid winters followed by arid summers, while across northern Australia the pattern is of humid summers followed by arid winters (Nix, 1981). Either way, the highly seasonal pattern of rainfall often leads to soil moisture deficits at certain times of the year. Secondly, many of the soils over the continent are derived from ancient geological substrates and have been leached for many millions of years. As a result, most soils are low to very low in mineral nutrients, with particular deficiencies in nutrients essential for plant growth such as phosphorus, as well as important trace elements (Specht, 1996). Thirdly, fire is a dominant feature of the Australian landscape, occurring frequently in all but the wettest terrestrial habitats (Recher, 1981). Natural fire has presumably occurred since at least the Late Tertiary, when climate patterns in the Southern Hemisphere became increasingly arid (Sturman and Tapper, 1996), while anthropogenic fire has been present since Aboriginal occupation of the continent at least 50–60 000 years ago (Nix, 1981).

The combination of aridity and nutrient-poor soils, together with likely long-term exposure to fire, has led to the evolution of a distinctive flora across much of the Australian continent. That flora is in large part exemplified by sclerophyllous forests and woodlands dominated by trees of the genus *Eucalyptus* with an associated flora of small-leafed, sclerophyllous shrubs (Recher, 1981). To help compensate for the poor nutritional status of the soils, most of these woody plants form mycorrhizae with a range of fungi. Ec-

* E-mail: andrew.claridge@npws.nsw.gov.au

tomycorrhizal fungi, in particular, form a significant yet poorly documented component of the fungal biodiversity present within sclerophyllous Australian vegetation communities (Castellano and Bougher, 1994). The fruit-bodies of epigeous species emerge aboveground and spores are dispersed by wind, often with the aid of mechanisms for forcible discharge of spores or percussive discharge through ruptures of the outer peridium (Johnson, 1996). In hypogeous species, the fruit-bodies develop below-ground and have no way to liberate spores directly into the air (Bougher and Lebel, 2001). Instead they largely rely for dispersal on vertebrates, which excavate and eat their fruit-bodies, later disseminating spores in faecal pellets (Claridge and May, 1994; Fogel and Trappe, 1978; see below also).

Hypogeous ectomycorrhizal fungi perform a number of beneficial roles in Australian vegetation. Functions relating to their symbiotic status, such as nutrient acquisition and cycling, have been reviewed at length elsewhere (i.e., Bougher and Lebel, 2001; Tommerup and Bougher, 1999) and are only dealt with briefly here. The primary purpose of this paper instead is to explore how hypogeous fungi function as a food resource for animals in forests and woodlands. In doing so, the wide range of fauna known to feed on these fungi, particularly terrestrial mammals, is highlighted. The role of hypogeous fungi in influencing the distribution and foraging patterns of fungus-dependent animals is demonstrated, across time and space, and also in relation to patterns of forest disturbance. Likely nutritional benefits of mycophagy (fungus-feeding) for different mammals are also discussed, as is the potentially important role these animals play in dispersal of spores in forest and woodland habitats. In conclusion, several priority areas for future research on the ecological relationships among hypogeous fungi and mycophagous mammals are identified.

Taxonomy and species diversity

The hypogeous fungi known to occur in Australian forests and woodlands are represented by species within three major classes of fungi, principally Basidiomycetes but also Ascomycetes and Zygomycetes (Bougher and Lebel, 2001; Castellano and Bougher, 1994; Claridge and May, 1994; Warcup, 1990b, 1991). The total diversity of species continent-wide is unknown, because individual taxa can only be reliably identified from their fruit-bodies and only a few areas have been intensively sampled. Fruit-bodies can be difficult to sample because they are generally ephemeral, patchily distributed and hidden at under the soil–litter interface or in the mineral soil. Consequently, building a profile of the number of hypogeous fungi within any given site usually takes repeated and intensive sampling over time (i.e., Claridge et al., 1993a; Johnson, 1994a). Despite the difficulties of inventoring hypogeous fungi, nearly 300 species from among 73 genera have been formally described for Australia, with a similar number of undescribed species known but awaiting taxonomic monographing (Bougher and Lebel, 2001; see Table 1). It is apparent that most major groups of macrofungi are represented by hypogeous forms in Australia, and further that the continent is likely a centre of diversity for certain families (Bougher and Lebel, 2001).

The list of species provided by Bougher and Lebel (2001) for Australia exceeds the total number of hypogeous fungi described from either Europe (Montecchi and Sarasini, 2000) or North America (J. M. Trappe, personal communication). Based on the total number of described Australian taxa, the rate at which new taxa were being monographed, and the proportion of new species being found in recent ecological studies (Claridge et al., 2000a,b; Johnson, 1994a), Bougher and Lebel (2001) estimated that between 12 and 23% of hypogeous species had so far been detected and that the number of Australian hypogeous species would total between 1278 and 2450, an extraordinary figure. Recent studies by Claridge et al. (2000a,b) in the forests and woodlands of the south-eastern Australian mainland provide insight into the incredible diversity of hypogeous fungi across the broader landscape. Across an area of approximately 32 281 km^2, these workers established 136 sample sites, each 50 $\times$ 20 m (totalling approximately 14 ha), representing a range of major environmental gradients for the study region. Although fruit-bodies of hypogeous fungi were collected from these sites on only two occasions (autumn and spring) from within a single year, over 200 species were identified, nearly three-quarters of which were undescribed. Subsequent sampling of the same plots, on two further autumn occasions since, have likely resulted in a further 5–10% increase in number of species detected (A. W. Claridge and J. M. Trappe, unpublished data). The total number of species so far detected is likely an underestimate, since the survey effort concentrated only on those species which produced easily visible fruit-bodies. Despite this, similar systematic collection of hypogeous fungi in other parts

Table 1. Summary of number of hypogeous fungal taxa either formally described or known but currently undescribed (in brackets) from Australia. Data principally from Bougher and Lebel (2001) with subsequent updates

Taxon	No. of genera	No. of species	Taxon	No. of genera	No. of species
BASIDIOMYCETES			Phallaceae	2	2
AGARICALES			RUSSULALES		
Agaricaceae	1	2	Leucogastraceae	1	1
Amanitaceae	2	6	Russulaceae	5	68 (98)
Coprinaceae	4	6	SCLERODERMATALES		
Cortinariaceae	8 (1)	29 (65)	Sclerodermataceae	2	2 (7)
Cribbeaceae	1	3	STEPHANOSPORALES		
Entolomataceae	2	3	Stephanosporaceae	1	1 (1)
Octavianaceae	2	3 (6)	UNCERTAIN STATUS	2 (2)	6 (7)
Tricholomataceae	2	6	Sub-total Basidiomycetes	58 (4)	236 (236)
BOLETALES			**ASCOMYCETES**		
Austropaxillaceae	1	3	ELAPHOMYCETALES		
Boletaceae	7 (1)	24 (10)	Elaphomycetaceae	0 (1)	0 (15)
Boletellaceae	1	3	PEZIZALES		
Strobilomycetaceae	3	6	Ascobolaceae	1	3
GOMPHALES			Humariaceae	1 (1)	1 (3)
Ramariaceae	2	11 (13)	Discineaceae	1	1
PHALLALES			Otidaceae	3 (1)	3 (1)
Clathraceae	1	1 (2)	Pezizaceae	5	8 (6)
Hysterangiaceae	3	18 (24)	Tuberaceae	6	12
Mesophelliaceae	6	32 (2)	Sub-total Ascomycetes	17 (3)	28 (25)
ZYGOMYCETES					
ENDOGONALES					
Densospora	1	4 (1)			
Endogonoceae	1	7 (1)			
GLOMALES	2	16 (11)			
Sub-total Zygomycetes	4	27 (13)			

of the continent would prove useful in approximating the total number of Australian species (Bougher and Lebel, 2001).

The mechanism(s) reponsible for causing the proliferation of hypogeous ectomycorrhizal fungi in Australia remains speculative, but may principally have resulted from long-term climatic change (Johnson, 1996). Thiers (1984) and Trappe (1988) have postulated that the hypogeous fruiting habit is favoured under two sets of environmental conditions: insulated in the soil, hypogeous fungi are protected from warm, dry air, particularly winds that dessicate epigeous mushrooms, or frost that freezes them. Under dry conditions the fertile tissue of epigeous species might quickly dessicate but in hypogeous species the fertile tissue is protected by an outer skin or peridium, which is further buffered against dessication by the surrounding soil (Johnson, 1996). Also, the formation of many small chambers within the fruit-bodies of hypogeous fungi presumably facilitates moisture retention. The selective advantages of remaining hypogeous in an arid environment are illustrated in modern-day times in Australian dry woodlands by various epigeous fungi which produce fruit-bodies with a long rooting stem and a cap that remains below ground or barely emerges (see Bougher and Tommerup, 2000). Ejected spores from such fruit-bodies do not have the opportunity to disperse in air currents and instead accumulate in a dense mass immediately below the cap (Bougher and Lebel, 2001). Further evidence for the advantages of the underground-fruiting form in an arid environment is provided by recent studies from south-eastern Australia. There, hypogeous fungi have been found to comprise a diverse and abundant component of the ectomycorrhizal fungal flora in forests subject to seasonal drought (Claridge et al., 1993a; Johnson, 1994a).

Environmental conditions thought favourable to the hypogeous fruiting habit likely began in the Late Tertiary, when the Australian continent became progressively more arid (Specht, 1981). Such conditions would presumably have arisen more recently on multiple occasions during the Pleistocene Epoch when Australia experienced repeated cooling and drying associated with glacial cycles (Sturman and Tapper, 1996). Johnson (1996) has speculated that these glacial cycles caused episodes of forest fragmentation, promoting multiple independent evolutions of hypogeous fruiting in closely related taxa. Testing this hypothesis is problematic, although molecular phylogenetic analyses of a broad array of fungi may shed further light on various lineages of taxa and timing of evolution of the below-ground fruiting form (Bougher and Lebel, 2001).

Another hypothesis relating to the proliferation of hypogeous fungi relates to their generally obligate ectomycorrhizal associations and dependence on animals for dispersal (Bougher and Lebel, 2001). Production of spores of hypogeous fungi at the root zone and maintenance of spores in the soil for long periods have been suggested to be evolutionarily advantageous for mycorrhizal fungi (Fogel, 1992; Miller et al., 1994). Mycophagous animals concentrate the spores of hypogeous fungi in faecal pellets and deposit them back into the environment, usually near established host root systems. Alternatively, seeds may also be present in the same faecal pellets, allowing for spores to be in direct contact with germinating plants. This notion is supported by the predominant mycorrhizal nature of hypogeous fungi, in contrast to epigeous fungi which have a broader array of nutritional modes, and the dependence on mycophagy for spore dispersal (Bougher and Lebel, 2001).

Role as ectomycorrhizal associates of plants

Most species of Australian hypogeous fungi are thought to be ectomycorrhizal. This presumption is mainly based on the observation of their fruit-bodies only in the presence of ectomycorrhizal hosts (i.e., Claridge et al., 1993a; Johnson, 1994a), but also the phylogenetic relationships to demonstrated ectomycorrhizal genera of epigeous fungi. Based on the co-occurrence of fruit-bodies, likely plant hosts include predominant trees such as species of *Eucalyptus* and *Leptospermum* (Myrtaceae), *Allocasuarina* and *Casuarina* (Casuarinaceae), *Acacia* (Mimosaceae) and *Nothofagus* (Nothofagaceae) (Bougher and Lebel, 2001). Associations may also occur with other plant families reported to form ectomycorrhizae, such as the Euphorbiaceae, Fabaceae, Goodeniaceae, Rhamnaceae and Sterculiaceae (Warcup, 1980). Ectomycorrhizal status has been confirmed for a variety of known hypogeous taxa through one of three means: (i) microscopic examination of plant roots in association with either hyphae and/or fruit-bodies (i.e., Ashton, 1976; Dell and Malajczuk, 1985; Dell et al., 1990; Malajczuk et al., 1987), (ii) formation of ectomycorrhizae on plant seedlings following inoculation with vegetative mycelium (i.e., Malajczuk et al., 1982; Thomson et al., 1994; Warcup, 1990b, 1991), or (iii) formation of ectomycorrhizae following inoculation with spores isolated from fruit-bodies (i.e., Lu et al., 1998; McGee, 1996; Warcup, 1985) (see Table 2).

Because the various functional roles of ectomycorrhizal fungi in plant communities have been reviewed at length elsewhere (e.g., Brundrett, 1991; Smith and Read, 1997; Tommerup and Bougher, 1999), the likely benefits that Australian hypogeous fungi confer to their respective plant hosts are only touched on briefly here. First and foremost, studies of specific fungus–plant combinations under controlled conditions highlight that ectomycorrhizal fungi can increase early growth of their plant host, primarily through the acquisition of phosphorus (Burgess et al., 1993; Malajczuk et al., 1975), which is otherwise immobile in the soil environment (Bougher et al., 1990), and organic sources of nitrogen (Read, 1987). The ability to do this varies between different fungal species (i.e., Burgess et al., 1993), but also within species where the edaphic environment changes (i.e., Bougher et al., 1990). In addition to phosphorus, ectomycorrhizae stemming from hypogeous fungi are probably also involved in the uptake of more mobile ions such as ammonium, nitrate, sulfate and potassium (Grove and LeTacon, 1993).

In the field the benefits of individual species of hypogeous fungi to plant growth and nutrition have been less well documented, as is true of other types of ectomycorrhizal fungi (Cairney, 1999). In Victoria, Ashton (1976) reported increased growth of Moun-

Table 2. Examples of taxa of hypogeous fungi shown to be ectomycorrhizal, and their respective plant associate (fungal names updated as appropriate from the originally cited papers)

Taxon	Plant Associate	Reference Source
ASCOMYCETES		
Hydnoplicata whitei	*Eucalyptus obliqua*	Warcup (1990a)
Labyrinthomyces varius	*Eucalyptus obliqua*	Warcup (1990a)
Muciturbo reticulatus	*Eucalyptus obliqua*	Warcup (1990a)
Ruhlandiella beroliensis	*Eucalyptus obliqua*	Warcup (1991)
BASIDIOMYCETES		
Cortinarius globuliformis	*Eucalyptus marginata*	Malajczuk et al. (1987b)
Descomyces albus	*Eucalyptus saligna*	Chu Chou and Grace (1982)
Descomyces albellus	*Eucalyptus camaldulensis*	Malajczuk et al. (1982)
	Eucalyptus sieberi	
Hydnangium carneum	*Eucalyptus delegatensis*	Chu Chou and Grace (1982)
	Eucalyptus globulus	Burgess et al. (1993)
Hysterangium inflatum	*Eucalyptus saligna*	Chu Chou and Grace (1982)
Hysterangium sp.	*Eucalyptus marginata*	Reddell and Malajczuk (1984)
Mesophellia trabalis	*Eucalyptus diversicolor*	Malajczuk et al. (1987b)
Protoglossum viscidus	*Eucalyptus globulus*	Burgess et al. (1993)
Scleroderma sp.	*Eucalyptus pellita*	Reddell and Milnes (1992)
Setchelliogaster sp.	*Eucalyptus globulus*	Thomson et al. (1996)
ZYGOMYCETES		
Densocarpa tubaeformis	*Eucalyptus obliqua*	Warcup (1991)
Endogone aggregata	*Brunonia australis*	Warcup (1990b)
	Eucalyptus obliqua	
	Melaleuca uncinata	
	Stylidium granifolium	
Endogone tuberculata	*Brunonia australis*	Warcup (1990b)
	Eucalyptus obliqua	
	Melaleuca uncinata	

tain Ash (*Eucalyptus regnans*) seedlings inoculated with *Mesophellia arenaria* and subsequent outplanting in a plantation. Malajczuk (1988) demonstrated 50% increases in mean height and diameter of 2-year-old Karri (*E. diversicolor*) seedlings inoculated with *Protubera canescens*, outplanted on a recently clear-felled forest site in Western Australia. Grove et al. (1991) reported 66–88% biomass increases in 1-year-old outplanted Karri seedlings inoculated with *Setchelliogaster* sp. Other potential functions of hypogeous fungi as ectomycorrhizal associates include: (i) acquisition of water from the soil environment and protection for the host plant from moisture stress (see Dixon et al., 1980; Parke et al., 1983), (ii) protection against soil-borne root pathogens such as *Phytophthora cinnamomi* (Marx, 1973), and (iii) protection against soil toxicity and salinity (Smith and Read, 1997). These functions have not yet been experimentally demonstrated for hypogeous species (Bougher and Lebel, 2001).

Hypogeous fungi apparently vary in both their spatial and temporal distribution (i.e., Anderson and Cairney, 2001). At a landscape-scale, Claridge et al. (2000a,b) found that the probability of detecting the fruit-bodies of certain hypogeous species either increased or decreased in relation to major climatic gradients in south-eastern mainland Australia. Similarly, at a more localised (within-site) spatial scale, fruit-bodies of some species commonly associate with key micro-habitats such as deep leaf litter, exposed mineral soil, or fallen trees and coarse woody debris (Amaranthus et al., 1994; Claridge et al., 2000b; Johnson, 1994a). Several studies have also shown that the occurrence and relative abundance of fruit-bodies of different hypogeous fungi may vary in relation to successional age of vegetation (i.e., Chu Chou and Grace, 1982; Claridge et al., 1993b; Johnson, 1997; Lu et al., 1999). Assuming that patterns in fruit-body distribution and abundance such as those described here in part reflect the distribution of these species in the environment, this suggests that individual species differ in their relative importance as mycorrhizal associates for plants in both space and time. For this reason, maintaining a diversity of hypogeous species in any landscape may be essential for ongoing and efficient nutrient cycling in forests and woodlands. Further, there is some evidence that these systems may better recover from perturbations such as fire or intensive logging with increasing diversity of ectomycorrhizal fungi (Amaranthus et al., 1989).

Role as a food resource for animals

The fruit-bodies of hypogeous fungi are a key food resource for a wide variety of Australian mammals and invertebrates but also possibly birds. Mycophagy (fungus-feeding) by terrestrial mammals has been reviewed by Claridge and May (1994) and Claridge et al. (1996), with additional recent records provided by Reddell et al. (1997). These reviews indicate that at least 38 native mammal species and three introduced mammal species in Australia eat hypogeous fungi. The existing list of mycophagous species is not exhaustive: the basic ecology of many mammals is undocumented, particularly for species in remote parts of the country. As more work is conducted the list of mammal species known to eat fungi will undoubtedly increase (Claridge et al., 1996).

Relating each of the species so far found to eat hypogeous fungi to their respective mammalian families reveals some distinctive patterns (Table 3). Mycophagy is particularly common among herbivorous and omnivorous mammals weighing less than about 3000 gms (Claridge et al., 1996; Johnson, 1996), most notably among the Potoroidae (rat-kangaroos), with all extant species known to eat hypogeous fungi. Similarly, more than half of the extant species within the Peramelidae (bandicoots and bilbies), and a third of the species within the Muridae (rodents), consume hypogeous fungi. In contrast, mycophagy among the larger terrestrial marsupials such as wallabies and kangaroos (Macroprodidae), wombats (Vombatidae) and carnivorous marsupials (i.e., Daysuridae) is less prevalent. While mycophagy is mainly restricted to ground-dwelling mammals, some arboreal (tree-dwelling) species, such as Common Brushtail Possum (*Trichosurus vulpecula*) and Mountain Brushtail Possum (*T. caninus*), also forage on the ground and eat fungi (Claridge and May, 1994).

Mycophagy by mammals is widespread across many habitat types, including rainforest, wet and damp sclerophyll forest, dry sclerophyll forest, open and closed woodland, heathland and to a much lesser extent desert (Claridge et al., 1996). Further, multiple species within any given habitat and locality may be mycophagous (Tory et al., 1997). In such instances the fungal food resource is usually partitioned in some way to minimise direct competition (Claridge and May, 1994). In south-eastern New South Wales, for example, Claridge et al. (1991) determined that sympatric populations of Southern Brown (*Isoodon obesulus*) and Long-nosed (*Perameles nasuta*) Bandicoots both fed on fungi. However, the main types of fungi eaten differed because the former species occurred more frequently in vegetation types on ridges and upper slopes and the latter in adjacent drainage lines. Whether such resource partitioning occurs among other mammals is unknown.

In other cases, the overlap in range of hypogeous fungi consumed by co-occurring mammal species may be considerable, but the degree of dependence on the fungal food resource differs temporally. In south-western Victoria, Tory et al. (1997) observed that the range of hypogeous fungi consumed by Long-nosed Potoroos (*Potorous tridactylus*) and Bush Rats (*Rattus fuscipes*) was generally similar across seasons, with certain groups of fungi dominating both their fungal diets within any season. However, the proportion of fungus in the diet of the Bush Rat was more highly seasonal than that of the Long-nosed Potoroo, with distinctive peaks in consumption during autumn and winter and troughs in spring and summer. This probably relates to the overall availability of fruit-bodies. In southern temperate Australia, fruit-body production of many hypogeous fungi is higher during autumn and winter than spring and summer (Claridge et al., 2000a; Johnson, 1994a), so a temporary excess of fungal resources may be available in autumn and winter such that several species of mammal become equally mycophagous. In contrast, during spring and summer, with diminished fungal productivity, food partitioning may be necessary for co-existence since harvesting pressure on fruit-bodies may be significant (see Johnson, 1994d).

Table 3. Summary of Australian mammal taxa known to consume fruit-bodies of hypogeous fungi. Data shown here mainly derived from Claridge and May (1994) and Claridge et al. (1996)

Taxon	No. of extant genera	No. of extant species	No. of mycophagous species	% Mycophagous
PROTOTHERIA				
(MONOTREMES)				
Ornithorhynchidaea	1	1	0	0
Tachyglossidae	1	1	0	0
MARSUPIALIA				
(MARSUPIALS)				
DASYUROMORPHIA				
(Carnivorous Marsupials)				
Dasyuridae	13	51	1	2
Myrmecobiidae	1	1	0	0
Thylacinidae	0	0	0	0
PERAMELOMORPHIA				
(Bandicoots and Bilbies)				
Peramelidae	3	7	4	57
Perorcytidae	1	1	0	0
DIPROTODONTIA				
(Koala, Wombat, Possums, Macropods)				
Phascolarctidae	1	1	0	0
Vombatidae	1	3	0	0
Burramyidae	2	5	1	20
Petauridae	3	6	0	0
Pseudocheridae	6	8	0	0
Tarsipedidae	1	1	0	0
Acrobatidae	1	1	0	0
Phalangeridae	4	5	2	40
Potoroidae	4	9	9	100
Macropodidae	9	43	4	9
Notoryctidae	1	1	0	0
EUTHERIA				
(PLACENTALS)				
RODENTIA (Rodents)				
Muridae*	15	59	19	32
CARNIVORA				
(Carnivorous Placentals)				
Canidae	2	2	0	0
Felidae	1	1	0	0
LAGOMORPHA				
(Rabbit and Hare)				
Leporidae	2	2	0	0
PERISSODACTYLA				
(Odd-toed Ungulates)				
Equidae	1	2	0	0
ARTIODACTYLA				
(Even-toed Ungulates)				
Suidae**	1	1	1	100
Camelidae	1	1	0	0
Bovidae	3	3	0	0
Cervidae	3	5	0	0

Key: *Includes two introduced species. **Introduced species.

Four broad trends in the level of dependence on fungi as a food resource can be identified among mammals. Firstly, for species like the Long-footed Potoroo (*Potorous longipes*), hypogeous fungi form the single most important food resource throughout the year, ranging between 90 and 100% of overall diet (Scotts and Seebeck, 1989). Secondly, for animals such as the Long-nosed Potoroo and Tasmanian Bettong (*Bettongia gaimardi*), fungi form the single most important dietary item during most but not all times of the year (Claridge et al., 1993b; Johnson, 1994c). Thirdly, as demonstrated by bandicoots and rodents, fungi may be highly seasonally important, with low levels of fungi consumed at all other times (Claridge et al., 1996; Tory et al., 1997). Fourthly, fungi may always be a minor dietary component, as seen with possums and wallabies (Claridge and May, 1994).

A diversity of non-mammalian fauna may also feed on fruit-bodies of hypogeous fungi (Bougher and Lebel, 2001; Claridge et al., 1996). Mycophagy by invertebrates is common and some groups, particularly the coleoptera, have been shown to associate closely with hypogeous taxa (Lawrence 1988). Birds, too, are known from elsewhere in the world to actively feed on hypogeous fungi (i.e., Alsheikh and Trappe, 1983), but current published reports of mycophagy by Australian avifauna are mostly restricted to consumption of epigeous fungi (i.e., Reichelt and May, 1997; Simpson, 2001). Nevertheless, recent anecdotal observations indicate that the forest-dwelling Yellow Robin (*Eopsaltria australis*) can be hand-fed fruit-bodies of some hypogeous species (Simpson, 2001). In future, closer inspection of the diet of birds that forage mainly at the soil–litter interface, such as the Superb Lyrebird (*Menura novaehollandiae*), may be informative. Even if the Lyrebird does not eat fungi, its extensive scratching can bring hypogeous fruit-bodies to the surface of the ground, where they are accessible to other birds (A.W. Claridge and J.M. Trappe, unpublished observations).

Fungi as a food source in environments subject to fire

Fire, either natural or anthropogenic in source, has been commonplace in most Australian plant communities for thousands of years (Gill, 1975; Recher, 1981). In habitats recently disturbed by fire the fruit-bodies of at least a few species of hypogeous fungi may provide an essential food resource for a range of ground-dwelling mammals where other dietary items become at least temporarily scarce. In the Jarrah (*Eucalyptus marginata*) forests of south-west Western Australia, Christensen (1980) observed that the Brush-tailed Bettong (*Bettongia penicillata*) fed almost exclusively on fruit-bodies of *Mesophellia* after deliberately applied fire. Similarly, in East Gippsland, Victoria, Claridge et al. (2001) noted that the normally herbivorous Swamp Wallaby (*Wallabia bicolor*) also fed on the fruit-bodies of *Mesophellia*, in the absence of plant browse on recently burnt sites.

In northern Tasmania, Taylor (1992) compared the foraging activity of Tasmanian Bettongs in plots within unburned and recently burned woodland. The density of forage-diggings of animals on burned plots on was 10-fold greater than elsewhere across the study area. Most forage-diggings in the burned plots had discarded fragments of fungal fruit-bodies, again entirely from species within the family Mesophelliaceae. At least four alternative hypotheses could account for this phenomenon. Firstly, fire removes ground cover, improving access to the soil and hence food resources. Secondly, animals may dig more often for fungi in recently burned habitats because other foods are unavailable. Thirdly, fire may actually stimulate the fungi to fruit. And fourthly, heating of certain hypogeous fruit-bodies may change their odour to make them more detectable. To establish the true nature of the observed pattern, Johnson (1995) set up two separate experimental burns in Tasmanian dry woodland habitat, the preferred habitat of the highly mycophagous Tasmanian Bettong. A series of matched sites was established, some subject to deliberately applied (prescribed) fire, others remaining as unburned control sites. In the first set of small-scale fires Johnson (1995) found that the density of forage-diggings of bettongs was 3- to 9-fold higher in burnt than unburned plots 1 month post-fire, before returning to pre-fire levels and matching forage-digging densities on control sites. A similar difference in foraging effort by bettongs was recorded during the second experiment. The relative abundance of fungal fruit-bodies also differed by treatment and over time. In the first set of small-scale burns, the relative abundance of fruit-bodies from plots in control sites did not differ from plots situated around trees burned at low intensity. However, in plots around trees burned at high intensity the density of fungal fruit-bodies was significantly greater than within plots burned at low intensity and control plots. This difference only occurred one month post-fire, and within 2 months densities of fruit-bodies equalled those around trees burned at lower intensity and trees in unburned sites. Differences in productivity were attributable to species within the Mesophelliaceae, in particular within the genera *Castoreum* and *Mesophellia*. In the second experiment, in which fruit-bodies were collected from within exclosures that bettongs could not access, numbers of fruit-bodies around trees on burned plots did not differ with time since fire.

Johnson (1995) concluded that fire had actually triggered the fungi to fruit, but the data presented indicate that the fungal response is complicated and perhaps unpredictable. What might trigger such a rapid fruiting response is unknown. Causal factors might include: (i) the soil being heated and dried, (ii) (unknown) chemical effects on plant roots and fungi, (iii) the rapid release of nutrients held in litter, (iv) actual damage to the fungi or above-ground plant parts, resulting in some (unknown) phenological response, or (v) some (again unknown) physiological change in the plants (see Johnson, 1995). Whatever the cause, given that increases in foraging activity occurred mostly within 2–3 days post-fire, the response is extremely quick, by far quicker than the *Mesophellia* fruit-body can normally form (Dell et al., 1990). Further studies examining the possible mechanisms behind this phenomenon would be valuable. Regardless, the ability of mycophagous mammals to rapidly enter burned areas and consume fungi may lead to quick dispersal of fungal spores which may later form mycorrhizal associations on host plant roots (Claridge et al., 1992).

Recent experiments by Vernes and Haydon (2001) at Davies Creek in north-eastern Queensland largely corroborates the patterns observed by Johnson (1995) in Tasmania, but firm conclusions about the stimulating influence of fire on fungal fruiting is hampered because sample plots were placed on or immediately adjacent to sites where Northern Bettongs (*Bettongia tropica*) dug. This observation may reflect foraging preferences of animals rather than any response by the fungi. Not withstanding this sampling bias, the biomass of fruit-bodies under forage-diggings was greater on burnt sites than under forage-diggings on unburnt sites shortly after fire. The initial increased response was again mainly due to *Mesophellia*. During that time, Vernes and Haydon (2001) found that significantly more animals foraged in burnt habitat than in unburnt habitat.

One alternative explanation for the results observed by Vernes and Haydon (2001) is that the fruit-bodies of *Mesophellia* survived intact after fire and were more attractive to bettongs than other fungal food resources in adjacent unburned habitat. The fruit-bodies of *Mesophellia* are certainly well equipped to survive fire. Firstly, they usually fruit at depth in the mineral soil and are therefore insulated against immediate heating effects (Claridge et al., 1993b). Secondly, the outer peridium comprises a thickened structure of interwoven tree roots and hyphae, which protects the inner spore-bearing tissue and central sterile core from scorching. Furthermore, heat from fire may change the aromatic compounds found in mature fruit-bodies (Millington et al., 1997): whether these changes act as a foraging stimulus to mycophagous mammals deserves further study.

The longer-term influence of fire on hypogeous fungi is largely undocumented. In one relevant study, Johnson (1997) compared diversity and relative abundance of fruit-bodies of hypogeous fungi at six sites matched for soil, vegetation and climate but differing in time since last fire. No clear relationship was found between fungal diversity, productivity and time since fire. However, the density and biomass of fungal fruit-bodies was lowest at 2 years post-fire and highest at 4 and 10 years post-fire. Fungal species diversity and richness was similarly lowest at 2 and 4 years post-fire, and highest at 10 years post-fire. Notably, all fungal taxa present in recently burned sites were also present at sites long unburned. Differences in fungal diversity between sites varying in time since disturbance were also noted by Claridge et al. (1993a) in south-eastern mainland Australia. There, Long-nosed Potoroos fed on a greater diversity of fungi in a site never logged than in a site logged 10 years previously. The relative abundance of different fungi also varied between the two stands, with the spores of some taxa (Mesophelliaceae again) more commonly represented in the scats of animals from the recently disturbed site. To the extent that the nature of fungal populations can be inferred from the diet of mycophagous animals, this suggests that logging may impact on some fungi at least in the short-term. This may or not influence the animals that rely on the fungal food resource. Recent studies by Green et al. (1998) on the ecology of the Long-footed Potoroo, an obligate mycophagist, imply that animals in sites with minimal recent disturbance may produce more offspring per annum and forage for significantly shorter periods than animals in habitats with more recent evidence of disturbance.

Patterns of fruiting influence distribution of mycophagous mammals

Patterns in the spatial and temporal distribution of hypogeous fungi may influence the distribution and foraging behaviour of mycophagous mammals at various spatial scales. For example, at a single study site in northern Tasmania, Taylor (1992) found that the density of forage-diggings left by the Tasmanian Bettong differed according to major vegetation types, being lowest in sedgelands and highest in heathlands, and forests and woodlands on low-nutrient status soils. Since the sedglands contained mainly non-mycorrhizal plant species, fungal food resources may have been limiting. Johnson (1994b) similarly observed variations in foraging effort according to habitat type by the same bettong species in south-eastern Tasmania. In a one-off sample, the density of forage-diggings was higher where Silver Peppermint (*Eucalyptus tenuiramus*) and *Acacia* dominated.

Changes in seasonal availability of fungal food resources may also influence micro-habitat use by mycophagous mammals. At Cabbage Tree Creek in East Gippsland, Victoria, Claridge et al. (1993b) found that Long-nosed Potoroos moved variously throughout the topographic sequence in a forest catchment, depending on time of year. During wet late-autumn and winter periods most foraging activity occurred in gullies and lower slopes, while in spring and summer, foraging increased, relatively, on upper slopes and ridges. These seasonal changes in foraging activity mirrored changes in the availability of different suites of hypogeous fungi. In late-autumn and winter, fungi which produced soft, desiccation-prone fruit-bodies fruited abundantly in gullies and lower slopes. Because these types of fungi usually fruited at or immediately beneath the soil–litter interface, they were readily accessible and energetically inexpensive to harvest. Once these habitats had dried out, production of the soft-bodied fungi decreased (Claridge et al., 1993c) and animals fed primarily on different suites of fungi that produced hard, desiccation-resistant fruit-bodies. These latter fungi fruited mainly in the mineral soil on upper slopes and ridges. In comparison with the soft-bodied fungi, they were available year-round in equivalent abundance.

Hypogeous fungi may also influence the distribution of mycophagous mammals over the broader landscape. In south-eastern mainland Australia, Claridge and Barry (2000) found that the probability of occurrence of potoroos and bandicoots increased as soil fertility decreased. This pattern presumably reflects differences in the relative productivity of hypogeous ectomycorrhizal fungi in relation to soil nutrient status. Similar findings have also been made in tropical Queensland for the Northern Bettong (*Bettongia tropica*), a rare rat-kangaroo that appears to be mainly restricted to forests and woodlands on low fertility soils (Laurance, 1997). Further research on the role of soil nutrient status in influencing the abundance of hypogeous fungi is urgently needed (Claridge and Barry, 2000). Ideally, fungal productivity should be estimated over several years.

Nutritional value of fungal fruit-bodies

The prevalence of mycophagy among Australian terrestrial mammals implies that fungi are nutritious. At face value this would appear to be the case. Fungal fruit-bodies accumulate significantly higher concentrations of phosphorus, nitrogen and potassium than other foods such as leaves and seeds (see Cork and Kenagy, 1989). They may also accumulate other non-metallic and metallic elements, with concentrations of copper, zinc, calcium, iron, magnesium, sodium and other elements reported. Organic compounds including vitamins, amino acids, steroids, triterpenes, amines, indoles and phenols are also present and of potential use in the mammalian diet. Vitamins such as niacin and riboflavin may reach concentrations greater than those in most nuts, fruits and leaf vegetables and be present in levels equivalent to most meat products.

Despite the above, experiments so far conducted with fungi suggest poor nutritional quality, at least in terms of provision of carbohydrates and nitrogen. Cork and Kenagy (1989) determined that fruit-bodies of several species of hypogeous fungi from North America contained indigestible complex carbohydrates, fungal nitrogen incorporated largely in indigestible cell wall constituents and part of the available nitrogen in non-protein forms. In a feeding trial, they fed captive Golden-Mantled Ground Squirrels (*Spermophilus lateralis*) the fruit-bodies of *Elaphomyces granulatus*, a common hypogeous fungus, and compared the digestibility of that fungus to other major foodstuffs. Despite consuming a high daily intake of fungus, squirrels lost weight. Further, the digestibility of nitrogen and energy from *Elaphomyces* was lower than that recorded for nearly all the other diets. Although chemical analyses revealed that the nitrogen content of fruit-bodies was relatively high, 80%

of it was bound in totally indigestible spores which the squirrels rarely ate. Of the remaining 20%, only half was present as protein nitrogen. Sources of energy were tied up in complex, relatively indigestible cell-wall tissue. More recent feeding trials conducted by Claridge et al. (1999), on the Northern Flying Squirrel (*Glaucomys sabrinus*) and Red-backed Vole (*Clethrionomys californicus*), further support the notion that the fruit-bodies of hypogeous fungi are of only moderate sources of carbohydrates and nitrogen, even though both animals subsist primarily on this food source.

Little is known about the nutritional value of fungi for Australian mammals. In Western Australia, Kinnear et al. (1979) assessed the chemical components of fruit-bodies of *Mesophellia*, a significant food for the Brush-tailed Bettong. Analyses of the inner core of the fruit-bodies (the portion largely consumed by Bettongs) indicated a rich source of lipids (around 40% by dry weight) and crude protein levels of 8–10%. Although protein levels were high, analyses revealed they were deficient in certain essential amino acids such as lycine. Other amino acids, particularly cysteine and methionine, were present in large quantities. It was suggested that the imbalances in amino acids in fruit-bodies could be largely corrected during digestion. Bettongs and other rat-kangaroos have special adaptations to the gut, including a large sacciform forestomach. The hind-gut is reduced to a well-developed though simple caecum and proximal colon. The enlarged sacculated foregut cultures anaerobic microbes, which ferment food and convert fungal nitrogen to a form more available for the host animal. This process is termed pre-gastric fermentation. Claridge and Cork (1994) provided the first clear indication that hypogeous fungi were nutritious for animals with such a digestive capacity. They fed captive Long-nosed Potoroos fruit-bodies of *Mesophellia glauca* and *Rhizopogon luteolus* and found that animals lost little weight and digested much of the dry matter, nitrogen and energy present in the pure fungal diets. This enabled them to maintain positive nitrogen and energy balances. Most other mycophagous mammals in Australia lack an enlarged foregut and most food is digested in the hindgut. The absence of this digestive system may help explain why hindgut-fermenters such as rodents and bandicoots seldom rely wholly upon fungi but commonly eat other foods such as seeds and invertebrates.

Field observations of Rat-kangaroos also suggest that fungi are highly nutritious. Johnson (1994c) found Tasmanian Bettongs were almost entirely mycophagous when production of hypogeous fruit-bodies was highest, whereas at times of low fruit-body production other foods such as leaves and fruits were consumed. Adult body condition improved with increasing amounts of fungus in the diet. When production of fruit-bodies increased, energy turnover in adult females and growth rates of pouch young increased concomitantly, suggesting that the fungi provided animals with a surplus of energy, perhaps used in lactation. Later, McIlwee and Johnson (1997) used stable isotopes to determine that nearly all nitrogen assimilated into body tissue by Northern Bettongs was from fungi. In contrast, the sympatric Northern Brown Bandicoot (*Isoodon macrourus*) derived much of its nitrogen from invertebrates and practically none from fungi.

Other adaptations for fungal diet

Mycophagous mammals may possess other morphological and physiological attributes that better enable them to successfully exploit the potential food resource offered by hypogeous fungi. Morphologically, well-developed forelimbs and claws better enable excavation and manipulation of fruit-bodies, as demonstrated by marsupials such as the Long-footed Potoroo and Long-nosed Bandicoot (Claridge and May, 1994). These animals also have elongate, tapered heads, which likely improve access into tight spaces within the mineral soil when in search of fruit-bodies. Physiologically, a keen sense of smell is undoubtedly important for detecting hypogeous fruit-bodies, which otherwise remain hidden within the soil (see Donaldson and Stoddart, 1994). To that end, some Australian mycophagous mammals, such as the rat-kangaroos, have exceptional olfactory acuity. This ability is likely due to their extensive olfactory cells and prominent olfactory brain lobes (Seebeck and Rose, 1980). In addition, a vomeronasal organ is present in at least the Rufous Bettong (*Aepyprymnus rufescens*) and probably occurs in all other rat-kangaroos: the functions of this organ are unknown but is thought principally to be used in detection, by males, of sexually responsive females. It may also have some role in the detection of food by adults and the acquisition of olfactory information by young developing *in utero*.

Detection of hypogeous fruit-bodies is made easier because they become increasingly odiferous as they approach maturity. Lack of odour in immature fungal fruit-bodies likely acts as a safeguard against being consumed prior to the maturation of the spore-bearing tissue (Trappe, 1988). The odours produced by mature hypogeous fruit-bodies can be species-specific. Examples include: *Mesophellia clelandii*, which produces an aroma resembling aniseed; *M. glauca*, which produces a sweet, pungent odour, and; *M. trabalis*, which has odours resembling crushed onion or garlic. The volatile compounds responsible for producing odours in fruit-bodies of Australian hypogeous fungi have been little studied, although the few studies so far conducted indicate that they are mainly alcohols and esters (see Donaldson and Stoddart, 1994; Millington et al., 1997).

Spore dispersal

In eating the fruit-bodies of hypogeous fungi, mycophagous animals inadvertently play a role in spore dispersal. Spores from consumed fruit-bodies, being indigestible (Cork and Kenagy, 1989), pass through the digestive tract of the mycophagist unscathed and are voided in the faeces of the animal. Spore viability is apparently unaffected by this process, as evidenced by germination of spores isolated from faeces (Trappe and Maser, 1976), or by successfully culturing mycorrhizae on plant seedlings by inoculating with spore-laden faecal pellets (i.e., Claridge et al., 1992; Lamont et al., 1985; Reddell and Spain, 1991). A few workers have even suggested that passage through the gut of mycophagous mammals may sometimes be necessary for spore germination (Cork and Kenagy, 1989; Lamont et al., 1995). The generality of this hypothesis is weakened, however, because at least several species of hypogeous fungi readily form mycorrhizae with their plant host without being eaten or dispersed by an animal vector (see Colgan and Claridge, 2002). Whatever the case, mycophagy is seemingly a highly efficient means for dispersal of hypogeous fungi, particularly in forest and woodland habitats. Movement studies of some Australian mycophagous mammals demonstrate that spores can potentially be disseminated across areas exceeding 100 ha, or radii of at least several hundred metres (Johnson, 1996). These distances may potentially be even greater in instances where mycophagous mammals fall prey to raptors or terrestrial carnivores with larger home ranges. This contrasts with the short effective dispersal distances of spores from epigeous fungi, particularly in forests where air turbulence close to the ground may be minimal (Allen, 1991). The foraging patterns of some mycophagous mammals, which focus on areas within the rooting zone of potential host plant species, enhance prospects for successful dissemination of fungal spores where faeces are deposited (Claridge et al., 1992; Johnson, 1996). Movement of spores back into the soil profile and into contact with fine roots can then be effected at such sites through abiotic means such as precipitation (Maser et al., 1978). Alternatively, secondary biotic vectors may be involved. In Australian forests, for instance, dung beetles may associate closely with heavily mycophagous mammals such as rat-kangaroos and play a key role in moving spore-laden faeces below-ground and into contact with roots of host plants (see Claridge et al. 1992).

Some fungal taxa may have additional strategies that facilitate the chances of dispersal. The genus *Mesophellia*, endemic to Australia, provides a useful example. Fruit-bodies of *Mesophellia* are complex structures, comprising a hard, multi-layered peridium or outer shell of soil particles, fine tree roots and fungal mycelium, within which lies an internal spore-bearing mass and central sterile core (Dell et al., 1990). Upon maturation, pungent odours are released by fruit-bodies, serving to attract a wide range of mycophagous mammals (Millington et al., 1997). Once excavated, fruit-bodies are processed and consumed within a few seconds (Claridge and May, 1994). Rat-kangaroos manipulate fruit-bodies in their forepaws, opening the outer peridium with their teeth and discarding much of it as well as some of the powdery spore mass to the forest floor. During this time, spores adhere to the fur of the animal and brush on to vegetation as the animal moves away or on to soil where the animal next digs. The inner sterile core, together with any further adhering spores, is then ingested, providing the animal with a nutritious food reward (Claridge and Cork, 1994). Spores passing through the gut of the animal are voided in faeces and form ectomycorrhizae on eucalypt seedlings (Claridge et al., 1992; Lamont et al., 1995). Finally, any spores left adhering to the discarded outer peridium, which are hydrophobic, can be aerially dispersed from raindrop splash, at a later date, or washed into the immediate soil environment (Claridge and May, 1994).

The extent to which spore dispersal by mycophagists contributes to maintenance and/or enhancement of fungal communities in nature is unknown, largely because the fate of spores in faecal pellets of animals has not been followed in a field situation (Colgan and Claridge, 2002). The relative extent to which hypogeous fungi principally colonise host plant root systems, either through vegetative growth or by dispersal of sexual spores, also needs to be better understood (Johnson, 1996). Once these pathways are defined, the relative impact on populations of hypogeous fungi of removing mycophagous mammals from forest and woodland systems may be better evaluated.

Future research

Studies of the ecological role of hypogeous fungi in Australian forests and woodlands are in their infancy. Critical research challenges in the future include, but are not restricted to, the following themes: (i) establishing a complete taxonomic inventory of hypogeous fungi, preferably through broadscale and systematic surveys across the broader Australian landscape which enable patterns in the distribution of individual taxa and groups of taxa to be elucidated; (ii) examining the mycorrhizal status of various hypogeous fungi and the respective role they might play in benefitting growth and survival of their plant host. Where feasible, these studies should be applied to the field situation; (iii) better understanding how patterns in the distribution and abundance of fruit-bodies of hypogeous fungi influence the distribution and foraging patterns of fungus-dependent animals, at various spatial and temporal scales; (iv) clarifying the temporal response of various hypogeous fungi to fire, and the implications of those changes for fungus-dependent animals; (v) further examining the nutritional value of fruit-bodies of hypogeous fungi for mycophagous animals with different digestive capabilities; (vi) evaluating the importance of hypogeous fungal diversity in diets of mammalian mycophagists, and; (vii) exploring the relative importance of spore dispersal by mammals in creating, maintaining and enhancing mycorrhizal associations in Australian vegetation.

Acknowledgements

The author wishes to acknowledge the many people who have collaborated with him on various projects relating to the ecology of Australian hypogeous fungi and their role as a food resource for mammals. Particular thanks go to: Prof. Jim Trappe, Drs Simon Barry, Wes Colgan III, Steven Cork, Ari Jumpponen, Teresa Lebel and Tom May. Funding to carry out some of the field activities cited in this paper was provided, in part, by the Australian Biological Resources Study. Comments from Prof. Trappe improved an earlier version of this manuscript.

References

Allen M F 1991 The Ecology of Mycorrhizae. Cambridge University Press, Cambridge.

Alsheikh A M and Trappe J M 1983 Taxonomy of *Phaengium lefebvrei*, a desert truffle eaten by birds. Can. J. Bot. 61, 1919–1925.

Amaranthus M P, Trappe J M and Molina R J 1989 Long-term forest productivity and the living soil. *In* Maintaining the Long-Term Productivity of Pacific Northwest Forest Ecosystems. Eds. D A Perry, R Meurisse, B Thomas, R Miller , J Boyle, J Means, C R Perry and R F Powers. pp. 36–52. Timber Press, Portland, OR.

Amaranthus M P, Trappe J M, Bednar L and Arthur D 1994 Hypogeous fungal production in mature Douglas-fir forest fragments and surrounding plantations and its relation to coarse woody debris and animal mycophagy. Can. J. For. Res. 24, 2157–2165.

Anderson I C, Chambers S M and Cairney J W G 2001 Distribution and persistence of Australian *Pisolithus* spp. genets at native sclerophyll forest field sites. Mycol. Res. 105, 971–976.

Ashton D H 1976 Studies of the mycorrhizae of *Eucalyptus regnans* F. Muell. Aust. J. Bot. 24, 723–741.

Bougher N L and Lebel T 2001 Sequestrate (truffle-like) fungi of Australia and New Zealand. Aust. Syst. Bot. 14, 1–47.

Bougher N, Grove T S and Malajczuk N 1990 Growth and phosphorus acquistion of karri (*Eucalyptus diversicolor* F. Muell.) seedlings inoculated with ectomycorrhizal fungi in relation to phosphorus supply. New Phytol. 114, 77–85.

Brundrett, M. 1991 Mycorrhizas in natural ecosystems. Adv. Ecol. Res. 21, 72–313.

Burgess T I, Malajczuk N and Grove T S 1993 The ability of 16 ectomycorrhizal fungi to increase growth and phosphorus uptake of *Eucalyptus globulus* Labill. and *E. diversicolor* F. Muell. Plant Soil 153, 155–164.

Cairney J W G 1999 Intraspecific physiological variation: implications for understanding functional diversity in ectomycorrhizal fungi. Mycorrhiza 9, 125–135.

Castellano M A and Bougher N L 1994 Consideration of the taxonomy and biodiversity of Australian ectomycorrhizal fungi. Plant Soil 159, 37–46.

Christensen P E S 1980 The biology of *Bettongia penicillatta* (Gray 1837) and *Macropus eugenii* (Desmarest 1817) in relation to fire. Bulletin of the Woods and Forests Department of Western Australia 91, 90 pp.

Chu Chou M and Grace L J 1982 Mycorrhizal fungi of *Eucalyptus* in the North Island of New Zealand. Soil Biol. Biochem. 14, 122–137.

Claridge A W and Barry S C 2000 Factors influencing the distribution of medium-sized ground-dwelling mammals in south-eastern mainland Australia. Aust. Ecol. 25, 676–688.

Claridge A W and Cork S J 1994 Nutritional value of hypogeous fungal sporocarps for the long-nosed potoroo (*Potorous tridactylus*), a forest-dwelling mycophagous marsupial. Aust. J. Zool. 42, 701–710.

Claridge A W, Tanton M T, Seebeck J H, Cork S J and Cunningham R B 1992 Establishment of ectomycorrhizae on the roots of two species *Eucalyptus* from fungal spores contained in the faeces of the long-nosed potoroo (*Potorous tridactylus*). Aust. J. Ecol. 17, 207–217.

Claridge A W, Robinson A P, Tanton M T and Cunningham R B 1993a Seasonal production of hypogeous fungal sporocarps in a mixed-species eucalypt forest stand in south-eastern Australia. Aust. J. Bot. 41, 145–167.

Claridge A W, Tanton M T and Cunningham R B 1993b Hypogeal fungi in the diet of the long-nosed potoroo (*Potorous tridactylus*) in mixed-species and regrowth eucalypt stands in south-eastern Australia. Wildl. Res. 20, 321–337.

Claridge A W, Cunningham R B and Tanton M T 1993c Foraging patterns of the long-nosed potoroo (*Potorous tridactylus*) for hypogeal fungi in mixed-species and regrowth eucalypt forest stands in south-eastern Australia. For. Ecol. Manage. 61, 75–90.

Claridge A W, Castellano M A and Trappe J M 1996 Fungi as food resource for mammals. *In* Fungi of Australia Volume 1B: Introduction — Fungi in the Environment. Eds. K Mallett and C Grgurinovic. pp. 239–267. Australian Biological Resources Study: Canberra.

Claridge A W, Trappe J M, Cork S J and Claridge D L 1999 Mycophagy by small mammals in the coniferous forests of North America: nutritional value of sporocarps of *Rhizopogon vinicolor*, a common hypogeous fungus. J. Comp. Physiol. B 169, 172–178.

Claridge A W, Cork S J and Trappe J M 2000a Diversity and habitat relationships of hypogeous fungi. I. Study design, sampling techniques and general survey results. Biodiv. Conserv. 9, 151–173.

Claridge A W, Barry S C, Cork S J and Trappe J M 2000b Diversity and habitat relationships of hypogeous fungi. II. Factors influencing the occurrence and number of taxa. Biodiv. Conserv. 9, 175–199.

Claridge A W, Trappe J M and Claridge D L 2001 Mycophagy by the swamp wallaby (*Wallabia bicolor*): a disperser of hypogeous fungi in recently burned habitats. Wildl. Res. 28, 643–645.

Colgan W III and Claridge A W 2002 Mycorrhizal effectiveness of *Rhizopogon* spores recovered from faecal pellets of small forest-dwelling mammals. Mycol. Res. (in press).

Dell B and Malajczuk N 1985 Structure of *Hysterangium-Eucalyptus* ectomycorrhizas. *In* Proceedings of the 6th North American Conference on Mycorrhizae. pp. 446. 25–29th June 1984, Bend, OR, USA. Ed. R. Molina. Published by the Forestry Research Laboratory, Corvallis, OR.

Dell B, Malajczuk N, Grove T S and Thomson G 1990 Ectomycorrhiza formation in *Eucalyptus*. IV. Ectomycorrhizas in the sporocarps of the hypogeous fungi *Mesophellia* and *Castoreum* in Eucalypt forests of Western Australia. New Phytol. 114, 449–456.

Dixon R K, Wright G M, Behrns G T, Teskey R O and Hinckley T M 1980 Water deficits and root growth of ectomycorrhizal white oak seedlings. Can. J. For. Res. 10, 545–548.

Donaldson R and Stoddart M 1994 Detection of hypogeous fungi by the Tasmanian bettong (*Bettongia gaimardi*: Marsupialia; Macropodoidea). J. Chem. Ecol. 20, 1201–1207.

Gill A M 1975 Fire and the Australian flora: a review. Aust. For. 38, 4–25.

Green K, Mitchell A T and Tennant P 1998 Home range and micro-habitat use by the long-footed potoroo, *Potorous longipes*. Wildl. Res. 25, 357–372.

Grove T S and LeTacon F 1993 Mycorrhiza in plantation forestry. Adv. Plant Pathol. 8, 191–227.

Grove T S, Malajczuk N, Burgess T, Thomson B D and Hardy G 1991 Growth responses of plantation eucalypts to inoculation with selected ectomycorrhizal fungi. Proceedings of IUFRO Symposium on Intensive Forestry: The Role of Eucalypts, 86–93.

Fogel R 1992 Evolutionary processes in truffles and false-truffles: evidence from distribution of hypogeous fungi in the Great Basin, USA. Micol. Veg. Mediterranea 7, 13–30.

Johnson C N 1994a Fruiting of hypogeous fungi in dry sclerophyll forest in Tasmania, Australia: seasonal variation and annual production. Mycol. Res. 98, 1173–1182.

Johnson C N 1994b Feeding activity of the Tasmanian bettong (*Bettongia gaimardi*) in relation to vegetation patterns. Wildl. Res. 21, 249–255.

Johnson C N 1994c Nutritional ecology of a mycophagous marsupial in relation to production of hypogeous fungi. Ecology 75, 2015–2021.

Johnson C N 1994d Mycophagy and spore dispersal by a rat-kangaroo: consumption of ectomycorrhizal taxa in relation to their abundance. Funct. Ecol. 8, 464–468.

Johnson C N 1995 Interactions between fire, mycophagous mammals, and dispersal of ectomycorrhizal fungi in *Eucalyptus* forests. Oecologia 104, 467–475.

Johnson C N 1997 Fire and habitat management for a mycophagous marsupial, the Tasmanian bettong *Bettongia gaimardi*. Aust. J. Ecol. 22, 101–105.

Kinnear J E, Cockson A, Christensen P E S and Main A R 1979 The nutritional biology of ruminants and ruminant-like mammals — a new approach. Comp. Biochem. Physiol. 64A, 357–365.

Lamont B B, Ralph C S and Christensen P E S 1985 Mycophagous marsupials as dispersal agents for ectomycorrhizal fungi on *Eucalyptus calophylla* and *Gastrolobium bilobum*. New Phytol. 101, 651–656.

Laurance W F 1997 A distributional survey and habitat model for the endangered northern bettong *Bettongia tropica* in tropical Queensland. Biol. Conserv. 82, 47–60.

Lawrence J F 1988 Mycophagy in coleoptera: feeding strategies and morphological adaptations. *In* Insect–Fungus Interactions. Eds. N Wilding, N M Collins, P M Hammond and J F Webber. pp. 1–23. Royal Entomological Society of London, 14th Symposium. Academic Press, London.

Lu X, Malajczuk N and Dell B 1998 Mycorrhiza formation and growth *of Eucalyptus globulus* seedlings with spores of various ectomycorrhizal fungi. Mycorrhiza 8, 81–86.

Lu X, Malajczuk N, Brundrett M and Dell B 1999 Fruiting of putative ectomycorrhizal fungi under blue gum (*Eucalyptus globulus*) plantations of different ages in Western Australia. Mycorrhiza 8, 255–261.

Malajczuk N 1988 Ecology and management of ectomycorrhiza in regenerated ecosystems in Australia. *In* Mycorrhizae in the Next Decade. Practical Applications and Research Priorities. Eds. D M Sylvia, L L Hung and J H Graham. pp. 290–292. University of Florida, Gainesville, FL.

Malajczuk N, McComb A J and Loneragan J F 1975 Phosphorus uptake and growth of mycorrhizal and uninfected seedlings of *Eucalyptus calophylla* R.Br. Aust. J. Bot. 23, 231–238.

Malajczuk N, Molina R and Trappe J M 1982 Ectomycorrhiza formation in *Eucalyptus*. I. Pure culture synthesis, host specificity and mycorrhizal compatability with *Pinus radiata*. New Phytol. 91, 467–482.

Malajczuk N, Dell B and Bougher N 1987a Ectomycorrhiza formation *in Eucalyptus*. III. Superficial ectomycorrhizas initiated by *Hysterangium* and *Cortinarius* species. New Phytol. 105, 421–428.

Malajczuk N, Trappe J M and Molina R 1987b Interrelationships among some ectomycorrhizal trees, hypogeous fungi and small mammals: Western Australian and northwestern American parallels. Aust. J. Ecol. 12, 53–55.

Marx, D H 1973 Ectomycorrhizae as biological deterrants to pathogenic root infection. Annu. Rev. Phytopath. 10, 429–454.

Maser C, Trappe J M and Nussbaum R A 1978 Fungal-small mammal interrelationships with emphasis on Oregon coniferous forests. Ecology 59, 799–809.

McGee P A 1996 The Australia Zygomycetous mycorrhizal fungi: the genus *Densospora* gen. nov. Aust. Syst. Bot. 9, 329–336.

McIlwee A P and Johnson C N 1997 The contribution of fungus to the diets of three mycophagous marsupials in *Eucalyptus* forests, revealed by stable isotope analysis. Funct. Ecol. 12, 223–231.

Miller S L, Torres P and McClean T M 1994 Persistence of basidiospores and sclerotia of ectomycorrhizal fungi and *Morchella* in soil. Mycologia 86, 89–95.

Millington S, Leach D N, Wyllie S G and Claridge A W 1997 Aroma profile of the Australian truffle-like fungus *Mesophellia glauca*. *In* Flavor Analysis — Developments in Isolation and Characterisation. Eds. C J Mussinana and M J Morello. pp. 331–342. American Chemical Society, Washington, DC.

Montecchi A and Sarasini M 2000 Fungi Ipogei d'Europa. A.M.B. Fondazione. Centro Studi Micologici, 714 pp.

Nix H A 1981 The environment of *Terra Australis*. *In* Ecological Biogeography of Australia. Ed. A Keast. pp. 105–133. Dr W Junk, The Hague.

Parke J L, Linderman R G and Black R H 1983 The role of ectomycorrhizas in drought tolerance of Douglas-fir seedlings. New Phytol. 95, 83–95.

Read D J 1987 In support of Frank's organic nitrogen theory. Angew. Bot. 61, 25–37.

Recher H F and Christensen P E S 1981 Fire and the evolution of the Australian biota. *In* Ecological Biogeography of Australia. Ed. A. Keast. pp. 137–161. Dr W Junk, The Hague.

Reddell P and Malajczuk N 1984 Formation of mycorrhizae by Jarrah (*Eucalyptus marginata* Donn ex Smith) in litter and soil. Aust. J. Bot. 32, 511–520.

Reddell, P. and Milnes, A.R. 1992 Mycorrhizas and other specialised nutrient-acquisition strategies: their occurrence in woodland plants from Kakadu and their role in rehabilitation of waste rock dumps at a local uranium mine. Aust. J. Bot. 40, 223–242.

Reddell P and Spain A V 1991 Earthworms as vectors for viable propagules of mycorrhizal fungi. Soil Biol. Biochem. 23, 767–774.

Reddell P, Spain A V and Hopkins M 1997 Dispersal of spores of mycorrhiza fungi in scats of native mammals in tropical forests of northeastern Australia. Biotropica 29, 184–192.

Reichelt R C and May T W 1997 Malleefowl eating fungi and orchid tubers. Vic. Nat. 114, 98.

Scotts, D J and Seebeck J H 1989 Ecology of *Potorous longipes* (Marsupialia: Potoroidae): and preliminary recommendations for management of its habitat in Victoria. Arthur Rylah Institute Technical Report Series 62, 129 pp.

Simpson J A 2001 More on mycophagous birds. Australasian Mycol. 19, 49–51.

Smith S E and Read D J 1997 Mycorrhizal Symbiosis. Academic Press, London.

Specht R L 1996 The influence of soils on the evolution of eucalypts. *In* Nutrition of Eucalypts. P M Attiwill and M A Adams. pp. 31–60. CSIRO, Collingwood.

Taylor R J 1992 Distribution and abundance of fungal sporocarps and diggings of the Tasmanian bettong, *Bettongia gaimardi*. Aust. J. Ecol. 17, 155–160.

Thiers H D 1984 The secotioid syndrome. Mycologia 76, 1–8.

Thomson B D, Grove T S, Malajczuk N and Hardy, G E StJ. 1994 The effectiveness of ectomycorrhizal fungi in increasing growth of *Eucalyptus globulus* Labill. in relation to root colonisation and hyphal development in soil. New Phytol. 126, 517–524.

Thomson, B.D., Hardy, G.E. StJ., Malajczuk, N. and Grove, T.S. 1996 The survival and development of inoculant ectomycorrhizal fungi on roots of outplanted *Eucalyptus globulus* Labill. Plant Soil 178, 247–253.

Tommerup I C and Bougher N L 1999 The role of ectomycorrhizal fungi in nutrient cycling in temperate Australian woodlands. *In* Temperate Eucalypt Woodlands in Australia: Biology, Conservation, Management and Restoration. Eds. R J Hobbs and C J Yates. pp. 190–224. Beatty and Sons, Chipping Norton, Surrey.

Tory M K, May T W, Keane P J and Bennett A F 1997 Mycophagy in small mammals: a comparison of the occurrence and diversity of hypogeal fungi in the diet of the long-nosed potoroo *Potorous tridactylus* and the bush rat *Rattus fuscipes* from southwestern Victoria, Australia. Aust. J. Ecol. 22, 460–470.

Trappe J M 1988 Lessons from Alpine fungi. Mycologia 80, 1–10.

Trappe J M and Maser C 1976 Germination of spores of *Glomus macrocarpa* (Endogonaceae) after passage through a rodent digestive tract. Mycologia 68, 433–436.

Vernes K and Haydon D T 2001 Effects of fire on northern bettong (*Bettongia tropica*) foraging behavious. Austral Ecol. 26, 649–659.

Warcup J H 1980 Ectomycorrhizal associations of Australian indigenous plants. New Phytol. 85, 531–535.

Warcup J H 1985 Ectomycorrhiza formation by *Glomus tubiforme*. New Phytol. 99, 267–272.

Warcup, J H 1990a Occurrence of ectomycorrhizal and saprophytic discomycetes after a wild fire in a eucalypt forest. Mycol. Res. 94, 1065–1069.

Warcup, J H 1990b Taxonomy, culture and mycorrhizal associations of some zygosporic Endogonaceae. Mycol. Res. 94, 173–178.

Warcup, J H 1991 The fungi forming mycorrhizas on eucalypt seedlings in regeneration coupes in Tasmania. Mycol. Res. 95, 329–332.

Plant and Soil **244:** 307–317, 2002.

Effects of mycorrhizal fungi on plant populations

Roger T. Koide[1,3] & Ian A. Dickie[1,2]
[1]*Department of Horticulture, The Pennsylvania State University, University Park, PA 16802 USA;* [2]*Current address: Department of Forest Resources, 115 Green Hall, 1530 N. Cleveland Ave., St. Paul, MN 55108, USA;* [3]*Corresponding author**

Received 21 August 2001. Accepted in revised form 21 january 2002

Key words: adult–seedling interactions, density-dependence, offspring quality, reproduction, size hierarchy

Abstract

We discuss four potentially important interactions between mycorrhizal fungi and populations of plants. First, vesicular-arbuscular mycorrhizal colonization has been shown to increase reproduction (via both male and female functions) and offspring survival, and thus it can increase population size, at least in the short term. This is undoubtedly important to wild plant species and especially to those whose success depends on high rates of reproduction such as early successional annuals. Second, the positive response in growth and reproduction to vesicular-arbuscular mycorrhizal colonization may be inversely related to plant population density. All else being equal, this would tend to stabilize the density of natural plant populations over time. It may also explain why positive responses to mycorrhizal inoculation of dense crops are rare. Third, vesicular-arbuscular mycorrhizal fungi can increase inequality in size and reproduction among plants within a population. Mycorrhizal fungi may thus exaggerate the genetic overrepresentation in the next generation of the most robust individuals in the current generation. Fourth, established mycorrhizal plants may serve as important sources of inoculum for initially non-mycorrhizal, conspecific seedlings. This may affect regeneration, and could contribute to patchy distributions of species within the community.

Introduction

A great deal of research has been conducted to determine the effects of mycorrhizal fungi on individual host plants. Much has been learned about the mycorrhizal symbiosis and much will continue to be learned by studying individual plants, particularly about the genetics, biochemistry and physiology of the symbiosis. In natural and agricultural field settings, however, plants often interact with other members of the same species, which comprise a population. Plant–plant interactions within populations strongly influence basic processes such as growth and reproduction. Plant populations, as crops, are of fundamental importance to the welfare of the human population. Finally, the population is the fundamental unit of genetic adaptation to environmental heterogeneity (Raven, 1979 and references therein). For these reasons, studies of the effects of mycorrhizal fungi on plant populations are of considerable interest to the ecologist, evolutionary biologist and agronomist. We review here some of the important research illustrating four important ways in which mycorrhizal fungi influence plant populations. We include discussions on the positive effect of vesicular-arbuscular mycorrhizal colonization on plant reproduction and offspring survival, the relationship between plant population density and response to vesicular-arbuscular mycorrhizal colonization, the effect of vesicular-arbuscular mycorrhizal colonization on size and reproductive hierarchies within plant populations, and the role of established vesicular-arbuscular and ectomycorrhizal plants as important sources of inoculum for conspecific seedlings.

* FAX No: 814-863-6139. E-mail: rxk13@psu.edu

Effects of mycorrhizal colonization on plant reproduction and offspring survival

Relatively little is known about the effects of mycorrhizal colonization on plant reproduction and, for technical reasons, this is especially true for perennial plant species. It is very difficult in most circumstances to maintain plants in a nonmycorrhizal condition for long, particularly in the field. Even when starting from sterile soil, chance inoculation by windblown or animal-dispersed inocula will eventually produce mycorrhizas. The necessity of life-long comparisons between mycorrhizal and nonmycorrhizal plants, however, is easily appreciated; the magnitude of mycorrhizal effects on reproduction early in the reproductive period may differ from that later in the period, even for annual species (Bryla and Koide, 1990a; Shumway and Koide, 1994b). Moreover, the effects of mycorrhizal fungi on reproduction of perennials may vary among years as nutrient availability varies among years. Extrapolation of lifetime reproduction from a single year, therefore, may not be particularly meaningful. Most hosts of ectomycorrhizal fungi are perennials and so we know essentially nothing about the effects of ectomycorrhizal fungi on the reproduction of their hosts.

Our knowledge of the effects of mycorrhizal fungi on plant reproduction is based almost entirely on annual, vesicular-arbuscular mycorrhizal plant species. For such species, a robust body of literature collectively shows that mycorrhizal colonization can significantly increase seed production (summarized in Koide, 2000). This is not surprising. Nitrogen and phosphorus frequently limit both vegetative growth and reproduction in terrestrial plants, and vesicular-arbuscular mycorrhizal fungi can improve host phosphorus status. Because ectomycorrhizal fungi can improve both nitrogen and phosphorus status of their hosts (Smith and Read, 1997), it is logical to suppose that colonization by ectomycorrhizal fungi would also increase plant reproduction, but this has not been demonstrated. Non-nutritional benefits from mycorrhizal colonization, such as a reduction in fungal pathogenesis, may also positively influence reproduction (Carey et al., 1992).

What is surprising, however, is that while vegetative size and reproductive output are often correlated (Bryla and Koide, 1990a and references therein), they may be influenced to very different extents by mycorrhizal colonization. In one experiment, *Abutilon theophrasti* plants were grown at a low density (12 m^{-2})in the field. Mycorrhizal colonization significantly increased vegetative growth when fertilizer phosphorus was not added to the soil, but failed to when fertilizer phosphorus was added. However, seed output was significantly increased by mycorrhizal colonization in both phosphorus treatments (Stanley et al., 1993). For two cultivars of *Lycopersicon esculentum*, reproductive trait responses to mycorrhizal colonization were greater than leaf area response (Poulton, 2000; Poulton et al., 2002). Other researchers have obtained similar results (Bryla and Koide, 1990b; Dodd et al., 1983; Koide et al., 1988). In many cases, the ability of vesicular-arbuscular mycorrhizal colonization to affect plants may be related to the plant phosphorus deficit, the difference between the demand for phosphorus by the plant and the phosphorus actually absorbed from the soil (Koide, 1991b). Thus, when there is a greater phosphorus deficit during reproduction than during vegetative growth, reproduction may be more responsive to mycorrhizal colonization (Fitter, 1989, 1991).

In a field study, Stanley et al. (1993) showed that as a consequence of significantly greater seed production in vesicular-arbuscular mycorrhizal swards of *Abutilon theophrasti*, seedling recruitment the following year was significantly greater in mycorrhizal plots than in nonmycorrhizal plots. The density of *Abutilon theophrasti* in mycorrhizal plots was about 175 plants m^{-2}, while the density in nonmycorrhizal plots was about 75 plants m^{-2}. Thus, early in succession, mycorrhizal colonization may immediately lead to greater densities. This may be particularly important for some early successional annual species for which rapid exploitation of space by the population is an important strategy.

Sexual reproduction consists of both male and female functions, and these may be influenced by mycorrhizal colonization independently. For example, in one cultivar of *Lycopersicon esculentum*, total pollen production per plant was significantly more responsive to mycorrhizal colonization than total seed production per plant (Poulton, 2000; Poulton et al., 2002). Thus, we discuss the male and female functions separately.

Female function

While the diploid embryo within a seed is genetically the product of both male and female gametophytes, we consider seed production to be primarily a female function. Provisioning of the endosperm and embryo with nutrients occurs within the ovary and, in angiosperms, the female contributes two-thirds of the genes to the triploid endosperm, the male one-third. Seed production can be significantly improved by mycorrhizal colonization. Presumably the major reason for this is the greater amount of resources possessed by mycorrhizal plants compared to nonmycorrhizal plants, but the proportional allocations of phosphorus or other resources to reproduction and vegetative growth may be altered by mycorrhizal colonization (Bryla and Koide, 1990b; Koide et al., 1988). Thus, the efficiency with which absorbed phosphorus is utilized to produce seeds may be different in mycorrhizal and nonmycorrhizal plants (Bryla and Koide, 1990a). Several components of seed production may be influenced by mycorrhizal colonization. For example, seed production by *Abutilon theophrasti* was increased by mycorrhizal colonization because mycorrhizal plants branched more extensively and consequently produced more flowers (Koide et al., 1994). Mycorrhizal colonization of *Abutilon theophrasti* also caused plants to flower for a longer period of time, and reduced the levels of fruit and seed abortion (Lu and Koide, 1994).

Male function

During pollen development, sporophyte tissues within the anther endow pollen grains with various storage products to be used during pollen tube growth. For example, phytate stored in the pollen grain is hydrolysed into phosphate and myo-inositol, which are used by the pollen tube for cell wall and membrane synthesis (Jackson and Linskens, 1982). The quantity and quality of pollen nutrient storage products can affect pollen germinability, tube growth rate, and the probability of siring seeds in competition with other pollen (Stephenson et al., 1994). It seems reasonable to expect mycorrhizal colonization to influence pollen quality in much the same way that it influences seed quality (see references in Koide, 2000). Indeed, we have shown recently for tomato (*Lycopersicon esculentum*) that mycorrhizal colonization can significantly increase *in vitro* pollen tube growth rate, which can allow pollen produced by mycorrhizal plants to outcompete slower growing pollen from nonmycorrhizal plants (Poulton et al., 2001a). This alone may result in mycorrhizal plants siring more seed than nonmycorrhizal plants (Poulton et al., 2001a). Moreover, mycorrhizal plants may produce more flowers (Poulton et al., 2001a), each of which may produce more pollen (Poulton et al., 2001b). Thus mycorrhizal colonization can improve male function in at least three ways.

Offspring vigor

Mycorrhizal colonization may result in enhanced offspring vigor and thus enhanced offspring survival. We first documented an effect of mycorrhizal colonization on seed quality in wild oat, *Avena fatua*. In this species, mycorrhizal colonization of the parent generation caused the offspring to grow faster (Koide and Lu, 1992; Lu and Koide, 1991). We therefore investigated the nutrient economy of the seeds and resultant seedlings. Mycorrhizal colonization of *Avena fatua* did not increase individual seed weight or seed nitrogen content, but increased seed phosphorus content by approximately 62% (Koide et al., 1988; Lu and Koide, 1991). Upon germination into a nonmycorrhizal soil, seeds produced by mycorrhizal plants were able to mobilize more phosphorus from the endosperm (about 50% of which was phytate) than seeds from nonmycorrhizal plants. Seedlings from mycorrhizal plants consequently contained significantly more phosphorus than seedlings from nonmycorrhizal plants.

Mycorrhizal colonization of *Abutilon theophrasti* increased weight, height, leaf area, lateral root formation and specific root length of offspring (Koide and Lu, 1995; Lewis and Koide, 1990) and the proportion of offspring plants that flowered (Shumway and Koide, 1994a). Seeds produced early in the reproductive phase were found to benefit more from mycorrhizal colonization than seeds produced later in the season (Shumway and Koide, 1994b), which is consistent with the observation that the number of arbuscules within a root system and thus the benefit from mycorrhizal colonization may be greatest in the earliest phases of growth (Smith and Dickson, 1991). By growing offspring plants in water, we showed that the superior growth of offspring produced by mycorrhizal parents was not due to a superior ability to take up nutrients (Koide and Lu, 1995). Instead, there was an inherent difference in initial growth rate (Koide and Lu, 1995). It is possible that the amount of phytate, the storage form of phosphorus in seeds, may have been more important than total phosphorus content in determining offspring vigor (Lewis and Koide, 1990).

A small but significant mycorrhizal effect on offspring vigor may not be particularly important when microsite variation is large or when germination is asynchronous. However, the effect may become very important when intraspecific competition among

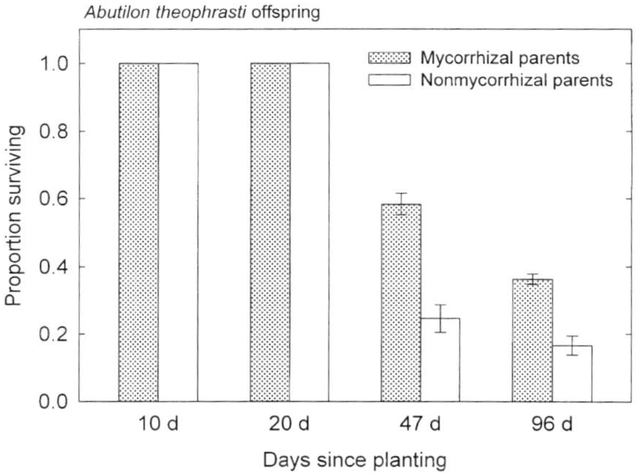

Figure 1. The mean (± S.E.M.) proportion of offspring from mycorrhizal and nonmycorrhizal *Abutilon theophrasti* parents surviving in dense, mixed populations of initially equal proportions. $n = 8$. See Heppell et al. (1998) for details. The differences are significant at 47 and 96 d after planting.

even-aged offspring is intense, as competition may greatly magnify small differences in size (Weiner and Thomas, 1986). Indeed, Solbrig (1981) indicates that during competition offspring size and fitness may be positively correlated. We therefore studied the consequences of mycorrhizal colonization of the parent generation of *Abutilon theophrasti* to the success of offspring under competitive conditions (Heppell et al., 1998). Seeds produced by mycorrhizal plants had 37% greater phosphorus concentrations, 6% greater weights, but did not differ significantly in nitrogen concentration. Seeds from mycorrhizal and nonmycorrhizal *Abutilon theophrasti* parents were germinated, and seedlings were planted in alternate 1.5 × 1.5 cm cells in large wooden boxes filled with a soil lacking mycorrhizal fungus propagules. The interior dimensions of the boxes were 50 cm × 50 cm × 32 cm high. All boxes were watered daily, but plants were not given fertilizer. Emergence did not differ significantly between offspring from mycorrhizal and nonmycorrhizal parents. However, after 20 d there were fewer offspring of nonmycorrhizal parents represented in the largest size classes and more offspring of nonmycorrhizal parents represented in the smallest size classes. As a result of self-thinning (mortality), only 42% of seedlings survived to 47 days. Mortality was not random with respect to parental treatment, however, as significantly more offspring from mycorrhizal parents survived compared to offspring from nonmycorrhizal parents (Figure 1). Offspring of mycorrhizal parents accounted for nearly all individuals in the larger size classes as well as the majority in the smallest size class, suggesting that many of the smallest offspring of nonmycorrhizal parents died. Thus, mycorrhizal colonization may influence more than one generation of plants.

Density-dependent responses to mycorrhizal colonization

The density of natural populations of plants varies widely for many reasons including varying levels of seed production and seedling mortality due to variation in resource availability. In any case, there is now substantial evidence that as density increases, the degree to which plants respond to vesicular-arbuscular mycorrhizal colonization decreases (Allsopp and Stock, 1992; Bååth and Hayman, 1984; Carey et al., 1992; Facelli et al., 1999; Hartnett et al., 1993; Hayman, 1983,; Koide, 1991a).

Three mechanisms have been proposed to account for this density-dependent effect of mycorrhizal colonization. The most likely explanation for this phenomenon is that at higher plant densities an increasing overlap of phosphorus depletion zones created by roots and root hairs results in a decreasing usefulness of hyphae for the acquisition of phosphorus (Figure 2; Hayman, 1983; Bååth and Facelli et al., 1999; Hartnett et al., 1993; Hayman, 1984; Koide, 1991a). If this is true, the effect of plant density on response to mycorrhizal colonization may differ among plant species. For example, species with a very sparse root system such as leek or onion may never have root densities high enough to render mycorrhizal hyphae superfluous, while species with dense root systems such as perennial ryegrass may, even at low plant densities, so effectively exploit soil that mycorrhizal fungi are nearly always superfluous (Koide, 1991b). In principle, a similar relationship between density and mycorrhiza responsiveness could occur for ectomycorrhizal hosts, but this has not yet been demonstrated.

A second possible reason for reduced plant response to mycorrhizal colonization at higher densities may be a reduction in mycorrhizal colonization (Allsopp and Stock, 1992; Bååth and Hayman, 1984; Koide, 1991a). Both the rate at which roots become colonized and the extent of colonization may have important consequences for the plant. For example, extensive colonization and increased phosphorus concentration of a seedling may increase its chances for

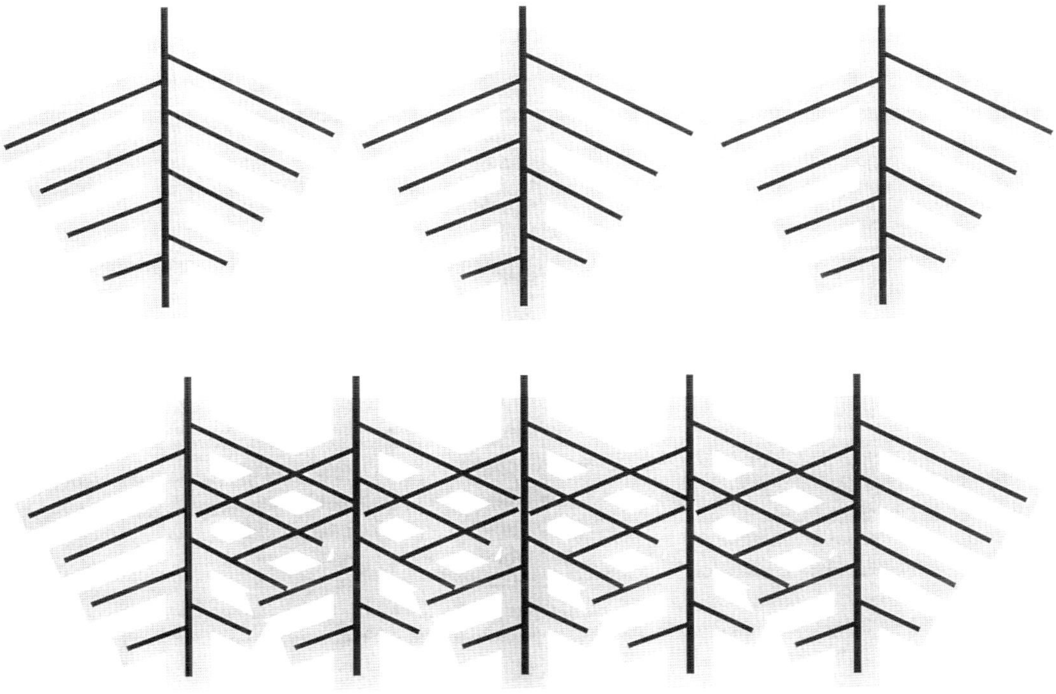

Figure 2. Thick black lines represent roots. Shading represents the phosphate depletion zone in the soil created by absorption of phosphate by roots and root hairs. Upper: Three widely spaced root systems do not very effectively exploit the soil volume. Thus, mycorrhizal hyphae can be effective in acquiring substantial phosphorus from unexploited soil. Lower: Five root systems occupy the same soil volume as the three above. In this case, roots and root hairs exploit much of the soil volume and mycorrhizal hyphae are largely superfluous.

survival and, remarkably, its reproductive effort (Barry and Miller, 1989). A given quantity of mycorrhiza inoculum in the soil can only colonize a given length of root in a given period of time. Thus, as root density increases, a smaller and smaller proportion of the total root length may become colonized in a given period of time. This may explain the results from experiments on *Abutilon theophrasti* (Koide, 1991a), *Aspalathus linearis* (Allsopp and Stock, 1992) and *Trifolium subterraneum* (Facelli et al., 1999) in which reduced proportions of the root systems were colonized at increased plant densities. It may also be possible that at higher densities, greater competition for light results in less carbohydrate available for mycorrhizal fungi (Abbott and Robson, 1984). Facelli et al. (1999) found, however, that reduced light levels do not necessarily result in less colonization.

In some cases, increasing plant density has been associated with *increasing* mycorrhizal colonization such as with the legume *Otholobium hirtum* (Allsopp and Stock, 1992) and with *Plantago lanceolata* (Eissenstat and Newman, 1990). The mechanism for this has not been shown, but the phenomenon is consistent with known positive effects of some root exudates and volatiles on hypha growth, hypha tropism and colonization of roots as shown by St. John et al. (1983), Gemma and Koske (1988) and Bécard and Piché (1989).

A third explanation involves the increase of the cost: benefit ratio of mycorrhizal colonization with increasing plant density (Allsopp and Stock, 1992). As plant density increases, competition for light increases and photosynthesis per plant declines. Plants may become limited more by carbon than by P. Thus, as the value of carbon to the plant increases and the value of phosphorus decreases, the beneficial effects of mycorrhizal colonization decrease. The effect of density on the cost:benefit ratio may differ among plant species. For example, grasses with upright leaves may suffer less from competition for light when densely packed than forbs with largely horizontal leaves (Begon et al., 1986). Therefore, grasses may suffer less proportional decline in responsiveness to mycorrhizal colonization than forbs as density increases.

In an experiment with onion described by Bååth and Hayman (1984), mycorrhizal fungi colonized a lower proportion of the root system at higher plant densities, despite using an amount of inoculum at each density that was proportional to the number of plants. This suggests that a simple 'dilution' of colonization at higher root densities is not the only cause of reduced colonization. Koide and Li (1991) showed that leachate from pots containing live roots of *Abutilon theophrasti* markedly reduced mycorrhizal colonization in other *Abutilon theophrasti* plants. This kind of density-dependent regulation of colonization may be important in maintaining a cost:benefit ratio that is favorable to the host. As root density increases and, therefore, as roots and root hairs more effectively explore the soil volume, mycorrhizal hyphae would become less and less useful (Bååth and Hayman, 1984; Koide, 1991a), increasing the cost:benefit ratio of maintaining them (Allsopp and Stock, 1992). Inherent mechanisms such as suggested by Koide and Li (1991) may maintain a lower cost:benefit ratio of the symbiosis to the plant.

The consequences of density-dependent responses to mycorrhizal colonization are potentially large. The lack of response of plants to mycorrhizal colonization in a number of field trials (McGonigle, 1988) may have many causes, but it may be that at high densities, common to many field crops, mycorrhizal colonization is not effective for one of the reasons discussed above (also see Facelli et al., 1999). Because at high densities crop growth may be largely unresponsive to mycorrhizal colonization, growers might improve economic efficiency by planting dense crops in soils of low inoculum density, but less dense crops in soils of high inoculum density. For wild species such as *Abutilon theophrasti*, colonization of one generation of plants at a comparatively low population density may lead to greater reproduction and thus a higher population density in the subsequent year (Stanley et al., 1993). Thus, members of the second year cohort would not benefit from colonization as much as the previous generation (Koide et al., 1994). Mycorrhizal fungi may thus serve to stabilize the density of natural plant populations through time.

The development of size and reproductive hierarchies within populations

The individuals comprising plant populations naturally vary in size. Some of this is due to microsite variation, some due to genetic variation. Grime et al. (1987) suggested that movement of assimilate from larger plants of one species to smaller plants of another species via vesicular-arbuscular mycorrhizal hyphae could account for greater community evenness (but see Robinson and Fitter, 1999). Such transport could have significant consequences for plant populations, resulting in a reduction in size disparity among individuals.

Most evidence, however, suggests that a reduction in size disparity within populations does not occur as a consequence of mycorrhizal colonization. Eissenstat and Newman (1990) indicated that the balance of competition between large and small *Plantago lanceolata* plants was not significantly influenced by mycorrhizal colonization. In other cases, mycorrhizal colonization resulted in a magnification rather than a diminution of inter-plant differences in size. For example, mycorrhizal colonization shifted the balance of competition in favor of larger *Prunella vulgaris* plants, not the smaller ones (Moora and Zobel, 1996), and mycorrhizal colonization resulted in an increase in the coefficient of variability (for size) in *Otholobium hirtum*, a species that is highly responsive to mycorrhizal colonization (Allsopp and Stock, 1992). Facelli et al. (1999) showed clearly that mycorrhizal colonization increased the intensity of intraspecific competition in *Trifolium subterraneum*, which resulted in greater size inequality. Carey et al. (1992) showed that among untreated (heavily mycorrhizal) *Vulpia ciliata* plants there were significantly more highly reproductive individuals than among the fungicide-treated (less mycorrhizal) plants, again suggesting that mycorrhizal colonization accentuates differences among individuals rather than reducing them. Shumway and Koide (1995) created mycorrhizal and nonmycorrhizal field populations of *Abutilon theophrasti* by soil fumigation and reinoculation, and obtained the same result as Carey et al. (1992); mycorrhizal populations had significantly higher inequality in terms of reproduction than nonmycorrhizal populations because the mycorrhizal populations had a higher frequency of highly reproductive individuals than nonmycorrhizal populations. Thus, it appears that slightly more robust plants are better able to capitalize on mycorrhizal colonization (Figure 3), possibly because phosphorus is transferred more freely to plants from which carbon is more available. We conclude, as did Facelli et al. (1999), that resource transfer via a common mycelial network does not appear to equalize resource distribution among intraspecific neighbours, and that movement of resources from larger to smaller plants via the mycorrhizal mycelium is not likely to have

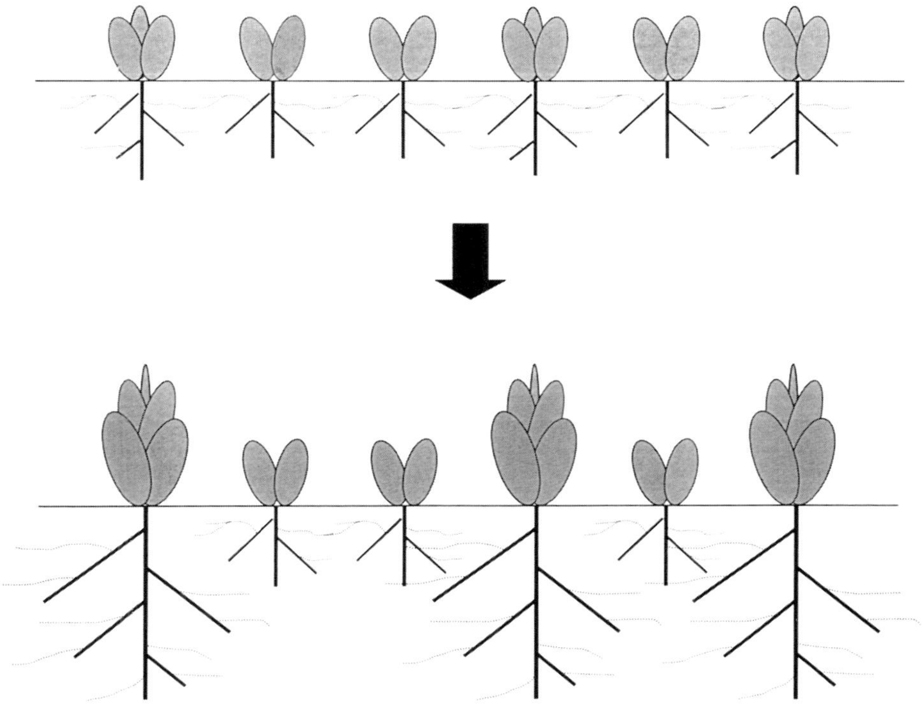

Figure 3. Thick black lines represent roots. Thin, dotted lines represent mycorrhizal hyphae. In most experiments to date, small, inherent differences among plants (note that the first, fourth and sixth plants are slightly larger than the others) are magnified by mycorrhizal colonization. This is contrary to the hypothesis that greater uniformity among plants should occur as a consequence of net resource transfer from dominant to suppressed individuals via a common mycelial network. Instead, the magnification of differences may result from greater phosphorus transfer to plants from which more carbon is available.

major ecological consequences. Instead, mycorrhizal fungi may result in an increased skewing of genetic representation in the next generation in favor of the dominant plants. If dominance has a genetic basis, the result could be a significant effect on population genetic structure (Facelli et al., 1999).

Mycorrhizal interactions between established plants and seedlings

Mycorrhizal networks and nutrient transfer

When seedlings establish within an existing mycorrhizal network, mycorrhizal linkages between established plants and seedlings may be formed (Newman, 1988). Autoradiographic evidence suggests that ^{32}P taken up by a fungal mycelium in microcosms is transferred both throughout the mycelium and into multiple host plants (Finlay and Read, 1986). A pre-existing mycelium may thus allow compatible seedlings to take advantage of a more extensive mycelial network than they otherwise could, and established plants may subsidize the costs of supporting this network (Newman, 1988).

Interest in mycorrhizal links has been raised by the possibility of net C transfer between plants. This may be particularly important to understory seedlings connected to mature canopy trees. Robinson and Fitter (1999) recently reviewed the literature on this subject and concluded that there is no evidence for significant net interplant C transfer through ectomycorrhizal or vesicular-arbuscular mycorrhizal links. Indeed, Fleming (1984) found that severing linkages between trees and seedlings actually increased seedling growth, and Simard et al. (1997) found that severing linkages caused no change in seedling height or dry mass. Again, it appears that if there is movement of resources between plants via mycorrhizal hyphae, it is not likely to have large consequences.

Seedling colonization

In many plant populations, seedlings grow near established plants and may thus become colonized by a preexisting mycorrhizal mycelium (Newman, 1988). This does not require that the established plants be living at the time of seedling growth, as shown by the positive influence of vesicular-arbuscular mycorrhizal winter cover crops on the mycorrhizal colonization of subsequent maize crops (e.g., Boswell et al., 1998; Kabir and Koide, 2000). Where vesicular-arbuscular mycorrhizal plants are absent, as in primary succession, lack of vesicular-arbuscular inoculum may limit succession (Allen, 1987). Low levels of mycorrhizal colonization also may reduce yields of agronomic crops (Boswell et al., 1998; Kabir and Koide, 2000). Thus, the provision of inoculum by previously established plants may be very important to newly establishing seedlings.

Because many ectomycorrhizal fungi produce abundant wind-dispersed spores, seedlings of ectomycorrhizal species may become rapidly colonized even in the absence of established plants (Allen, 1987). This is illustrated by 'contaminant' mycorrhizal colonization of seedlings grown in pasteurised soil in greenhouse studies (e.g., Pilz and Perry, 1984), the rapid ectomycorrhizal colonization of fumigated nursery beds by *Thelephora terrestris* (Marx and Cordell, 1987), and the establishment of ectomycorrhizal plants in primary succession (Allen, 1987; Trappe and Luoma, 1992). Nonetheless, in recent studies we found that *Quercus rubra* seedlings in a site surrounded by largely ectomycorrhizal forest were colonized to a much greater extent when planted near ectomycorrhizal trees than when planted near vesicular-arbuscular mycorrhizal trees (Dickie, 2000; Dickie et al., 2001). Lack of, or low extent of, colonization by ectomycorrhizal fungi may be an important factor preventing or reducing tree establishment following disturbance (Perry et al., 1989; Terwilliger and Pastor, 1999).

In addition to increasing the extent of seedling colonization, established plants may also influence the species of fungi that colonize seedlings (Last et al., 1992 and citations therein). Seedlings establishing in forest clearcuts may be colonized by different species of fungi than seedlings establishing at forest edges or within the forest (e.g., Kranabetter and Wylie, 1999). Even where seedlings establish within the root zone of established trees, the distance to the nearest tree may influence the composition of mycorrhizal fungal species (Gibson and Deacon, 1988). It appears that some species of fungi are largely incapable of colonizing seedlings except through hyphae extending from established tree roots, as trenching can substantially alter the community composition of fungi colonizing seedlings (Fleming, 1984; Simard et al., 1997).

Established plants may influence succession and regeneration if they control to a large extent either the degree of mycorrhizal colonization of seedlings or the species composition of the colonizing fungi. Succession may be hastened when at least some mycorrhizal fungal species colonizing the established individuals of one species are capable of colonizing the seedlings of the succeeding plant species, and it may be hindered when there is no such compatibility. Cross-species compatibility is obviously not necessary for regeneration of a given plant species, however, since this would only require that the fungi supported by the established individuals are capable of colonizing conspecific seedlings. An inability of the fungi to colonize both established plants and seedlings could negatively affect regeneration. Established plants could conceivably influence the distribution of other plants by providing the appropriate mycorrhizal fungi. One might, therefore, hypothesize that ectomycorrhizal tree species would be patchily distributed in primarily vesicular-arbuscular mycorrhizal forest, and vice versa.

Influences of established plants on the species composition of vesicular-arbuscular mycorrhizal fungi colonizing seedlings have been less well studied than in ectomycorrhiza. Vesicular-arbuscular mycorrhizal fungi are generally believed to be less diverse than ECM fungi, and to have lower levels of host specificity. Nonetheless, communities of vesicular-arbuscular mycorrhizal fungi in soils may differ depending on the presence of different host plant species (Johnson et al., 1992; McGonigle and Fitter, 1990), and this could result in different species of fungi infecting seedlings. The recent development of molecular approaches to identify vesicular-arbuscular mycorrhizal fungi in individual roots may permit further advances in this area.

Net influence of established plants on seedlings

We have found that *Quercus rubra* seedlings planted near established ectomycorrhizal trees had increased ectomycorrhizal colonization, nitrogen and phosphorus uptake, and growth compared with seedlings planted near vesicular-arbuscular mycorrhizal trees (Dickie, 2000; Dickie et al., in press). Nonetheless, although established plants may have beneficial influences on seedlings in terms of mycorrhizal colonization, the net influence of established plants on seedlings may be negative. For example, Zhou and Sharik (1997) found that *Quercus rubra* seedling growth was maximized in forest clearcuts, although mycorrhizal colonization was highest at intermediate canopy levels. Mycorrhizal colonization was reduced under a full canopy, perhaps due to reduced seedling photosynthesis (Zhou et al., 1997). Negative influences of established plants on seedlings may occur either directly, through competition, or indirectly through support of pathogenic organisms (Packer and Clay, 2000). It appears that the net influence of established plants on seedlings is most likely to be positive in sites where lack of mycorrhizal colonization would limit seedling growth; net negative interactions are more likely as plant density increases.

Conclusions

Populations possess properties that individuals do not. Thus, the effects of mycorrhizal fungi on populations of plants are not simply the sum of their effects on the individuals within a population. The study of mycorrhizas at this higher level of complexity is therefore of great importance to ecologists who seek to understand their role in natural ecosystems, and to agronomists who seek practical applications of them in agricultural systems. The research reviewed herein indicates that mycorrhizal fungi can interact with plant populations in several important ways. Reproduction and offspring survival can be significantly increased by vesicular-arbuscular mycorrhizal colonization, and this can increase plant population size, at least in the short term. The effects of mycorrhizal colonization on reproduction can be manifested through both the male and female functions. This effect on reproduction, however, may not occur generation after generation because of the density dependence of response to mycorrhizal colonization. Because low-density populations may benefit more from mycorrhizal colonization than high-density populations, mycorrhizal fungi may contribute to the stabilization of population density through time. There is no evidence that variation among individuals within populations is reduced by vesicular-arbuscular mycorrhizal colonization as predicted from the few studies suggesting that significant quantities of materials may be transferred from more vigorous to less vigorous individuals. Instead, differences among individuals may actually be exacerbated by mycorrhizal colonization. Finally, previously established plants may influence seedlings of the same species by providing an important source of mycorrhiza inoculum - living hyphae. In agricultural systems, winter cover crops that support vesicular-arbuscular mycorrhizal fungi thus benefit subsequent mycorrhizal crops. In natural ecosystems in which inoculum is limited, this pattern of colonization may naturally result in patchy distributions of vesicular-arbuscular, ecto- or ericoid mycorrhizal plants. These interactions between mycorrhizal fungi and plant populations suggest that further research on the effects of mycorrhizal fungi at the level of the plant population will continue to be valuable in both ecological and agricultural contexts.

Acknowledgements

We thank the US National Science Foundation and the A W Mellon Foundation for financial support, and the editors and two anonymous reviewers for helpful suggestions in the preparation of this contribution.

References

Abbott L K and Robson A D 1984 The effect of root density, inoculum placement and infectivity of inoculum on the development of vesicular-arbuscular mycorrhizas. New Phytol. 97, 285–299.

Allen M F 1987 Re-establishment of mycorrhizas on Mount St. Helens: Migration vectors. Trans. Br. Mycol. Soc. 88, 413–417.

Allsopp N and Stock W D 1992 Density dependent interactions between VA mycorrhizal fungi and even-aged seedlings of two perennial Fabaceae species. Oecologia 91, 281–287.

Barry D A J and Miller M H 1989 Phosphorus nutritional requirement of maize seedlings for maximum yield. Agronomy J 81, 95–99.

Bååth E and Hayman D S 1984 Effect of soil volume and plant density on mycorrhizal infection and growth response. Plant Soil 77, 373–376.

Bécard G and Piché Y 1989 Fungal growth stimulation by CO_2 and root exudates in vesicular-arbuscular mycorrhizal symbiosis. Appl. Environ. Microbiol. 55, 2320–2325.

Begon M, Harper J L and Townsend C R 1986. Ecology, Individuals, Populations, and Communities. Sinauer Associates, Sunderland, MA, USA. 876 p.

Boswell E P, Koide R T, Shumway D L and Addy H D 1998. Winter wheat cover cropping, VA mycorrhizal fungi and maize growth and yield. Agric. Ecosys. Environ. 67, 55–65.

Bryla D R and Koide R T 1990a Regulation of reproduction in wild and cultivated *Lycopersicon esculentum* Mill. by vesicular-arbuscular mycorrhizal infection. Oecologia 84, 74–81.

Bryla D R and Koide R T 1990b Role of mycorrhizal infection in the growth and reproduction of wild vs. cultivated plants. II. Eight wild accessions and two cultivars of *Lycopersicon esculentum* Mill. Oecologia 84, 82–92.

Carey P D, Fitter A H and Watkinson A R 1992 A field study using the fungicide benomyl to investigate the effect of mycorrhizal fungi on plant fitness. Oecologia 90, 550–555.

Dickie I A 2000. Mycorrhizal infection, nutrition, and growth of *Quercus rubra* L. seedlings as influenced by overstory trees on a disturbed forest site. Ph. D. Thesis, The Pennsylvania State University Intercollege Graduate Degree Program in Ecology.

Dickie I A, Koide R T and Fayish A C 2001 Vesicular-arbuscular mycorrhizal infection of *Quercus rubra* seedlings. New Phytol. 151, 257–264.

Dickie I A, Koide R T and Steiner K C Influence of established trees on mycorrhizas, nutrition, and growth of *Quercus rubra* seedlings. Ecol Monogr, in press.

Dodd J, Krikun J and Haas J 1983 Relative effectiveness of indigenous populations of vesicular-arbuscular mycorrhizal fungi from four sites in the Negev. Israel J. Bot. 32, 10–21.

Eissenstat D M and Newman E I 1990 Seedling establishment near large plants: Effects of vesicular-arbuscular mycorrhizas on the intensity of plant competition. Funct. Ecol. 4, 95–99.

Facelli E, Facelli J M, Smith S E and McLaughlin M J 1999 Interactive effects of arbuscular mycorrhizal symbiosis, intraspecific competition and resource availability on *Trifolium subterraneum* cv. Mt. Barker. New Phytol. 141, 535–547.

Finlay R D and Read D J 1986 The structure and function of the vegetative mycelium of ectomycorrhizal plants. II. The uptake and distribution of phosphorus by mycelium inter-connecting host plants. New Phytol. 103, 157–165.

Fitter A H 1989 The role and ecological significance of vesicular-arbuscular mycorrhizas in temperate ecosystems. Agric. Ecosys. Environ. 29, 137–151.

Fitter A H 1991 Costs and benefits of mycorrhizas: Implications for functioning under natural conditions. Experientia 47, 350–355.

Fleming L V 1984 Effects of soil trenching and coring on the formation of ectomycorrhizas on birch seedlings grown around mature trees. New Phytol. 98, 143–153.

Gemma J N and Koske R E 1988 Pre-infection interactions between roots and the mycorrhizal fungus *Gigaspora gigantean*: Chemotropism of germ-tubes and root growth response. Trans. Br. Mycol. Soc. 91, 123–132.

Gibson F and Deacon J W 1988 Experimental study of establishment of ectomycorrhizas in different regions of birch root systems. Trans. Br. Mycol. Soc. 91, 239–251.

Grime J P, Mackey J M L, Hillier S H and Read D J 1987 Floristic diversity in a model system using experimental microcosms. Nature 328, 420–422.

Hartnett D C, Hetrick B A D, Wilson G W T and Gibson D J 1993 Mycorrhizal influence on intra-and interspecific neighbour interactions among co-occurring prairie grasses. J. Ecol. 81, 787–795.

Hayman D S 1983 The physiology of vesicular-arbuscular endomycorrhizal symbiosis. Can. J. Bot. 61, 944–963.

Heppell K B, Shumway D L and Koide R T 1998 The effect of mycorrhizal infection of *Abutilon theophrasti* on competitiveness of offspring. Funct. Ecol. 12, 171–175.

Jackson J F and Linskens H F 1982 Phytic acid in *Petunia hybrida* pollen is hydrolysed during germination by a phytase. Acta Bot. Neerlandica 31, 441–447.

Johnson N C, Tilman D and Wedin D 1992 Plant and soil controls on mycorrhizal fungal communities. Ecology 73, 2034–2042.

Kabir Z and Koide R T 2000 The effect of dandelion or a cover crop on mycorrhiza inoculum potential, soil aggregation and yield of maize. Agric. Ecosys. Environ. 78, 167–174.

Koide R T 1991a Density-dependent response to mycorrhizal infection in *Abutilon theophrasti* Medic. Oecologia 85, 389–395.

Koide R T 1991b Nutrient supply, nutrient demand and plant response to mycorrhizal infection. New Phytol. 117, 365–386.

Koide R T 2000 Mycorrhizal symbiosis and plant reproduction. *In* Arbuscular Mycorrhizas: Physiology and Function. Eds. Kapulnik Y and Douds D D Jr. pp. 19–46. Kluwer Academic Publishers, Dordrecht, The Netherlands.

Koide R T and Li M 1991 Mycorrhizal fungi and the nutrient ecology of three oldfield annual plant species. Oecologia 85, 403–412.

Koide R T, Li M, Lewis J and Irby C 1988 Role of mycorrhizal infection in the growth and reproduction of wild vs. cultivated plants. Oecologia 77, 537–543.

Koide R T and Lu X 1992 Mycorrhizal infection of wild oats: maternal effects on offspring growth and reproduction. Oecologia 90, 218–226.

Koide R T and Lu X 1995 On the cause of offspring superiority conferred by mycorrhizal infection of *Abutilon theophrasti*. New Phytol. 131, 435–441.

Koide R T, Shumway D L and Mabon S A 1994 Mycorrhizal fungi and reproduction of field populations of *Abutilon theophrasti* Medic. (Malvaceae). New Phytol. 126, 123–130.

Kranabetter J M and Wylie T T 1998 Ectomycorrhizal community structure across forest openings on naturally regenerated western hemlock seedlings. Can. J. Bot. 76, 189–196.

Last F T, Natarajan K, Mohan V and Mason P A 1992 Sequences of sheathing (ecto-) mycorrhizal fungi associated with man-made forests, temperate and tropical. *In* Mycorrhizas in Ecosystems. Eds. Read D J, Lewis D H, Fitter A H and Alexander I J. pp. 214–219. CAB International, Wallingford, UK.

Lewis J D and Koide R T 1990 Phosphorus supply, mycorrhizal infection and plant offspring vigour. Funct. Ecol. 4, 695–702.

Lu X and Koide R T 1991 *Avena fatua* L. seed and seedling nutrient dynamics as influenced by mycorrhizal infection of the maternal generation. Plant Cell Environ. 14, 931–939.

Lu X and Koide R T 1994 The effects of mycorrhizal infection on components of plant growth and reproduction. New Phytol. 128, 211–218.

Marx D H and Cordell C E 1987 Ecology and management of ectomycorrhizal fungi in regenerating forests in the eastern United States. *In* Mycorrhizae in the Next Decade: Practical Applications and Research Priorities. Eds Sylvia D M, Hung L L and Graham J H. pp. 69–71. Institute of food and agricultural sciences, University of Florida, Gainesville, FL, USA.

McGonigle T P 1988 A numerical analysis of published field trials with vesicular-arbuscular mycorrhizal fungi. Funct. Ecol. 2, 473–478.

McGonigle T P and Fitter AH 1990 Ecological specificity of vesicular-arbuscular mycorrhizal associations. Mycol. Res. 94, 120–122.

Moora M and Zobel M 1996 Effect of arbuscular mycorrhiza on inter- and intraspecific competition of two grassland species. Oecologia 108, 79–84.

Newman E I 1988 Mycorrhizal links between plants: their functioning and ecological significance. Adv. Ecol. Res. 18, 243–270.

Packer A and Clay K 2000 Soil pathogens and spatial patterns of seedling mortality in a temperate tree. Nature 404, 278–281.

Perry D A, Amaranthus M P, Borchers J G, Borchers S L and Brainerd R E 1989 Bootstrapping in ecosystems. BioScience 39, 230–237.

Pilz D P and Perry D A 1984 Impact of clearcutting and slash burning on ectomycorrhizal associations of Douglas-fir seedlings. Can. J. For. Res. 14: 94–100.

Poulton 2000 Effects of mycorrhizal infection and soil phosphorus availability on the male function of reproduction in *Lycopersicon esculentum* Mill., PhD Dissertation, The Pennsylvania State University.

Poulton J L, Bryla D, Koide R T and Stephenson A G 2002 Effects of mycorrhizal infection and soil phosphorus availability on vegetative growth and the female and male functions in two cultivars of *Lycopersicon esculentum*. New Phytol. in press.

Poulton J L, Koide R T and Stephenson A G 2001a Effects of mycorrhizal infection and soil phosphorus availability on *in vitro* and *in vivo* pollen performance in *Lycopersicon esculentum* (Solanaceae). Am. J. Bot. 88, 1786–1793.

Poulton J L, Koide R T and Stephenson A G 2001b Effects of mycorrhizal infection, soil phosphorus availability, and fruit production on the male function in two cultivars of *Lycopersicon esculentum* Mill. Plant Cell Environ. 24, 841–849.

Raven P H 1979 Future directions in plant population biology. *In* Topics in Plant Population Biology. Eds. Solbrig O T, Jain S, Johnson G B and Raven P H. pp. 461–481. Columbia University Press, New York.

Robinson D and Fitter A H 1999 The magnitude and control of carbon transfer between plants linked by a common mycorrhizal network. J. Exp. Bot. 50, 9–13.

Shumway D L and Koide R T 1994a Reproductive responses to mycorrhizal colonization of *Abutilon theophrasti* Medic. plants grown for two generations in the field. New Phytol. 128, 219–224.

Shumway D L and Koide R T 1994b Within-season variability in mycorrhizal benefit to reproduction in *Abutilon theophrasti* Medic. Plant Cell Environ. 17, 821–827.

Shumway D L and Koide R T 1995 Size and reproductive inequality in mycorrhizal and nonmycorrhizal populations of *Abutilon theophrasti*. J. Ecol. 83, 613–620.

Simard S W, Perry D A, Smith J E and Molina R 1997 Effects of soil trenching on occurrence of ectomycorrhizas on *Pseudotsuga menziesii* seedlings grown in mature forests of *Betula papyrifera* and *Pseudotsuga menziesii*. New Phytol. 136, 327–340.

Smith S E and Dickson S 1991 Quantification of active vesicular-arbuscular mycorrhizal infection using image analysis and other techniques. Aust. J. Plant Phys. 18, 637–648.

Smith S E and Read D J 1997 Mycorrhizal Symbiosis, second edition. Academic Press, San Diego. 605 p.

Solbrig O T 1981 Studies on the population biology of the genus *Viola*. II. The effect of plant size on fitness in *Viola sororia*. Evolution 35, 1080–1093.

St John T V, Hays R I and Reid C P P 1983 Influence of a volatile compound on formation of vesicular-arbuscular mycorrhizas. Trans. Br. Mycol. Soc. 81, 153–154.

Stanley M R, Koide R T and Shumway D L 1993 Mycorrhizal symbiosis increases growth, reproduction and recruitment of *Abutilon theophrasti* Medic. in the field. Oecologia 94, 30–35.

Stephenson A G, Erickson C W, Lau T-C, Quesada M R and Winsor J A 1994 Effects of growing conditions on the male gametophyte. *In* Pollen–Pistil Interactions and Pollen Tube Growth. Eds. Stephenson A G and Kao T H. Vol 12, pp. 220–229. Current Topics in Plant Physiology.

Terwilliger J and Pastor J 1999 Small mammals, ectomycorrhizae, and conifer succession in beaver meadows. Oikos 85, 83–94.

Trappe J M and Luoma D L 1992. The ties that bind: fungi in ecosystems. *In* The Fungal Community: Its Organization and Role in the Ecosystem, 2nd edn. Eds. Carroll G C and Wicklow D T. pp. 17–27. Marcel Dekker, New York.

Weiner J and Thomas S C 1986 Size variability and competition on plant monocultures. Oikos 47, 211–222.

Zhou M and Sharik T L 1997 Ectomycorrhizal associations of northern red oak (*Quercus rubra*) seedlings along an environmental gradient. Can. J. For. Res. 27, 1605–1713.

Zhou M, Sharik T L, Jurgensen M F and Richter D L 1997 Ectomycorrhizal colonization of *Quercus rubra* seedlings in response to vegetation removals in oak and pine stands. For. Ecol. Manag. 93, 91–99.

Plant and Soil **244:** 319–331, 2002.

The role of mycorrhizas in plant community structure and dynamics: lessons from grasslands

David C. Hartnett[1] & Gail W. T. Wilson
Division of Biology, Kansas State University, Ackert Hall, Manhattan, KS 66506, U.S.A. [1]*Corresponding author**

Introduction

Research on the mycorrhizal associations over the past several decades has yielded increased understanding and appreciation of the important role of this symbiosis in the functioning and performance of plants in a wide array of terrestrial ecosystems. We now understand that the role of mycorrhizal fungi extends beyond the symbiotic acquisition of phosphorus for the host plant and reciprocal carbon provision from the host to fungus. Additional effects of mycorrhizal fungi on the functioning of their host plants including increased disease resistance, improved water relations, acquisition of other soil nutrients, and alterations in other soil physico-chemical properties have been documented. Other aspects of the ecology of mycorrhizas, including variation in the costs and benefits of carbon and nutrient exchange, the ecological significance of mycelial networks, the role of mycorrhizal symbiosis in multi-species interactions, and the extent and consequences of host-specificity in these associations have also recently been explored.

The effects of mycorrhizal fungi on the function and growth of individual plants have numerous consequences that extend to the plant population and community levels. The costs and benefits of colonization by mycorrhizal fungi for plant resource availability can strongly influence patterns of plant reproduction and demography, responses to competitors, herbivores, and other biotic interactions, and patterns of species composition, diversity, and succession. This ancient mutualistic symbiosis may be one of the most important but least understood biotic interactions regulating plant community structure and dynamics. Hopefully, an increased understanding of these patterns, their underlying mechanisms, and the costs and benefits of this symbiosis will lead in the future to the formulation of general predictive theory of mycorrhizal effects on plant communities.

* E-mail: dchart@ksu.edu

In this paper, we examine the consequences of mycorrhizal symbiosis at the plant community level, with a primary focus on grasslands. Mycorrhizal fungi are ubiquitous and abundant in grasslands and savannas worldwide (e.g., Chiariello et al., 1982; Davidson and Christensen, 1977; Fitter, 1986; Koide et al., 1988; Miller, 1987; Newsham and Watkinson, 1998; O'Connor et al., 2001; Perez and Frangi, 2000; Smilauer and Smilaveraer, 2000; van Auken and Brown, 1998). Indeed, our sampling of >100 herbaceous species in North American tallgrass prairie and southern African low veld grasslands revealed that all species thus far examined are colonized by mycorrhizal fungi and that most dominant grasses in these systems are obligate mycotrophs. In addition, studies over the past decade have demonstrated that mycorrhizas have particularly large influences on several aspects of grassland community and ecosystem functioning, and experiments in grasslands have contributed much to the lessons learned and new questions raised about the ecological role of this symbiosis. Here we review evidence for the varied effects of mycorrhizas on the composition and diversity of plant communities, discuss the various hypothesized mechanisms by which mycorrhizal fungi influence plant community structure, and explore the role of this symbiosis in the broader context of the complex array of biotic interactions characteristic of grasslands and many other terrestrial ecosystems.

Mycorrhizas and plant community structure

The processes regulating the composition and diversity natural communities has been a central focus in ecology. The roles of species interactions, natural disturbances operating at different scales, and historical and biogeographic factors have all been explored (e.g., Brown, 1995; Huston, 1994; Pickett and White, 1985; Ricklefs and Schluter, 1993). Studies of the role of biotic interactions influencing plant community structure have focused largely on plant competition and plant-herbivore interactions (e.g., Frank et al., 1998; Grace and Tilman, 1990). The important role of mycorrhizal symbiosis in shaping plant community structure has recently become recognized and gained significant attention. This has been a result of both a general increase in research on mycorrhizas, and an increasing recognition that positive interactions (e.g., direct and indirect facilitation) are common in natural communities and may have large effects on plant community structure (Bengtsson et al., 1994; Callaway, 1995; Clay and Holah, 1999).

Recent studies have shown that mycorrhizal fungi can have large but quite variable effects on plant species diversity, composition, and successional dynamics (e.g., Allen and Allen 1984; Connell and Lowman, 1989; Grime et al., 1987; Hartnett and Wilson, 1999). There is also increasing evidence that the effects of mycorrhizal fungi on their host plant communities are context-dependent, varying with host species, plant life history stage, resource availability, and abiotic conditions (e.g., Bethlenfalvay et al., 1982; Fitter, 1986; Francis and Read, 1995; Hartnett et al., 1994; Johnson et al., 1997). In some herbaceous communities, mycorrhizas enhance plant species diversity by increasing the establishment and abundance of subordinate species relative to the community dominants (Gange et al., 1993; Grime et al., 1987; Newman and Reddle, 1988), and plant diversity may be positively correlated with the species diversity of mycorrhizal fungi (van der Heijden et al., 1998a). In other communities, evidence indicates that mycorrhizal fungi decrease plant species coexistence and diversity (e.g., Hartnett and Wilson, 1999; Newsham et al., 1995; O'Connor et al., 2001). Experiments in both North American tallgrass prairie (Figure 1) and European calcareous grasslands (Zobel and Moora, 1995) indicate that mycorrhizal symbiosis decreases plant species diversity by increasing the competitive success and abundance of the community dominants relative to subordinate species. Studies in other grasslands and grassland restorations indicate that mycorrhizal fungi may also alter species composition with no net effect on species richness (Smilauer and Smilauerova, 2000), or may alter rates of succession by hastening the displacement of ruderal species by native grasses (Smith et al., 1998).

Thus, it is clear that at the local community (alpha) scale, mycorrhizal fungi have the potential either to decrease or increase plant species diversity, alter plant species composition, and/or influence the rate and trajectories of community succession. The hypothesized underlying mechanisms causing these observed effects of mycorrhizas on plant communities are similarly varied and include: (1) direct and differential effects on plant species traits such as colonization, establishment, and competitive ability, (2) effects of mycorrhizal mycelial networks and interplant resource transfer through hyphal connections on plant growth and relative abundances, and (3) mycorrhizal mediation of other biotic interactions such as plant–herbivore or plant–pathogen interactions. In the following sections, we review the evidence and discuss various proposed mechanisms for mycorrhizal influence on plant communities.

Patterns of plant species diversity in nature are scale-dependent, and conservation of biodiversity at the regional or global scale should not be confused with maximization of species richness and diversity at the local community or patch scale (Poiani et al., 2000). Many natural grassland communities are and have historically been characterized by low diversity and high dominance at the local community (alpha) scale. In these systems, this low local patch diversity, coupled with significant heterogeneity at larger scales, results in a pattern of higher diversity at the landscape or regional scales. Furthermore, high alpha diversity at the local community scale is often associated with a larger number of invasive exotic or ruderal species, whereas stable, undisturbed communities are characterized by lower richness. Thus, high alpha diversity or species richness at the local scale is not necessarily a good indicator of the biotic integrity or stability of a community. The study of mycorrhizas or any other factors that influence patterns of diversity must consider spatial scale, and focusing solely on the maximization of local species richness as a conservation goal is an unwarranted oversimplification and generalization that ignores scale and overemphasizes species numbers relative to species composition and relative abundances. It is clear from the various studies reviewed above that mycorrhizas have large influences on plant community structure and are an important factor in the stability of plant species composition, as evidenced by the large magnitude of changes in plant communities reported in response to experimental alteration of mycorrhizal activity.

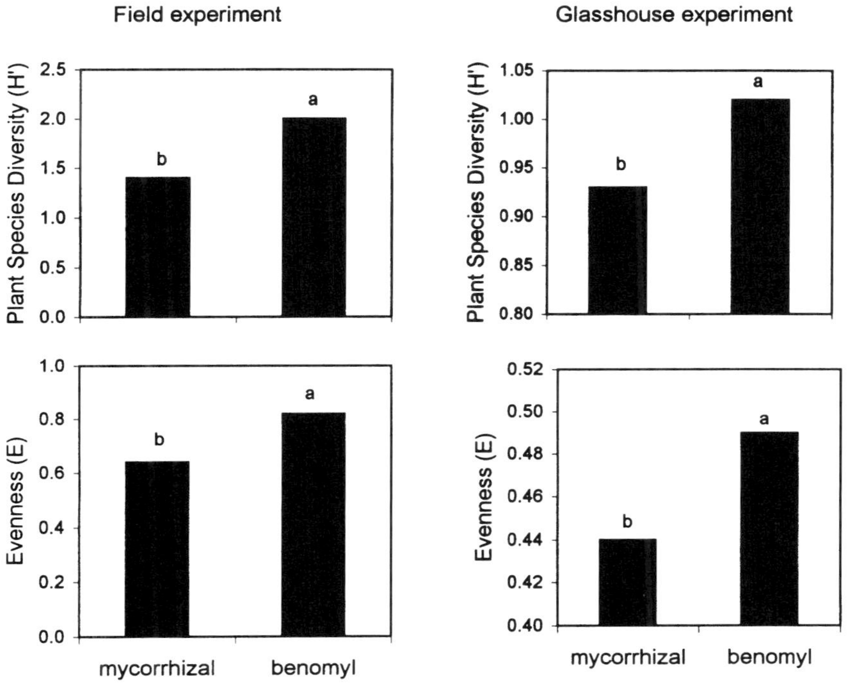

Figure 1. Field experiment: Plant species diversity and evenness of mycorrhizal (control) and mycorrhizal-suppressed (Benomyl application) plots in North American tallgrass prairie after five years of treatment application. Mycorrhizal and mycorrhizal-suppressed plots were not different prior to fungicide applications (year 1). Bars represent means of six transects (10 plots/transect). Different letters indicate that control plots are significantly different ($P \leq 0.05$) from mycorrhizal-suppressed plots. Data from Hartnett and Wilson (1999). Glasshouse experiment: plant species diversity and evenness of mycorrhizal (control) and mycorrhizal-suppressed (Benomyl application) tallgrass prairie microcosms maintained for 42 weeks under glasshouse conditions. Each microcosm contained warm- and cool-season grasses and native prairie forbs planted at approximate field densities (eight species total). Different letters indicate that control plots are significantly different ($P \leq 0.05$) from mycorrhizal-suppressed plots. Data from Wilson and Hartnett (1997).

The spatial heterogeneity hypothesis

Allen (1991) suggested that spatial heterogeneity of mycorrhizal fungal infectivity may increase plant species diversity, allowing nonmycotrophic and mycotrophic species to coexist in patches of low and high inoculum, respectively. This mechanism depends strongly on the existence of interspecific differences in responses to mycorrhizal colonization among co-occurring plant species. We extend Allen's hypothesis further and suggest that, in habitats with significant spatial heterogeneity in fungal infectivity, there will be a positive relationship between interspecific variability in host plant mycotrophy and plant species diversity.

Allen's hypothesis also assumes that fungal infectivity or inoculum potential is limiting to the growth of at least some plant species in the community. We know little about if, and under what conditions low mycorrhizal fungal populations and low root colonization levels may actually be limiting to their host plant populations. Variation in spore population densities or root colonization rates may have little effect on host plant populations if these factors are not limiting to the maintenance of functional mycorrhizal symbiosis in the field.

A related hypothesis is that spatial heterogeneity in infectivity, coupled with variation in host species dependency on mycorrhiza for colonization and establishment (e.g., regeneration niche differences) may enhance species diversity in plant communities. A field experiment in tallgrass prairie showed that plant species differed considerably in their requirement for mycorrhizal colonization for successful germination and establishment, and that some species which were highly dependent upon the symbiosis for establishment showed low dependency on mycorrhizas for adult plant growth, and *vice versa* (Hartnett et al., 1994). Effects of mycorrhizal colonization on plant growth may also differ from its effect on seed and vegetative reproduction within prairie species (Koide, 2000, and this volume; Wilson et al., 2001a). Thus, in these grasslands, mycorrhizal dependency is an important aspect of the regeneration niche of plants, and the dependency of the plant on the symbiosis may vary considerably among different plant life history stages. The mycorrhizal enhancement of germination and establishment of many tallgrass prairie plant species may not only promote species diversity, but may also increase the rate of succession by allowing highly mycotrophic species to more rapidly displace ruderal species (Smith et al., 1998).

Mycorrhizal regulation of plant competition

A second hypothesis is that mycorrhizal fungi influence plant species coexistence, relative abundances, and diversity by altering the intensity and outcomes of interspecific plant competition. Given interspecific differences among plants in mycorrhizal effects on growth and competitive ability, any changes in mycorrhizal colonization or symbiotic function will alter plant species competitive effects and responses, resulting in shifts in species dominance, coexistence, and diversity. We predict that, in general, mycorrhizal symbiosis will strongly influence plant species diversity in any community in which the dominant competitors are significantly more strongly or more weakly mycotrophic than their neighbors.

In turfgrass microcosm experiments conducted by Grime et al., (1987), the dominant competitors (C_3 grasses) were more weakly mycotrophic than their neighbors. Thus, active mycorrhizal associations conferred greater benefit to subordinate competitors and increased species diversity. By contrast, in our tallgrass prairie experiment (Hartnett and Wilson, 1999), the dominant competitors (C_4 perennial grasses) were more strongly mycotrophic than their interspecific neighbors. Thus, AM fungal colonization further enhanced their dominance and reduced local species diversity in these grasslands. The abundance of C_4 grasses decreased with mycorrhizal suppression, with a concomitant increase in abundance of C_3 grasses, but with no effect on total aboveground plant production. This clearly indicates that the subordinate prairie species experience compensatory increases in growth and abundance as the competitive effects of the dominant grasses are greatly reduced in the absence of their mycorrhizal symbionts.

These conclusions from this 5-year field experiment are further supported by a series of pairwise competition experiments conducted in the glasshouse (Hartnett et al., 1993; Hetrick et al., 1989, 1994), and a neighbor removal experiment conducted in the field (Smith et al., 1999), which showed that mycorrhizas regulated the outcomes of intra- and interspecific competition among tallgrass prairie plant species. Connell and Lowman (1989) suggested that this mechanism of mycorrhizal alteration of plant competitive relationships may also explain low diversity and high dominance in some tropical forest communities.

Interspecific variability in plant response to mycorrhizal symbiosis

Both the 'spatial heterogeneity hypothesis' and the 'mycorrhizal regulation of plant competition hypothesis' depend strongly on a key assumption of significant variability among host plant species in their responses to mycorrhizal colonization, and indeed there is considerable evidence for such variability in grassland plant communities. For example, a study that assessed the growth responses to arbuscular mycorrhizal colonization of 95 species of tallgrass prairie grasses and forbs found high interspecific variation in host-plant benefit from the symbiosis. Species varied from large positive growth responses (e.g., non-mycorrhizal plant biomass $< 2\%$ of mycorrhizal plant biomass) to negative effects of AMF colonization on plant growth. The patterns of mycorrhizal dependence varied among plant life history and taxonomic groups (e.g., grass, forb, legume, annual, perennial) and phenological guilds (Wilson and Hartnett, 1998). In general, perennial warm-season C_4 grasses and forbs benefited significantly from the symbiosis, whereas growth of the perennial cool-season C_3 grasses and annual species were minimally affected. Mycorrhizal responsiveness of the prairie forbs was strongly related to root system architecture (lower colonization in species with highly branched, fine root structure) and, to a lesser extent, flowering phenology.

Variation among host plant species in mycorrhizal dependency may not always provide reliable predictions of mycorrhizal effects on community structure if host species responses are strongly influenced by the particular species of AM fungi present (van der Heijden et al., 1998a,b). While different AM fungal taxa have been shown to produce varying growth responses, it should also be noted that the dominant plant species in the C_3-dominated grasslands that have been studied thus far did not respond significantly to *any* AM fungi (e.g., van der Heijden et al., 1998a,b). Also, while plant growth benefits varied among fungal isolates, eight of 11 sub-dominant plant species failed to grow in the absence of the AM fungi. Therefore, suppression of the symbiosis necessarily reduces diversity due to the loss of obligately mycotrophic plant species from the system, with a concomitant increase in dominance of the non-responsive dominant grasses. In contrast, the dominant plant species of the C_4-dominated grasslands studied are obligately mycorrhizal, and although different AM fungi may differ in their benefit to host plants (e.g., Hetrick et al., 1986), these host species fail to grow at all in the absence of the symbiosis. Because many sub-dominant plant species of these grasslands are facultative mycotrophs (low responsiveness), elimination of mycorrhizal fungi differentially suppresses the dominant C_4 grasses and results in competitive release and increased abundance of the sub-dominants, thereby increasing plant species diversity. Therefore, while AM fungal community composition has the potential to differentially influence different plant species (van der Heijden et al., 1998b), the general patterns of mycorrhizal responsiveness of the dominant plant species appears to be the overriding factor influencing plant species diversity in these systems. In tallgrass prairie communities, both field and glasshouse microcosm studies comparing fungicide-treatments with non-sterile native prairie soil containing a natural AM fungal community produced the same effects on the plant communities as glasshouse experiments using steam-pasteurized soil and re-inoculation with one or several fungal taxa (Hartnett and Wilson, 1999; Wilson and Hartnett, 1997, 1998; Wilson et al., 2001b). This further suggests that the presence or absence of AM fungi, and general differences in mycorrhizal dependency among dominant and subordinate plant species is the primary mechanism accounting for mycorrhizal effects on plant species diversity in tallgrass prairie. Interspecific variation in mycorrhizal dependency is also a key factor influencing patterns of succession in many communities, as more weakly or non-mycotrophic species are replaced by obligate mycotrophs with time since disturbance (Miller, 1987).

Overall, evidence to date suggests that the presence or absence of mycorrhizal fungi, the growth responses or mycorrhizal dependency of host plant species, and/or differential plant species responses to the particular species of AM fungi present may all be potentially important factors influencing the performance and relative abundance of plant species within local herbaceous communities.

The resource sharing hypothesis

Grime et al. (1987) hypothesized that their observed increase in plant species diversity in turfgrass microcosms in response to the presence of mycorrhizas was due to extensive mycelial networks facilitating interplant resource transfer via hyphal connections. This, they argued, allowed small, suppressed plants to obtain carbohydrates from the larger, dominant species via shared mycorrhizal hyphae, ultimately increasing the equitability of species abundances. These workers suggested that the 'export of assimilate from 'source' (canopy dominants) to 'sink' (understory components) through a common mycelial network may be an important mechanism maintaining species-rich communities in infertile soils'. A fundamental distinction between this hypothesis and the mechanisms discussed above is that it suggests that a key role of mycorrhizal fungi is to *decrease* rather than *increase* interspecific differences in resource acquisition and competitive ability among co-occurring plant species. Secondly, this hypothesis invokes a network of interplant hyphal connections and interplant resource transfer as an important process operating in natural plant communities.

Observations that mycorrhizas can form hyphal connections between the roots of related or unrelated hosts (e.g., Chiariello et al., 1982; Heap and Newman, 1980; Newman et al., 1994) have stimulated debate and experimentation on the potential role of interplant nutrient transfer in natural and agricultural ecosystems (Eissenstat, 1990; Francis and Read, 1984; Francis et al., 1986; Frey and Schuepp, 1992; Hamel and Smith, 1992; Ritz and Newman, 1984; Whitingham and Read, 1982; Zobel et al., 1997) and on the evolution of functional mycelial networks (Wilkinson, 1998). If mycorrhizal fungi transfer significant amounts of nutrients among neighboring plants, they may potentially act as significant integrators of plant community dynamics, affecting seedling establishment near mature plants, nutrient cycling from dying roots, and the acquisition and allocation of plant resources. Ultimately, such resource movement may influence patterns of local neighborhood competition, succession, and species diversity (Allen and Allen, 1990; Eissenstat and Newman, 1990; Grime et al., 1987; Miller and Allen 1992; Newman, 1988; Read et al., 1985; Ritz and Newman, 1985).

The potential role and importance of interplant hyphal connections in grasslands and other plant communities remains a matter of debate and deserves much further study. It is unknown whether interplant transfer of resources occurs in sufficient quantities and over sufficient spatial scales to influence the relative performance and interactions of neighboring plants, and whether a net movement of resources occurs toward particular plant species. Furthermore, if a net movement of resources via hyphal connections does indeed occur, it remains unclear whether such resource movement results in a more or less equitable distribution of limiting resources among neighbors. Currently, there is little empirical support for the assumption that subordinate plant species will be stronger sinks and that the movement of resources through mycelial networks will necessarily result in an equitable sharing of resources and higher species diversity. Furthermore, processes such as hyphal grazing by soil invertebrates and high hyphal turnover rates may preclude the maintenance of interplant hyphal connections for a long enough period to influence the outcome of neighbor interactions in nature.

A number of studies in grasslands have addressed some of these issues and suggest that interplant nutrient transfer via hyphae may be important in these communities. Glasshouse studies have shown that carbon and mineral nutrients can move from root to root directly via AM hyphal links (Fitter et al., 1998; Francis and Read, 1984; Read et al., 1985; Robinson and Fitter, 1999) or may be transferred among plants via leakage and subsequent uptake (Newman and Ritz, 1986). Several studies have also examined the patterns and controls of nutrient transfer between plants under glasshouse conditions (Frey and Schuepp, 1992; Haystead et al., 1988; Mader et al., 2000; Ritz and Newman, 1986; Rogers et al., 2001; Whittingham and Read, 1982).

Few studies have examined the patterns or controls of interplant nutrient transfer under natural field conditions. Chiariello et al., (1982) documented intra- and interspecific transfer of radio-labeled phosphorus between plants in a California annual grassland, but found no type of pattern that might be predicted based on a diffusion model, source-sink interactions, taxonomic specificity, or other factors. By contrast, in Kansas tallgrass prairie, significant interplant transfer of phosphorus occurs, but in a clearly non-random pattern. A field ^{32}P labeling experiment showed that transfer of phosphorus from the dominant grass *Andropogon gerardii* to its interspecific neighbors occurs over distances up to 50 cm, but with significantly different amounts of phosphorus transferred to different species within the local neighborhood, in patterns

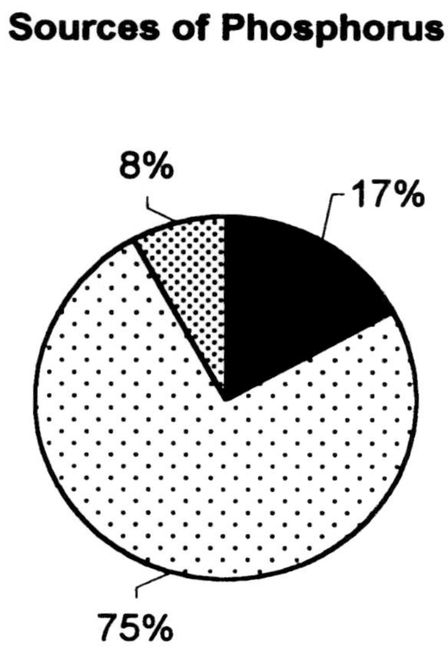

Figure 2. Differentiation of the sources of phosphorus taken up by *Andropogon gerardii*. (□) Mycorrhizal mediated uptake within the plant's own rhizosphere. (■) Mycorrhizal interplant transfer of phosphorus. (▩) Nonmycorrhizal uptake of phosphorus within the plant's own rhizosphere.

unrelated to relative plant size (Fisher-Walter et al., 1996). The results of this tallgrass prairie study suggest that hyphal interconnections and mycelial networks do not result in an equitable distribution of limiting nutrients among neighbors, as has been hypothesized for other grasslands. However, in this experiment, there was more phosphorus transferred from *A. gerardii* to the sub-dominant forbs and C_3 grasses than to other competitively dominant C_4 perennial grasses (Fisher-Walter et al., 1996). This pattern is consistent with the general hypothesis of Grime et al., (1987) that canopy dominants and subordinate species may function differentially as source and sink for resources moving through hyphal networks. Fisher-Walter et al., (1996) caution that their study did not demonstrate that mycorrhizal hyphal interconnections played a principal role in the interplant transfer process. Other mechanisms such as phosphorus leakage and uptake in the context of extensive and differential rhizosphere overlap among neighboring species were not ruled out.

Although occurring over sufficient spatial scales, whether such transfer occurs in sufficient quantities to influence neighbor interactions in tallgrass prairie remains unknown. However, a glasshouse experiment examining phosphorus transfer between tallgrass prairie plant species via hyphal interconnections suggests that interplant transfer may make a significant contribution to the total phosphorus acquired by a plant. Using a dual-pot and dual-labeling (^{32}P and ^{33}P) experiment, we grew pairs of tallgrass prairie plant species together and variously manipulated the presence or absence of AM fungi and hyphal connections between plants. The use of two P isotopes enabled the differentiation of the sources of phosphorus taken up by a given plant. Through these experiments, and a series of simple difference equations, we were able to quantify the phosphorus budget of an individual *A. gerardii* plant and to determine what proportion of its phosphorus was acquired from different sources (Hartnett and Wilson, unpubl.). Although the results were variable, this study showed that a significant amount (an average of 17%) of the phosphorus taken up by this grass was acquired directly from its neighbor via hyphal interconnections (Figure 2). Whether similar quantities of nutrients are transferred among neighboring plants in the field is unknown. This study showed further that approximately 75% of the P acquired by an *A. gerardii* plant was acquired from its own rhizosphere via mycorrhizal hyphae, and less than 10% of its P was acquired directly from the soil without the aid of mycorrhizal fungi. These patterns underscore the overriding importance of arbuscular mycorrhizas in P acquisition by these dominant grassland plants.

The results of these various experiments indicate the potential that the movement of plant resources through hyphal interconnections may be an important mechanism influencing plant species interactions and community structure in grasslands and other plant communities. However, results to date are equivocal and much further study is needed to determine whether this mechanism plays a significant role in patterns of species abundances and diversity in natural communities. The methodological limitations and difficulties in measuring the patterns and consequences of this phenomenon in the field will be the greatest challenge.

Complex biotic interactions and indirect effects

In addition to their direct effects on their host plants, mycorrhizal fungi may influence plant communities indirectly through their effects on interactions between plants and their herbivores, pathogens, pollinators, and other microbial mutualists (e.g., Finlay and Soderstrom, 1989; Fitter and Garbaye, 1994; Gehring and Whitham, 1994; Hodge, 2000; Lussenhop, 1996; Newsham et al., 1994). In turn, these species interactions can also influence mycorrhizal fungal communities and their function (Eom et al., 2001). Because a given host plant interacts simultaneously with mycorrhizal fungi and with its herbivores, pathogens, and other symbionts, and all of these organisms jointly influence important plant resources and costs and benefits, strong interactions and indirect effects of mycorrhizal fungi in plant communities should be expected. Furthermore, because many interactions such as those between mycorrhizal fungi or belowground invertebrate herbivores and their host plants are relatively non-host-specific, many communities may be characterized by 'diffuse' mutualisms or plant-consumer interactions in which interactions occur and are best defined at the functional group level rather than between particular pairs of species (e.g., Setala, 1995). Thus, as argued for the structure of competitive relationships in natural communities (e.g., Connell, 1990), 'diffuse' and 'apparent' interactions may be a common characteristic of plant communities associated with mycorrhizal fungi.

Although the direct effects of mycorrhizas through symbiotic uptake of mineral nutrients have been well studied, accumulating evidence suggests that these strong indirect effects of mycorrhizas may be of great importance in plant communities and should not be ignored. These multiple, complex interactions remain poorly understood, as most studies of species interactions have examined them independently as binary interactions (e.g., plant–plant, plant–herbivore, plant–microbe), without considering their function and effects in the multi-species context of natural communities. A further complexity of these relationships is that effects of mycorrhizas and other organisms on their host plant are not universally positive or negative, but may vary in time and space along a mutualism-antagonism continuum (e.g., Johnson et al., 1997).

Every approach to study in ecology involves a trade-off between simplicity, realism, precision, and generality (Harper, 1982). Most studies to date have involved simple pairwise experimental study of plant–mycorrhizal fungi interactions under a single or limited set of environmental conditions. Such experimental approaches are tractable, offer simplicity and precision, and have contributed to our understanding of the range of potential effects of mycorrhizal symbiosis. However, they lack the necessary realism to draw strong conclusions about the role and function of mycorrhizas in natural communities. To accurately assess the role of mycorrhizal symbiosis in plant communities in the context of natural multi-species interactions and under the range of environmental conditions found in nature will be a significant challenge. Mycorrhizal associations play a large role in influencing other species interactions and plant community dynamics, however their population and community-level effects appear to be quite varied and their underlying mechanisms are still not well understood (Bergelson and Crawley, 1988; Crowell and Boerner, 1988; Grime et al., 1987; Hartnett and Wilson, 1999; Miller, 1987).

Interactions between herbivores and mycorrhizal fungi are expected because both depend upon and influence important plant resources. Aboveground consumers may reduce photosynthate translocated to the root system and available to mycorrhizal fungi, resulting in a reduction in mycorrhizal colonization and reduced development of the symbiosis (Gehring et al., 1997; Hetrick et al., 1990a; Trent et al., 1988). Mycorrhizas, in turn, have many potential effects on plant–herbivore interactions. Under certain conditions, up to 40–50% of a plant's net production may be allocated to its fungal symbiont (Fogel and Hunt, 1979; Harris and Paul, 1987). Because mycorrhizal fungi both consume photosynthate and at the same time enhance mineral nutrient acquisition and growth capacity, the cost–benefit relationships among mycorrhizal fungi, herbivores and host plants are likely to be complex. Mycorrhizas may affect herbivores through alteration of plant growth or foliar chemistry (e.g., Goverde et al., 2000; Koide, 2000), and they may have large effects on plant responses to herbivores by influencing anti-herbivore defenses and/or herbivory tolerance (regrowth capacity). Mycorrhizas may significantly enhance grazing tolerance in grassland plants because their regrowth capacity following defoliation is often limited by the availability of nutrients such as phosphorus or nitrogen, or because mycorrhizas alter biomass allocation patterns in ways that promote persistence under grazing (Chapin and McNaughton, 1989; Coughenour et al., 1985; McNaughton et al., 1983).

Evidence from several grasslands indicates a strong relationship between mycorrhizal symbiosis and plant–grazer interactions. In east African grasslands dominated by *Panicum coloratum*, Wallace (1981) found a positive correlation between grazing intensity and mycorrhizal colonization, and in North American grasslands Reece and Bonham (1978) found that grazing increased vesicle density in *Bouteloua gracilis* roots. In tallgrass prairie, mycorrhizal *Andropogon gerardii* plants showed higher clipping tolerance than non-mycorrhizal plants, although repeated

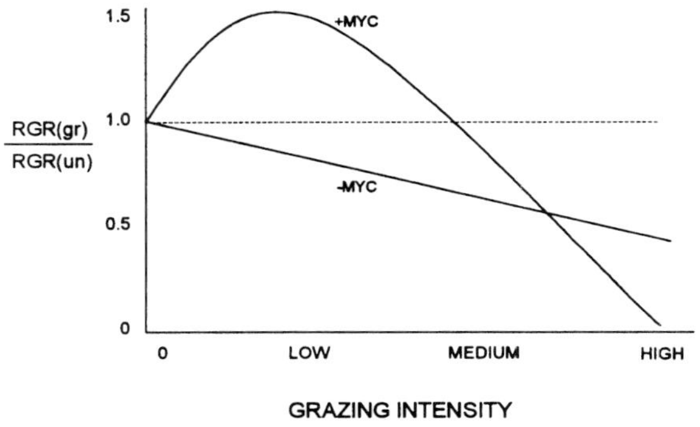

Figure 3. Hypothesized effects of mycorrhizas on the grazing tolerance of dominant tallgrass prairie grasses. +MYC, mycorrhizal symbiosis present; −MYC, mycorrhizal symbiosis absent or suppressed; gr, grazed; un, ungrazed.

defoliation decreased mycorrhizal root colonization (Hetrick et al., 1990a). Gange and Brown (1992) similarly reported an interaction of mycorrhizal fungi and plant responses to belowground herbivores in a successional community. Effects of experimental herbivore reductions on the abundance of perennial grasses disappeared when mycorrhizal fungi were also suppressed. Based on patterns observed in various grasses, we hypothesize that the effect of mycorrhizas on plant responses to defoliation will vary in both direction and magnitude with the intensity of herbivory (Figure 3). Under light to moderate herbivory, when mineral nutrient availability limits plant regrowth rates, mycorrhizal symbiosis will increase grazing tolerance (compensatory growth capacity) by increasing nutrient supply. However, under intense or repeated defoliation, regrowth of aboveground tissue and mycorrhizal fungi may become competing sinks for limited photosynthate. This switch from nutrient limitation to carbon limitation as herbivory increases results in a switch in the cost–benefit balance to the plant and mycorrhizas may become detrimental rather than beneficial by reducing the plants ability to tolerate herbivory (Figure 3).

We further hypothesize that this change in cost–benefit balance of mycorrhizal symbiosis as herbivory intensity increases, coupled with interspecific variation in plant responses to mycorrhizal colonization, will alter the effects of herbivore intensity on plant species diversity (Figure 4). Conceptual models such as the intermediate disturbance hypothesis and the model of Milchunas et al., (1988) predict a maximization of plant species diversity at intermediate intensities of herbivory in many grassland communities.

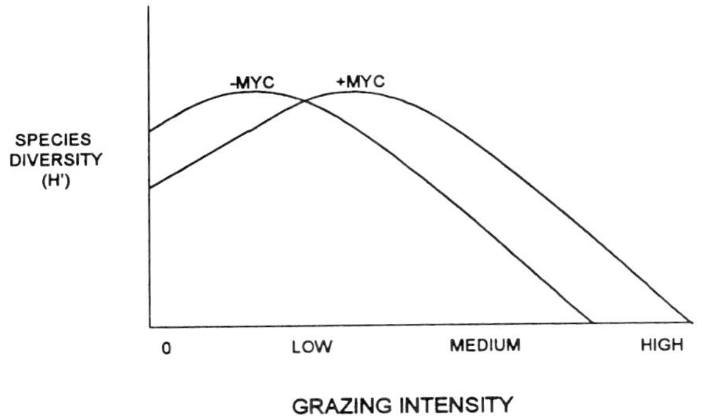

Figure 4. Hypothesized relationship between tallgrass prairie plant species diversity and grazing intensity under mycorrhizal and non-mycorrhizal conditions. +MYC, mycorrhizal symbiosis present; −MYC, mycorrhizal symbiosis absent.

The interaction of mycorrhizal symbiosis and plant–herbivore interactions is predicted to shift this relationship and influence the grazing intensity at which diversity is maximized (Figure 4).

Mycorrhizal fungi may also alter effects of belowground grazers of plant roots (Bakhtiar et al., 2001; Bonkowski et al., 2000). Currently, we are examining effects of above- and belowground grazing and their potential interactions on plant community composition, using experimental microcosms (0.4 × 0.52 × 0.32 m) planted with an assemblage of eight tallgrass prairie grass and forb species growing in native prairie soil. These microcosms are maintained under mycorrhizal and nonmycorrhizal conditions, with and without native soil nematodes, and with and without aboveground herbivory (clipping treatments and grasshopper herbivory) to elucidate the interactions and effects of these key species interactions. Preliminary results indicate that belowground herbivory by nematodes reduces biomass production of mycorrhizal C_4 grasses, and nonmycorrhizal C_3 grasses, while having variable effects on the forb species. Greater compensatory above- and belowground biomass production was observed in defoliated plants under mycorrhizal conditions than under nonmycorrhizal conditions, whether foliage removal was accomplished by clipping or by grasshopper herbivory. These preliminary results support our hypothesis that mycorrhizal symbiosis may enhance grazing tolerance under light to moderate grazing in grasslands. These results also underscore the importance of above-and belowground linkages and indicate that alterations in mycorrhizal and rhizosphere processes can have large indirect effects on plant communities through their effects on plant responses to above- and belowground consumers.

A comment on experimental approaches

One of the greatest difficulties in attempting to assess the role of mycorrhizas in natural plant communities is that all of the various approaches for experimentally manipulating mycorrhizal colonization and symbiotic function have limitations. Suppression of mycorrhizas with fungicides may result in non-target effects on other fungi or soil processes, however both target and potential non-target effects of the fungicide can be quantified and dealt with as co-variates in statistical analyses (e.g., Smith et al., 2000). The fungicide approach also provides greater realism than approaches using soil sterilization (which alters numerous soil processes), artificial growing media, and/or the use of one or only a few AM fungal species obtained from a source other than the native community under study. These latter approaches provide some insight into potential effects and mechanisms, but they lack the realism necessary to draw strong conclusions about the role and functioning of mycorrhizas in nature.

The fungicide benomyl has been used successfully in our studies and by several other workers to experimentally vary mycorrhizal activity (Carey et al., 1992; Fitter, 1986; Fitter and Nichols, 1988; Jacobson et al., 2001; Kahiluoto et al., 2000; Merryweather and Fitter, 1986; Schweiger and Jakobson, 1999). Several studies suggest that it is effective in reducing AM fungal colonization, has no direct effects on a wide range of plants in the absence of fungi, and has the fewest drawbacks of all potential methods of AM fungal suppression in the field (Fitter and Nichols, 1988; Paul et al., 1989; Sukarno et al., 1993; Wilson and Hartnett, 1997). In addition, several lines of evidence indicate that, in tallgrass prairie ecosystems, the use of benomyl is a conservative approach and its *primary*, although not exclusive effect is suppression of mutualistic mycorrhizal associations and their symbiotic function (Carey et al., 1992; Hartnett and Wilson, 1999; Hetrick et al., 1990b; Smith et al., 2000). Thus, in our view suppression of mycorrhizal fungi with fungicides remains the best approach currently available to assess the roles and function of mycorrhizas in natural communities, and indeed the only approach practical for field situations and large-scale experiments.

Summary

Although their specific effects vary, it is clear that mycorrhizas exert a significant influence on plant community structure and dynamics in grasslands and other terrestrial ecosystems. The effects of mycorrhizal fungi on patterns of plant diversity and their underlying mechanisms are quite varied. In one community type AM fungi may increase local plant diversity, whereas in another they may maintain the stability of the local community by maintaining the competitive balance among dominant and subordinate species. The structure of some plant communities may be regulated simply by the presence and abundance of mycorrhizal fungi and the patterns of variation in mycorrhizal dependency among co-occurring species. In others the taxonomic or genotypic composition of the fungal community may also be important. The role of mycorrhizas in patterns of diversity at larger scales has not been examined.

Each of the various methodological approaches to studying the ecological roles of mycorrhizal symbiosis has limitations, and each represents a compromise of generality, simplicity, precision and realism. Finding solutions to some of these methodological limitations is a significant challenge. Glasshouse experiments typically lack the realism necessary to draw conclusions about the functioning of mycorrhizas in natural communities, and field studies alone rarely reveal underlying mechanisms and controls. We suggest that both field studies and complementary glasshouse experiments, conducted together in different communities, have the greatest potential to provide a better understanding of the patterns and mechanisms of mycorrhizal effects and the role of this symbiosis in natural plant communities.

Recent empirical and theoretical approaches indicate that the greatest or most frequent effects of mycorrhizal fungi may be indirect rather than direct, as mycorrhizas influence interactions between plants and their competitors, herbivores, and other microorganisms. Because mycorrhizal fungi and these other organisms all influence important plant resources, they are unlikely to have independent effects on their host plants, and each is likely to alter the cost-benefit relationships of the other to the plant. Furthermore, in many grassland ecosystems which are characterized by non-equilibrium dynamics, the relative importance and limitation of different plant resources changes greatly over space and time (Knapp et al., 1998). Thus, the effects of mycorrhizas, their interactions at the community level, and the cost–benefit relationships among mycorrhizal fungi, other organisms, and host plants are likely to be complex and are likely to vary considerably spatially and from year to year. Long-term field studies will be essential to elucidate the dynamic consequences of mycorrhizal symbiosis in native plant communities.

References

Allen E B and Allen M F 1984 Competition between plants of different successional stages: mycorrhizae as regulators. Can. J. Bot. 62, 2625–2629.

Allen E B and Allen M F 1990 The mediation of competition by mycorrhizae in successional and patchy environments. *In* Perspectives on Plant Competition. Ed. J B Grace and D Tilman. pp. 367-389. Academic Press, San Diego, CA.

Allen M F 1991 The Ecology of Mycorrhizae. Cambridge University Press, Cambridge, UK.

Bakhtier Y, Miller D, Cavagnaro T and Smith S 2001. Interactions between two arbuscular mycorrhizal fungi and fungivorous nematodes and control of the nematode with fenamifos. Appl. Soil Ecol. 17, 107–117.

Bengtsson J, Fagerstrom T and Rydin H 1994 Competition and coexistence in plant communities. Trends Ecol. Evol. 9, 246–250.

Bergelson J M and Crawley J M 1988 Mycorrhizal infection and plant species diversity. Nature 334, 202.

Bethlenfalvay G J, Ulrich J M and Brown M S 1982 Plant response to mycorrhizal fungi: host, endophyte, and soil effects. Soil Sci. Soc. Am. J. 49, 1164–1168.

Bonkowski M, Cheng W, Griffiths B S, Alphei J and Scheu S 2000 Microbial-faunal interactions in the rhizosphere and effects on plant growth. Eur. J. Soil Biol. 36, 135–147.

Brown J H 1995 Macroecology. University of Chicago Press, Chicago, IL.

Callaway R M 1995 Positive interactions among plants. Bot. Rev. 61, 306–349.

Carey P D, Fitter A H and Watkinson A R 1992 A field study using the fungicide benomyl to investigate the effect of mycorrhizal fungi on plant fitness. Oecologia 90, 550–555.

Chapin F S III and McNaughton S J 1989 Lack of compensatory growth under phosphorus deficiency in grazing-adapted grasses from Serengeti Plains. Oecologia. 79, 551–557.

Chiariello N, Hickman J C and Mooney H A 1982 Endomycorrhizal role for interspecific transfer of phosphorus in a community of annual plants. Science 217, 941–943.

Clay K and Holah J 1999 Fungal endophyte symbiosis and plant diversity in successional fields. Science 285, 1742–1744.

Connell J H 1990 Apparent versus 'real' competition in plants. *In* Perspectives on Plant Competition. Eds. J B Grace and D Tilman. pp. 9–26. Academic Press, San Diego, CA.

Connell J H and Lowman M D 1989 Low-diversity tropical rain forests: some possible mechanisms for their existence. Am. Nat. 134, 88–119.

Coughenour M B, McNaughton S J and Wallace L L 1985 Responses of an African graminoid (*Themeda traindra* Forsk.) to frequent defoliation, nitrogen and water: a limit of adaptation to herbivory. Oecologia 68, 80–86.

Crowell H F and Boerner R E J 1988 Influences of mycorrhizae and phosphorus on belowground competition between two old-field annuals. Environ. Exp. Bot. 28, 381–392.

Davidson D E and Christiansen M 1977 Root-micofungal and mycorrhizal associations in a shortgrass prairie. *In* The Below-ground Ecosystem: a Synthesis of Plant-Associated Processes. Ed. J K Marshall. pp. 279–287. Colorado State University Press, Fort Collins, CO.

Eissenstat D M 1990 A comparison of phosphorus and nitrogen transfer between plants of different phosphorus status. Oecologia 82, 342–347.

Eissenstat D M and Newman E I 1990 Seedling establishment near large plants: effects of vesicular-arbuscular mycorrhizas on the intensity of plant competition. Funct. Ecol. 4, 95–100.

Eom A H, Wilson G W T and Hartnett D C 2001 Effects of ungulate grazers on arbuscular mycorrhizal symbiosis and fungal community composition in tallgrass prairie. Mycologia 93, 233–242.

Finlay R D and Soderstrom B 1989 Mycorrhizal mycelia and their role in soil and plant communities. *In* Ecology of Arable Land. Ed. M Clarholm and L. Bergstrom. pp. 139–148. Kluwer, Amsterdam.

Fischer Walter L E, Hartnett D C, Hetrick B A D and Schwab A P 1996 Interspecific nutrient transfer in a tallgrass prairie plant community. Am. J. Bot. 83, 180–184.

Fitter A H 1986 Effect of Benomyl on leaf phosphorus concentration in alpine grassland: a test of mycorrhizal benefit. New Phytol. 103, 767–776.

Fitter A H and Garbaye J 1994 Interactions between mycorrhizal fungi and other soil organisms. Plant Soil 159, 123–132.

Fitter A H and Nichols R 1988 The use of Benomyl to control infection by vesicular-arbuscular mycorrhizal fungi. New Phytol. 110, 201–206.

Fitter A H, Graves J D, Watkins N K, Robinson D and Scrimgeour C M 1998 Carbon transfer between plants and its control in networks of arbuscular mycorrhizas. Funct. Ecol. 12, 406–412.

Fogel R and Hunt G 1979 Fungal and arboreal biomass in a western Oregon Douglas-fir ecosystem: distribution patterns and turnover. Can. J. Forest Res. 9, 245–256.

Francis R and Read D J 1984 Direct transfer of carbon between plants connected by vesicular-arbuscular mycorrhizal mycelium. Nature 307, 53–56.

Francis R and Read D J 1995 Mutualism and antagonism in the mycorrhizal symbiosis, with special reference to impacts on plant community structure. Can. J. Bot. 73, 1301–1309.

Francis R, Finlay R D and Read D J 1986 Vesicular-arbuscular mycorrhiza in natural vegetation systems IV. Transfer of nutrients in inter- and intra-specific combinations of host plants. New Phytol. 102, 103–111.

Frank D A, McNaughton S J and Tracy B 1998 The ecology of the earth's grazing ecosystems. Bioscience 48: 513–521.

Frey B and Schuepp H 1992 Transfer of symbiotically fixed nitrogen from berseem (*Trifolium alexandrinum* L.) to maize via vesicular-arbuscular mycorrhizal hyphae. New Phytol. 122, 447–454.

Gange A C and Brown V K 1992 Interactions between soil-dwelling insects and mycorrhizas during early plant succession. *In* Mycorrhizas in Ecosystems. Eds. D J Read, D H Lewis, A H Fitter and I J Alexander. pp. 177–182. C A B Int. Cambridge University Press, Cambridge.

Gange A C, Brown V K and Sinclair G S 1993 Vesicular-arbuscular mycorrhizal fungi: a determinant of plant community structure in early succession. Funct. Ecol. 7, 616–622.

Gehring C A and Whitham T G 1994 Interactions between above-ground herbivores and the mycorrhizal mutualists of plants. TREE 9, 251–255.

Gehring C A, Cobb N S and Whitham T G 1997 Three-way interactions among ectomycorrhizal mutualists, scale insects, and resistant and susceptible pinyon pines. Am. Nat. 149, 824–841.

Goverdr M, van der Heijden M G A, Wiemken A, Sandres I R and Erhardt A 2000 Arbuscular mycorrhizal fungi influence life history traits of a lepidopteran herbivore. Oecologia 125, 362–369.

Grace J B and Tilman D G (Eds.) 1990 Perspectives on Plant Competition. 484 pp. Academic Press, San Diego, CA.

Grime J P, MacKey J M Hillier S H and Read D J 1987 Floristic diversity in a model system using experimental microcosms. Nature 328, 420–422.

Hamel C and Smith D L 1992 Mycorrhizae-mediated ^{15}N transfer from soybean to corn in field-grown inter-crops: effect of component crop spatial relationships. Soil Biol. Biochem. 24, 499–501.

Harper J L 1982 After description. *In* The Plant Community as a Working Mechanism. Ed. E I Newman. pp. 11–26. Blackwell Scientific Publications, Oxford, UK.

Harris K K and Paul E A 1987 Carbon requirements of vesicular-arbuscular mycorrhizae. *In* Ecophysiology of VA Mycorrhizal Plants. Ed. G R Safir. pp. 93–105. CRC Press, Boca Raton, FL.

Hartnett D C and Wilson G W T 1999 Mycorrhizal influence plant community structure and diversity in tallgrass prairie. Ecology 80, 1187–1195.

Hartnett D C, Hetrick B A D Wilson G W T and Gibson D J 1993 VA-mycorrhizal influence on intra- and interspecific neighbour interactions among co-occurring prairie grasses. J. Ecol. 81, 787–795.

Hartnett D C, Samenus R J Fischer L E and Hetrick B A D 1994 Plant demographic responses to mycorrhizal symbiosis in tallgrass prairie. Oecologia 99, 21–26.

Haystead A, Malajczuk N and Grove T S 1988 Underground transfer of nitrogen between pasture plants infected with vesicular-arbuscular mycorrhizal fungi. New Phytol. 108, 417–423.

Heap A J and Newman E I 1980 Links between roots by hyphae of vesicular-arbuscular mycorrhizas. New Phytol. 85, 169–171.

Hetrick B A D, Kitt D G and Wilson G W T 1986 The influence of phosphorus fertilization, drought, fungal species, and nonsterile soil on mycorrhizal growth response in tallgrass prairie plants. Can. J. Bot. 64, 1199–1203.

Hetrick B A D, Wilson G W T and Hartnett D C 1989 Relationship between mycorrhizal dependence and competitive ability of two tallgrass prairie grasses. Can. J. Bot. 67, 2608–2615.

Hetrick B A D, Wilson G W T and Owensby C E 1990a Mycorrhizal influences on big bluestem rhizome regrowth and clipping tolerance. J. Range Manag. 43, 286–290.

Hetrick B A D, Wilson G W T and Todd T C 1990b Differential responses of C_3 and C_4 grasses to mycorrhizal symbiosis, phosphorus fertilization, and soil microorganisms. Can. J. Bot. 68, 461–467.

Hetrick B A D, Hartnett D C Wilson G W T and Gibson D J 1994 Effects of mycorrhizae, phosphorus availability, and plant density on yield relationships among competing tallgrass prairie grasses. Can. J. Bot. 72, 168–176.

Hodge A 2000 Microbial ecology of the arbuscular mycorrhiza. Micro. Ecol. 32, 91–96.

Huston M A 1994 Biological Diversity: The Coexistence of Species on Changing Landscapes. 681 pp. Cambridge University Press, Cambridge.

Jakobsen I, Gazey C and Abbott L K 2001 Phosphate transport by communities of arbuscular mycorrhizal fungi in intact soil cores. New Phytol. 149, 95–103.

Johnson N C, Graham J H and Smith F A 1997 Functioning of mycorrhizal associations along the mutualism-parasitism continuum. New Phytol. 135, 575–585.

Kahiluoto H, Ketoka E and Vestberg M 2000 Creation of a non-mycorrhizal control for a bioassay of AM effectiveness. Mycorrhiza 9, 241–258.

Knapp A K, Briggs J M, Hartnett D C and Collins S L (eds) 1998 Grassland Dynamics: Long-Term Ecological Research in Tallgrass Prairie. 364 pp. Oxford University Press, New York.

Koide R T 2000 Mycorrhizal symbiosis and plant reproduction. *In* Arbuscular Mycorrhizas: Physiology and Function. Eds. Y Kapulnik and D D Douds. pp. 19–46. Kluwer Academic Publishers, Dordrecht.

Koide R T, Huenneke L F, Hamburg S P and Mooney H A 1988 Effects of applications of fungicide, phosphorus and nitrogen on the structure and productivity of an annual serpentine plant community. Funct. Ecol. 2, 335–344.

Lussenhop J 1996 Collembola as mediators of microbial symbiont effects upon soybean. Soil Biol. Biochem. 28, 363–369.

Mader P, Vierheilig H, Streitwolf-Engel R, Boller T, Frey B, Christie P and Wiemken A 2000 Transport of ^{15}N from a soil compartment separated by a polytetrafluoroethylene membrane to plant roots via the hyphae of arbuscular mycorrhizal fungi. New Phytol. 146, 155–161.

McNaughton S J, Wallace L L and Coughenour M B 1983 Plant adaptation in an ecosystem context: effects of defoliation, nitrogen, and water on growth of an African C_4 sedge. Ecology 64, 307–318.

Merryweather J and Fitter A H 1996 Phosphorus nutrition of an obligately mycorrhizal plant treated with the fungicide benomyl in the field. New Phytol. 132, 307–311.

Milchunas D G, Sala O E and Lauenroth W K 1988 A generalized model of the effects of grazing by large herbivores on grassland community structure. Am. Nat. 132, 87–106.

Miller R M 1987 The ecology of vesicular-arbuscular mycorrhizae in grass- and shrublands. *In* The Ecophysiology of VA Mycorrhizal Plants. Ed. G R Safir. pp. 136–163. CRC Press, Boca Raton, FL.

Miller S L and Allen E B 1992 Mycorrhizae, nutrient translocation, and interactions between plants. *In* Mycorrhizal Functioning. Ed. M F Allen. pp. 301–332. Chapman and Hall, New York.

Newman E I 1988 Mycorrhizal links between plants: their functioning and ecological significance. Adv. Ecol. Res. 18, 243–270.

Newman E I and Reddell P 1988 Relationship between mycorrhizal infection and diversity in vegetation: evidence from the Great Smokey Mountains. Funct. Ecol. 2, 259–262.

Newman E I and Ritz K 1986 Evidence on the pathways of phosphorus transfer between vesicular-arbuscular mycorrhizal plants. New Phytol. 104, 77–87.

Newman E I, Devoy C L N Easen N J and Fowles K J 1994 Plant species that can be linked by VA mycorrhizal fungi. New Phytol. 126, 691–693.

Newsham K K and Watkinson A R 1998 Arbuscular mycorrhizas and the population biology of grasses. *In* Population Biology of Grasses. Ed. G P Cheplick. pp. 286–312. Cambridge University Press, Cambridge.

Newsham K K, Fitter A H and Watkinson A R 1995 Multi-functionality and biodiversity in arbuscular mycorrhizas. TREE. 10, 407–411.

O'Connor P J, Smith S E and Smith F A 2001 Arbuscular mycorrhizas influence diversity and structure in a semi-arid plant community. *In* Diversity and Integration in Mycorrhizas. Proc. 3rd International Conference on Mycorrhizas. Ed. S E Smith. Adelaide, South Australia.

Paul N D, Aryes P G and Wyness L E 1989 On the use of fungicides for experimentation in natural vegetation. Funct. Ecol. 3, 759–769.

Perez C A and Frangi J L 2000 Grassland biomass dynamics along an altitudinal gradient in the Pampa. J. Range Manag. 53, 518–528.

Pickett S T A and White P S (Eds.) 1985 The Ecology of Natural Disturbance and Patch Dynamics. 472 pp. Academic Press, Orlando, FL.

Poiani K A, Richter B D, Anderson M G and Richter H E 2000 Biodiversity conservation at multiple scales: functional sites, landscapes, and networks. Bioscience 50, 133–146.

Read D J, Francis R and Finlay R D 1985 Mycorrhizal mycelia and nutrient cycling in plant communities. *In* Ecological Interactions in Soil. Eds. A H Fitter, D Atkinson D J Read and M B Usher. pp. 193–217. Blackwell, Oxford UK.

Reece P E and Bonham C D 1978 Frequency of endomycorrhizal infection in grazed and ungrazed blue grama plants. J. Range Manage. 31, 149–154.

Ricklefs RE and Schluter D (Eds.) 1993 Species Diversity in Ecological Communities. University of Chicago Press, Chicago, IL.

Ritz K and Newman E I 1984 Movement of ^{32}P between intact grassland plants of the same age. Oikos, 43, 138–142.

Ritz K and Newman E I 1985 Evidence for rapid cycling of phosphorus from dying roots to living plants. Oikos 45, 174–180.

Ritz K and Newman E I 1986 Nutrient transport between ryegrass plants differing in nutrient status. Oecologia 70, 128–131.

Robinson D and Fitter A H 1999 the magnitude and control of carbon transfer between plants linked by a common mycorrhizal network. J. Exp. Bot. 50, 9–13.

Rogers J B, Laidlaw A S and Christie P 2001 The role of arbuscular fungi in the transfer of nutrients between white clover and perennial ryegrass. Chemosphere 42, 153–159.

Schweiger P F and Jakobsen I 1999 Direct measurement of phosphorus uptake by native arbuscular mycorrhizal fungi associated with field-grown winter wheat. Agron. J. 91, 998–1002.

Setala H 1995 Growth of birch and pine seedlings in relation to grazing by soil fauna on ectomycorrhizal fungi. Ecology 76, 1844–1851.

Smilauer P and Smilauerova M 2000 Effect of AM symbiosis exclusion on grassland community composition. Folia Geobot. 35, 13–25.

Smith M D, Hartnett D C and Wilson G W T 1999 Interacting influence of mycorrhizal symbiosis and competition on plant diversity in tallgrass prairie. Oecologia 121, 574–582.

Smith M D, Hartnett D C and Rice C W 2000 Effects of long-term fungicide application on microbial processes in tallgrass prairie soils. Soil Biol. Biochem. 32, 935–946.

Smith M R, Charvat I and Jacobson R L 1998 Arbuscular mycorrhizae promote establishment of prairie species in a tallgrass prairie restoration. Can. J. Bot. 76, 1947–1954.

Sukarno N, Smith S E and Scott E S 1993 The effect of fungicides on vesicular-arbuscular mycorrhizal symbiosis. I. The effects on vesicular-arbuscular mycorrhizal fungi and plant growth. New Phytol. 25, 139–147.

Trent J D, Wallace L L Svejcar T J and Christiansen S 1988 Effect of grazing on growth, carbohydrate pools, and mycorrhizae in winter wheat. Can. J. Plant Sci. 68, 15–120.

Van Auken O W and Brown S C 1998 Importance of arbuscular mycorrhizae to drymass production of a native Texas C_3 and C_4 grass. Texas J. Sci. 50, 291–304.

Van der Heijden M G A, Boller T Wiemken A and Sanders I R 1998a Different arbuscular mycorrhizal fungal species are potential determinants of plant community structure. Ecology 79, 2082–2091.

Van der Heijden M G A, Klironomos J N Ursic M Moutoglis P Streitwolf-Engel R Boller T Wiemken A and Sanders I R 1998b Mycorrhizal fungal diversity determines plant biodiversity, ecosystem variability and productivity. Nature 396, 69–72.

Wallace L L 1981 Growth, morphology and gas exchange of mycorrhizal and nonmycorrhizal *Panicum coloratum* L. a C_4 grass species, under different clipping and fertilization regimes. Oecologia 49, 272–276.

Whittingham J and Read D J 1982 Vesicular-arbuscular mycorrhiza in natural vegetation systems. III. Nutrient transfer between plants with mycorrhizal interconnections. New Phytol. 90, 277–284.

Wilkinson D M 1998 The evolutionary ecology of mycorrhizal networks. Oikos 82, 407–410.

Wilson G W T and Hartnett D C 1997 Effects of mycorrhizae on plant growth and dynamics in experimental tallgrass prairie microcosms. Am. J. Bot. 84, 478–482.

Wilson G W T and Hartnett D C 1998 Interspecific variation in plant responses to mycorrhizal colonization in prairie grasses and forbs. Am. J. Bot. 85, 1732–1738.

Wilson G W T, Hartnett D C, Smith M D and Kobbeman K 2001a Effects of mycorrhizal suppression on growth dynamics of selected tallgrass prairie forbs. Am. J. Bot. 88: 1452–1457.

Wilson G W T, Hartnett D C and Todd T C 2001b Interactions of mycorrhizal symbiosis and herbivory in tallgrass prairie. *In* Diversity and Integration in Mycorrhizas. Proc. 3rd International Conference on Mycorrhizas. Ed. S E Smith. Adelaide, South Australia.

Zobel M and Moora M 1995 Interspecific competition and arbuscular mycorrhiza: importance for the coexistance of two calcareous grassland species. Folia Geobot. Phytotax. Praha 30, 223–230.

Zobel M, Moora M and Haukioja E 1997 Plant coexistance in the interactive environment: arbuscular mycorrhizae should not be out of mind. Oikos 78, 202–208.

Contents Volume 244 2002

Volume 244 Nos. 1–2

DIVERSITY AND INTEGRATION IN MYCORRHIZAS
Proceedings of the 3rd International Conference on Mycorrhizas (ICO3)

Adelaide, 8–13 July 2001

Guest editors:
S.E. Smith & F.A. Smith

Author index Volume 244